BUSINESS Algebra

Third Edition

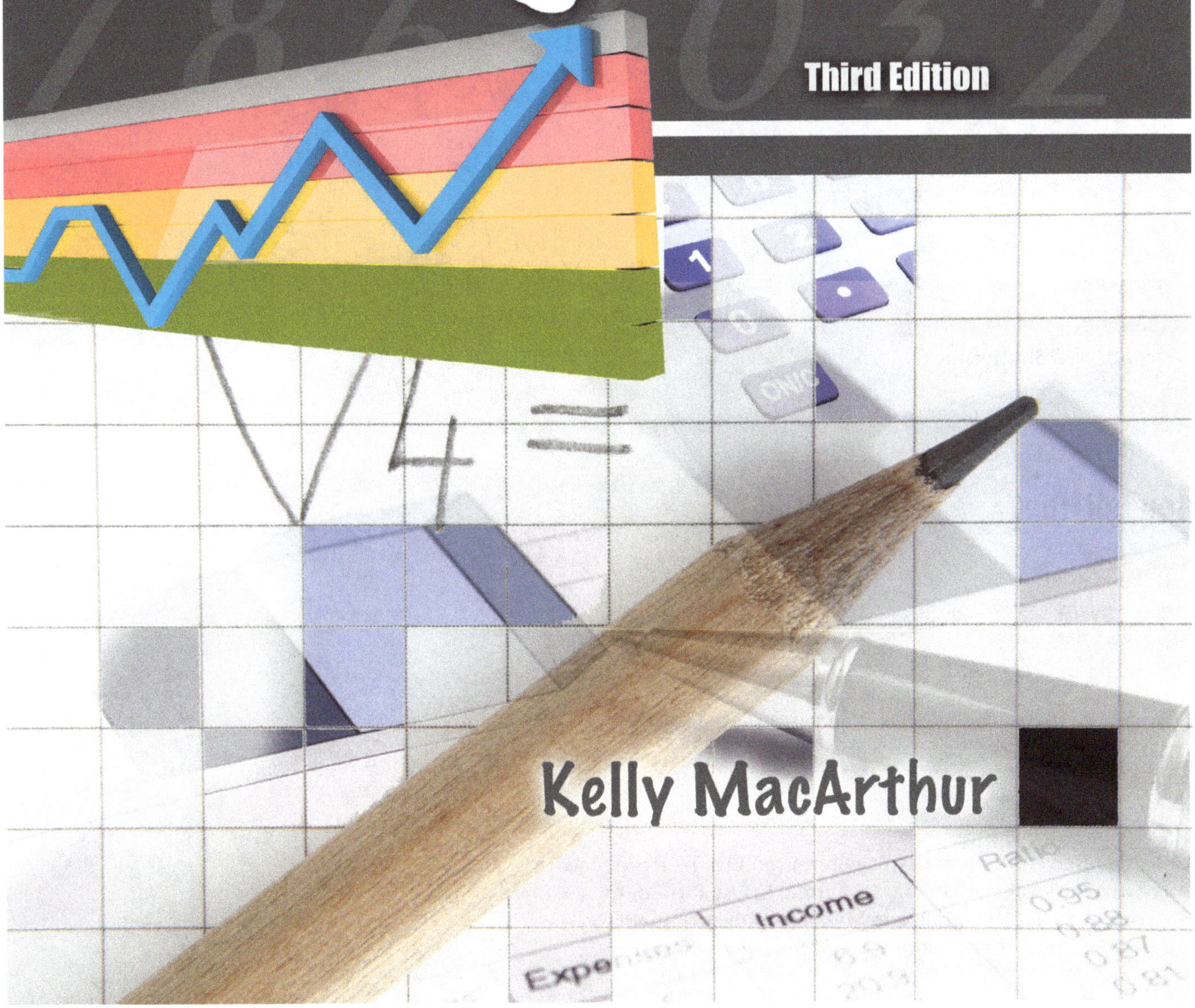

Kelly MacArthur

Kendall Hunt
publishing company

Kendall Hunt
publishing company

www.kendallhunt.com
Send all inquiries to:
4050 Westmark Drive
Dubuque, IA 52004-1840

ISBN 978-1-5249-7699-6

Printed in the United States of America
10 9 8 7 6 5 4 3 2

Contents

Preface and Acknowledgments

ACKNOWLEDGMENTS

I must first thank my colleagues Rod Millar, Clayton Palmer and Dave Burkhard for all their help and creativity with the homework problems in this book! This entire project would not have been possible without their efforts, since it was too big a project to undertake alone.

Secondly, I will always have gratitude in my heart for my sweet son, Jackson MacArthur Stafford, who gave me great reason and motivation to keep getting up early to work on my book (wink, wink). There were, of course, many other close friends, family and colleagues who supported my extra hours of work by giving me pep talks, kind bits of wisdom along the way and giving me the freedom to write instead of playing, skiing or biking with them. Everyone at Kendall Hunt Publishers were also excellent to work with, especially you, Kimberly. Overall, I quite enjoyed this first experience of writing a text book.

TEACHING PHILOSOPHY

I believe strongly that mathematics, at its core, is the art/experience/science of problem solving and pattern recognition. It is inherently a creative process, one to be struggled with, repeated, and ultimately enjoyed. The process requires imagination, persistence, courage, processing time, much repetition, and eventually produces experiential, mathematical skill. It is from this perspective that I teach and write. I'm not as concerned with the destination, i.e. the answer, as I am about the journey of problem-solving and mathematical exploration since it is exactly the entirety of the journey that creates understanding and creation of the answer and therefore brings one to the destination as a consequence of the process. Self-confidence and mastery are then natural by-products of the mathematical journey.

ABOUT THIS BOOK

After teaching this course at the University of Utah for a number of years, and trying many different text books, I had written about three binders of my own notes that I used in class for my students. On several of my course evaluations over those years, students would frequently comment that they relied more on my notes than the text book and some boldly challenged me to write my own text. I laughed at the notion, and even initially laughed when Kendall Hunt Publishers came to my office one day and asked me if I'd be interested in writing this book. I didn't want to take the time, and it seemed like an overwhelmingly huge venture. Clearly I was convinced to go against my initial reaction, mainly because I was tired of not having a reliable book, from a reliable company that would aid in my instruction of our Business Algebra course.

The primary goal of writing this text book was to keep it as simple, organized and straight forward as possible. It seems many of our students are somewhat mathemaphobic when taking this course and are intimidated by mathematics text books. I've noticed over the years that students typically don't have a great desire (or any at all) to read all the text of the sections in a mathematics book, but they desperately want to review the examples in detail in order to increase their own understanding of the material. Thus, I have attempted to write only what's necessary and give plenty of examples with every step outlined so students may get maximum use from the text. I also decided to provide the answers to every homework problem in the end of the book to provide feedback for all homework, not just half of the problems.

One aspect of my teaching that's harder to bring into writing is my strong opinion that there are always infinitely many ways to solve any problem. Many of those ways will be inefficient, and one path may be the most efficient for any particular problem. But, to suggest that there is "the" way to solve a problem is ludicrous indeed. What I present in this book are efficient ways to solve the problems posed, but by no means the only way to do so. In class when teaching, I can take the time to follow the lead of students many times and purposely take us on a longer journey than is absolutely necessary to answer a mathematical question. The benefit of that process is to remind them that as long as we're following mathematical rules properly, we always have lots of choices, and it encourages us to stay creative. The main goal is to get to the answer, secondarily we can worry about doing so efficiently. I typically prompt my students by asking "is it mathematically legal" when they suggest a next step. If the answer is yes, then it's usually at least worth pursuing. The next question to ask is "is it useful?" and that's harder to answer until we get a little further along in the process. It's like driving a car around a city. To get to your destination, you make legal turns and follow the legal rules of the road. If you're in a new city and you're not really sure of your way around, then you may not take the most efficient path to get to your destination the first time traveling there. But, if you travel to that destination again and again, you'll figure out an efficient way to travel. Most importantly, you don't get stuck, throw up your arms in disgust with the process and just stop in the middle of traffic, giving up your journey (at least I hope you don't). Treat the process of learning mathematics the same way. Don't get stuck, just keep trying different ideas and practicing legal mathematical maneuvers until you reach your destination. With enough practice, you'll get there.

Happy teaching and learning to all!
Kelly MacArthur, University of Utah, macarthur@math.utah.edu

How to Succeed in Math

The overriding tip I can give you is that mathematics requires PRACTICE!

1. **Do the homework.**
   - It's much better to do the homework in about two-hour time blocks. I would recommend you do the homework after each class so that you stay caught up and you are only doing small chunks (like one section) at a time.
   - Your mind actually needs processing time in between "sets" of doing homework for everything to sink in.
   - Think of it like physical exercise—it's much better to work out five days per week for one hour than once per week for five hours. Brainy exercise is the same way!

2. **Read the book.**
   - I would highly recommend that you read through the section before attending the lecture on it, even if it's confusing. Just read through it once to get some idea of what's happening. Then, when it's covered in class, it will make a lot more sense. You don't need to spend a lot of time on this to soak in every detail in the book. The idea is to have an overview of what's coming.

3. **Go to class.**
   - This sounds obvious to some people. There is really no substitute for attending class and picking up the extra bits of information that you'll get from the lectures. It's likely the instructor will give you additional strategies not found in the book. Also, repetition is a key for learning mathematics, so if you come to class, it helps embed the lessons in your head.

4. **Be courageous!**
   - This is by far the most important success tip I can give you. Math requires courage and persistence. Don't create a negative story about yourself if you can't get some problems finished. Just pick up the pieces, try again and look back through your notes or examples in the book to see if that can help you. Go get help from a tutor. Form a study group and get help from your peers. Ask questions in class. Trust me, if you have a question, many other students are wondering about the same thing.

Tips for Taking Math Exams

1. Keep up with the homework throughout the semester. This is truly the biggest success tip for the exams. Cramming the night before an exam is analogous to running 25 miles the day before you actually run a marathon. You'll just be exhausted for the actual event and not do as well as you could have! Most of the "training" for a math exam is done in the weeks prior to the exam, with the homework.

2. Do the exam review problems given to you by your instructor within three days preceding the exam to refresh your skills.

3. Get a good night's sleep before an exam to make sure you are well rested and ready to think.

4. When you get the exam, skim through the entire test FIRST.
   - Do the problems that seem easiest to you. This makes sure you get those points on your exam score, and it also boosts your emotional feelings and confidence about the test.
   - Then go through and do the medium-level problems.
   - Save the hardest problems for the end.
   - Know yourself! In other words, if you tend to second-guess yourself and change correct answers to incorrect answers, then don't look at the problems after you've finished. On the other hand, if you tend to have mental "aha" moments after you've struggled a bit, then go back through the exam a second or third time to make sure you sufficiently checked your work.
   - If you get frustrated and you find that your mind is going crazy with negative thoughts, take a deep breath and force yourself to replace those thoughts with something positive. Be prepared for this when you enter the exam room. In other words, have a positive affirmation already memorized to fall back on, in case you fall into negative thinking. Some of my past students have even turned their positive thoughts into some sort of an artistic symbol, like a rainbow or a funny face, which they can draw quickly. As soon as they get the exam, they draw that symbol on their test and go back to the symbol for positive reinforcement whenever necessary during the exam. (Besides helping them, seeing the various symbols drawn on the tests makes grading the exams entertaining.)

5. Finally, keep in mind that it's just an exam, not a statement about your worth as a human being.

Review Assignment

1. Evaluate/simplify these expressions and place the letter corresponding to each one on the number line below. (*Place a dot on the number line and the letter above it, as shown in the example.*) If the number does not simplify to some defined value, i.e., it is undefined, then just state that and you will not be able to plot that on the number line. Also fill in the table telling the smallest set the number belongs to. Choose from these number sets: **R** = real numbers, **Q** = rational numbers, **I** = irrational numbers, **Z** = integers, **W** = whole numbers.

	NUMBER TO EVALUATE AND PLOT:	SET:
A.	$-\sqrt{2}$	
B.	-2^2	
C.	5^0	
D.	$\pi - 1$	
E.	-0.3	
F.	$0.\overline{3}$	
G.	$\sqrt{7-3}$	
H.	$\lvert 3-6 \rvert$	
I.	$\frac{\lvert x \rvert}{x}$ if x is negative	
J.	$\frac{\lvert x \rvert}{x}$ if x is postive	
K.	Additive Identity	
L.	Additive Inverse of $\frac{3}{4}$	
M.	Multiplicative Identity	
N.	Multiplicative Inverse of $\frac{3}{4}$	
O.	$\frac{0}{4}$	
P.	$\frac{4}{0}$	
Q.	$\frac{0}{0}$	
R.	$0.\overline{9}$	
EXAMPLE:	S. $\frac{8}{2} = 4$	W

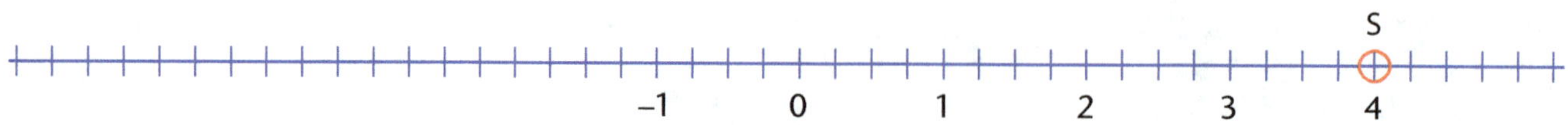

2. List all the integers in these intervals.
   (a) (–3, 4]
   (b) [1, 5]
   (c) (5, ∞)
   (d) (3, 4)
   (e) (–∞, 1]

3. Using $5x^3 - 2x + 4 = 0$, fill in an example of each of these questions to practice your understanding of the vocabulary words. (For instance, if I asked for "degree," you'd say 3 is the degree.)

VOCABULARY WORD	EXAMPLE
Equation	
Expression	
Term	
Factor	
Constant	
Coefficient	
Exponent	

4. Use the order of operations to evaluate these expressions.
   (a) $3 \cdot 5 - 6 \div 4 + 2$
   (b) $4 + 3 \cdot 2^3 \div 4 - 2$
   (c) $\dfrac{2x^3 - x}{yx + y}$ if $x = -2, y = 3, z = -6$
   (d) $(2 \cdot (-2)^3 - (-2)) \div (5 \cdot (-2) + 3)$

5. Evaluate these power (exponent) expressions.
   (a) $64^{\frac{2}{3}}$
   (b) $64^{\frac{2}{3}}$
   (c) $64^{\frac{-2}{3}}$
   (d) $64^{\frac{-3}{2}}$
   (d) $-64^{\frac{3}{2}}$
   (f) $(-64)^{\frac{2}{3}}$

6. Simplify, by rationalizing the denominator.

   (a) $\dfrac{5}{\sqrt{10}}$

   (b) $\dfrac{3}{\sqrt{5}-2}$

   (c) $\dfrac{\sqrt{2x^3}}{\sqrt{8x^6}}$

7. Simplify and write each expression with positive, rational exponents.

   (a) $\dfrac{5^{\frac{-1}{2}}\left(5x^{\frac{5}{2}}\right)}{(5x)^{\frac{3}{2}}}$

   (b) $\sqrt[3]{\sqrt{8x^3y^6}}$

   (c) $\dfrac{32 \cdot 8 \cdot 2^4}{64 \cdot 16 \cdot 2^{-3}}$ (*Hint:* Rewrite this as 2 to some power.)

8. Decide which expression, from each pair, cannot be evaluated and explain why not.

   (a) $\dfrac{4}{0}$ $\dfrac{0}{4}$

   (b) $\sqrt{2}$ $\sqrt{-2}$

   (c) $(-4)^2$ $(-4)^{\frac{1}{2}}$

   (d) $-16^{\frac{1}{4}}$ $(-16)^{\frac{1}{4}}$

   (e) $\sqrt{-64}$ $\sqrt[3]{-64}$

9. Simplify each expression, if possible.

   (a) $\dfrac{3}{4}+\dfrac{2}{3}$

   (b) $\dfrac{3}{4}-\dfrac{2}{3}$

   (c) $\dfrac{3}{4}\cdot\dfrac{2}{3}$

   (d) $\dfrac{3}{4}\div\dfrac{2}{3}$

   (e) $\sqrt{2}+\sqrt{8}$

   (f) $\sqrt{2}-\sqrt{8}$

   (g) $\sqrt{2}\cdot\sqrt{8}$

   (h) $\sqrt{2}\div\sqrt{8}$

10. Factor these quadratic polynomials.

   (a) $x^2 - 2x - 15$

   (b) $x^2 - 13x + 36$

   (c) $2x^2 + 17x - 9$

   (d) $3x^2 - 3x - 60$

   (e) $4x^2 + 14x + 6$

   (f) $12x^2 - 7x - 10$

Review Assignment Key

1. A. **I**
   B. –4; **Z**
   C. 1; **W**
   D. **I**
   E. **Q**
   F. 1/3; **Q**
   G. 2; **W**
   H. 3; **W**
   I. –1; **Z**
   J. 1; **W**
   K. 0; **W**
   L. –3/4; **Q**
   M. 1; **W**
   N. 4/3; **Q**
   O. 0; **W**
   P. undefined
   Q. undefined
   R. 1; **W**

2. (a) –2, –1, 0, 1, 2, 3, 4
   (b) 1, 2, 3, 4, 5
   (c) 6, 7, 8, 9, . . .
   (d) there are none
   (e) . . ., –2, –1, 0, 1

3. Equation $5x3 - 2x + 4 = 0$
   Expression $5x3 - 2x + 4$
   Term $-2x$
   Factor 5 (or x^3)
   Constant 4
   Coefficient –2
   Exponent 3

4. (a) 15½
   (b) 8
   (c) 14/3
   (d) 2

5. (a) 16
   (b) 512
   (c) 1/16
   (d) 1/512
   (e) –512
   (f) 16

6. (a) $\frac{\sqrt{10}}{2}$
   (b) $3(\sqrt{5} + 2)$
   (c) $\frac{1}{2}x^{-\frac{3}{2}}$

7. (a) $\frac{x}{5}$
   (b) $(2\,x)^{\frac{1}{2}}y$
   (c) 2^5

8. (a) $\frac{4}{0}$; cannot divide by zero
   (b) $\sqrt{-2}$; no real number squared gives –2
   (c) $(-4)^{\frac{1}{2}}$; because $(-4)^{\frac{1}{2}} = \sqrt{-4}$ and no real number squared gives –4
   (d) $(-16)^{\frac{1}{4}}$; because $(-16)^{\frac{1}{4}} = \sqrt[4]{-16}$ and no real number raised to the fourth power gives –16
   (e) $\sqrt{-64}$; no real number squared gives –64

9. (a) $\frac{17}{12}$
   (b) $\frac{1}{12}$
   (c) $\frac{1}{2}$
   (d) $\frac{9}{8}$
   (e) $3\sqrt{2}$
   (f) $-\sqrt{2}$
   (g) 4
   (h) $\frac{1}{2}$

10. (a) $(x - 5)(x + 3)$
    (b) $(x - 4)(x - 9)$
    (c) $(2x - 1)(x + 9)$
    (d) $3(x + 4)(x - 5)$
    (e) $2(2x + 1)(x + 3)$
    (f) $(4x - 5)(3x + 2)$

chapter 1

Linear Equations and Inequalities

1.1 Linear Equations in One Variable

VOCABULARY

Term—A mathematical entity (i.e. number or variable) that is participating in addition or subtraction.

Factor—A mathematical entity (i.e. number or variable) that is participating in multiplication or division.

Variable—An unknown quantity. We typically give it a letter name since we don't yet know its value. We can name a variable anything we'd like, even though x is the most common name for a variable.

Expression—Similar to a fragment of a sentence in English. An expression is a collection of terms and/or factors that need simplifying or evaluating. EXAMPLE: $5(x - 2) + 1 - 6x$. We can simplify this expression by gathering like terms. We would get: $5x - 10 + 1 - 6x$ and then $-x - 9$.

Equation—Notice the same root here as "equals." An equation is a complete mathematical statement that equates two expressions. An equation must have an equals sign and "stuff" on both sides of the equals sign from the beginning. We typically solve equations for a particular value of our variable, thereby finding a solution for the equation. EXAMPLE: $3x - 4 = 2$.

Identity—An identity is a particular type of equation. It's an equation that's true for ALL values of the variable. For instance, $x + 3 = 2 + x + 1$ is an identity because, regardless of the value chosen for x, this equation is always true.

We cannot magically transform an expression into an equation. If something starts as an expression, it remains so, as is also true for equations. The equals sign used in simplifying an expression is merely stating that the following expression is equal to the previous expression. It does not now mean there is something to solve.

LINEAR EQUATIONS

When solving any type of equation, we can think of the process as the same as keeping a balance scale (or a teeter-totter) balanced. An equation essentially states that we are starting with a balanced scale. The equals sign behaves as the fulcrum (balance point) of the scale. Our goal is to do legal maneuvers that maintain the balance of that scale while figuring out what the value of the variable, or the unknown quantity, is. We can maintain the balance if we constantly do the same thing to both sides of the scale. For instance, if we add five pounds to each side of the scale, it will stay balanced. If we double both sides of the scale, again it will keep its balance.

Here's an example:

$$3x + 24 = 7x + 4$$

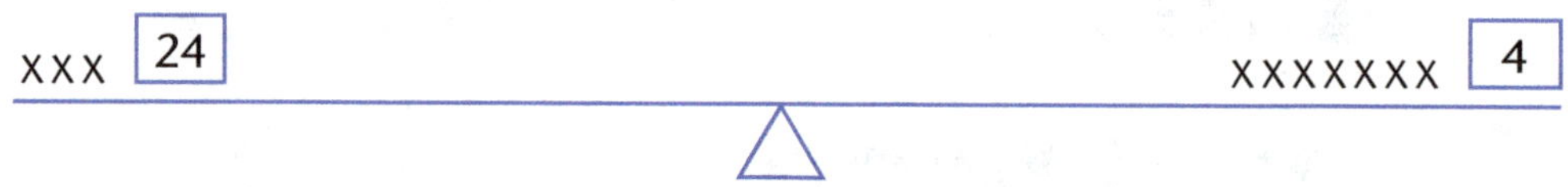

If we remove three of x from each side of the scale, we're left with:

$$24 = 4x + 4 \qquad (\text{or} \quad 20 + 4 = 4x + 4)$$

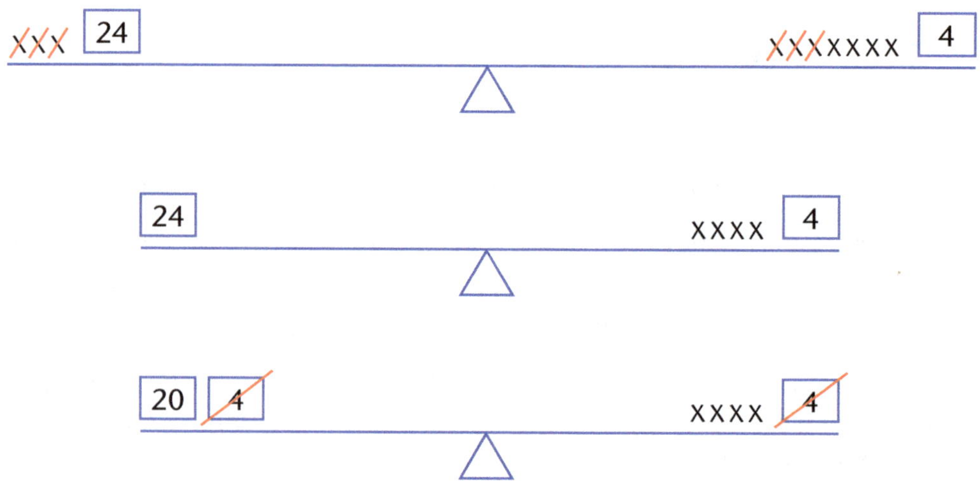

Now, we can remove the constant four from each side. You can think of each of these as a fixed four pounds. We have the following equation now:

$$20 = 4x$$

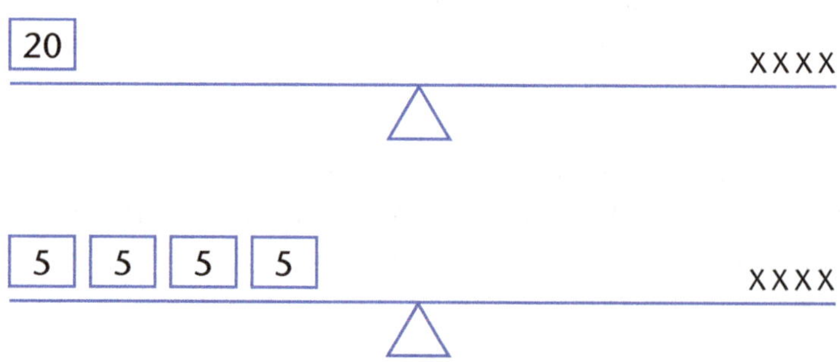

If we ask ourselves, "Four times what gives 20?" we get the answer that $x = 5$.

We will not solve all of our equations on the balance scale, as that would take up too much space. However, as we go about solving equations in this book, keep the idea of the balanced scale in mind. We want to constantly take only those actions that maintain the balance of the scale when solving equations.

A linear equation is a type of equation that only has x's and constants in it. We can move things around by adding or subtracting the same quantity from both sides of the equals sign.

> **LINEAR EQUATION**
> An equation that can be written in the form
> $$ax + b = 0,$$
> where $a \neq 0$ and $a, b \in \mathbb{R}$.

EXAMPLE 1

Solve the equation to find the value of x.

$$\frac{2x}{3} - 2 = \frac{x-3}{2}$$

Notice that it has an equals sign and expressions on each side of it. Thus, this is an equation. We can find the least common denominator, which is 6, and multiply both sides of the equation by it.

$$6\left(\frac{2x}{3} - 2\right) = \left(\frac{x-3}{2}\right)6$$

Now we need to distribute the 6 to each term in the parentheses.

$$\frac{6(2x)}{3} - 6(2) = \frac{6(x-3)}{2}$$

Simplify all terms.

$$2(2x) - 12 = 3(x - 3)$$
$$4x - 12 = 3x - 9$$

Each side is completely simplified. Move all x terms to one side of the equation by subtracting $3x$ from both sides of the equals sign.

$$4x - 3x - 12 = 3x - 3x - 9$$
$$x - 12 = -9$$

$$x - 12 + 12 = -9 + 12$$
$$x = 3$$

To finish getting x by itself, we need to add 12 to both sides of the equals sign.

When solving an equation, you can always check your solution by using that value of x in the original equation.

$$\frac{2(3)}{3} - 2 = \frac{3-3}{2}$$
$$\frac{6}{3} - 2 = \frac{0}{2}$$
$$2 - 2 = 0$$
$$0 = 0$$

CHECK: Since we got a true statement, namely $0 = 0$, our solution is correct.

VOCABULARY

Polynomial—An expression made up of variable terms of the form ax^n where $a \in \mathbb{R}$ and $n \in \{0, 1, 2, 3, \ldots\}$. For instance, $3x^2 - 5x + 1$ is a polynomial because we can think of each term as a number times x to some whole power. (1 can be thought of as x^0.)

Rational Equation—An equation that contains rational expressions, which are fractions that have polynomials in the numerators and/or denominators.

Domain—The set of all allowable inputs or variable values. Typical problems come from the fact that we cannot let the denominator be zero.

EXAMPLE 2

Solve this rational equation, which turns into a linear equation. *Note:* Check the domain.

$$\frac{3x}{x-1} = 4 + \frac{5}{x-1}$$

The LCD (least common denominator) is $(x - 1)$. Multiply both sides of the equation by the LCD to get rid of the fractions. (*Note:* We can only use this strategy in an equation, not an expression.)

$$(x-1)\left(\frac{3x}{(x-1)}\right) = \left(4 + \frac{5}{(x-1)}\right)(x-1)$$

Distribute the $(x - 1)$ to all terms.

$$\frac{3x(x-1)}{(x-1)} = 4(x-1) + \frac{5(x-1)}{(x-1)}$$

Simplify all terms.

$$3x = 4x - 4 + 5$$

$$3x = 4x + 1$$

Subtract $4x$ from both sides of the equation.

$$-x = 1$$

Multiply both sides of the equation by -1.

$$x = -1$$

Before we can count on our solution, we need to check the domain of this equation. We start with a default domain: $x \in \mathbb{R}$ (i.e., x can be any real number). Then we can add the restrictions on x. For this equation, we cannot allow the denominator to be zero, since dividing by zero is not defined. Thus, we cannot let $(x - 1) = 0$, which means we cannot let $x = 1$. The complete domain is: $x \in \mathbb{R}$, $x \neq 1$. Since our answer is -1 (not 1), it is allowed.

EXAMPLE 3

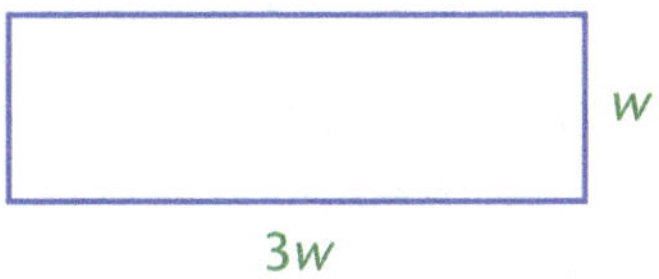

The perimeter of a rectangle is 400 feet and the length is three times as long as the width. Find the dimensions of the rectangle.

Perimeter $P = 2 * \text{width} + 2 * \text{length}$, but length $= 3 * \text{width}$

Let $w = \text{width}$.

Then $P = 2w + 2(3w)$, and we were given that $P = 400$ ft.

So, we have $400 = 2w + 2(3w)$

$400 = 2w + 6w$ Simplify.

$400 = 8w$ Divide both sides of the equation by 8.

$50 = w$

$\Rightarrow \text{length} = 3(50) = 150$

We know our measurements are in feet, which means the width is 50 feet and the length is 150 feet.

EXAMPLE 4

An investor spent $500,000 on two homes, Home A and Home B. After one year, those homes were sold. Home A was sold for a 6% profit and Home B was sold for a 10% profit. The total profit was $43,000. What was the purchase price of each home?

Let x = purchase price of Home A.

We can organize the information in a table that might help us out.

	HOME A	HOME B	TOTALS
Purchase Price	x	$500{,}000 - x$	$500,000
Profit	$0.06x$	$0.1(500{,}000 - x)$	$43,000

If we add the profit from Home A to the profit from Home B, that will equal the total profit.

$0.06x + 0.1(500{,}000 - x) = 43{,}000$	Simplify the left side.
$0.06x + 50{,}000 - 0.1x = 43{,}000$	If we don't like working with decimals in our equation, we can multiply both sides of the equals sign by 100 to eliminate them.
$100(0.06x + 50{,}000 - 0.1x) = (43{,}000)100$	Distribute the 100 to each term.
$6x + 5{,}000{,}000 - 10x = 4{,}300{,}000$	Simplify the left side.
$-4x + 5{,}000{,}000 = 4{,}300{,}000$	Subtract 5,000,000 from both sides of the equation.
$-4x = -700{,}000$ $x = 175{,}000$	Divide both sides of the equation by –4 to get x alone.

⇒ Home B purchase price = 500,000 – 175,000 = 325,000

This means that Home A was purchased for $175,000 and Home B was purchased for $325,000.

1.1 Homework

Solve these linear equations, check your solutions:

1. $2x + 5 = 9$
2. $5x + 8 = 3$
3. $5 - 3x = -4$
4. $3 - 4x = 7$
5. $5x - 18 = 7$
6. $5 - 7x = -9$
7. $5x - 7 = 9$
8. $4x - 2 = 3x + 4$
9. $5x + 13 = 3x + 7$
10. $3x + 7 = 12 - 2x$
11. $7 - x = 21 - 8x$
12. $5x - 9 = 2x + 17$
13. $4x - 7 = 8x + 2$
14. $3x + 2 = 7x + 2$
15. $8(x - 2) = 6(3x - 4)$
16. $\frac{5x}{2} - 4 = \frac{(2x - 7)}{6}$
17. $\frac{3x}{4} = 24$
18. $3x + 3x - 3 = 7$
19. $2(x - 7) = 5(x + 3) - x$
20. $\frac{(5x - 1)}{9} = \frac{5(x - 1)}{6}$
21. $\frac{1}{3}x + 7 = \frac{1}{5}x + 13$
22. $\frac{1}{6}x - 4 = 10 - \frac{1}{8}x$
23. $\frac{x}{7} + 6 = 5 - \frac{x}{5}$
24. $\frac{x + 2}{5} = \frac{x - 2}{7} - 2$
25. $\frac{x + 3}{2} - 2 = \frac{x - 2}{3} - 4$
26. $\frac{x - 5}{6} - 1 = \frac{x - 3}{8} - 2$
27. $\frac{x + 3}{6} - 1 = \frac{x - 3}{4} - 1$
28. $\frac{x - 7}{5} + 1 = \frac{x - 4}{3} - 1$

Solve these linear or rational equations, check the domain to ensure that the solution is valid.

29. $\frac{x}{3} - \frac{1}{2} = \frac{1}{3}$

30. $\frac{x}{5} - 2 = \frac{3}{5}$

31. $10x + 25(x - 3) = -5$

32. $-26 = x - 19$

33. $x + \frac{1}{12} = \frac{-5}{6}$

34. $\frac{3}{2}x = \frac{-1}{4}$

35. $\frac{5}{6}x = \frac{-7}{8}$

36. $4x - 12 = 60$

37. $\frac{6}{5}x + \frac{4}{10}x = \frac{32}{10}$

38. $0.9x - 0.7x = 42$

39. $3x - 7 = -7 - 5x$

40. $2(x + 6) = 8x$

41. $80 = 10(3x + 2)$

42. $\frac{2x}{x-3} - 1 = 3 - \frac{4}{x-3}$

43. $\frac{2x+1}{x-5} = 4 - \frac{3}{x-5}$

44. $\frac{5x-4}{x+7} = \frac{3}{x+7} + 2$

45. $\frac{x+4}{x-4} = \frac{x}{x-4} + 1$

46. $\frac{x-3}{x-5} = \frac{3x}{x-5} - 4$

47. $\frac{4}{(x-6)} + 1 = \frac{6}{(x-6)}$

48. $\frac{2x}{(x-3)} = 4 - \frac{6}{(x-3)}$

49. $\frac{5}{(3x+8)} + \frac{2}{(x-1)} = 0$

50. $\frac{3}{x} + \frac{1}{4} = \frac{2}{3} + \frac{1}{x}$

51. $\frac{(33-x)}{5x} = 2$

52. $\frac{1}{2x} + \frac{1}{x} = \frac{1}{2}$

53. $\frac{(x+3)}{(x-2)} = \frac{8}{3}$

Which of the following are polynomials and which are not?

54. $3x^2 + 2x - 7$

55. $\frac{3}{x^2}$

56. -7

57. $5x + 2$

58. $\frac{7}{x+1}$

59. $\sqrt{3}x^2 + 5x - 1$

60. $3\sqrt{x} + 5x - 1$

61. Five more than two times a number is 39. What is the number?

62. Three less than 4 times a number is 25. What is the number?

63. Ten more than three times a number is the same as five times that same number. What is the number?

64. Dan opens a savings account with $500. He plans to add $10 a week. How many weeks will it take for him to have a total of $810 in this savings account?

65. While vacationing at a beach, Mike rents a bicycle. The rental fee for the bicycle includes $5.00 to rent the bicycle and 20 cents per mile that he rides. Write a linear equation which shows how much Mike will owe for renting the bicycle and riding m miles. Then use this equation to determine how much he will owe after riding 12 miles.

66. A car rental company rents cars for $80.00 plus 15 cents per mile. Write a linear equation which shows how much renting a car will cost if the car is driven x miles. Then use this equation to determine the total rental cost if the car is driven 220 miles.

67. Sam has been building a fence to enclose his property. He now has 25 sections of fencing completed. He will be able to build another 8 sections of fencing every week. Write a linear equation which shows how many sections of fencing he will have built after w weeks. Use this equation to find out how many sections of fencing he will have built after 15 weeks.

68. The perimeter of a rectangle is 700 feet and the length of the rectangle is four times as long as the width. Find the dimensions of the rectangle.

69. The perimeter of a rectangle is 24 inches and the length of the rectangle is five times as long as the width. Find the dimensions of the rectangle.

70. A collector bought two rare coins for $10,000. A year later, the collector sold these coins for $10,400 or a $400 profit. One of the coins was sold for a 3% profit and the other was sold for a 7% profit. What was the purchase price of each coin?

71. A car dealership has two vehicles which were bought for $14,000 (total). The dealership was able to sell the first vehicle for an 11% profit and the second for a 6% profit. The total profit for the dealership was $1,100. What did the dealership originally pay for each vehicle?

72. A real estate investor bought two parcels of land for $1,000,000. Several years later, the investor sold these parcels for a $10,000 profit. One of the parcels was sold for a 1% loss and the other was sold for a 4% profit. What was the purchase price of each parcel of land?

Temperatures. The relationship between Fahrenheit and Celsius temperature readings is: $F = \frac{9}{5}C + 32$, where F is the temperature in degrees Fahrenheit and C is the temperature in degrees Celsius.

73. Find the temperature in degrees Fahrenheit when the temperature is 35° Celsius.

74. Find the temperature in degrees Celsius when the temperature is 50° Fahrenheit.

75. Determine the temperature where both of these temperature scales have the same numerical reading. *Hint:* Change the equation $F = \frac{9}{5}C + 32$ by replacing the F with C since both F and C will have the same numerical value for this particular temperature. Then solve the resulting equation to find C.

Future Value of an Investment. When money is invested and will receive simple interest, the future value of the investment is given by $S = P + Prt$, where S is the future value of the investment, P is the principal amount invested, r is the interest rate per time interval, and t is the length of time measured in units compatible with. (Usually the time interval is in years so the interest rate will be the annual interest rate.)

76. At what annual interest rate must \$1,000 be invested so that its Future Value is \$2,300 after 5 years?

77. For how many years would \$5,000 need to be invested at a 5% annual interest rate to have an investment with a Future Value of \$7,000?

78. How much of a principal amount would need to be invested at a 2% annual interest rate for 10 years to have an investment with a Future Value of \$9,000?

79. For how many years would \$1,000 need to be invested at a 1% annual interest rate to have an investment with a Future Value of \$2,000?

80. Sierra has 2 liters of a mixture containing 45% of boric acid. How much water must be added to make the mixture 36% boric acid?

81. There are three consecutive even integers. Half the sum of the second and third numbers is 55. What are the integers?

82. A hardware store employee is paid a base salary of \$2,000 per month plus an 8% commission on sales over \$7,000. How much must the employee sell to earn \$4,000 for the month?

83. If the low temperature for a 24-hour period at an arctic outpost was –49F°, what was the temperature in degrees Celsius?

84. A 30-foot piece of cable is cut into two pieces. One piece is 2 feet longer than the other. How long are the pieces of cable?

85. Five plus twice a number is seven times the number. What is the number?

86. The perimeter of a basketball court is 104 ft. and the length is 15 ft. more than the width. What are the dimensions of the court?

87. A rectangular field has a perimeter of 170 m. The length is 5m more than the width. Find the dimensions of the field.

88. A car rents for \$45 per day with unlimited mileage, or for \$25 per day plus 18¢ per mile. How many miles need to be traveled for the rates to be equal?

89. Find three consecutive integers whose sum is 36.

90. A mixture of silver and copper alloy weighs 128 pounds and contains 12 pounds of silver. How many pounds of silver must be added in order that every 8 pounds of the resulting alloy will contain 1 pound of silver. (*Hint:* find the constant quantity.)

91. IQ (Intelligence Quotient) is measured by dividing Mental Age (MA) by chronological age (CA) and multiplying by 100. A person with a mental age of 12 and a chronological age of 8 has an IQ of 150. If a 9-year-old girl has an IQ of 140, what is her mental age?

92. If a group of 14-year-old children have IQs that range from 85 to 137, what is their range of mental ages?

93. A 30-foot piece of rope is cut into two pieces. One piece is 5 feet longer than the other. How long are the pieces of rope?

94. A student has test scores of 93, 89, 72, 80, and 96. What must the sixth test score be to have an average of 88?

95. A loan is obtained at 7% simple interest. After one year $856.00 pays off the loan. How much was the loan?

96. A car sells for $22,461.00 and this price includes 6.75% sales tax. How much is the sales tax?

97. A small business reported profits during 2008 of $170,000. If this was 147% of the 2007 year profits, what was the profit in 2007?

98. Morgan received her paycheck this week. From her gross pay, 37% was deducted for federal taxes, 8% for FICA, 8% for state taxes, and 3% for her 401K. She received $660 after the deductions. What is her gross pay?

1.2 Linear Inequalities in One Variable

A linear inequality is a type of inequality that only has x's (variables) and constants in it. Our basic strategy for solving linear inequalities is the same as that for solving linear equations. (The main difference occurs when we multiply or divide both sides of the inequality by a negative number.)

We can move things around by adding or subtracting the same quantity from both sides of the inequality sign. If we think about an inequality in the context of the balanced scale as in the last section, we see that the scale is imbalanced from the beginning. It's like a parent sitting on one side of the teeter-totter and the child sitting on the other side. When solving an inequality, the goal is to maintain that imbalance with every step. Again, if we perform the same operation on both sides of the inequality, we'll definitely keep the imbalance in place, as it was at the beginning.

> **LINEAR INEQUALITY**
> An inequality that can be written in the form
> $$ax + b \le c \quad \text{or} \quad ax + b \ge c,$$
> where $a \ne 0$ and $a, b, c \in \mathbb{R}$.

What happens when we multiply or divide both sides of an inequality by a negative number? Let's look at an example to see.

We'll start with an inequality that we know is true: $3 < 5$. If we multiply both sides of the inequality by -1, we get $-3 < -5$. Is that a true statement? No, because the order of numbers, as we can see on a number line, is reversed or a mirror image on either side of zero.

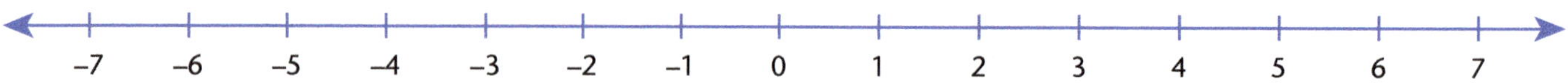

For our example, we should get $-3 > -5$, instead of our original guess.

> When we multiply or divide both sides of an inequality by a negative number, we must remember to switch the inequality sign.

EXAMPLE 1

Solve the inequality to find all the values of x that are acceptable.

$4(x - 1) + 3 < 5 - 2x + 2$ $4x - 4 + 3 < 7 - 2x$	Distribute the 4 through the parentheses on the left and start simplifying both sides of the inequality.

$4x - 1 < 7 - 2x$	Now, add $2x$ to both sides of the inequality.
$6x - 1 < 7$	Add 1 to both sides of the inequality.
$6x < 8$	To get x by itself, we need to divide both sides of the inequality by 6.
$x < \frac{8}{6}$	Finally, simplify the fraction on the right by dividing out two on both the top and the bottom of the fraction.
$x < \frac{4}{3}$	

This says that we can plug in any value of x less than $\frac{4}{3}$ into the original inequality and we'll get a true statement. To graph the solutions of this inequality, we draw a real number line and shade in all the acceptable values of x.

Notice that there are infinitely many variable values (or x-values) that will make a true statement for a linear inequality, i.e., infinitely many solutions, usually congregated in a particular interval of the real number line. We can write these solutions a few different ways. Let's go over the various ways to write interval notation.

$a < x \leq b$		$x \in (a, b]$
$a \leq x < b$		$x \in [a, b)$
$a < x < b$		$x \in (a, b)$
$a \leq x \leq b$		$x \in [a, b]$
$x > a$	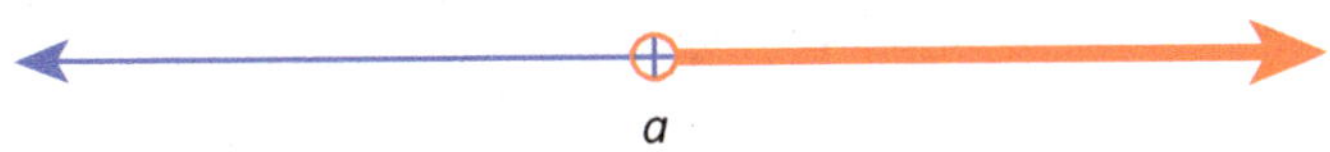	$x \in (a, \infty)$
$x < b$	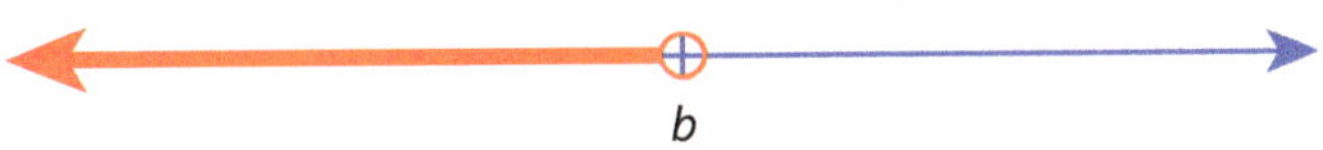	$x \in (-\infty, b)$
$x \geq a$	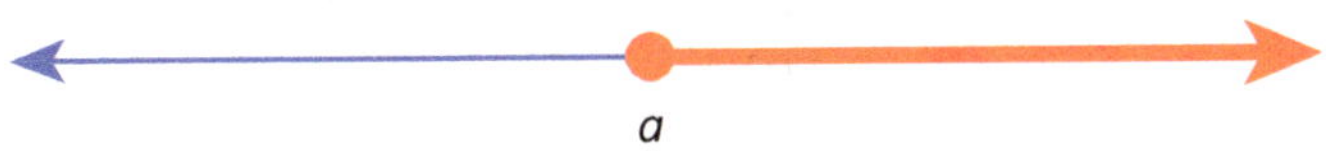	$x \in [a, \infty)$
$x \leq b$	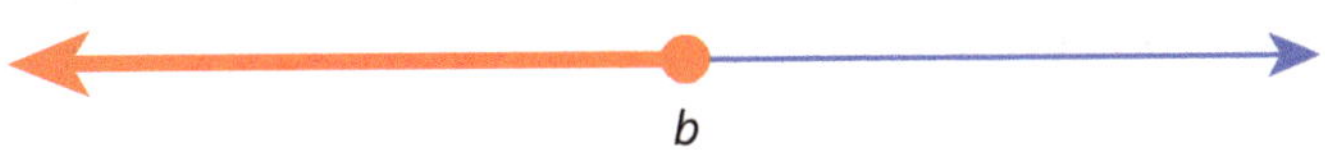	$x \in (-\infty, b]$
$-\infty < x \leq \infty$		$x \in (-\infty, \infty)$ or $x \in \mathbb{R}$

For the graphs of the solutions, we can either use a circle that's empty if we're not including that value of x or a (sign. If we are including that value for x, then we can use a circle that's filled (shaded) in or a] sign.

EXAMPLE 2

Solve the inequality to find all the solutions.

$$\frac{-3x}{5} + 1 \geq -10 - \frac{x}{2}$$

Notice the LCD of all fractions here is 10. If we multiply both sides of the inequality by 10, we should be able to eliminate the fractions and make this easier to handle.

$$10\left(\frac{-3x}{5} + 1\right) \geq \left(-10 - \frac{x}{2}\right)10$$

Distribute the 10 to every term.

$$\frac{10(-3x)}{5} + 10(1) \geq -10(10) - \frac{10x}{2}$$

Simplify each term.

$$2(-3x) + 10 \geq -100 - 5x$$

Keep simplifying on each side of the inequality.

$$-6x + 10 \geq -100 - 5x$$

Add $5x$ to each side of the inequality.

$$-x + 10 \geq -100$$

Subtract 10 from each side of the inequality.

$$-x \geq -110$$

To get x by itself (meaning only one of them instead of -1 of x), multiply both sides of the inequality by -1.

$$x \leq 110$$

Remember that since we multiplied by a negative number, we had to switch the inequality sign.

Graph the solutions on a real number line.

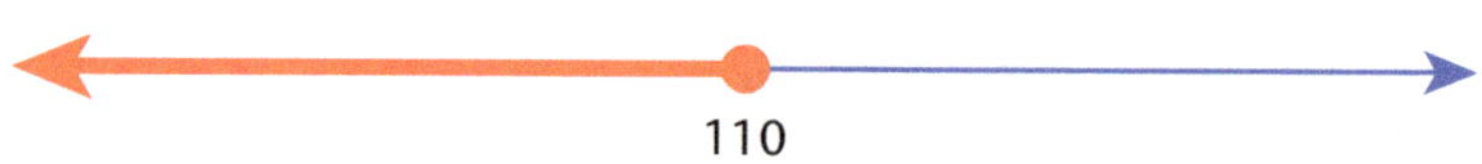

SOLUTIONS: $x \leq 110$ or we can write it as $x \in (-\infty, 110]$.

EXAMPLE 3

A math professor counts the final exam as 30% of the total grade in an algebra course. There are also four midterm exam grades in the total grade calculation. The top three midterm scores are each worth 20% of the overall grade and the lowest midterm score counts for only 10% of the overall grade. If a student's scores on the midterms were 82, 71, 66 and 85 (out of 100), what is the lowest score the student can earn on the final exam to get at least an 80 in the class?

Let x = final exam score. We want to know all the values of x that will get at least an 80 in the course. It's okay if we get a higher grade overall.

$0.3x$ = actual grade points earned from the final exam since it's worth 30%.
$0.2(82 + 85 + 71)$ = actual grade points earned from the three highest midterm grades.
$0.1(66)$ = actual grade points earned from the lowest midterm grade.

$0.3x + 0.2(82 + 85 + 71) + 0.1(66) \geq 80$	We want to add up all the grade points earned (with the midterms) and to be earned (with the final exam) and that quantity has to be at least 80 total.
$0.3x + 0.2(238) + 6.6 \geq 80$	We started simplifying on the left side.
$0.3x + 47.6 + 6.6 \geq 80$	If we want to eliminate the decimals, we can multiply both sides of the inequality by 10.
$10(0.3x + 47.6 + 6.6) \geq (80)10$	Distribute and simplify.
$3x + 476 + 66 \geq 800$	Keep simplifying on the left side.
$3x + 542 \geq 800$	Subtract 542 from both sides of the inequality.
$3x \geq 258$	To finish getting x alone, divide both sides of the inequality by 3.
$x \geq \frac{258}{3}$	Simplify the fraction.
$x \geq 86$	

This means that the student needs to get at least an 86% on the final exam in order to achieve the goal of getting an 80 as the overall total grade.

EXAMPLE 4

A company invests a total of $50,000 in two different accounts. The riskier investment yields an annual average of 6.5% profit, and the safer investment has an annual yield of 4.8%. How much money should be invested in each account in order to earn at least $2,600 profit in a year?

Let x = amount of money invested in the 6.5% account.

This means that $50{,}000 - x$ = amount of money invested in the 4.8% account.

We want the sum of the annual interest from the two accounts to be at least $2,600.

$0.065x + 0.048(50{,}000 - x) \geq 2{,}600$	If we want to eliminate the decimals, then we can multiply both sides of the inequality by 1,000.
$1000(0.065x + 0.048(50{,}000 - x)) \geq (2{,}600)1000$	Distribute and simplify.
$1000(0.065x) + (1000 \cdot 0.048)(50{,}000 - x) \geq 2{,}600{,}000$	Continue simplifying.
$65x + 48(50{,}000 - x) \geq 2{,}600{,}000$	Distribute and simplify again.
$65x + 2{,}400{,}000 - 48x \geq 2{,}600{,}000$	Subtract 2,400,000 from both sides of the inequality and simplify.
$17x \geq 200{,}000$	To get x by itself, divide both sides of the inequality by 17.
$x \geq \dfrac{200{,}000}{17}$	Convert the fraction into a decimal dollar amount.
$x \geq 11{,}764.71$	
$\Rightarrow 50{,}000 - x = 38{,}235.29$	

This means that the company should invest at least $11,764.71 in the 6.5% account and at most $38,235.29 in the safer 4.8% account to meet its goal.

1.2 Homework

For problems 1–45, solve the linear inequalities, and graph the solutions on a real number line:

1. $2x + 5 < 9$

2. $x + 2 \geq 8$

3. $1 \leq 1 + x$

4. $5x + 6 > 21$

5. $4x + 7 > -17$

6. $5 - 3x \leq -10$

7. $9 - 2x \geq 3$

8. $4 - 6x < 2$

9. $5x + 3 \leq -3$

10. $\frac{3}{2}x + 3 \geq -6$

11. $\frac{4}{5}x - 2 < 1$

12. $\frac{5}{3}x + 1 > \frac{4}{5}$

13. $\frac{2}{7}x + 2 > -2$

14. $\frac{7}{4}x - 2 \geq 5$

15. $1 - \frac{5}{2}x \leq -4$

16. $5 - \frac{4}{3}x > -5$

17. $5 - \frac{3}{5}x < 6$

18. $1 - \frac{11}{13}x \leq -3$

19. $\frac{3}{4}x + 1 \leq \frac{2}{3}x - 3$

20. $\frac{5x - 1}{2} - 2 < \frac{2}{3}x + 3$

21. $\frac{3x + 2}{4} + 3 \geq \frac{3x}{2} - 1$

22. $\frac{2x + 4}{3} - 2 < \frac{x + 5}{4} + 1$

23. $\frac{3x - 3}{4} + 3 \leq \frac{x + 1}{6} - 2$

24. $\frac{5x + 3}{8} - 1 \geq \frac{x + 4}{6} + 1$

25. $\frac{x + 10}{3} + 1 < \frac{3x + 11}{4} - 3$

26. $2 \leq x + 3 \leq 5$

27. $-4 < 2x - 3 < 9$

28. $\frac{x}{2} - \frac{2}{3} < \frac{x}{3} + 2$

29. $2 - x \le 4(x - 2)$

30. $-1 \le \frac{2x}{3} + 5 \le 10$

31. $11 < x - 6 \le 31$

32. $\frac{3}{2} \le \frac{4x}{3} + 7 < 9$

33. $5(x + 3) + 9 < 3(x - 2) + 6$

34. $\frac{1}{3}(6x + 24) - 20 \ge -\frac{1}{4}(12x - 72)$

35. $y + (y - 3) \le (y + 2) - (y + 1)$

36. $\frac{7}{8}(5 + 4x) + 17 \le 38$

37. $3 \le x + 7 < 11$

38. $\frac{2}{3}(2w - 1) > 10$

39. $\frac{3}{5}(3x + 4) < 25$

40. $\frac{x}{-12} - 5 \le -4$

41. $-5(y + 3) \le 26$

42. $20 > 2(q + 11)$

43. $-22 > -7x + 15$

44. $\frac{y}{-4} + 3 < 12$

45. $v + 24 < 3v + 29$

For problems 46–50, translate to an inequality:

46. The price of a movie ticket is less than $8.

47. A number is greater than –4.3.

48. A number is less than 12 and greater than 7.

49. A number is less than –4 and greater than –11.

50. Three times a number is less than 13 and greater than –3.

For problems 51–70, answer the given question.

51. Four less than three times a number is no more than eleven. What range of values can this number have?

52. Five more than four times a number is at least 21. What range of values can this number have?

53. A number is less than 10. Twice that number is at least 6. What range of values can this number have?

54. Six more than two times a number is at least as much as four times that same number. What range of values can this number have?

55. Maria has gotten scores of 75, 77, 81 and 84 (out of a possible 100) on the four midterm examinations for a math class. If each of these midterms counts as 15% of the total grade in the class and the final exam counts as 40% of the total grade, what must she get as a score (out of 100 possible) on the final examination to get at least 85% as her total grade in the class?

56. An investor wants to invest a total of \$10,000 in two different accounts. The riskier investment yields an annual average of 9.5% profit and the safer investment has an annual yield of 4.5%. How much money should be invested in each account in order to earn at least \$600 profit in a year?

57. A company invests a total of \$100,000 in two different accounts. The riskier investment yields an annual average of 7.5% profit and the safer investment has an annual yield of 4.0%. How much money should be invested in each account in order to earn at least \$6,000 profit in a year?

58. During a 24-hour period at an Arctic station, the temperature range was –51F° to 11F° ($-51 \le F \le 11$). What is this range in degrees Celsius?

59. An employee is given a base salary of \$2,000 a month plus an 8% commission on sales over \$7,000 during the month. How much in total sales must the employee have to make at least \$4,000 for the month?

60. Your new job allows you to choose between two ways of being paid:

	Monthly Salary	*Sales Commission*
(1)	\$800	4% of all sales
(2)	\$1000	6% of sales over \$10,000

When is plan (1) better than plan (2)?

61. If the revenue function is $R(x) = 40x$ and the cost function is $C(x) = 20x + 1600$, how many items must be sold to produce a profit?

62. The wind chill factor is a temperature in still air that is the equivalent of the ambient temperature plus the depressing effect a wind has on temperature due to evaporation of water on skin. For a wind speed of 25 mph the wind chill can be expressed by:

$$WC = 1.479t - 43.821, \text{ where } t = \text{temperature.}$$

What range of temperatures does the wind chill factor of 30°F represent? (*Hint:* Find t such that $WC \le t - 30$.)

63. A video camera sells for \$695 and tapes for \$5.75 each. How many tapes can be purchased along with the camera for \$900?

64. A line of t-shirts sells for \$3.50 each. The fixed costs to manufacture them is \$6500 and it costs \$1.30 to make each t-shirt. How many t-shirts must be sold to realize a profit?

65. A magazine costs \$0.60 to produce and is sold to dealers for \$0.55. The publisher also gets advertising fees of 10% of the amount the dealers pay for the magazines, but only on an amount over 10,000 magazines. How many magazines must be published to make a profit?

66. The relationship between height, H, in inches and age, A, in boys 4 to 16 years of age is related by:

$$H = 2.31A + 31.26$$

Normal variation at these ages is ± 5% of H. Express the variability, as an inequality, in height of boys 10.5 years old.

67. A report states that, of 18,000 pieces of advertising mail sent out by a company, less than 20% is not opened by the recipients. Write an inequality that expresses the amount of unopened mail.

68. A product sells for $20 and has a unit cost of $15, and fixed costs of $200,000. Find the least number of products that must be sold to have a profit.

69. You are offered a job with two options:

(a) A salary of $45,000 per year, or
(b) A salary of $2,500 per month plus a commission of 6% on gross sales.

For what range of sales is the second plan better than the first?

1.3 Equations of Lines

Now that we have seen linear equations and inequalities in one variable, we'll next look at linear equations in two variables, also known as linear functions. The solutions to these equations will graph into a line in two dimensions, i.e., in a Cartesian coordinate system.

> **LINEAR EQUATION IN TWO VARIABLES**
>
> Any equation which can be written in the form
>
> $$y = ax + b,$$
>
> $a, b \in \mathbb{R}$.

VOCABULARY

Input Variable—The independent variable graphed on the horizontal axis in a Cartesian coordinate system (usually called x).

Output Variable—The dependent variable graphed on the vertical axis in a Cartesian coordinate system (usually called y or $f(x)$).

y-intercept—The point where the line crosses the y-axis.

x-intercept—The point where the line crosses the x-axis.

Slope—The steepness of the line, which is constant at any point on the line. (*Note:* Common synonyms for slope include (i) rate or rate of change, (ii) speed and (iii) velocity.)

Parallel—Two lines are parallel if they have the same steepness and never intersect each other.

Perpendicular—Two lines are perpendicular if they intersect in a right angle.

We only need two points to connect together and determine a line. Let's call those points (x_1, y_1) and (x_2, y_2). Then the slope of the line through those points will be the ratio of the vertical change to the horizontal change between the two given points.

> **SLOPE OF A LINE**
>
> $$m = \frac{y_2 - y_1}{x_2 - x_1} = \frac{y_1 - y_2}{x_1 - x_2}$$
>
> *Note:* For a horizontal line, the slope is 0. For a vertical line, the slope is undefined.

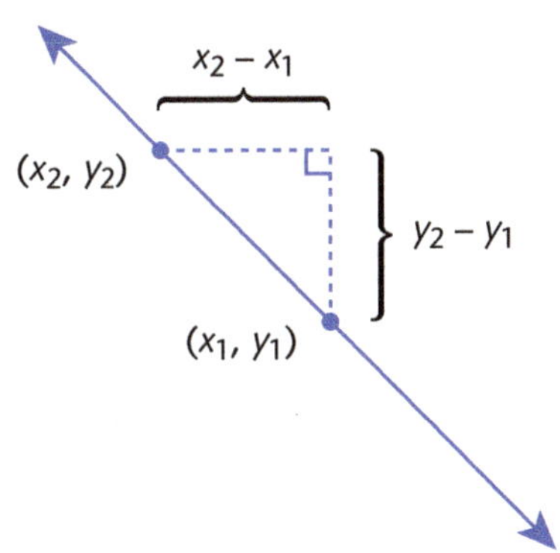

If the slope is positive, it means that the line is increasing from left to right. Likewise, a negative slope means that, looking from left to right, the line is decreasing.

EXAMPLE 1

Find the slope of the line through the points (2, 3) and (1, –5).

Let's call $(x_1, y_1) = (2, 3)$ and $(x_2, y_2) = (1, -5)$. Then the slope is

$$m = \frac{-5-3}{1-2} = \frac{-8}{-1} = 8.$$

PARALLEL LINES

Two lines are parallel if and only if the slopes of the lines are the same, i.e., $m_1 = m_2$.

PERPENDICULAR LINES

Two lines are perpendicular if and only if the slopes of the lines are the negative reciprocals of one another, i.e., $m_1 = \frac{-1}{m_2}$ and $m_2 = \frac{-1}{m_1}$. (*Note:* Horizontal and vertical lines are also perpendicular to each other, even though they do not fit this definition because vertical line slopes are undefined.)

EXAMPLE 2

Determine if the lines with the given slopes are (i) parallel, (ii) perpendicular or (iii) neither.

(a) $m_1 = 5$, $m_2 = \frac{1}{5}$ These two lines both have positive slope, but they're not the same slope. So they're neither parallel nor perpendicular.

(b) $m_1 = -3$, $m_2 = -3$ These two lines have exactly the same slope, which means they're parallel.

(c) $m_1 = \frac{-3}{4}$, $m_2 = \frac{4}{3}$ These two lines have slopes that are negative reciprocals of each other, so they're perpendicular. Another way to think of this is that $m_1 \cdot m_2 = -1$, which means they're perpendicular.

EQUATION OF A LINE

Point-Slope Form: Given a point (x_1, y_1) and the slope, m, of the line, the line equation is

$$y - y_1 = m(x - x_1).$$

Slope-Intercept Form: Given the slope, m, of the line and the y-intercept, $(0, b)$, the line equation is

$$y = mx + b.$$

In fact, we can write any line in its slope-intercept form, regardless of how the line equation is originally given. In the slope-intercept form, we can more easily see how to graph the line quickly. That will be the preferred format for the line equation for this text.

EXAMPLE 3

Find the slope and y-intercept for the line given by $-3x + 2y = 8$.

We need to legally move things around in the line equation in order to get y by itself. At that point, we'll be able to immediately find the slope and y-intercept.

$-3x + 2y = 8$ Add $3x$ to both sides of the equation.

$2y = 3x + 8$ To finish getting y by itself, divide both sides of the equation by 2.

$y = \frac{3}{2}x + 4$

In this format, we can see that the slope is $\frac{3}{2}$ and the y-intercept is the point $(0, 4)$.

(*Note:* The y-intercept is NOT the number 4. It is the entire point that needs the appropriate designation with two coordinates. The x-coordinate will always be zero for the y-intercept.)

EXAMPLE 4

Write the equation of the line given the following information.

(a) $m = 4$, y-intercept: $(0, -2)$

For this line, we need to substitute 4 for m and -2 for b in the slope-intercept equation of the line.
$\Rightarrow y = 4x + -2$ or $y = 4x - 2$

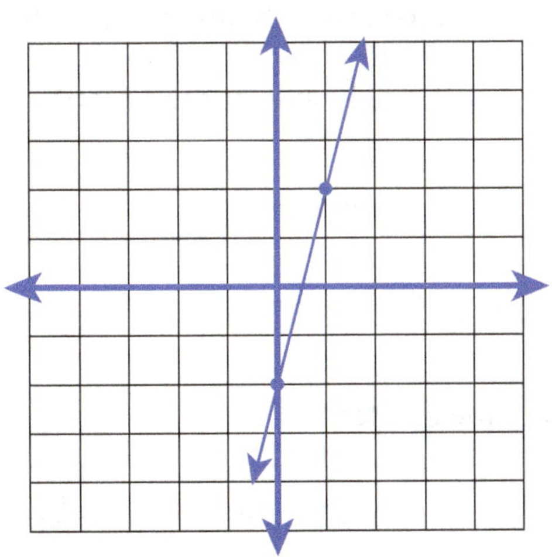

(b) $m = -3$, and the line goes through the point $(-2, 1)$

With this information, we'll use the point-slope formula of a line and substitute -3 for m, -2 for x_1 and 1 for y_1.

$\Rightarrow y - 1 = -3(x - (-2))$ Now, simplify on the right side of the equation.

$y - 1 = -3x - 6$ To get y by itself, add 1 to both sides of the equation.

$y = -3x - 5$

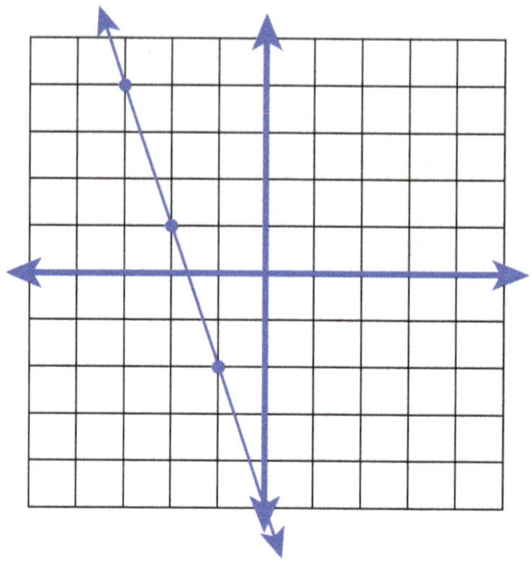

To graph this line, we can first plot the y-intercept, $(0, -5)$. From there, we can get another point using the slope information. We can think of the slope of -3 as $\frac{-3}{1}$ or $\frac{3}{-1}$. From the y-intercept, we'd go down 3 and to the right 1 or up 3 and to the left 1 to plot another point on the graph. We can continue to use the slope to plot more points from any other point on the line.

(c) The line goes through the points $(-1, 4)$ and $(2, 3)$.

Given this information, we need to first compute the slope of the line. We'll let $(x_1, y_1) = (-1, 4)$ and $(x_2, y_2) = (2, 3)$.

$$\Rightarrow m = \frac{4-3}{-1-2} = \frac{1}{-3} = \frac{-1}{3}$$

Now we have enough information to substitute into the point-slope formula for a line.

$y - 4 = \frac{-1}{3}(x - (-1))$ Simplify the right side of the equation.

$y - 4 = \frac{-1}{3}x - \frac{1}{3}$ Add 4 to both sides of the equation, to get y by itself.

$y = -\frac{1}{3}x - \frac{1}{3} + 4\left(\frac{3}{3}\right)$ Multiply 4 by a form of one to get a common denominator.

$y = \frac{-1}{3}x - \frac{1}{3} + \frac{12}{3}$ Combine like terms.

$y = \frac{-1}{3}x + \frac{11}{3}$ This is now simplified in slope-intercept form.

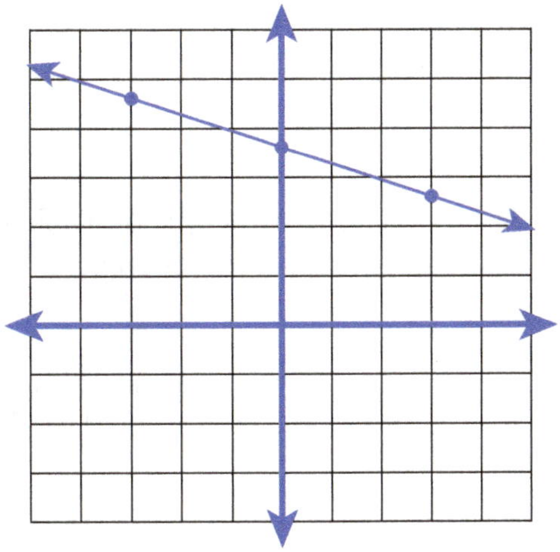

To graph this, we can plot the y-intercept: $(0, \frac{11}{3})$. Then, we can use the slope and plot more points by going down one and to the right 3.

EXAMPLE 5

Write the equation of the line through the point (8, –4) and perpendicular to the line $4x - 5y = 6$.

We have a point on the line, $(x_1, y_1) = (8, -4)$, but we need to know the slope of the line in order to use our point-slope formula to determine the line equation. We can gather the slope information from the line given, since our desired line is perpendicular to that line. The given line is

$4x - 5y = 6$ Subtract $4x$ from both sides of the equation. (We're trying to get y by itself.)

$-5y = -4x + 6$ To finish getting y by itself, divide both sides of the equation by –5.

$y = \frac{-4}{-5}x + \frac{6}{-5}$ Simplify one last step.

$y = \frac{4}{5}x - \frac{6}{5}$ This was the given line, not the line we are looking for. What we need from this line is the slope. In this slope-intercept form, we can see that the slope is $\frac{4}{5}$.

The line we are finding is perpendicular to this given line, which means its slope is $\frac{-5}{4}$.

To summarize, we now have $(x_1, y_1) = (8, -4)$ and $m = \frac{-5}{4}$, so we're ready to find the equation of the desired line.

$\Rightarrow$ $y - (-4) = \frac{-5}{4}(x - 8)$ Simplify the right side of the equation by distributing the $\frac{-5}{4}$ to the terms in the parentheses.

$y + 4 = \frac{-5}{4}x + \frac{5(8)}{4}$ Simplify the fraction on the right side of the equation.

$y + 4 = \frac{-5}{4}x + 10$ To finish getting y by itself, subtract 4 from both sides of the equation.

$y = \frac{-5}{4}x + 6$

To graph this equation, we can first plot the y-intercept (0, 6). From there, again, we'll use the slope information to plot another point, by going down 5 and to the right 4.

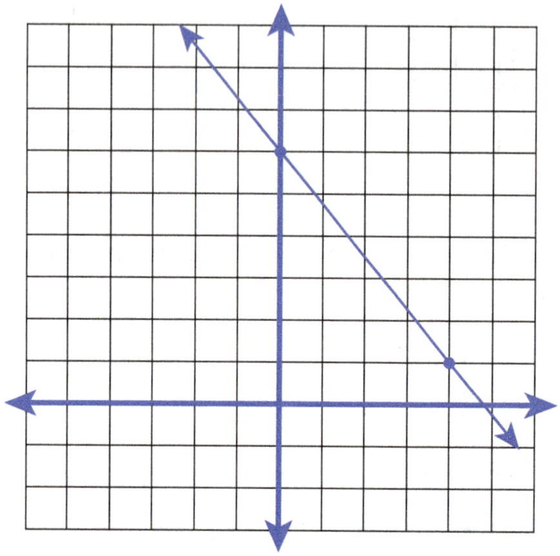

1.3 Homework

For problems 1–15, find the slope of the line through each set of two points:

1. (1, 2) and (3, 4)
2. (–8, 8) and (5, –5)
3. (3, 7) and (8, –8)
4. (3, –5) and (2, 5)
5. (–1, –5) and (5, –8)
6. (1, –1) and (–2, –3)
7. (8, –8) and (8, 7)
8. (–2, –10) and (6, –6)
9. (–3, 3) and (3, 3)
10. $(\frac{3}{4}, \frac{1}{2})$ and (2, 3)
11. (–16.3, 12.4) and (–5.2, 8.7)
12. $(\frac{7}{2}, \frac{1}{4})$ and $(-\frac{3}{4}, \frac{1}{2})$
13. (14, 3) and (2, 12)
14. (3, 2) and (–7, –5)
15. (7, 1) and (10, –3)

Determine whether lines with these slopes are parallel, perpendicular or neither:

16. $m_1 = 1,\ m_2 = 1$
17. $m_1 = 2,\ m_2 = -\frac{1}{2}$
18. $m_1 = \frac{3}{2},\ m_2 = -\frac{3}{2}$
19. $m_1 = \frac{1}{2},\ m_2 = -2$
20. $m_1 = -\frac{1}{3},\ m_2 = -3$
21. $m_1 = -\frac{1}{3},\ m_2 = 3$

Find the slope and y-intercept for these lines, if they exist:

22. $x + y = 2$
23. $6x + y = -1$
24. $3x - 2y = 6$
25. $4y - 5x = 12$
26. $8x + 3y + 9 = 0$
27. $7x - 10y = 4$

28. $y - 8x = 12$

29. $3x = 12 + y$

30. $5x - 6 = 15$

31. $2x - 17 = -y$

32. $4 - 5y + 7x = -10$

33. $17y + 4x + 3 = 7 + 4x$

34. $3y - 2x = 5 + 9y - 3x$

35. $4x = 9y + 36$

36. $-6x = 4y + 3$

37. $5x + 4y = 16$

38. $y = 3.8x$

39. $-8x - 7y = 24$

Write the equation of the line given the slope and y-intercept, and sketch the line.

40. $m = 1$, y-intercept is $(0, -2)$

41. $m = -1$, y-intercept is $(0, 9)$

42. $m = 3$, y-intercept is $(0, -8)$

43. $m = 0$, y-intercept is $(0, 2)$

44. $m = \frac{2}{3}$, y-intercept is $(0, 1)$

45. $m = -\frac{1}{2}$, y-intercept is $(0, -4)$

46. $m = -\frac{7}{3}$, y-intercept is $(0, -5)$

47. $m = 5$, y-intercept is $(0, 8)$

48. $m = \frac{4}{5}$, y-intercept is $(0, 28)$

49. $m = 3$, y-intercept is $(0, -3)$

50. $m = 0$, y-intercept is $(0, 5)$

51. m is undefined, x-intercept is $(2, 0)$

Write the equation of the line given the slope and a point that the line passes through, and sketch the line.

52. $m = 7$, point is $(0, 6)$

53. $m = 2$, point is $(1, 1)$

54. $m = -1$, point is $(1, 9)$

55. $m = -3$, point is $(-2, 4)$

56. $m = 1$, point is $(-2, -1)$

57. $m = -\frac{1}{3}$, point is $(10, -4)$

58. $m = \frac{5}{2}$, point is $(1, 2)$

59. $m = 4$, point is $(3, 2)$

60. $m = 5$, point is $(5, 4)$

61. $m = \frac{3}{2}$, point is $(6, 7)$

62. $m = -\frac{1}{3}$, point is $(4, 5)$

Write the equation of the line that passes through the two given points, and sketch the line.

63. (1, 3) and (–1, 5)

64. (1, –6) and (–1, 3)

65. (–9, –1) and (6, –6)

66. (–7, –2) and (–5, 1)

67. (–7, –2) and (5, 1)

68. (–8, 1) and (4, 5)

69. (–4, 3) and (8, 10)

70. (1, 4) and (5, 6)

71. (–1, –1) and (2, 2)

72. (0, –3) and (7, 0)

73. (0, 0) and (–3, 4)

74. $\left(\frac{1}{4}, -\frac{1}{2}\right)$ and $\left(\frac{3}{4}, 6\right)$

Determine whether the lines are parallel or perpendicular to each other:

75. $y = x + 6, \quad y - x = -2$

76. $y + 7x = -9, \quad y = -7x - 1$

77. $2x - 3y = 7, \quad 3x = -2y - 6$

78. $y = 4x - 5, \quad 4y = 8 - x$

79. $2x - 7 = y, \quad y - 2x = 8$

Write the equation of the line that passes through the given point and is parallel to the given line:

80. The point (2, –1) and the line $3x + 4y = 10$

81. The point (–3, 8) and the line $2x + 7y = 3$

82. The point (2, –8) and the line $4x - y = 4$

83. The point (3, 7) and the line $x + 2y = 6$

84. The point (0, 3) and the line $2x - y = 7$

85. The point (–7, 0) and the line $2y + 5x = 6$

Write the equation of the line that passes through the given point and is perpendicular to the given line:

86. The point (2, –1) and the line $3x + 4y = 10$

87. The point (7, 8) and the line $x + 2y = 7$

88. The point (–3, –1) and the line $8x - y = 5$

89. The point (–3, 4) and the line $2x - y = 5$

90. The point (5, 2) and the line $3x - 2y = 6$

91. The point (–2, 2) and the line $x + 5y = 6$

92. While vacationing at a beach, Carla rents a bicycle. The rental fee for the bicycle includes $7.00 to rent the bicycle and 25 cents per mile that she rides. Write a linear equation which shows how much Carla would owe for renting the bicycle and riding x miles. Identify the slope value in this equation and what this slope value means.

93. A car rental company rents cars for $100.00 plus 38 cents per mile. Write a linear equation which shows how much renting a car will cost if the car is driven x miles. Then identify the y-intercept of this equation and explain what it means.

94. Jorge is going bowling by himself. The cost of his bowling will be $2.00 to rent a pair of shoes and then $1.50 per game that he bowls. Write a linear equation which shows how much the bowling will cost Jorge if he bowls x games. Then identify the slope and y-intercept of this equation and explain what each means.

95. A manufacturing plant can build widgets which can be sold at a profit of $50.00 per widget over the cost of the raw materials used to create the widget. The overhead of running the manufacturing plant (electricity, employee salaries, etc.) is $4500.00 per day. If the manufacturing plant can create 100 widgets in a day, what is the total profit from its operations in a day? (*Hint:* Write a linear equation to represent the total profit for manufacturing x widgets in a day, and then substitute 100 for x.)

96. An investor can invest money through a broker at a guaranteed rate of return of 5%. The broker's fee for making the investment will be $100.00. Write a linear equation to represent the investor's profit from the investment. How much will the investor make if he invests $10,000.00? (*Hint:* Write a linear equation for the money returned from the investment of x dollars, then substitute 10,000 for x.)

97. The cost to rent a car and drive it 100 miles is $30. If you drive it 150 miles the cost is $37.50. Write a linear equation that fits these data.

98. Water freezes at 32°F which is 0°C. Water boils at 212°F which is the same as 100°C. What celsius temperature corresponds to 70°F ? Write a linear equation that fits these data.

99. An office machine is purchased for $5200 and after two years its value has depreciated to $4225. Write a linear equation that fits this data.

100. If a revenue function is $R(x) = 34x$ and the cost function is $C(x) = 17x + 3400$, write a linear equation for the profit function.

101. The r-value for 3.5 inches of insulation is 11 and the r-value for 6 inches is 19. Write a linear equation that relates r-value to thickness of insulation.

102. A building is purchased for $360,000 and after 5 years its value has depreciated to $270,000. Write a linear depreciation equation for this situation, put its value units in thousands ($).

103. In 1970 the life expectancy of males in the U.S. was 68 years. In 2005 it was 78 years. Write a linear equation to represent this information.

1.4 Systems of Linear Equations

VOCABULARY

System of Equations—Two or more equations, of two or more variables.

Solution—A solution of a system of equations is the point of intersection between two or more equations (of two or more variables). It's the point that lies on both lines or curves.

Let's consider first a system of two linear equations of two variables. Basically that means we have two line equations and we're trying to find the point of intersection between those two lines. Let's look at the three possible ways two lines might intersect.

One solution	No solution (i.e., the lines are parallel)	Infinitely many solutions (i.e., they're the same line)
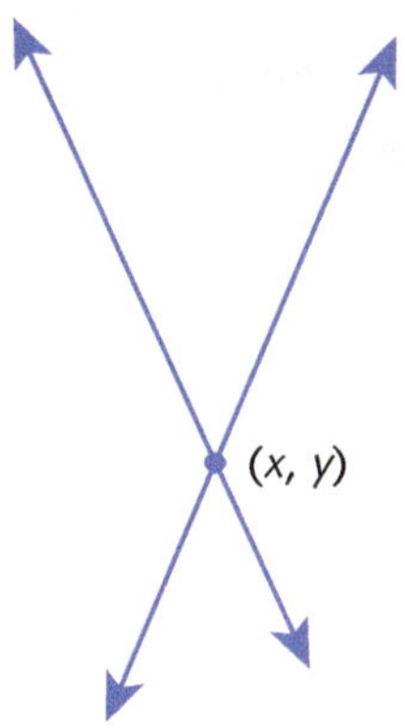	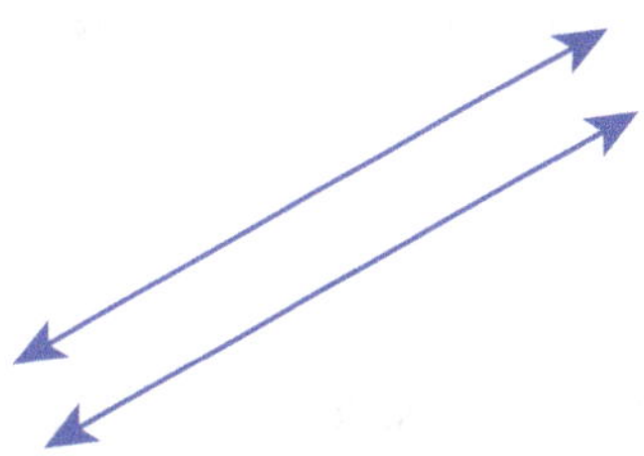	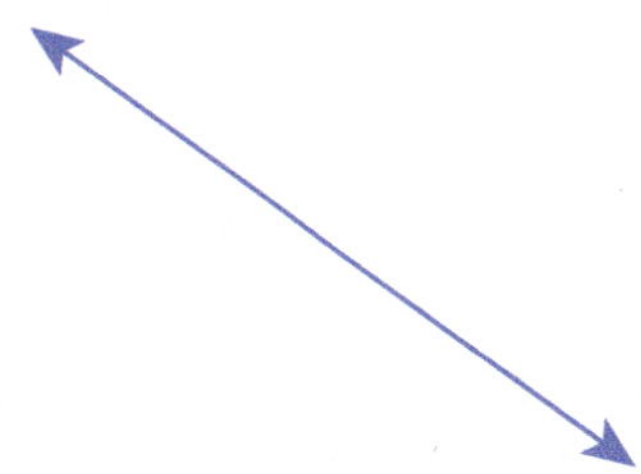

METHODS FOR SOLVING A SYSTEM OF TWO LINEAR EQUATIONS (IN TWO VARIABLES)

1. **Graphing**—This is the least reliable method because our graph may be slightly off when we draw it by hand, and we would have a hard time finding an intersection point like $(\frac{1}{3}, \frac{2}{7})$.
2. **Substitution**—This method will always work and is useful even beyond a linear system of equations.
3. **Elimination**—This method also will always work for solving a linear system of equations.

SUBSTITUTION METHOD

1. Solve for one of the variables in one of the two equations.
2. Substitute that variable's expression into the OTHER equation. This will yield a linear equation in one variable that we already know how to solve.
3. Solve for the variable.
4. Substitute that variable's value into the equation used in step 1 to get the value of the other variable.

EXAMPLE 1

Solve this system of equations. (We'll use the Substitution Method.)

(A) $2x - y = -3$
(B) $-x + 2y = 9$

I labeled the equations (A) and (B) for ease of use and reference.

STEP 1: I'll choose to solve for y in equation (A).

$-y = -2x - 3$	We subtracted $2x$ on each side of the equation.
$y = 2x + 3$	We multiplied both sides of the equation by -1.

STEP 2: Substitute that expression for y into equation (B).

$-x + 2(2x + 3) = 9$	Now, simplify.

STEP 3: Solve for x.

$-x + 4x + 6 = 9$	Collect like terms and subtract 6 from both sides of the equation.
$3x = 3$	To get x by itself, divide both sides of the equation by 3.
$x = 1$	

STEP 4: Substitute 1 for x in equation (A).

$2(1) - y = -3$	Simplify.
$2 - y = -3$	Add y and 3 to both sides of the equation.
$5 = y$	

⇒ Our point of intersection, a.k.a. the solution, is (1, 5).

ELIMINATION METHOD

1. Arrange both equations so that everything lines up vertically the same. That is, place both the x terms above each other, the y terms above each other, the = signs above each other and the constants above each other, in whatever order you like.
2. Decide which variable to eliminate.
3. Multiply one or both equations by a nonzero constant in order to make the coefficients of one of the variables be the opposite of each other, so that they add up to zero.
4. Add the equations straight down, thereby eliminating one of the variables. Solve that remaining equation for the first variable value.
5. Substitute that variable value into either equation to get the other variable value.

EXAMPLE 2

Solve this system of equations. (We'll use the Elimination Method.)

$$\begin{aligned} 2x - 3y &= 5 \\ 4x + y &= -11 \end{aligned}$$

STEP 1: This step has already been accomplished, so we have nothing extra to do here.

STEP 2: I will choose to eliminate y.

STEP 3: Multiply the bottom equation (both sides) by 3 in order to get $3y$.

$$\begin{aligned} 2x - 3y &= 5 \\ 3(4x + y) &= (-11)3 \end{aligned} \quad \Leftrightarrow \quad \begin{aligned} 2x - 3y &= 5 \\ 12x + 3y &= -33 \end{aligned}$$

We simplified the bottom equation.

STEP 4: Add straight down now to eliminate y.

$$\begin{aligned} 2x - 3y &= 5 \\ +\ 12x + 3y &= -33 \end{aligned}$$ Add straight down.

$14x = -28$ Now, solve for x by dividing both sides of the equation by 14.

$x = -2$

STEP 5: Substitute -2 in for x into either of the original equations.

$2(-2) - 3y = 5$ Simplify.

$-4 - 3y = 5$ Add 4 to both sides of the equation.

$-3y = 9$ To finish getting y by itself, divide both sides of the equation by -3.

$y = -3$

$\Rightarrow$ The solution for our system of linear equations is the point $(-2, -3)$.

Let's look at some examples where we don't get one nice point of intersection as the solution.

EXAMPLE 3

Solve the system of equations.

(A) $x + 3y = 3$
(B) $5x + 15y = 12$

We'll choose to use the substitution method for this example. Solve for x in equation (A).

(A) $x = 3 - 3y$

Now, substitute the expression for x into equation (B).

(B) $5(3 - 3y) + 15y = 12$	Simplify by distributing the 5 through each term in the parentheses.
(B) $15 - 15y + 15y = 12$	Simplify by collecting like terms.
(B) $15 = 12$	Hmmm, the y terms added to zero, and now we're left with a statement that 15 is the same as 12. We know that's not a true statement, which means that our system of equations has no solution because there is no value of x or y that will ever make 12 equal 15.

⇒ There is No Solution (N.S.). This means the lines are parallel to one another.

EXAMPLE 4

Solve the system of equations.

$$\begin{aligned} 4x - 3y &= 5 \\ -8x + 6y &= -10 \end{aligned}$$

This time around, we'll use the method of elimination to solve this system of equations. I'll choose to eliminate x.

$$\begin{aligned} 2(4x - 3y) &= (5)2 \\ -8x + 6y &= -10 \end{aligned}$$

Multiply the top equation by 2 (on both sides).

Now, we have

$$\begin{array}{r} 8x - 6y = 10 \\ + \ -8x + 6y = -10 \\ \hline 0 = 0 \end{array}$$

We can add straight down to eliminate the x terms.

What happened, though, was that both the x terms and the y terms added to zero, as did the right hand side of the equations. Once again we're left with an equation that is missing both variables. However, this equation that $0 = 0$ is true, as opposed to the last example. This means that any value of x and y will make that true, so there are infinitely many solutions.

⇒ These two lines are the same line and have all their points in common.

Next, let's solve a system of three equations with three variables. In this case, each equation represents a plane in three dimensions, and we're trying to find the intersection point between all three planes.

EXAMPLE 5

Solve the system of equations.

$$\begin{aligned} x + 3y - 8z &= 8 \\ y - 3z &= 1 \\ -2y + 7z &= -2 \end{aligned}$$

Because the last two equations only have two variables represented in them, for this problem we'll solve that system of two equations for y and z. Then, we'll substitute those y and z values into the first equation to find the value of x.

We'll use the substitution method to solve the middle equation for y.

$y - 3z = 1$ Add $3z$ to both sides of the equation.

$y = 1 + 3z$ (Call this equation A.)

Now, substitute that expression for y into the bottom equation.

$-2(1 + 3z) + 7z = -2$ Simplify.

$-2 - 6z + 7z = -2$ Collect like terms on the left side of the equation.

$-2 + z = -2$ Add 2 to both sides of the equation.

$z = 0$

Next, substitute 0 in for z in equation A.

$y = 1 + 3(0) = 1 + 0 = 1$

Last, substitute the values for y and z into the top original equation to solve for x.

$x + 3(1) - 8(0) = 8$ Simplify.

$x + 3 = 8$ Subtract 3 from both sides of the equation.

$x = 5$

$\Rightarrow$ The solution, i.e., the point of intersection between the three planes, is the point (5, 1, 0).

EXAMPLE 6

Charlie has \$200,000 invested in two mutual funds. One yields an annual return of 10% of his investment and the other returns 8% per year on the investment. His total annual return from the two funds is \$19,000. Let x = the amount invested in the 10% fund and y = the amount invested in the 8% fund. How much is invested in each fund?

We need to set up two equations, since we have two different variables. We can get the first equation by knowing that he's invested a total of $200,000.

This means $x + y = 200{,}000$.

We can get another equation by looking for two ways to represent the profit from the investments.

$0.10x$ = the amount of money Charlie makes on the first fund in one year
$0.08y$ = the amount of money Charlie makes on the second fund in one year
and 19,000 = the total amount of money he makes in one year from the two investments

This means $0.10x + 0.08y = 19{,}000$.

Now, we have a system of equations to solve.

(A) $x + y = 200{,}000$
(B) $0.10x + 0.08y = 19{,}000$

If we don't really want the decimals in the second equation, we can multiply both sides of the equation (B) by 100 to eliminate them.

$100(0.10x + 0.08y) = (19{,}000)100$ Simplify.

$10x + 8y = 1{,}900{,}000$ We can divide both sides of the equation by 2 now to simplify further.

$\frac{1}{2}(10x + 8y) = \frac{1{,}900{,}000}{2}$ Simplify.

$5x + 4y = 950{,}000$

Now our system of equations is:

(A) $x + y = 200{,}000$
(B) $5x + 4y = 950{,}000$

We'll use the method of elimination this time to solve this system of equations. Let's eliminate x. Then we need to multiply both sides of equation (A) by –5.

$$\begin{array}{rr} \text{(A)}\ -5x - 5y = & -1{,}000{,}000 \\ +\ \text{(B)}\ \ 5x + 4y = & 950{,}000 \\ \hline -y = & -50{,}000 \\ y = & 50{,}000 \end{array}$$

Since all terms are properly aligned, add straight down.
Multiply both sides of the equation by –1.

We can substitute that value of y into equation (A) to find x.

$x + 50{,}000 = 200{,}000$ Subtract 50,000 from both sides of the equation.
$x = 150{,}000$

⇒ Charlie invested $150,000 in the 10% fund and $50,000 in the 8% fund.

1.4 Homework

Solve these systems of equations, using the substitution method.

1. $3x - y = 5$
   $x = 4 - 2y$

2. $\sqrt{2}x + y = 1$
   $x - y = \sqrt{2}$

3. $3x - y = 5$
   $5x - y = 7$

4. $2x + y = 0$
   $2x + y = 3y + 7 - x$

5. $2x + 3y = 13$
   $3x + y = 7 - y$

6. $x - 4y = 8$
   $4x + y = -2$

7. $x + y = 0$
   $3y - x = 8$

8. $2x + y = 9$
   $3x = y + 11$

9. $\frac{1}{3}y = -x - 19$
   $x + \frac{1}{3}y = 5$

10. $x + \frac{3}{10}y = -2$
    $2x + \frac{7}{6}y = -4$

11. $x + \frac{7}{2}y = 12$
    $x + \frac{5}{3}y = 1$

12. $x - \frac{3}{4}y = -9$
    $\frac{1}{3}x = \frac{1}{4}y - 3$

Solve these systems of equations, using the elimination method.

13. $\frac{1}{2}(3x - y) = x - \frac{1}{2}$
    $\frac{1}{(2x - y)} = -1$

14. $3(2x + 3y - 2) = -x + y$
    $x + 5 = 2 - 5y$

15. $3x - y = 5$
    $5x - y = 7$

16. $2x + y = 15$
    $\frac{3x}{4y} + \frac{1}{y} = \frac{4}{7}$

17. $5x + 7y - 3 = 0$
$6x - 5y - 17 = 0$

18. $4x - 6y = -2$
$7x + 2y = 9$

19. $5x + 7y = 3$
$4x + 3y = 5$

20. $3x + 2y = -7$
$10y - 3x = -53$

21. $2x + 9y = -19$
$9x + 8y = 12$

22. $3x + 9y = 3$
$x + 3y = 1$

23. $4y = 3x - 1$
$7x = 5y + 11$

24. $-\frac{5}{4}x + \frac{1}{3}y = -\frac{27}{4}$
$15x - 4y = 1$

Solve these systems of equations, using any method.

25. $x + \frac{3}{2}y = -\frac{1}{2}$
$\frac{7}{8}x + \frac{9}{8}y = -\frac{5}{2}$

26. $\frac{3}{2}x = y + 2$
$\frac{7}{6}x + y = 14$

27. $\frac{1}{3}x - \frac{5}{6}y = 9$
$\frac{7}{3}x + \frac{7}{6}y = -7$

28. $9x - \frac{5}{2}y = -19$
$x - \frac{3}{2}y = 15$

29. $0.2x - y = -6$
$0.9x + 0.5y = 3$

30. $0.04x = 0.03y - 4$
$0.2x + 0.12y = -3.8$

31. $0.3x - 0.5y = 2.4$
$1.1x + 0.05y = 3.15$

32. $0.06x - 0.07y = 0.28$
$0.04x + 0.06y = 0.4$

33. $4.1x - 1.1y = 4.9$
$20.5x - 5.5y = 24.5$

34. $3.6x + 0.4y = -24.4$
$4.6x + 2.3y = 4.6$

35. $2.2x + 2.7y = 5.6$
$1.8x - 2.8y = 14.6$

36. $0.9x + 0.5y = -9.4$
$1.8x = -y$

37. $x + 1.3y = -5.3$
$2.1x + 2.4 = 6y$

38. $5x + 8y = -11$
$9x + 10y = 3$

39. $x + 3y = 5$
$2x = -6y + 10$

40. $-x - 5y = 2$
$15y = -6 - 3x$

41. $6x = y + 3$
$2y = 8x + 3$

42. $6x - 8y = -3$
$8x + 4y = 7$

43. $2x - 4y = 9$
$x = 2y + 3$

44. $5x + 3y = 7$
$y = \frac{5}{6} - \frac{5}{3}x$

45. $x = 4$
$y = 6$

46. $x = -3$
$y = 1$

47. $5x + 2 = 3y$
$2x - 4y = -7$

48. $2x - 3y = 4$
$x + y = 0$

49. $2x + 3y = 7$
$6x = -9y + 9$

50. $x + 3y = -1$
$5x - 6y = 3$

51. $x + 7y = 1$
$-2x - 14y = 3$

52. $x + 7y = 1$
$-2x - 14y = -2$

53. $1 = 10x + y$
$x - y = 1$

54. $4x + 5y = -10$
$4x - 5y = -5$

Solve the system of three linear equations.

55. $5x + y + 4z = 18$
$6y = 2z$
$7y + z = 10$

56. $14x + y - z = 16$
$2x - z = -5$
$3x - z = -1$

57. $6x - 4y + 3z = -3$
$5x = y + 8$
$9x + y = 6$

58. $4y = z + 42$
$14x + 8y + 5z = 70$
$3y + z = 35$

59. $2x + y + z = 8$
$-3y + 4z = -15$
$y - 2z = 7$

60. $3x + y - z = -16$
$2x + 5z = 29$
$-4x + z = 19$

61. $y + 3z = -7$
$2x - y + 4z = -21$
$3y + 2z = 7$

62. $5z = 15$
$x - 2y + 3z = 17$
$2x + 3y + z = 12$

63. $x + y + z = 8$
$3y = 6$
$-2x - y + 5z = -14$

64. $2x - y = 4$
$-2x + y + z = 0$
$-3z = 21$

65. $3x - y + z = 4$
$-6y = 12$
$6x - 3y + 2z = 5$

66. At a large school play, 1435 tickets were sold. A student ticket cost \$1.50 and an adult ticket cost \$5.00. If the ticket sales totaled \$3552.50, how many of each type of ticket was sold?

67. A mother and her daughter frost cupcakes for a birthday party. The mom frosts 3 cupcakes for every 2 cupcakes finished by the daughter. If they frost a total of 80 cupcakes, how many did each person frost?

68. Kayla works at a juice company. The company's most popular fruit juice contains 15% real fruit juice and 85% of a non-fruit juice. How much fruit juice should be added to 51 liters of non-fruit juice to make the popular fruit juice?

69. The length of a rectangle is 7 feet more than its width. The perimeter of the rectangle is 260 feet. What are the dimensions of the rectangle?

70. Suppose regular gas costs \$1.29 per gallon and premium unleaded gas costs \$1.84 per gallon. If the station sold \$79.98 of regular gas, how many gallons of premium unleaded gas must be sold to have an average selling price of \$1.608 per gallon?

71. If Gabrielle were 9 years older, she'd be three times as old as Olivia. Olivia is 27 years younger than Gabrielle. How old is Olivia?

72. Alexandra is 65 less than three times the age of Madison. Charlie is 39 less than two times the age of Madison. The sum of all their ages is 58. How old is Madison?

73. Brittany lives 31 miles from work. During the summer, she rides her bike at 9 mph to the metro station. She then takes the metro to work, which travels at 40 mph. If she spends 14 minutes less on the bike than on the metro, what is the distance from her house to the metro?

74. It takes 36 days for 27 people, who are all working together at the same rate, to build a house. How long would it take 9 people to build the house?

75. Matt has two types of iodine in his lab. He has brand 206 iodine, which costs 28 cents per ounce. He also has brand 822 iodine which costs \$1.70 per ounce. How many ounces of each should he mix together if he wants to make 367 ounces of iodine that has a total cost of \$203.26?

76. Jack's basketball team scored 41 less than two times the number of points that Dylan's team scored. The sum of both teams' final points was 106. How many points did each team score?

77. A concert made \$165,000 on the sale of 8,000 tickets. If the tickets sold for \$15 and \$25 each, how many of each type were sold?

78. A collection of dimes and quarters contains 100 coins worth \$14.50. How many dimes and quarters are there in the collection?

1.5 Functions

DEFINITIONS

Relation—A relation is defined by a set of ordered pairs (x, y), such that x is an input and y is the output that goes with x.

Function—A function is a relation that satisfies the condition that every input has *exactly one* output. The graph of a function is the set of (x, y) points that satisfy the function, as it's defined analytically.

Notation: function is typically represented as $f(x)$ and read "f of x." The input is x, and f is the output. (*Note:* There is no multiplication between the f and the x. The f is the name of the function acting on the input x.)

VOCABULARY

Domain—The set of allowable inputs.

Range—The set of outputs.

EXAMPLE 1

For each of these pairs of inputs with outputs, decide if it describes a relation or function.

(a) Input—person, Output—bicycle owned by the person.

Since any person could have more than one bicycle, and some people may have no bicycle at all, then this would describe a relation, not a function.

(b) x = student in Business Algebra, y = grade in that class

Since each student receives exactly one grade, this then describes a function.

(c) Input—person, Output—biological father

This also describes a function since every person has exactly one biological father.

(d) Input—person, Output—that person's child(ren)

This describes a relation because every person can have more than one child or no children at all.

EXAMPLE 2

Decide if the following input and output sets describe a function or a relation.

(a) Domain: {red, blue, green, yellow}
Range: {1, 2, 3, 4, 5, 6, 7}
Ordered pairs: {(red, 5), (blue, 7), (green, 2), (yellow, 2)}

This set of ordered pairs is a function because every single input has exactly one output. It's okay that two of the inputs have the same output; what matters is that each input has only one output. It also doesn't matter if all of the elements in the range are used as outputs. However, it is indeed necessary that each element in the domain is used as an input.

(b) **Domain: {Mary, Joe, Bob, Sue, Laura, Mark}**
Range: {1, 2, 3, 4, . . . , 19, 20}
Ordered pairs: {(Mary, 1), (Mary, 4), (Joe, 9), (Bob, 11), (Sue, 11), (Laura, 13)}

This set of ordered pairs is only a relation, for two reasons. One, the input Mary has two different outputs, and, two, the input Mark has no output associated with him.

(c) 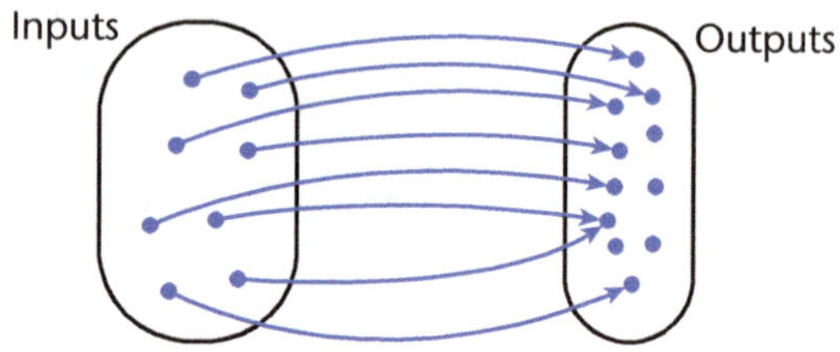

This is a function because every input is mapped to a single output.

VERTICAL LINE TEST

If we graph $f(x)$ in a Cartesian Coordinate System, then the input corresponds with the horizontal axis and the output corresponds with the vertical axis. One test to determine if the input-output relation is a function is the Vertical Line Test. This states that if every vertical line crosses the graph at most one time, then the graph represents a function. If there is any vertical line that crosses the graph more than once, then it's a relation (not a function.)

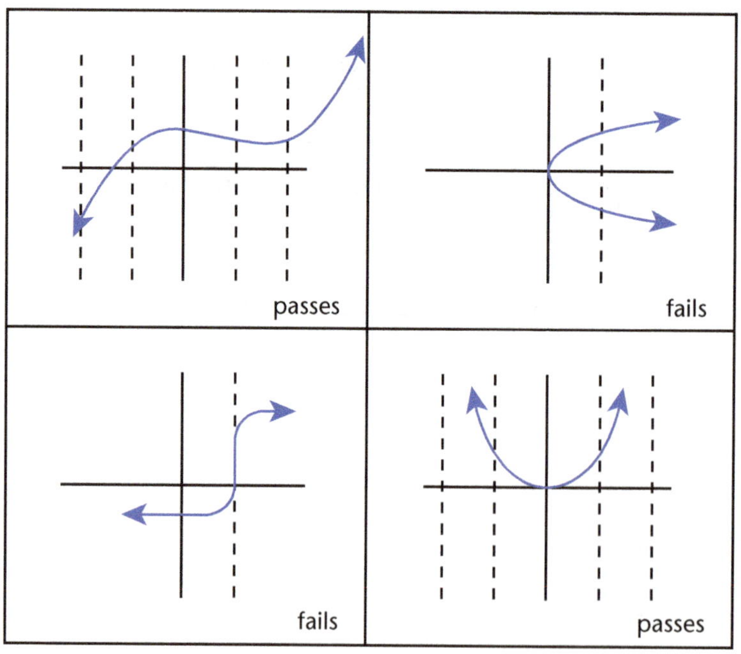

EXAMPLE 3

Determine if these are functions. Identify the domain.

(a) $y = f(x) = 3x^2 + x - 1$

The domain is the set of all allowable inputs, in this case, x-values. *In general at this level of mathematics, the restrictions in the domain will come from (1) making sure the denominator is not zero and/or (2) making sure we are taking the even root of only nonnegative numbers.* In this case, there is nothing happening to x that is ever disallowed. This means that the domain is all real numbers. We can notate that as $x \in \mathbb{R}$. That is basically our default domain, unless we have some particular restrictions we need to add for x.

So, is this a function? Yes, because regardless of what x (input) we put in, we will get back exactly one y-value.

(b) $y^2 = 3x^2$

The domain for this function again is asking for the set of allowable inputs. Since there is nothing happening to x that we have to be careful about, the domain is all real numbers, i.e., $x \in \mathbb{R}$.

Is this a function? No, because for every value of x we input, there are two possible y-values that will work. For example, if $x = 1$, then

$$y^2 = 3(1^2) \quad \Leftrightarrow \quad y^2 = 3 \quad \Leftrightarrow \quad y = \pm\sqrt{3}.$$

(c) $y = \dfrac{2x}{x^2 - 1}$

Is this a function? Yes, because for every x-value we put in, we get out only one y-value.

The domain of this function will be a little more interesting than the previous two examples. We need to restrict the values of x to ensure the denominator is never zero.

$$\Rightarrow \quad x^2 - 1 \neq 0 \quad \Leftrightarrow \quad x^2 \neq 1 \quad \Leftrightarrow \quad x \neq \pm 1$$

Thus, the domain is $x \in \mathbb{R}$, $x \neq \pm 1$.

EXAMPLE 4

Evaluate the following, given $f(x) = -2x^2 + 3x + 1$.

(a) $f(0)$

This is asking us to substitute 0 in for the input in our function definition given above.

$$\Rightarrow \quad f(0) = -2(0^2) + 3(0) + 1 = 0 + 0 + 1 = 1$$

That is, the output is 1 when the input is 0.

As a result, if we wanted to graph this function, we would plot the point (0, 1) and know the graph goes through that point.

(b) $f(3)$

For this problem, we need to substitute 3 in for x.

$$\Rightarrow \quad f(3) = -2(3^2) + 3(3) + 1 = -2(9) + 9 + 1 = -18 + 9 + 1 = -8$$

So, this function also goes through the point $(3, -8)$.

(c) $f(w)$

This is asking to substitute w in for x in our function.

$$\Rightarrow \quad f(w) = -2w^2 + 3w + 1$$

There is no more simplifying to do here.

(d) $f(♥)$

This may look a little absurd, but the point will be well taken. Regardless of what symbol or number we need to use for the input, the function, f, gives clear instructions what to do in order to get our output.

$$\Rightarrow \quad f(♥) = -2(♥)^2 + 3(♥) + 1$$

EXAMPLE 5

For the function $g(x) = x^2 + \sqrt{x+5} - 3$, answer the following questions.

(a) Find $g(x + h)$.

This is asking us to input $(x + h)$ (the whole thing) wherever we see the input x in our original function definition.

$$\Rightarrow \quad g(x+h) = (x+h)^2 + \sqrt{(x+h)+5} - 3$$

We can simplify this somewhat by multiplying everything out.

$$g(x+h) = (x+h)(x+h) + \sqrt{(x+h)+5} - 3 = x^2 + 2xh + h^2 - 3 + \sqrt{(x+h)+5}$$

(b) Find $g(x) + h$.

Notice that this is different than part (a). It boils down to the order of operations. Part (a) asks us to add h to x first, then apply the function g to that quantity. This question asks us to apply the function g to x first, and then add h to that.

$$\Rightarrow \quad g(x) + h = (x^2 + \sqrt{x+5} - 3) + h = x^2 + \sqrt{x+5} - 3 + h$$

Hopefully, you can see that the result of this is vastly different than the result for part (a).

(c) What is the domain?

Remember we have to constantly be on the lookout for possibly (1) dividing by zero and (2) taking an even root of a negative number, both of which are not allowed mathematically.

In this function, the square root causes some concern because we must limit the values of x so that we only take the square root of a non-negative number.

$\Rightarrow$ We need to force the stuff inside the square root to be greater than or equal to zero.

$\Leftrightarrow \quad x + 5 \geq 0 \quad \Leftrightarrow \quad x \geq -5$

Thus, the domain is $x \in \mathbb{R}$, $x \geq -5$.

1.5 Homework

For problems 1–10, indicate whether the set of ordered pairs represents a function or not.

1. {(2, 10), (8, 4), (1, 3), (9, 10), (6, 6), (7, 5)}
2. {(9, 5), (5, 10), (4, 7), (8, 2), (3, 4), (4, 9)}
3. {(1, 7), (9, 9), (5, 10), (3, 9), (2, 2), (7, 3)}
4. {(4, 10), (8, 8), (5, 10), (3, 2), (1, 4), (7, 2)}
5. {(8, 4), (6, 7), (1, 9), (6, 4), (7, 1), (9, 9)}
6. {(8, 4), (9, 6), (10, 4), (6, 5), (4, 1), (5, 7)}
7. {(10, 8), (1, 4), (5, 9), (2, 3), (6, 3), (7, 9)}
8. {(3, 6), (10, 1), (8, 4), (6, 2), (5, 10), (4, 2)}
9. {(5, 9), (1, 4), (8, 1), (8, 3), (9, 5), (2, 9)}
10. {(10, 10), (5, 9), (9, 2), (2, 1), (7, 5), (4, 2)}

For problems 11–20, indicate whether the relation represents a function of x; i.e. does each value of x yield a unique value for y.

11. $y = x$
12. $3x + 2y = 5$
13. $y = 2x^2 + 8x + 2$
14. $y = \sqrt{7x + 1}$
15. $3x^2 + 4y^2 = 7$
16. $y = \dfrac{x + 1}{x - 2}$
17. $y = \dfrac{3}{\sqrt{3x - 1}}$
18. $y = \dfrac{1}{x^2 + 2x - 15}$
19. $5 = y^4 - x^4$
20. $y = x^5 - 4x^4 + 2x^3 - 3x^2 + 5x - 1$

For problems 21–30, indicate whether the set of ordered pairs represents a function or not and then, if it is a function, find both the domain and the range of the function.

21. {(10, 2), (3, 4), (4, 6), (7, 10), (5, 9), (6, 3)}
22. {(9, 9), (4, 9), (1, 5), (5, 2), (7, 2), (8, 10)}
23. {(1, 2), (1, 5), (10, 8), (7, 8), (9, 1), (2, 3)}
24. {(1, 4), (4, 7), (6, 1), (8, 4), (6, 9), (4, 3)}
25. {(10, 2), (8, 1), (1, 3), (6, 9), (9, 1), (7, 10)}
26. {(4, 5), (5, 7), (6, 1), (7, 4), (8, 9), (2, 3)}
27. {(4, 9), (5, 10), (7, 2), (6, 6), (3, 8), (8, 5)}
28. {(5, 6), (7, 6), (9, 1), (1, 10), (6, 10), (10, 2)}
29. {(3, 2), (1, 9), (2, 4), (9, 5), (4, 5), (6, 8)}
30. {(7, 8), (2, 8), (8, 6), (1, 3), (3, 6), (4, 4)}

For problems 31–40, indicate whether the relation represents a function of x; i.e. does each value of x yield a unique value for y. Then, if it is a function, find both the domain and range of the function.

31. $10x - y = 2$
32. $y = 9x^3 + 2x - 5$
33. $y = \sqrt{x + 10}$
34. $x^2 + 2y^2 = 10$
35. $y = \dfrac{2}{x - 3}$
36. $y = \dfrac{1}{x - 5}$
37. $y = \dfrac{x + 1}{x}$
38. $y = \sqrt{10 - 5x}$
39. $y = \sqrt{3 - x} + \sqrt{1 + x}$
40. $y = \dfrac{1}{x - 4} - \dfrac{1}{x - 1}$

For problems 41–57, evaluate each function for the given input value.

41. $f(x) = 5x^2 + 6x + 10$, with $x = 5$
42. $f(x) = \sqrt{4x - 5}$, with $x = 10$
43. $f(x) = \sqrt{20 - 2x}$, with $x = -8$
44. $f(x) = \dfrac{x - 3}{x + 4}$, with $x = -5$
45. $f(x) = \dfrac{1}{\sqrt{5x - 3}}$, with $x = 3$
46. $f(x) = 9 - 7x$, with $x = 2u - 6$
47. $f(x) = x^2 - x + 7$, with $x = 8 - 2a$
48. $f(x) = x^2 - 4x + 4$, with $x = b + 2$
49. $f(x) = \dfrac{x + 1}{x + 3}$, with $x = \dfrac{3w - 1}{w - 1}$
50. $f(x) = x^3 - 7x + 6$, with $x = p + 2$
51. $g(x) = \sqrt{x^2 - 2}$, with $x = 4$
52. $h(x) = \dfrac{2x + 1}{x - 3}$, with $x = 2$

53. $h(y) = y^4 - 3y^2 + y + 2$, with $y = -2$

54. $g(p) = \dfrac{p^2 - 4}{p - 4}$, with $p = 5$

55. $f(x) = x^2 - 4$, with $x = y + h$

56. $g(y) = y^2 - 2y + 3$, with $y = w^2$

57. $h(x) = 4x - 3$, with $x = h(x)$

For the following functions, find the domain.

58. $f(x) = x^2 - 4$

59. $f(x) = \dfrac{x + 2}{x - 4}$

60. $g(x) = \dfrac{x^2 - 2x + 1}{\sqrt{x - 6}}$

61. $f(x) = \sqrt{x - 5}$

62. $g(x) = \dfrac{x}{x^2 - 16}$

63. $h(x) = \dfrac{x^4 - 4}{x^3 - 8}$

64. $h(x) = \sqrt{2x - 1}$

65. $g(x) = \dfrac{\sqrt{x + 5}}{\sqrt{5 - x}}$

66. $h(x) = \dfrac{\sqrt{x + 2}}{x^2 - 4}$

Sketch a graph of each relation by plotting points. Decide whether each relation is a function using the vertical line test.

67. {(0, 1), (2, 4), (–1, 3), (2, 1)}

68. {(–4, –2), (2, 3), (–1, 1), (3, 0), (0, 0)}

69. $x^2 = y$

70. $x = y^2$

Decide whether each relation pictured is a function.

71.

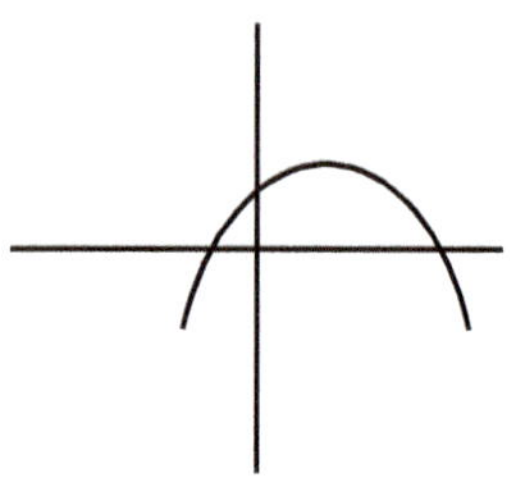

72.

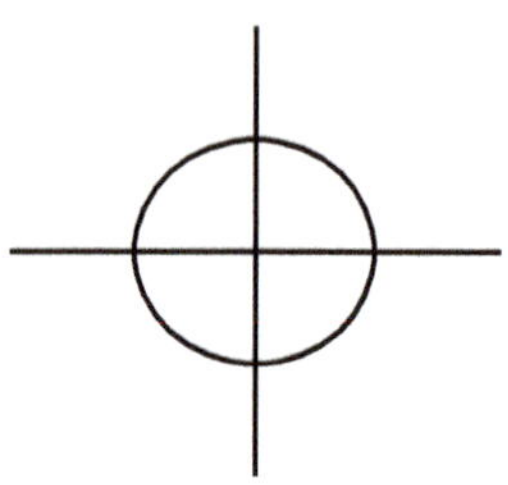

73.

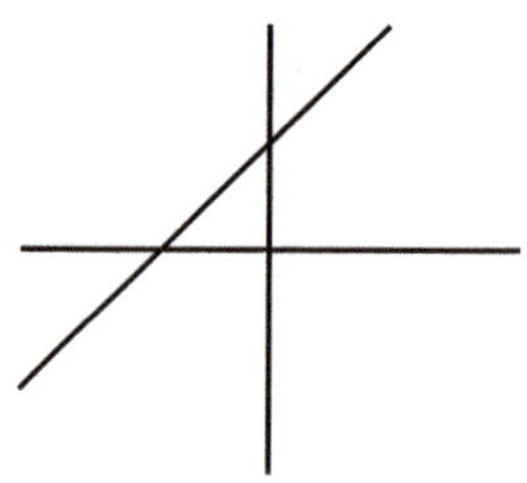

74.

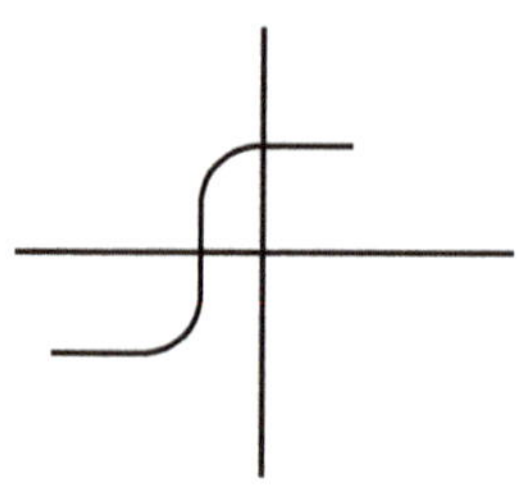

1.6 Linear Business Applications

In this section, we'll look at two types of business applications: (1) Profit, Revenue, Cost problems and (2) Supply/Demand problems.

PROFIT/REVENUE/COST

VOCABULARY

Revenue—The amount of money earned in a business, typically from selling a service or a product, usually denoted as $R(x)$.

Cost—The amount of money spent to produce and sell the product or service. Total costs consist of (a) Fixed Costs and (b) Variable Costs, and they are usually denoted as $C(x)$.

Fixed Costs—Those costs that need to be paid regardless of how much is produced (e.g., utilities or rent).

Variable Costs—Those costs that depend on how many items are produced.

Profit—The amount of money left after subtracting the costs from what money a business earned. That is, $P(x) = R(x) - C(x)$.

Break-Even Point—The point (x, y) where x = number of items produced and sold that will yield zero profit (i.e., where revenue = cost) and y = the revenue or cost associated with those x items.

Marginal—An adjective that describes how much something changes for one more unit sold. For example, we could ask for marginal profit, which would indicate how much profit changed for one more unit sold. It's basically the slope of the profit function!!

EXAMPLE 1

Suppose a watch manufacturer has a total cost function of $C(x) = 40x + 2500$ and a revenue function of $R(x) = 220x$.

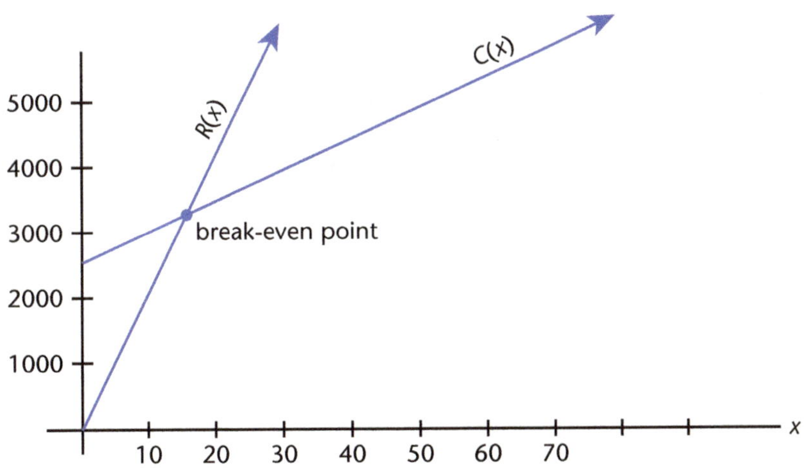

(a) What is the profit function, $P(x)$?

Remember that profit = revenue – cost. So,

$P(x) = 220x - (40x + 2500)$	Distribute the subtraction through the parentheses.
$P(x) = 220x - 40x - 2500$	Simplify further.
$P(x) = 180x - 2500$	

(b) What is the profit earned from producing and selling 30 watches?

$P(30) = 180(30) - 2500$	Substitute 30 in for x in the profit function from part (a).
$P(30) = 5400 - 2500 = 2900$	So, the total profit from selling 30 watches is \$2,900.

(c) How many watches must the business sell to break even?

For this question, we want to know what x is when $P = 0$. Set P equal to zero for our profit function in part (a).

$0 = 180x - 2500$	Now, solve for x. First, add 2,500 to both sides of the equation.
$2500 = 180x$	Divide both sides of the equation by 180 to finish getting x by itself.
$\frac{2500}{180} = x$	Simplify further.
$x = \frac{250}{18} = \frac{125}{9} = 13\frac{8}{9}$	This means the company needs to sell 13 and $\frac{8}{9}$ watches to break even, i.e., to cover all their costs without making any additional money. But, of course, $\frac{8}{9}$ of a watch cannot be sold. Thus, we'd round up and say that the business needs to sell 14 watches in order to break even.

(d) What is the marginal revenue?

We can denote marginal revenue as MR. The marginal revenue basically asks for the slope of the revenue line. We know $R(x) = 220x$ is the revenue, and we remember that $y = mx + b$ is the slope-intercept form of a line. In our case then, $y = R$, $m = 220$ and $b = 0$. That implies the slope of the revenue line is 220. So, the marginal revenue is 220, which means that if we sell one more watch, we'll bring in an additional \$220.

(e) What is the marginal profit?

We'll do the same analysis as in part (d). However, we need to use the profit function this time. We know from part (a) that $P(x) = 180x - 2500$. By comparing this equation to the slope-intercept form of a line again, we can see that the slope is 180, which is exactly the marginal profit. This means that for each additional watch sold, the company profits \$180.

EXAMPLE 2

My nephew, Jack, who lives in hot, sunny Arizona, runs a successful lemonade stand in the summer months. To stay in the shade, he rents a space in front of the grocery store down the street and has fixed costs of \$500 per summer. His variable costs, including the supplies, are \$0.40 per cup. He charges \$1.00 per cup for his yummy homemade lemonade.

(a) **What is the revenue function? The cost function?**

Let x = the number of cups of lemonade he sells per summer.

The revenue function represents how much money he brings in from selling his lemonade.

$$\text{Then} \quad R(x) = 1.00x = x.$$

The cost function represents the total costs incurred from operating his lemonade business.

$$C(x) = 500 + 0.40x \quad \text{The fixed costs plus variable costs.}$$

(b) **How many cups of lemonade does Jack need to sell every summer in order to break even?**

We want to know when profit is zero, or another way to say that is when revenue equals cost.

$R(x) = C(x)$

$x = 500 + 0.4x$ We want to solve for x. Subtract $0.4x$ from both sides of the equation.

$0.6x = 500$ To finish getting x by itself, divide both sides of the equation by 0.6.

$x = \dfrac{500}{0.6}$ Now, we need to simplify this quotient.

$$x = \frac{500}{0.6} = \frac{500}{0.6}\left(\frac{10}{10}\right) = \frac{5000}{6} = \frac{2500}{3} = 833\frac{1}{3}$$

So Jack needs to sell 833 and $\frac{1}{3}$ cups of lemonade per summer in order to break even. Since he can't sell $\frac{1}{3}$ of a cup of lemonade, we need to round up. Thus, he needs to sell 834 cups of lemonade per summer to break even.

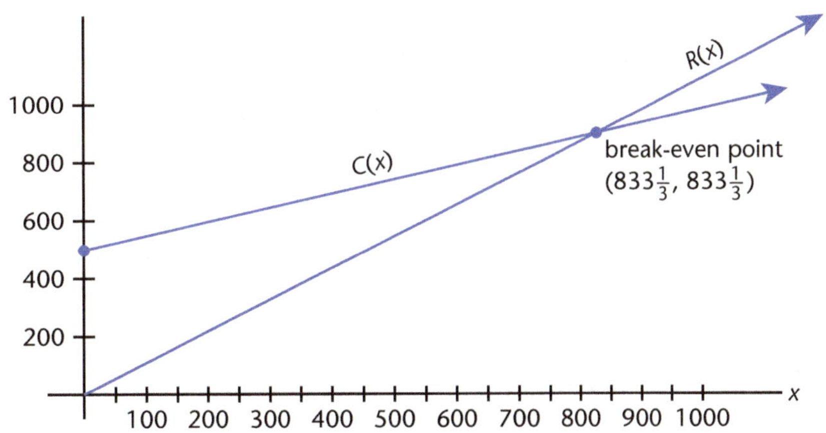

(c) What is the profit function?

$P(x) = R(x) - C(x) = x - (500 + 0.4x)$ Simplify further by distributing the subtraction through the parentheses.

$P(x) = x - 500 - 0.4x$ Continue simplifying.

$P(x) = 0.6x - 500$

(d) Last summer, Jack sold 3,000 cups of lemonade. What was his profit?

Substitute 3,000 in for x in the profit function from part (c).

$$P(3000) = 0.6(3000) - 500 = 1800 - 500 = 1300$$

So, Jack made \$1,300 last summer selling lemonade. That's a pretty good business for a ten-year-old.

SUPPLY/DEMAND

VOCABULARY

Price—The amount of money charged for a product; denoted as p; we will graph it as the vertical coordinate on a Cartesian Coordinate System.

Quantity—The number of products sold; denoted as q; we will graph it as the horizontal coordinate on a Cartesian Coordinate System.

Supply Equation—The equation that relates p and q, explaining mathematically how producers are willing to supply quantities at various prices.

Demand Equation—The equation that relates p and q, explaining mathematically how consumers demand products at various prices.

Equilibrium Point—The point (q, p) that represents where the supply and demand equations intersect in a Cartesian Coordinate System. This point indicates the quantity and price where both producers and consumers are willing to meet.

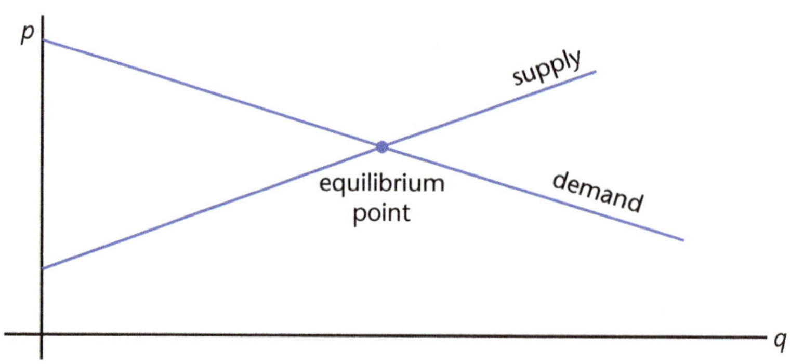

EXAMPLE 3

Given the supply equation for a cell phone is $45p - 30q = 600$ and the demand equation is $p = 200 - 2q$.

(a) What number of cell phones is the producer willing to supply at a price of \$25?

We need to use the supply equation here and substitute 25 in for p.

$45(25) - 30q = 600$ — Simplify.

$1125 - 30q = 600$ — Subtract 1,125 from both sides of the equation.

$-30q = -525$ — Divide both sides of the equation by –30.

$q = \frac{-525}{-30}$ — Simplify the quotient.

$$q = \frac{-525}{-30} = \frac{105}{6} = \frac{35}{2} = 17\frac{1}{2}$$

Since we can't supply part of a cell phone, we need to round this answer up to 18. Thus, the supplier is willing to supply 18 cell phones at a price of \$25. (Note: We could also round down, instead of rounding up.)

(b) What number of cell phones will consumers demand at a price of \$25?

Here, we need to use the demand equation and substitute 25 in for p.

$25 = 200 - 2q$ — To work toward solving for q, subtract 200 from both sides of the equation.

$-175 = -2q$ — To finish getting q by itself, divide both sides of the equation by –2.

$\frac{-175}{-2} = q$ — Simplify this quotient.

$$q = \frac{175}{2} = 87\frac{1}{2}$$

Again, since we can't demand half of a phone, we need to round this answer up to 88. Thus, consumers will demand 88 cell phones at a price of \$25.

(c) Discuss the results of parts (a) and (b).

Basically, we can see that consumers will demand many more cell phones at a price of \$25 than the supplier is willing to supply, probably because the supplier cannot profit on such a low price. It's unlikely, then, that \$25 will be a good price for the supplier.

(d) What is the equilibrium point?

Here, we want to know what price and quantity both the supplier and the consumers will agree upon. We basically have a system of two equations with two variables, and we need to solve the system of equations. We can use either the substitution or the elimination method.

(A) $45p - 30q = 600$
(B) $p = 200 - 2q$

This is already set up nicely to use the substitution method. Let's substitute the expression for p in B into equation A.

$\Rightarrow \quad 45(200 - 2q) - 30q = 600$	Distribute the 45 through the parentheses.
$9000 - 90q - 30q = 600$	Collect like terms on the left side of the equation.
$9000 - 120q = 600$	Subtract 9,000 from both sides of the equation.
$-120q = -8400$	To finish getting q by itself, divide both sides of the equation by -120.
$q = \frac{-8400}{-120} = \frac{840}{12} = 70$	

We now have a value for q. We need to substitute 70 in for q into equation B to find the value of p.

$p = 200 - 2(70)$	Now simplify to find the value of p.
$p = 200 - 140 = 60$	

Thus, the equilibrium point (q, p) is (70, 60). This means that at a price of \$60 the supplier will supply and the consumer will demand 70 cell phones.

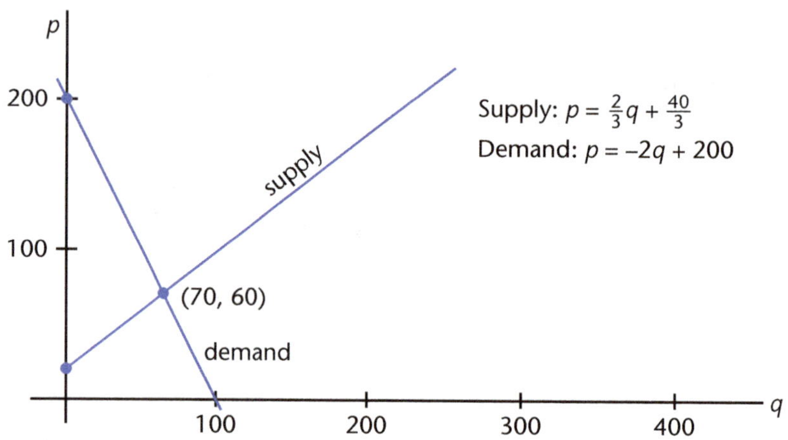

EXAMPLE 4

Retailers will buy 150 jackets from a wholesaler if the price is \$36 each, but only 75 jackets at a price of \$51. The wholesaler will supply 100 jackets at \$40 and 300 at \$60 each. Assuming the supply and demand equations are linear, find the market equilibrium point.

The supply and demand equations were not explicitly given for this problem. However, there is enough information to uncover the supply and demand equations. Let's work with the demand equation first.

We know two points that the demand line goes through, namely (150, 36) and (75, 51). Now we just need to find the equation of the line through those points.

$\text{Slope} = m = \dfrac{51-36}{75-150}$ This is just using the formula for slope of a line. Next, simplify.

$m = \dfrac{51-36}{75-150} = \dfrac{15}{-75} = \dfrac{-3}{15} = \dfrac{-1}{5}$

Equation of the Demand Line

Use the point-slope formula for a line $y - y_1 = m(x - x_1)$ (for our purposes $x = q$ and $y = p$).

$p - 51 = \dfrac{-1}{5}(q - 75)$ Distribute the slope through the parentheses.

$p - 51 = \dfrac{-1}{5}q + 15$ Add 51 to both sides of the equation.

$p = \dfrac{-1}{5}q + 66$ If we want to get rid of the fraction, multiply both sides of the equation by 5.

$5p = -q + 330$ This is the **demand equation**.

Do this exact same process to determine the supply equation.

We know two points that the supply line goes through, namely (100, 40) and (300, 60). Now we just need to find the equation of the line through those points.

$\text{Slope} = m = \dfrac{60-40}{300-100}$ This is just using the formula for slope of a line. Next, simplify.

$m = \dfrac{60-40}{300-100} = \dfrac{20}{200} = \dfrac{2}{20} = \dfrac{1}{10}$

Equation of the Supply Line

Use the point-slope formula for a line $y - y_1 = m(x - x_1)$ (for our purposes $x = q$ and $y = p$).

$p - 40 = \dfrac{1}{10}(q - 100)$ Distribute the slope through the parentheses.

$p - 40 = \dfrac{1}{10}q - 10$ Add 40 to both sides of the equation.

$p = \dfrac{1}{10}q + 30$ If we want to get rid of the fraction, multiply both sides of the equation by 10.

$10p = q + 300$ This is the **supply equation**.

Now we can use either the substitution or elimination method to determine the point of intersection between the supply and demand lines. Let's use elimination this time.

$$\begin{array}{r} 5p = -q + 330 \\ +\ 10p = q + 300 \\ \hline 15p = 630 \end{array}$$

Since all terms are in the same order in both equations, we can add straight down.

Divide both sides of the equation by 15.

$p = \dfrac{630}{15} = 42$ So, the equilibrium price is \$42.

Substitute 42 in for p into either the supply or the demand equation to determine the value of q.

$10(42) = q + 300$	Now, simplify the left side of the equation.
$420 = q + 300$	Subtract 300 from both sides of the equation.
$q = 120$	

⇒ The market equilibrium point is (120, 42). That is, both the wholesaler and retailer agree to exchange 120 jackets at a price of $42 each.

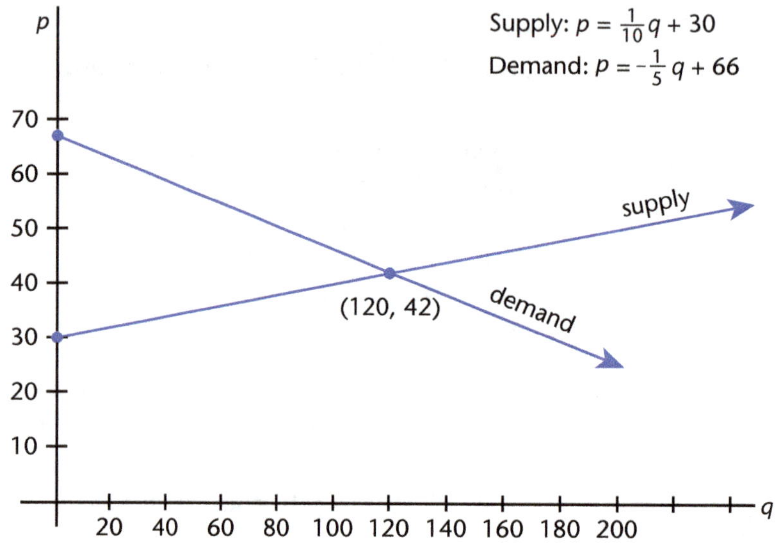

1.6 Homework

1. Printing a particular book has a cost function of $C(b) = 3.5b + 8900$ and a revenue function of $R(b) = 7.95b$ where these two functions are in dollars and b is the number of books printed.

   (a) What is the amount of the Fixed Costs?
   (b) What is the value of the Marginal Cost?
   (c) What is the value of the Marginal Revenue?
   (d) What is the fully simplified version of the Profit Function: $P(b)$?
   (e) How many books need to be printed to get to the break-even point?

2. Remanufacturing diesel engines has a cost function of $C(e) = 2e + 7500$ and a revenue function of $R(e) = 8e$ where these two functions are in dollars and e is the number of engines produced.

   (a) What is the amount of the Fixed Costs?
   (b) What is the value of the Marginal Cost?
   (c) What is the value of the Marginal Revenue?
   (d) What is the fully simplified version of the Profit Function: $P(e)$?
   (e) How many engines need to be remanufactured to break even?

3. A company will be selling a new type of chair. The demand equation for this chair is $80p = 625800 - 9q$, and the supply equation for this chair is $500p - 3q = 12600$.

   (a) How many chairs could be sold (demand) if the price were $330.00 ?
   (b) How many chairs would the company be willing to sell (supply) if the price were $600.00 ?
   (c) How many chairs could be sold if the price were $510.00 ?
   (d) How many chairs would the company be willing to sell if the price were $510.00 ?
   (e) What is the equilibrium point for this market for chairs?

4. Market research has shown that for a sporting event, the supply equation for tickets is $20p - q = 100$ and the demand equation is $4p = 6260 - 5q$.

   (a) How many tickets will the fans purchase if the price is $30.00 ?
   (b) How many tickets will the sponsors of the event be willing to sell if the ticket price is $30.00 ?
   (c) How many tickets will the fans purchase if the price is $100.00 ?
   (d) How many tickets will the sponsors of the event be willing to sell if the ticket price is $100.00 ?
   (e) What is the equilibrium point for this market for sporting event tickets?

5. A marketing research company has polled potential buyers of a new type of MP3 player and found that if the cost of the player is \$35.00 then there should be approximately 4,000,000 buyers nationally and that if the cost of the player is \$70.00 then there should be approximately 2,000,000 buyers nationally. The manufacturer of the new player is willing to sell these at a price of \$80.00 if 5,000,000 of the players can be sold, but would sell these for \$40.00 if only 2,000,000 could be sold.
   (a) What is the demand equation? (*Hint:* Calculate the slope and then use the Point-Slope form of a linear equation.)
   (b) Approximately how many buyers would buy this new player if its price was \$50.00?
   (c) What is the supply equation? (*Hint:* Calculate the slope and then use the Point-Slope form of a linear equation.)
   (d) What price would the manufacturer request if there were 4,400,000 players available for sale?
   (e) What is the equilibrium point for this market based on the research company's data?

6. A company will be marketing youth sports uniforms and found that if the cost of a uniform is \$35.00 then there should be approximately 3,750 uniforms purchased and that if the cost of a uniform is \$110.00 then there should be approximately 750 uniforms purchased. The company is willing to make 1000 uniforms if they can be sold at a price of \$50 per uniform and would make 3000 uniforms to be sold for \$90 each.
   (a) What is the demand equation?
   (b) How many uniforms could be sold if the price was \$50.00?
   (c) What is the supply equation?
   (d) What price would the company request per uniform if there were 1,200 uniforms available for sale?
   (e) What is the equilibrium point for this market based on the research company's data?

7. The KM Book Company wants to publish an adventure novel and has determined that fixed costs would be \$55,000 and variable costs are \$1.60 per book. The LB Book Distributing Company will buy them for \$11 each.
   (a) Write the cost function.
   (b) Write the revenue function.
   (c) Write the profit function.
   (d) How many books must be sold for the publisher to break even?
   (e) What profit would the publisher expect if the distributor purchased 12,000 books?

8. The KM Book Publishing Company was approached to publish a cookbook. Fixed costs were determined to be \$92,000 and the variable cost to produce each book was \$2.10.
   (a) One distributor will buy this cookbook for \$15 each. How many cookbooks must the publisher sell to break even?
   (b) If the cookbook goes into a second printing the variable costs will be \$1.95 per book. What is the new break even point for the publisher?
   (c) A second distributor offers to buy 15,000 copies from this second printing at a price of \$12 each. What profit will this produce for the publisher?
   (d) What is the marginal revenue using your results from the previous problem above?
   (e) What is the marginal profit from the problem above?

9. Fixed costs for a roller skate manufacturer are \$400 per day with a total cost of \$4,500 per day, where output is 100 pairs of skates per day.

   (a) What are the variable costs per day?
   (b) What is the marginal cost for a day's output?
   (c) If a pair of skates is sold to a retailer for \$60.00, what is the manufacturer's revenue for 200 pairs of skates?
   (d) What is the marginal revenue for problem (c)?
   (e) What is the manufacturer's profit for selling 600 pairs of skates?

10. A donut shop has fixed costs of \$175 per day and variable costs of \$0.15 per donut.

    (a) What is the shop's total cost function per day?
    (b) How many donuts can be produced for \$500 per day?

11. One motorcycle costs a total of \$7647 to manufacture. If variable costs are \$1250 per motorcycle what are the fixed costs?

12. If the motorcycles from problem (11) sell for \$8500 each, to a local retailer, what is the marginal profit per motorcycle?

13. If the motorcycle manufacturer in problem (11) above sells 5 motorcycles and all are sold at \$8500 each, what is the profit?

14. A craftsman makes and sells leather belts for \$15 each. If his tools and machines required \$300 in one-time setup costs and his variable costs are \$9 per belt, how many belts must he sell to break even?

15. A manufacturer sells pocket watches to retailers for \$45 each. Monthly fixed costs are \$12,000 and variable costs are \$28 each.

    (a) Write the cost and revenue functions for the pocket watch manufacturer.
    (b) Find the number of watches that need to be sold for the manufacturer to break even.
    (c) What is the marginal cost and marginal revenue of producing and selling 2500 pocket watches?
    (d) What is the marginal profit for making and selling 2500 pocket watches?
    (e) What is the break-even point if variable costs increase to \$30 per pocket watch and the price to retailers remains at \$45 per pocket watch?

16. A music player has a marginal cost of \$29 when 100 players are manufactured at a total cost of \$7,500. Write the cost function assuming it is linear.

17. Find the market equilibrium point for these demand and supply functions, where p is price and q is quantity.

    Demand: $p = -2q + 220$
    Supply: $p = 8q + 10$

18. Find the market equilibrium point for:

    Demand: $p = -4q + 300$
    Supply: $p = 21q + 50$

19. The demand for pine 2×4's is 2500 when the price is \$1.10 each, and the demand is 3000 when the price is \$1.00 each. Assuming a linear demand function write the demand equation. Use p for price and q for quantity.

20. A distributor will supply 10,000 calendars if the price is \$2.00 each or will supply only 8,000 calendars if the price is \$1.25 each. Write the linear supply function using p for price and q for quantity.

21. A certain commodity has fixed costs of \$1800 and variable costs of \$35 per unit. If the commodity is sold for \$60 per unit, how many units need to be sold in order to break even?

22. Suppose that for a product A, the demand equation is $p = \frac{-1q}{70} + 10$ and the supply equation is $180p - q = 550$ where q is the number of units sold per week and p is the price per unit. Solve the system of equations to find the point of equilibrium.

23. A manufacturer sells cell phones for \$50 each. The costs incurred in the production and sale of the cell phones are \$65,000 plus \$10 for each cell phone produced and sold.

    (a) Write the revenue function, cost function and profit function.
    (b) How many cell phones must be produced and sold to break even?

24. A manufacturer sells bracelets for \$50 per bracelet. The fixed costs are \$10,000 per month and the variable costs are \$20 per bracelet.

    (a) Write the revenue and cost functions.
    (b) What is the profit function?
    (c) How many bracelets need to be sold to break even?

25. Assume supply for a toy is given by $35p - 20q = 350$ and demand is $p + 2q = 100$.

    (a) Compare the quantity demanded and supplied when the price is \$12.
    (b) What is the equilibrium point?

26. Retailers will buy 45 books from a wholesaler if the price is \$10 each, but only 20 books if the price is \$60. The wholesaler will supply 35 books at \$30 each and 70 books at \$50 each. Assuming supply and demand are linear, find the market equilibrium point.

27. Cap'n Bob plans to outfit a pirate ship. The ship, supplies and crew salaries will cost him \$35,000. In addition, every sea battle will cost him \$2,700 in gunpowder and hospital bills. A sea battle yields \$5200 in booty.

    (a) What are Cap'n Bob's cost, revenue and profit functions, where x = number of battles fought?
    (b) How many battles must Cap'n Bob fight to break even?

28. You are considering buying a hot dog cart and selling hot dogs at the stadium during football games. The cart costs \$1000. You will also need to pay a one-time fee of \$500 to get a vendor's license. It will cost you \$0.35 per hot dog and \$0.15 per bun. (You may neglect the cost of condiments.) You plan to sell your hot dogs for \$3.00 each. Let x = the number of hot dogs sold.

    (a) Write the cost, revenue and profit functions.
    (b) How many hot dogs must you sell to break even?

1.7 Linear Inequalities in Two Variables

VOCABULARY

Linear Inequality—An inequality that can be written in the form $Ax + By < C$, where A, B, and $C \in \mathbb{R}$. (*Note:* The inequality sign could be any of these four choices: $<$, $>$, $\leq$, $\geq$.)

Linear System of Inequalities—Two or more linear inequalities of two variables for which we want to find the solution set.

Solution Set—The region in the two-dimensional plane that contains all the points that satisfy all inequalities simultaneously.

Half-plane—Part of the two-dimensional unbounded plane that a line has cut in half.

Test Point—When solving a system of linear inequalities, we can use an arbitrarily chosen test point (x, y) to determine whether or not it lies in the region comprised of solutions.

Here are some illustrations of possible solution sets.

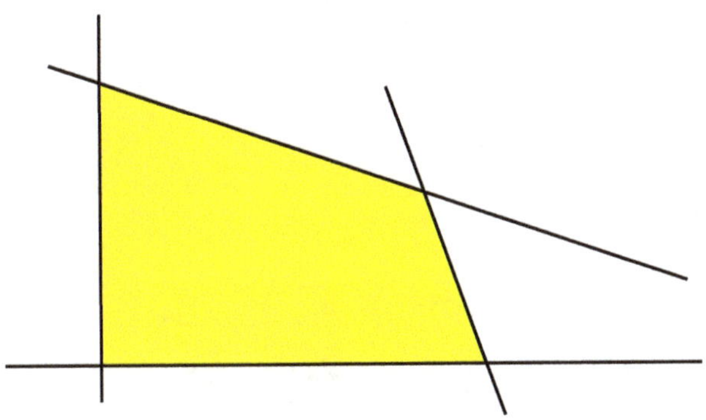

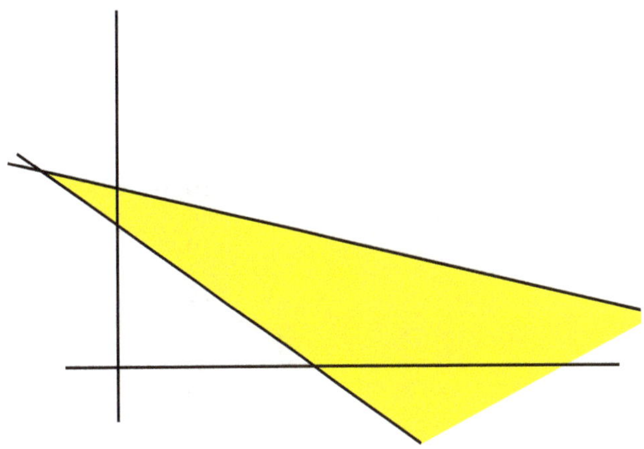

Let's start by graphing one linear inequality and then move on to a system of linear inequalities.

EXAMPLE 1

Graph the solutions for $2x - \frac{y}{3} < \frac{-4}{3}$.

Notice that this looks exactly like an equation of a line, except that an inequality sign has replaced the equals sign. Rather than graphing just the points on the line, we need to graph the line as a boundary and then shade in the half plane that contains all the points that satisfy this inequality. First, let's get this inequality in the slope-intercept form of a line so we can recognize its slope and y-intercept more readily.

$3\left(2x - \frac{y}{3}\right) < \left(\frac{-4}{3}\right)3$ To eliminate all the fractions, multiply both sides of the inequality by the least common denominator of 3.

$6x - y < -4$ We distributed the 3 through the parentheses on the left side and simplified both sides of the inequality.

$-y < -6x - 4$ We subtracted $6x$ from both sides of the inequality.

$y > 6x + 4$ We multiplied both sides of the inequality by -1 to finish getting y by itself. Remember that we had to switch the inequality sign since we multiplied both sides by a negative number!

With this linear inequality, we next graph the line as we normally would. However, if the inequality does not include the "or equal to" sign, then our line must be dotted rather than solid. The dotted line indicates that the solution set does not include the points on the line. But, the line represents where the boundary of the half-plane is. (*Note:* If the inequality sign includes the "or equal to" symbol, then the line would be drawn as a solid line, as we're used to.)

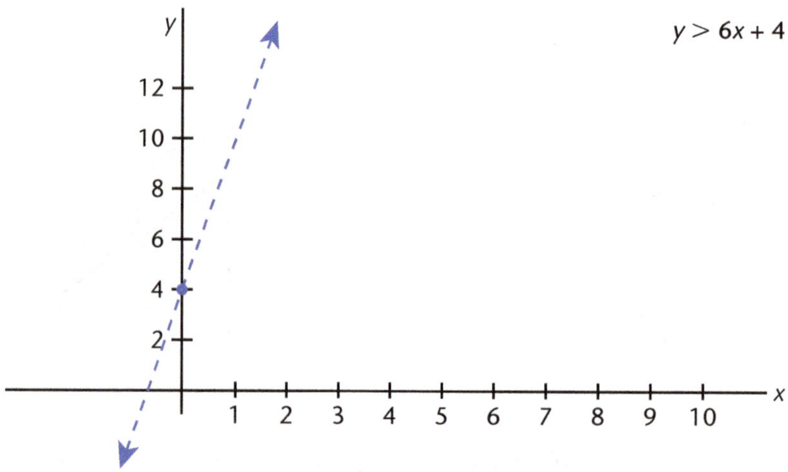

Next, we need to decide which of the two half planes to shade in, i.e., which of the two half planes contains the points that satisfy this inequality. Let's pick a very easy test point: (0, 0). In general, that's the easiest test point to choose. The only time we cannot choose the origin as our test point is when the line goes through the origin.

Test point: (0, 0) We substitute 0 in for x and 0 in for y in our linear inequality. We get $0 > 6(0) + 4$? We must decide if this is a true or a false statement. Is $0 > 4$? No. Thus, when we substituted our test point into the inequality, we got a false statement, which means that we do NOT want to shade in that side of our line. (If we had a true statement instead, we would shade in that side of the line containing the point (0, 0).)

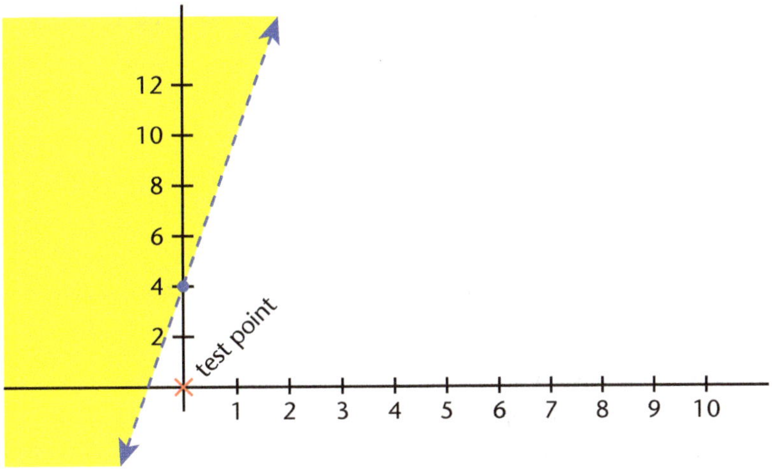

EXAMPLE 2

Graph the solution set of this system of linear inequalities.

(A) $4x + 3y \le 9$
(B) $5x - y \le 16$
(C) $x \ge 0$

First, let's change the first two inequalities around to be in slope-intercept form so that it's easier to graph those lines.

For inequality (A):

$4x + 3y \le 9$	Subtract $4x$ from both sides of the inequality.
$3y \le -4x + 9$	Divide both sides of the inequality by 3.
$y \le \frac{-4}{3}x + 3$	Now we have slope-intercept form.

For inequality (B):

$5x - y \le 16$	Subtract $5x$ from both sides of the inequality.
$-y \le -5x + 16$	Multiply (or divide) both sides of the inequality by –1. (Remember to switch the inequality sign since we're multiplying by a negative number.)
$y \ge 5x - 16$	Now we have slope-intercept form for inequality (B).

Let's graph all three lines on the same coordinate axes. (Notice that all three lines are solidly drawn since they each have the "or equal to" option in the inequalities.)

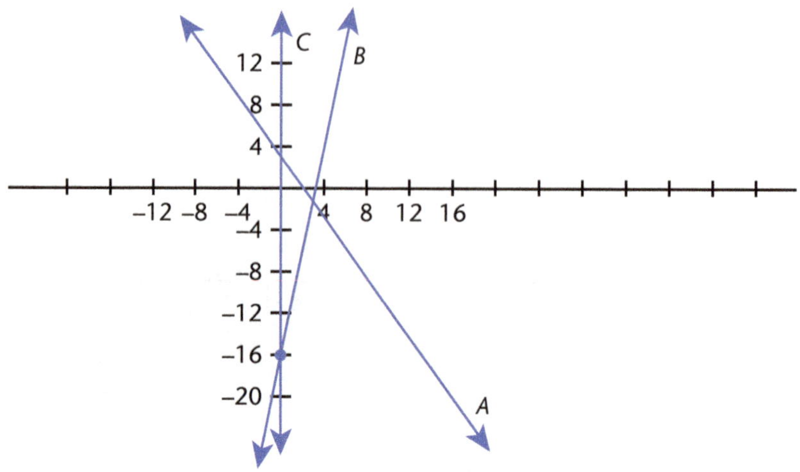

The last step for graphing the solution set is to shade in the region that satisfies all three inequalities. Let's choose the test point of (0, 1) this time since inequality (C) passes through the point (0, 0) which means we can't choose that as a test point for all three inequalities.

Test point: (0, 1) For each inequality, substitute $x = 0$ and $y = 1$ into the inequality to determine if it makes a true or a false statement.

(A): $1 \le \frac{-4}{3}(0) + 3 \quad \Leftrightarrow \quad 1 \le 0 + 3 \quad \Leftrightarrow \quad 1 \le 3$

Is this a true statement? Yes. So we want to shade in the side of line (A) that contains our test point.

(B): $1 \ge 5(0) - 16 \quad \Leftrightarrow \quad 1 \ge -16$

Is this a true statement? Yes. Again, we want to shade in the side of this line (B) that contains our test point.

(C): $0 \ge 0$

This is also a true statement. We'll shade in the right side of the line $x = 0$.

Let's put this all together in one graph.

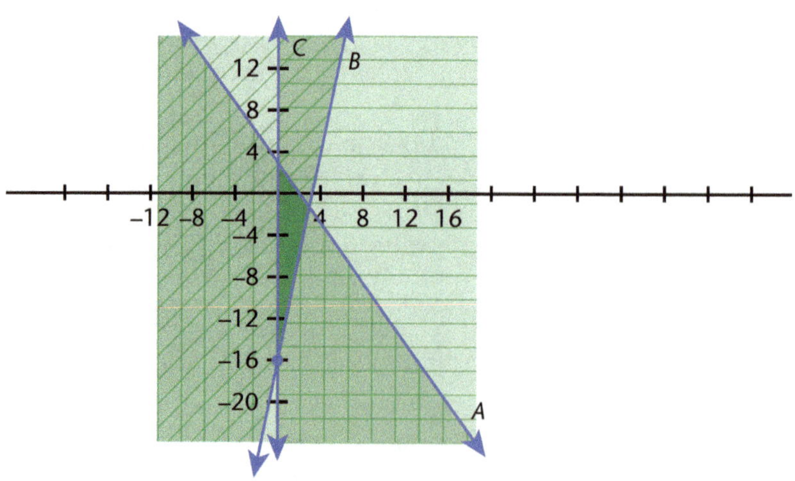

EXAMPLE 3

Graph the solution set for this system of linear inequalities.

(A) $3y \le -4x + 30$
(B) $x - y \le 4$
(C) $5x - 2y \le 10$
(D) $x \ge 0$
(E) $y \ge 0$

Again, our first step is to move things around in the first three inequalities in order to get them into slope-intercept form.

(A): $3y \le -4x + 30$ — To get y by itself, we only need to divide both sides of the inequality by 3.

$y \le \frac{-4}{3}x + 10$

(B): $x - y \le 4$ — Subtract x from both sides of the inequality.

$-y \le -x + 4$ — Multiply or divide both sides of the inequality by –1. (Remember to switch the inequality sign since we're multiplying by a negative number.)

$y \ge x - 4$

(C): $5x - 2y \le 10$ — Subtract $5x$ from both sides of the inequality.

$-2y \le -5x + 10$ — Divide both sides of the inequality by –2. (Remember to switch the sign.)

$y \ge \frac{-5}{2}x + 5$

Inequalities (D) and (E) are already in the form we want.

Next, we'll graph the lines. Notice that they're all solid lines since all the inequalities have the "or equal to" option in them.

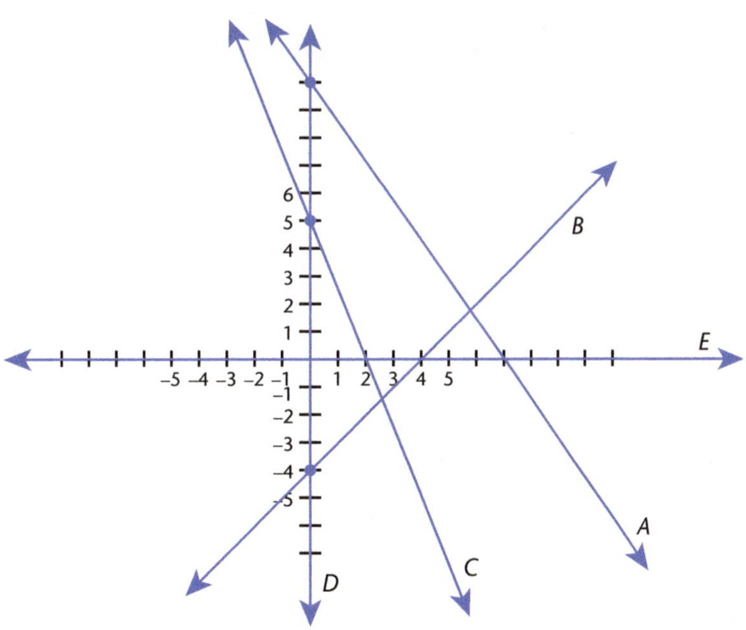

Test point: (1, 1) Use this test point to substitute $x = 1$ and $y = 1$ into every inequality to determine which side of the line gets shaded in.

(A): $1 \le \frac{-4}{3}(1) + 10 \Leftrightarrow 1 \le \frac{-4}{3} + \frac{30}{3} \Leftrightarrow 1 \le \frac{26}{3}$ True

(B): $1 \ge 1 - 4 \Leftrightarrow 1 \ge -3$ True

(C): $1 \ge \frac{-5}{2}(1) + 5 \Leftrightarrow 1 \ge \frac{-5}{2} + \frac{10}{2} \Leftrightarrow 1 \ge \frac{5}{2}$ False

(D): $1 \ge 0$ True

(E): $1 \ge 0$ True

Therefore, we shade in the sides of lines (A), (B), (D) and (E) that contain the point (1, 1). For line (C), we shade in the side of that line that does not contain the point (1, 1).

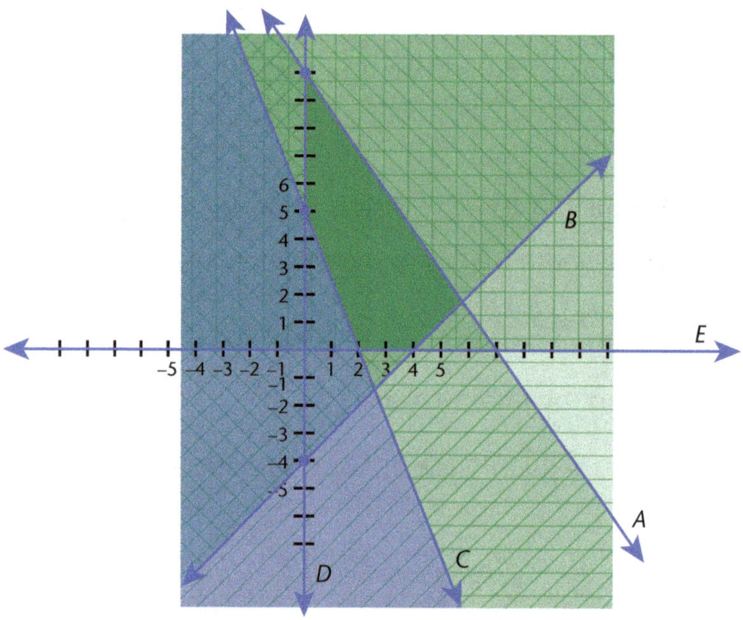

EXAMPLE 4

A company manufactures two types of hammers. One machine can produce 200 small hammers in an hour. Another machine can produce 140 large hammers in one hour. The packaging department can only package a total of 300 hammers per hour. Write out the system of inequalities that represents this problem and graph the solution set. (Assume we don't want any hammers stacking up to be packaged later.)

Let x = the number of small hammers produced and packaged in an hour and
y = the number of large hammers produced and packaged in an hour.

Since x and y represent something countable, it must be true that $x \geq 0$ and $y \geq 0$ since we can't produce a negative quantity of hammers.

We also know that we can produce at most 200 small hammers in an hour. This means that $x \leq 200$.

And, we can produce at most 140 large hammers in an hour. Therefore, $y \leq 140$.

Lastly, we're told that the packaging department can only keep up with a total maximum of 300 hammers per hour. So $x + y \leq 300$.

Putting that all together, we have the following system of inequalities.

(A) $x \leq 200$		(A) $x \leq 200$
(B) $y \leq 140$		(B) $y \leq 140$
(C) $x + y \leq 300$	You can write this as	(C) $y \leq -x + 300$ in slope-intercept form.
(D) $x \geq 0$		(D) $x \geq 0$
(E) $y \geq 0$		(E) $y \geq 0$

The graph of the lines looks like this.

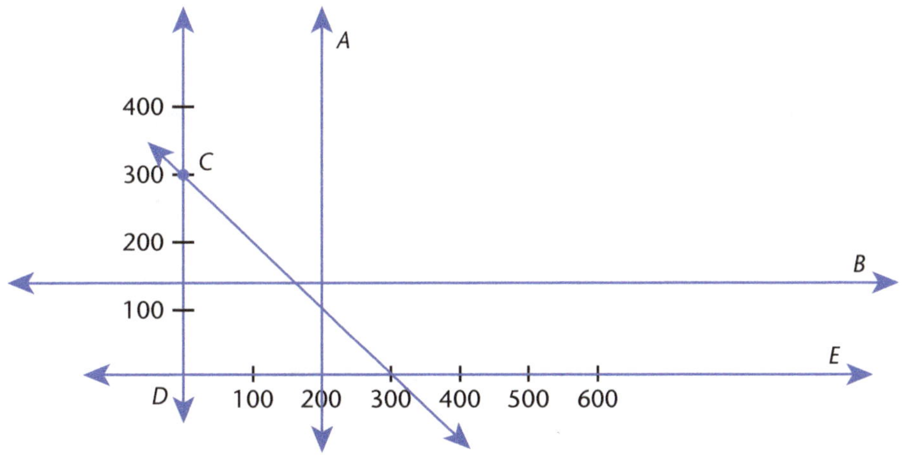

Let's choose a test point of (1,1) to determine which region is shaded in, representing the solution set.

Test point: (1, 1)	**(A):** $1 \leq 200$	True
	(B): $1 \leq 140$	True
	(C): $1 \leq -1 + 300$	True
	(D): $1 \geq 0$	True
	(E): $1 \geq 0$	True

Thus, we need to shade in the side of each line that contains the test point (1, 1).

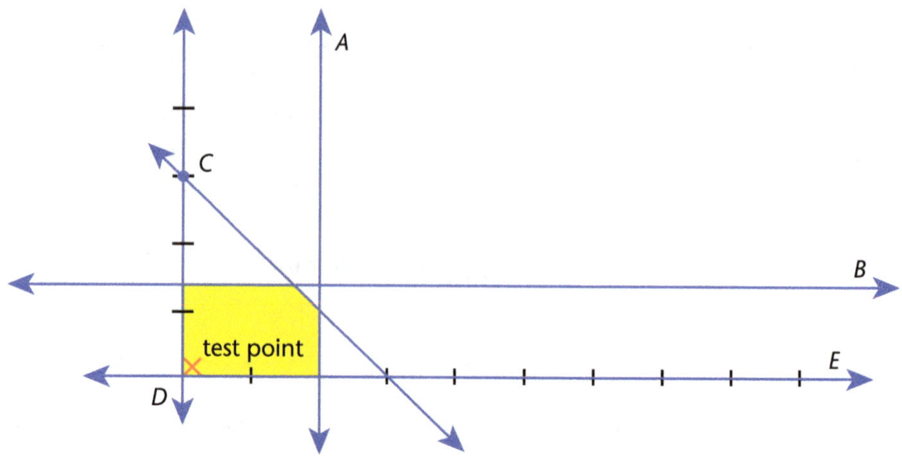

This shaded region represents all the x-values and y-values that will satisfy all the conditions of production and packaging.

1.7 Homework

Graph the solutions for these inequalities:

1. $y > 2x + 1$
2. $y < \frac{1}{3}x - 2$
3. $3x + 4y \leq 2$
4. $5x - 2y \geq 3$
5. $3y - 4x > 5$
6. $4x + 3y \leq 9$
7. $3x + 9y \leq -1$
8. $3x + \frac{7}{3}y \geq 2$
9. $-\frac{1}{2}x - \frac{1}{2}y \leq 1$
10. $\frac{6}{5}x + \frac{1}{10}y \geq -\frac{2}{5}$
11. $-2x - \frac{9}{5}y < \frac{9}{4}$
12. $-\frac{7}{2}x + \frac{3}{2}y < -\frac{3}{8}$
13. $-y - 7x \geq -\frac{5}{7}$
14. $2x - \frac{3}{5}y \geq -\frac{2}{5}$
15. $\frac{2}{3}x - \frac{7}{4}y < \frac{1}{2}$

Graph the solution set of each of these systems of inequalities:

16. $x + 2y < 4$
 $2x + y < 6$
 $x > -1$
 $y > -1$

17. $y > 1$
 $2x + y > -2$
 $\frac{2}{3}y - 4x < 9$
 $\frac{3}{2}x + \frac{1}{2}y < \frac{5}{2}$

18. $x + y < \frac{1}{2}$
 $2x + \frac{3}{4}y < 3$
 $\frac{1}{3}x + \frac{1}{2}y > -2$
 $\frac{1}{3}y - \frac{2}{3}x < 5$

19. $y > -3$
$7y < -6x + 21$
$y + 2x > -2$
$3y + 20x > -30$

20. $\frac{1}{3}x - \frac{1}{2}y > -\frac{5}{2}$
$y > 7x - 14$
$y > \frac{-1}{2}$
$\frac{9}{10}y > \frac{3}{5}x$

21. $x + 3y < 4$
$\frac{7}{3}y - \frac{1}{5}x > -4$
$\frac{2}{3}x + \frac{5}{9}y > \frac{3}{4}$
$\frac{8}{5}x - \frac{2}{3}y > -3$

22. $3y - \frac{1}{2}x > 4$
$\frac{9}{8}x - \frac{1}{2}y > -4$
$y < -4x - 1$
$\frac{1}{3}y - 4x > -5$

23. $3y > 2$
$\frac{4}{5}x - \frac{2}{3}y > \frac{2}{5}$
$\frac{3}{2}x - \frac{1}{2}y < 7$
$2x + \frac{9}{4}y > \frac{7}{3}$

24. $\frac{8}{9}x + \frac{8}{5}y < -\frac{3}{2}$
$\frac{3}{2}x + y < \frac{7}{2}$
$\frac{9}{10}x - \frac{5}{4}y > \frac{2}{9}$
$\frac{1}{3}y - x < \frac{7}{4}$

25. $2x + \frac{9}{7}y > \frac{12}{5}$
$\frac{4}{3}x - \frac{5}{3}y > -4$
$2y - \frac{5}{2}x > -\frac{13}{9}$
$\frac{2}{3}y - \frac{5}{8}x > -7$

26. $2x - 3y > -9$
$5x - y > -16$
$9x - 7y < -8$

27. $2y > -2$
$3x - 2y < -7$
$3x - 4y > -11$

28. $5x - 3y < 0$
$x - 5y > -22$
$5x - 3y > -22$
$x - 5y < 0$

29. $x - 3y < 16$
$2x + 3y < -4$
$4x - 3y > 10$

30. $x + 7y < -15$
$5x - y > -3$
$x - 2y < 12$

31. $8x - 7y > 19$
$3x + y < 18$
$3x - 2y > 4$
$x - 3y > 6$

32. $x - 2y > -4$
$x - 7y < -4$
$x + 3y < 6$

33. $7x - 4y < 7$
$2x - y > -6$
$x - 2y > -9$
$2x + 3y > 2$

34. A factory can manufacture 1,000 vehicles, either cars or trucks, per day. The factory must manufacture at least 200 more cars than trucks. Find the inequalities which represent this situation and then graph the solution set.

35. A farmer can plant 100 acres in either beans or corn. The farmer must plant at least 10 acres of corn. Find the inequalities which represent this situation and then graph the solution set.

36. A fuel delivery company must supply up to 10,000 gallons of gasoline to a gas station. By contract, the delivery company must purchase at least 1,000 gallons of this gasoline from Refinery X, and also 2,000 gallons from Refinery Y. The fuel delivery company must also purchase at least twice as much gasoline from Refinery Y as from Refinery X. Find the inequalities which represent this situation and then graph the solution set.

37. A manufacturing plant can produce 8,000 vehicles, either cars or trucks, per day. The Quality Assurance portion of the plant can check up to 12,000 vehicles per day, either 12,000 cars and no trucks or 6,000 trucks and no cars, or something in between. Find the inequalities which represent this situation and then graph the solution set.

38. A farmer can plant 9,000 acres in either beets or corn. Each acre of beets takes about 200 gallons of irrigation water per day while each acre of corn takes about 400 gallons of water per day. There is only 2,400,000 gallons per day of irrigation water available for these 9,000 acres of land. Find the inequalities which represent this situation and then graph the solution set.

39. For a concert, there are \$50 reserved seat tickets and \$35 general admission tickets. There are 2000 reserved seats available, and the fire code limits the total number of audience members to 3000. The promoter must take in \$80,000 in ticket sales to break even. Find a system of inequalities that represents this problem and graph all possible solutions.

40. A furniture company makes and sells two types of tables, one small and one large. Each large table requires two hours of assembly and two hours of finish work. Each small table requires three hours of assembly, due to the detailed craftsmanship, and $1\frac{1}{4}$ hours of finish work. The assembly shop is open a maximum of 12 hours per day. The finishing shop can be available for 10 hours per day. Find the system of linear inequalities to represent this situation and graph all possible solutions.

1.8 Graphical Linear Programming

VOCABULARY

Linear Programming—A technique used to solve problems that consist of a system of linear inequalities combined with a linear function that is to be maximized or minimized.

Constraints—The limits or conditions set on the variables in the problem, typically represented as inequalities.

Objective Function—The function that is to be maximized or minimized (also referred to as optimized).

Feasible Region—That shaded-in region that represents all the solutions that meet the constraint conditions, i.e., the solution set for the system of inequalities.

Optimal Solution—The solution that optimizes the objective function, subject to the constraints. The optimal solution must be one of the solutions (or points) in the feasible region.

Closed and Bounded Region—A region is closed and bounded if it is totally enclosed geometrically and includes the boundary.

For a linear programming problem, an optimal (maximum or minimum) solution is guaranteed to exist at one of the corners of the feasible region. (*Note:* This is assuming that our feasible region is convex. A convex region is one such that every line segment connecting any two points inside the region is also wholly contained in the region. Basically, a convex region has no "holes" or "divots" in it.) If the feasible region is open, not closed, then there is no guarantee that an optimal solution exists.

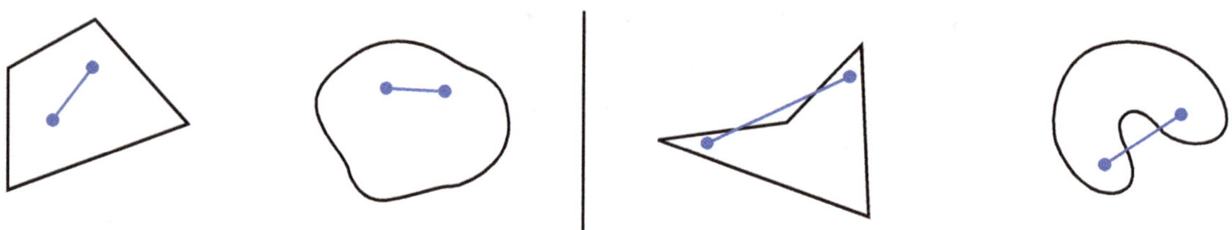

convex regions

concave regions (not convex)

Basically, to solve a linear programming problem, we employ the exact same technique as used in the last section. Namely, we graph the solution set (feasible region) for the system of linear inequalities. After that, we find and label all corners of the feasible region. Then we substitute all of those corner points into the objective function to determine which corner point yields optimal value.

GRAPHICAL LINEAR PROGRAMMING

1. Graph all constraints (linear inequalities) on the same coordinate axes.
2. Shade in the feasible region.
3. Find and label all the vertices (corner points) of the feasible region.
4. Plug in the corner points to the objective function to determine which vertex (corner point) gives the minimum and/or maximum values.

Warning: If the feasible region is not closed and bounded, then we're not guaranteed a minimum nor maximum of the objective function, but we can look for one and determine based on the problem if it will exist.

EXAMPLE 1

Find the maximum and minimum values of the objective function given by $P = 3x + 5y$, which is subject to these constraints.

(A) $5x + 8y \le 64$
(B) $y - x \ge -5$
(C) $x \ge 0$
(D) $y \ge 0$

First, let's graph the feasible region, i.e., the solution set for the system of inequalities. Changing things around to get all lines in slope-intercept form, our system of inequalities becomes:

(A) $y \le \frac{-5}{8}x + 8$

(B) $y \ge x - 5$

(C) $x \ge 0$

(D) $y \ge 0$

Next, choose a test point to help determine the shaded region. Since graphing inequalities (C) and (D) is the same as simply shading in the first quadrant of our Cartesian Coordinate system, we'll choose the easiest point, (0, 0), to test for inequalities (A) and (B) only.

Test point: (0, 0)

(A) $0 \le \frac{-5}{8}(0) + 8 \iff 0 \le 8$ True

(B) $0 \ge 0 - 5 \iff 0 \ge -5$ True

Thus, we want to shade in the region in the first quadrant that is also on the side of (A) and (B) that contains the origin.

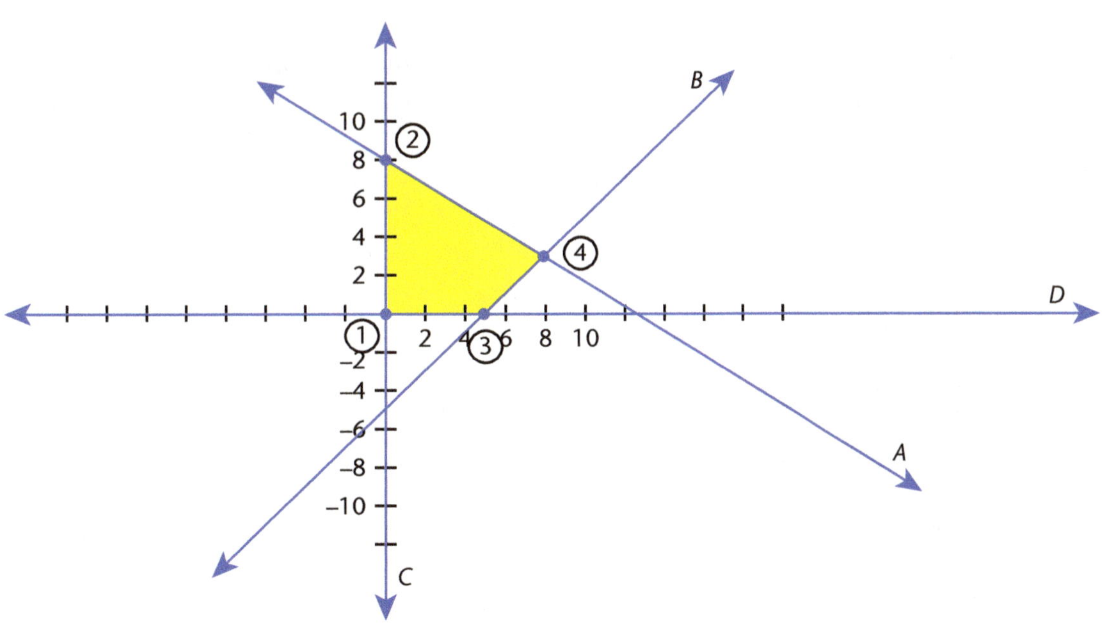

Now that we have the feasible region, we need to find and label the corner points.

1. This point is the origin, (0, 0).
2. This point is the y-intercept of line (A), which is (0, 8).
3. This point is the x-intercept of line (B). We can find that point by substituting $y = 0$ into (B) and solving for the x-value.

   $0 = x - 5$ Add 5 to both sides of the equation.

   $x = 5$

   So, the corner point here is (5, 0).

4. This point is the intersection between lines (A) and (B). We need to solve this system of equations to determine the point.

   (A) $y = \frac{-5}{8}x + 8$ This is nicely set up to use the substitution method. We'll substitute the expression for y from (B) into (A).

   (B) $y = x - 5$

   $\Rightarrow$ $x - 5 = \frac{-5}{8}x + 8$ Add $\frac{5}{8}x$ and 5 to both sides of the equation.

   $x + \frac{5}{8}x = 8 + 5$ Simplify.

   $\frac{8}{8}x + \frac{5}{8}x = 13$ Combine like terms on the left side of the equation.

   $\frac{13}{8}x = 13$ To get x by itself, multiply both sides of the equation by $\frac{8}{13}$.

   $x = 13\left(\frac{8}{13}\right) = 8$

Now we need to substitute $x = 8$ into either equation (A) or (B) to find the corresponding y-value.

$\Rightarrow$ (B) $y = 8 - 5 = 3$

The point of intersection between lines (A) and (B) is (8, 3).

Last, we need to substitute each of our four corner points into the objective function, $P = 3x + 5y$, to determine which point yields a maximum value and a minimum value.

1. (0, 0) $P = 3(0) + 5(0) = 0$
2. (0, 8) $P = 3(0) + 5(8) = 40$
3. (5, 0) $P = 3(5) + 5(0) = 15$
4. (8, 3) $P = 3(8) + 5(3) = 24 + 15 = 39$

Therefore, the maximum value of P is 40, which occurs at the point (0, 8), and the minimum value of P is 0, which occurs at the point (0, 0).

EXAMPLE 2

Minimize the function $G = -4x + 7y$ subject to the constraints below.

(A) $5y - 45 \geq -4x$
(B) $2y + x \geq 12$
(C) $x \geq 0$
(D) $y \geq 0$

Let's first find the feasible region by shading in the solution set of the system of inequalities. To make that easier, we'll change the inequalities (A) and (B) around to be in slope-intercept form. Then we'll have the following system.

(A) $y \geq \frac{-4}{5}x + 9$

(B) $y \geq \frac{-1}{2}x + 6$

(C) $x \geq 0$
(D) $y \geq 0$

When the lines are in slope-intercept form, we can choose to shade in "above" or "below" the lines, rather than choosing a test point each time. For (A), we need to shade above the line, since we want to shade in all the points that are greater than or equal to the y-values on the line. The same will be true for (B). Then, we'll have this graph of the feasible region.

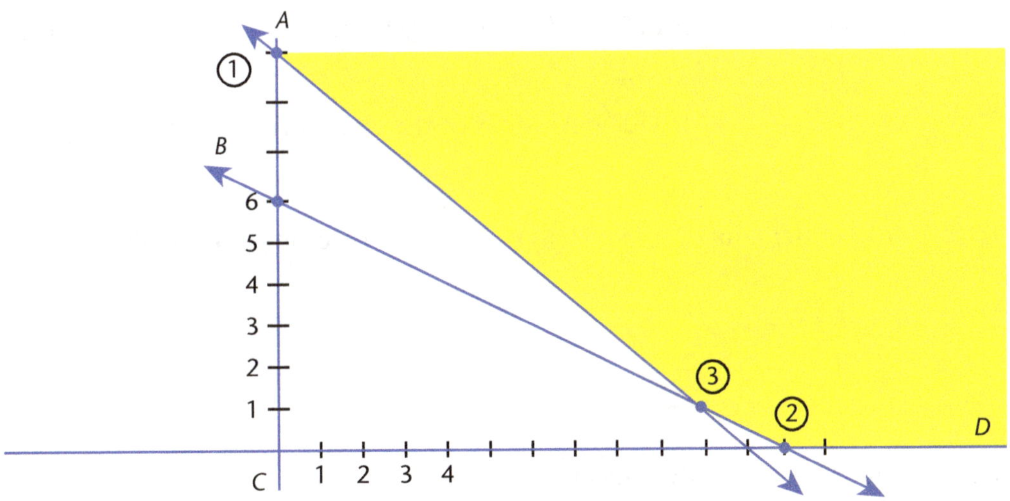

Notice that we have an unbounded shaded region this time. We are not guaranteed to find a maximum or minimum value for our objective function with this region. We can certainly check the three corner points that we do have.

We need to find those three corner points.

1. This point is the y-intercept for line (A), which is (0, 9).
2. This point is the x-intercept for line (B). We know every x-intercept has a y-value of 0. We can find the x-value for the point by substituting $y = 0$ into line (B).

$\Rightarrow$ $0 = \frac{-1}{2}x + 6$ Subtract 6 from both sides of the equation.

$-6 = \frac{-1}{2}x$ Multiply both sides of the equation by –2.

$-6(-2) = \frac{-1}{2}x - 2$ Simplify.

$12 = x$

$\Rightarrow$ The point is (12, 0).

3. This point is the intersection point between lines (A) and (B). We can solve the following system of equations to find the point.

(A) $y = \frac{-4}{5}x + 9$

(B) $y = \frac{-1}{2}x + 6$

We'll use the substitution method to solve this system of equations. Substitute the expression for y in (B) into (A).

$\frac{-1}{2}x + 6 = \frac{-4}{5}x + 9$ Multiply both sides of the equation by 10 (the least common denominator of the equation) in order to eliminate the fractions for now.

$10\left(\frac{-1}{2}x + 6\right) = 10\left(\frac{-4}{5}x + 9\right)$ Distribute and simplify.

$-5x + 60 = -8x + 90$ Add $8x$ and subtract 60 from both sides of the equation.

$3x = 30$

$x = 10$ Divide both sides of the equation by 10 to get x by itself.

Now, substitute 10 in for x into either equation (A) or (B) to find the corresponding y-value.

$$y = \frac{-1}{2}(10) + 6 = -5 + 6 = 1$$

So, the point is (10, 1).

Last, substitute the corner points into the objective function $G = -4x + 7y$ to find out if there is a possible minimum value.

1. (0, 9) $G = -4(0) + 7(9) = 63$
2. (12, 0) $G = -4(12) + 7(0) = -48$
3. (10, 1) $G = -4(10) + 7(1) = -40 + 7 = -33$

Certainly between these three values, –48 is the minimum that we have at the point (12, 0). However, if you look back at the shaded feasible region, the point (100, 0) would also be a candidate. If we substitute that point into G, we would get a value of –400, which is smaller still. We could continue in this fashion to find point after point that yields ever smaller values of G, which means there is no minimum! Again, we were not guaranteed to find a minimum value for the objective function G because we have an unbounded region.

EXAMPLE 3

An inventor, Doug, builds and sells two different types of prosthetic arms. The simple Hook Arm costs \$3,000 to make and requires 50 worker-days of labor. The more complex Real Arm costs \$4,500 and requires 80 worker-days of labor to make. Doug has \$333,000 capital to invest in his inventions and a total of 5,800 worker-days of labor to utilize. The profit from each Hook Arm sold is \$900, and the profit for the Real Arm is \$1,300. How many of each type of prosthetic arm should he create to maximize profit? What is the maximum profit?

For this problem, the first step that we must take is to define what our variables will represent.

Let x = the number of Hook Arms made and sold and y = the number of Real Arms made and sold. We know we can only produce a positive number of units. That is,

$$x \geq 0 \quad \text{and} \quad y \geq 0.$$

We can get an inequality from the costs given. It takes \$3,000x to make the Hook Arms and \$4,500y to make the Real Arms. The total cost can only be as big as \$333,000, which is the start-up money available.

$$3000x + 4500y \leq 333{,}000$$

Last, we know that we only have a maximum of 5,800 worker-days of labor available. This yields the final inequality.

$$50x + 80y \leq 5800$$

We want to maximize the profit function, given by $P = 900x + 1300y$.

Altogether, the following inequalities give our feasible region.

(A) $3000x + 4500y \leq 333{,}000$

(B) $50x + 80y \leq 5800$

$x \geq 0 \quad \text{and} \quad y \geq 0$

If we move things around in (A) and (B) to get those inequalities in slope-intercept form, we have:

$\Leftrightarrow$ (A) $y \leq \dfrac{-2}{3}x + 74$

(B) $y \leq \dfrac{-5}{8}x + \dfrac{145}{2}$

$x \geq 0 \quad \text{and} \quad y \geq 0$

Graph those inequalities on the same coordinate axes and shade below both lines in the first quadrant. That gives us the feasible region.

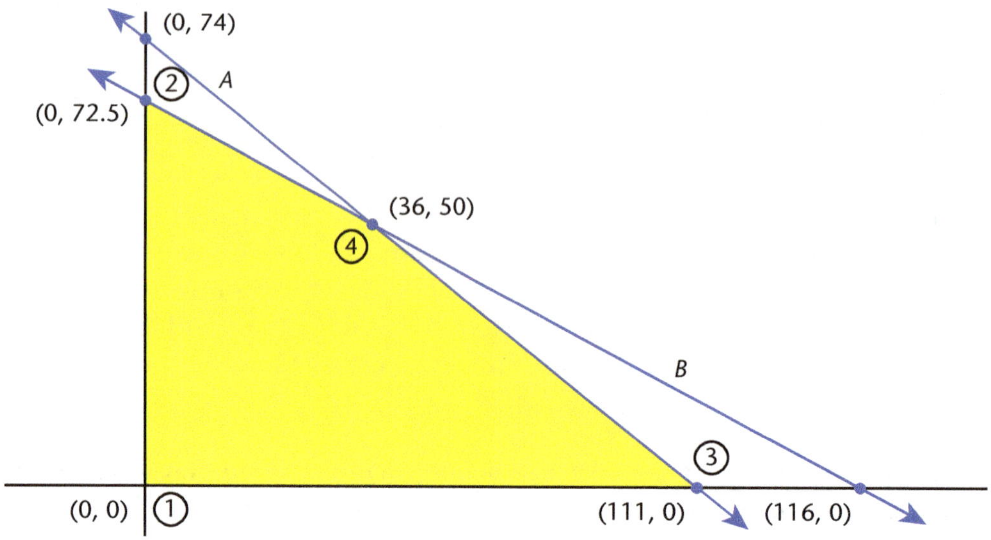

Now we need to determine the corner points.

1. The origin (0, 0).
2. The y-intercept of line (B): (0, 72.5).
3. The x-intercept of line (A). We can find this point by substituting $y = 0$ into the line (A) equation.

$$0 = \frac{-2}{3}x + 74 \iff \frac{2}{3}x = 74 \iff \left(\frac{3}{2}\right)\frac{2}{3}x = 74\left(\frac{3}{2}\right) \iff x = 111 \qquad (111, 0)$$

4. The intersection point between lines (A) and (B).
   We have this system of equations to solve.

(A) $y = \frac{-2}{3}x + 74$

(B) $y = \frac{-5}{8}x + \frac{145}{2}$

We'll use the substitution method to solve this system.

$\frac{-2}{3}x + 74 = \frac{-5}{8}x + \frac{145}{2}$	Multiply both sides of the equation by 24 to eliminate the fractions.
$\frac{-2(24)}{3}x + 74(24) = \frac{-5(24)}{8}x + \frac{145(24)}{2}$	Simplify.
$-16x + 1776 = -15x + 1740$	Add $16x$ and subtract 1,740 from both sides of the equation.
$36 = x$	

Now, substitute 36 in for x in either equation (A) or (B) to get the corresponding value for y.

$$y = \frac{-2}{3}(36) + 74 = -24 + 74 = 50$$

Thus, the intersection point is (36, 50).

Last, we need to find out which corner point yields the greatest profit. We need to substitute each corner point into our profit function.

$$P = 900x + 1300y$$

1. (0, 0) $P = 900(0) + 1300(0) = \$0$
2. (0, 72.5) $P = 900(0) + 1300(72.5) = \$94{,}250$
3. (111, 0) $P = 900(111) + 1300(0) = \$99{,}900$
4. (36, 50) $P = 900(36) + 1300(50) = \$97{,}400$

From these numbers, we can see that the greatest profit is achieved at the point (111, 0), i.e., from producing 111 Hook Arms and 0 Real Arms. The maximum profit for that point is $99,900.

1.8 Homework

1. Find both the maximum and minimum of the objective function $x + y$ given these constraints:

$$y \le x + 6$$
$$y \ge 9x - 18$$
$$y \ge 0$$
$$x \ge 0$$

2. Find both the maximum and minimum of the objective function $x - 3y$ given these constraints:

$$y \le 7 - x$$
$$y \ge \frac{1}{2}x - 2$$
$$y \ge 0$$
$$x \ge 0$$

3. Find both the maximum and minimum of the objective function $y - 2x$ given these constraints:

$$y \le \frac{1}{2}x + 1$$
$$y \le 6 - 2x$$
$$y \ge 0$$
$$x \ge 0$$

4. Find both the maximum and minimum of the objective function $10x - 7y$ given these constraints:

$$3x - 2y \ge -12$$
$$x + y \le 9$$
$$y \ge 0$$
$$x \ge 0$$

5. Find both the maximum and minimum of the objective function $6y - 8x$ given these constraints:

$$y - x \le 3$$
$$5y - 8x \ge -3$$
$$y \ge 0$$
$$x \ge 0$$

6. Find both the maximum and minimum of the objective function $4x + y$ given these constraints:

$$10x - y \le 12$$
$$x - 2y \le 5$$
$$2x + y \ge -5$$
$$4y - 9x \le 14$$

7. Find both the maximum and minimum of the objective function $5y - x$ given these constraints:

$$y - 5x \ge -7$$
$$6y \ge x - 13$$
$$2x + 2y \ge -9$$
$$13y \le 18x + 50$$

8. Find both the maximum and minimum of the objective function $3y - 2x$ given these constraints:

$$9x - y \le 11$$
$$x - 3y \le 7$$
$$2x + y \ge -7$$
$$5y - 8x \le 19$$

9. Find both the maximum and minimum of the objective function $2x + 9y$ given these constraints:

$$y \le 1$$
$$y - x \ge 0$$
$$x + y \ge -6$$
$$4y - 3x \le 4$$

10. Find both the maximum and minimum of the objective function $8y - 9x$ given these constraints:

$$3x + 4y \le 4$$
$$x - y \le 6$$
$$x + 2y \ge -6$$
$$2y - 3x \le 2$$

11. Find both the maximum and minimum of the objective function $y - 9x$ given these constraints:

$$5x - 2y \le 13$$
$$y \ge -4$$
$$y - 7x \le 31$$
$$2x + 7y \le 13$$

12. Find both the maximum and minimum of the objective function $4y - 2x$ given these constraints:

$$3x - 2y \le 4$$
$$2x - 3y \le 1$$
$$x + y \ge -2$$
$$3y - 2x \le 4$$

13. Find both the maximum and minimum of the objective function $7y - 6x$ given these constraints:

$$x + 2y \le 6$$
$$2x - y \le 12$$
$$15y \ge -2x - 36$$
$$3y - 5x \le 9$$

14. Find both the maximum and minimum of the objective function $9y - 7x$ given these constraints:

$$x + y \le 2$$
$$4x - 3y \le 22$$
$$4x + 3y \ge -14$$
$$y - 5x \le 8$$

15. Find both the maximum and minimum of the objective function $3x + 9y$ given these constraints:

$$5x + 9y \le 44$$
$$5x - 6y \le 29$$
$$7x + 5y \ge -13$$
$$2y - 3x \le 18$$

16. Find both the maximum and minimum of the objective function $9x + y$ given these constraints:

$$x + y \le -1$$
$$2x - y \le 1$$
$$x - 6y \le 17$$
$$2y - 3x \le 13$$

17. Find both the maximum and minimum of the objective function $9x + 2y$ given these constraints:

$$x + 2y \le 13$$
$$3x - y \le 11$$
$$5x - 9y \le 11$$
$$3y - 5x \le 13$$

18. Find both the maximum and minimum of the objective function $6y - 5x$ given these constraints:

$$2y - x \le 3$$
$$5x + y \le 7$$
$$4x - 5y \le 23$$
$$y - 4x \le 5$$

19. Find both the maximum and minimum of the objective function $x + y$ given these constraints:

$$2y - x \le 2$$
$$4x - 3y \le 7$$
$$x - 4y \le 5$$
$$5y - 4x \le 2$$

20. Find both the maximum and minimum of the objective function $4x + 5y$ given these constraints:

$$y \ge x - 2$$
$$y \ge -x - 1$$
$$y \le x + 2$$

21. Find both the maximum and minimum of the objective function $7x - 5y$ given these constraints:

$$x + 2y \le 6$$
$$2y \le 3x + 1$$
$$3x + y \ge -3$$

22. Find both the maximum and minimum of the objective function $4x - 3y$ given these constraints:

$$3y \le 2x - 15$$
$$3x + 2y \ge 2$$
$$2x + 3y \ge -9$$

23. Find both the maximum and minimum of the objective function $8y - 4x$ given these constraints:

$$4x + 3y \le 8$$
$$y \ge 2x - 4$$
$$2x + 3y \ge -9$$

24. Find both the maximum and minimum of the objective function $8y - 4x$ given these constraints:

$$2y \ge 4x + 7$$
$$2y \ge x + 4$$
$$10y \ge x - 8$$

25. Find both the maximum and minimum of the objective function $2.5x + y$ given these constraints:

$$x \ge 0$$
$$y \ge 0$$
$$3x + 5y \le 15$$
$$5x + 2y \le 10$$

26. Find both the maximum and minimum of the objective function $9x + 9y$ given these constraints:

$$x - 2y \ge 3$$
$$3x - y \ge -6$$
$$x - y \le 2$$

27. Find both the maximum and minimum of the objective function $-8x + 5y$ given these constraints:

$$2x + 11y \le 51$$
$$8y \ge 24$$
$$5y \le 25$$
$$-x - y \le -2$$

28. Find both the maximum and minimum of the objective function $3x + 4y$ given these constraints:

$$x \ge 0$$
$$y \ge 0$$
$$x + y \ge 1$$
$$2x + y \le 4$$

29. A factory can manufacture 1,000 vehicles, either cars or trucks, per day. The factory must manufacture at least 200 more cars than trucks. If the profit to be made from a car is $1,000 and the profit from a truck is $1,500, what numbers of cars and trucks should be manufactured to maximize the profits of the operation of the factory?

30. A farmer can plant 100 acres in either beans or corn. The farmer must plant at least 10 acres of corn. If the farmer can make a profit of $2,500 per acre of beans and $2,000 per acre of corn, how much of each should he plant to make the most profit?

31. A fuel delivery company must supply between 9,000 and 10,000 gallons of gasoline to a gas station. By contract, the delivery company must purchase at least 1,000 gallons of this gasoline from Refinery X, and also 2,000 gallons from Refinery Y. The fuel delivery company must also purchase at least twice as much gasoline from Refinery Y as from Refinery X. If the gasoline from Refinery X costs $1.10 per gallon and the gasoline from Refinery Y costs $1.15, how much gasoline should the delivery company purchase from each refinery in order to pay the least for the gasoline?

32. A manufacturing plant can produce 8,000 vehicles, either cars or trucks, per day. The Quality Assurance portion of the plant can check up to 12,000 cars per day or up to 6,000 trucks per day. The profit from manufacturing a car is $2,200 and the profit from manufacturing a truck is $2,800. How many cars and trucks should the plant manufacture in a day to maximize the profits?

33. A farmer can plant 9,000 acres in either beets or corn. Each acre of beets takes about 200 gallons of irrigation water per day while each acre of corn takes about 400 gallons of water per day. There is only 2,400,000 gallons per day of irrigation water available for these 9,000 acres of land. The profit for an acre of cultivation is the final sales revenue less the cost of irrigation and other expenses. If the profit from an acre of beets is $3,750 and the profit from an acre of corn is $4,150, how many acres of each should the farmer plant to maximize his income?

34. A manufacturer wants to maximize the profit for two toys. The first toy yields $3 profit and the second toy yields a profit of $4 per unit. The demand for toy 2 is no more than half the demand for toy 1. The manufacturer can only produce up to 1200 toys per month. The maximum number of type-1 toys produced is 600 plus three times the number of type-2 toys produced. What is the maximum profit and how many of each type of toy should be produced for that maximum profit?

Chapter 1 Review

1.1

Solve these equations.

1. $5(x - 1) + 3 = 4x - 7$

2. $\frac{3}{2}x - 6 = \frac{2x + 1}{5}$

3. $\frac{x - 9}{4} + 2 = \frac{x - 4}{3} + 3$

4. $\frac{x}{x - 2} + 5 = 3 - \frac{4}{x - 2}$

5. $\frac{2}{2x + 1} + \frac{3}{x - 5} = 0$

6. $\frac{3x - 1}{x + 5} = \frac{9}{2}$

7. $\frac{6 - x}{x + 2} = \frac{8}{5}$

8. $\frac{2x + 1}{x - 2} + 4 = \frac{5}{x - 2}$

9. Twenty more than five times a number is the same as seven times that same number. What is the number?

10. A car sells for $25,482.60 and this price includes 7.25% sales tax. How much is the sales tax?

11. Find three consecutive even numbers whose sum is 72.

1.2

Solve the linear inequalities and graph the solutions on a real number line.

12. $\frac{1}{3}x - 4 \geq 1$

13. $\frac{1}{4}x + 5 \leq \frac{1}{3}x + 1$

14. $\frac{5x + 2}{4} - 1 \leq \frac{x + 3}{6} + 1$

15. $-1 \leq 2x + 3 < 5$

16. $\frac{5}{3}(2w + 4) < 25 + 5w$

17. $\frac{1}{2}(8x - 12) + 5 < \frac{-1}{3}(18x - 6)$

18. Malik scored 69, 78, 85 and 71 (out of a possible 100 points) on the four midterm exams for a math class. If the three highest scoring exams each count as 20% of the total grade, the lowest exam counts as only 10% of the total grade, and the final exam counts as 30% of the total grade, what must Malik score on the final exam to get at least 75% as the overall grade in the class?

19. Jaleigh invests a total of $10,000 in two different accounts. The riskier investment yields an annual average of 11% profit and the safer investment typically yields 6% profit annually. How much money should be invested in the safer account if Jaleigh needs at least $850 profit per year?

1.3

Find the slope of the line through the two given points.

20. (–5, 3) and (1, 1)

21. (4, 2) and (–3, 2)

22. (1, 9) and (1, 0)

Write the equation of the line, in slope-intercept form, given two pieces of information about the line.

23. $m = 3$, goes through the point (–2, 5)

24. goes through the points (–1, 7) and (2, 1)

25. goes through the point (4, 0) and is parallel to the line $2x + 4y = 5$

26. goes through the point (3, –6) and is perpendicular to the line $x - 5y = 9$

27. goes through the point (10, 8) and has undefined slope

28. The plumber, Paul, charged Ann a flat fee of $20 just to come to her house. Then, his hourly labor rate was $45, prorated for parts of hours. If he was there for 45 minutes, how much was her plumbing bill?

1.4

Solve these systems of equations.

29. $2x + 5y = 7$
$x - 3y = -2$

30. $\frac{1}{3}x + 5y = -1$
$2x - \frac{1}{2}y = -6$

31. $4x - y = -1$
$5x + y = 19$

32. $10x - 4y = 3$
$\frac{5}{2}x - y = 1$

33. $-3x + y = 9$
$6x - 2y = -18$

34. $5x + 4y = 1$
$-15x - 12y = 0$

35. $2x - 3y = 2$
$5x + 6y = \frac{1}{2}$

36. $\frac{-3}{2}x + \frac{5}{2}y = 4$
$9x - 15y = -24$

37. $-x + y + 2z = -2$
$5y = 15$
$3x - 4y + z = 3$

38. $x + y + z = 3$
$3x - 2y = -1$
$2z = -6$

39. At Leslie's Party Store, you can purchase fancy party hats for $1.50 each and party favors for $0.80 each. Carmen spent $104.50 for a total of 100 items, including only the party hats and party favors. How many party hats and how many party favors did Carmen purchase?

1.5

Indicate whether or not the relation represents a function. If it is a function, find the domain.

40. $3x^2 + 4y^2 = 1$

41. $y = \dfrac{2x}{(x-5)(2x+1)}$

42. $y = (3 + x^2)\sqrt{x+5}$

43. $y = \dfrac{18}{\sqrt{x-4}}$

Evaluate each function at the given input values.

44. $f(x) = 2x^2 - 4x + 1$

a. $f(3)$

b. $f(h)$

45. $g(p) = \dfrac{9}{\sqrt{3p+2}}$

a. $g(0)$

b. $g(w + 1)$

46. $h(x) = \dfrac{2x+1}{x-8}$

a. $h(4)$

b. $h(x^2)$

47. $f(y) = \sqrt{2-y} + \sqrt{y+1}$

a. $f(2)$

b. $f(y + h)$

48. $y(x) = \sqrt{x^2 - 4}$

a. $y(3)$

b. $y(x^2 + 1)$

1.6

49. Find the market equilibrium point given a demand equation of $2p + 3q = 492$ and a supply equation of $p - 3q = -429$.

50. Van's Furniture Store sells a popular rocking chair. They've discovered, through market analysis, that the demand equation is $2p + q = 740$ for this chair, and their supply equation is $p = 2q - 530$. Given this information, what price should they charge for a rocking chair? How many chairs will they sell at this price?

51. Mohomodou is thinking of starting a business selling t-shirts with inspirational quotes on them. His target consumer group is comprised of wealthy clients, including celebrities. After some market research, he finds the consumers will buy 180 t-shirts if the price is \$42 and will buy 200 t-shirts for a price of \$26. He's willing to supply t-shirts according to the equation $\frac{1}{3}p + 204 = q$.

a. How many t-shirts will he sell if he prices each t-shirt at \$25?

b. What price must he charge to sell 230 shirts?

c. What is the market equilibrium point?

52. The demand for Ki's chocolate truffles, during the winter holiday season, is 400 when the cost is \$1.50 each, but demand goes up to 800 when the cost is only \$1 each. Assuming the demand equation is linear, find the demand equation. At what price would consumers be willing to buy 1000 of Ki's truffles?

53. Given a cost equation of $C = 18x + 10800$ and a revenue equation of $R = 45x$, where x is the quantity of books made and sold, how many books should be sold to break even? What is the cost of creating that many books? What is the revenue from selling those books?

54. In her retirement, my mother decided to sell her homemade jellies. Her fixed costs are \$250 per month and it costs an additional \$1.25, on average, to produce one jar of jelly. She can sell each jar for \$4.50. How many jars of jelly must she sell to break even? What will her profit be if she sells 150 jars this month? What will her profit be if she sells 300 jars of jelly?

55. Alkira realized there is a great need for helping students overcome their fear of mathematics. She's decided to go into business offering workshops to help students overcome this fear and fulfill this need. Her fixed costs will be $540, according to her calculations, and materials will cost $40 per attendee. She's planning on charging $95 per student for participating in her workshop. How many students must sign up for the workshop for Alkira to break even? She wants to keep the workshop enrollment to a maximum of 30 students. If it has a maximum capacity, what is Alkira's profit for the workshop?

56. For a cost equation given by $C = 3x + 310$ and a revenue equation of $R = 6.5x$, answer the following questions.

a. What is the profit equation?

b. What is the equilibrium point?

1.7

Graph the solution set of each of these systems of inequalities.

57. $x \geq 0, y \geq 0$
$3x + 2y \leq 20$
$x - y \leq 5$

58. $3y - 5x \leq 15$
$4y - 20 \leq -7x$
$2x + 7y \geq -6$

59. $x \geq 2$
$x \leq 6$
$y \leq x + 1$
$5x + 4y \geq 10$

60. $y \geq x - 4$
$5y \leq 4x + 8$
$5x - y \leq 32$
$3x + 2y \geq 17$

61. $3y + 2x \geq -6$
$3x + y \geq 5$
$3y \leq 5x + 15$

62. A Bikram yoga studio charges $10 per class for adults and only $6 per class for students. The studio can hold a maximum of 65 students for a class. The owner wants to bring in revenue of at least $590 for a special New Year's Eve class. Graph the possible solution set, providing also the system of inequalities for that solution set.

1.8

Find both the maximum and minimum of the objective function, f, given the inequality constraints on x and y.

63. $f = -2x + 5y$
$y \leq 8$
$6y - 5x \geq 8$
$2y \geq -5x + 16$

64. $f = 3x + y$
$x \geq 0, y \geq 0$
$7x + 3y \leq 21$
$9y \leq -4x + 36$

65. $f = 5x - 2y$
$3y \leq -8x + 24$
$4x - 3y \leq 12$
$5y \leq 8x + 40$
$-5y \leq 4x + 20$

66. (Refer to problem 62.) If the yoga studio owner has a profit equation of $P = 5x + 2y$, where x is the number of adult participants and y is the number of student participants, what is the maximum profit for the special New Year's Eve class? How many adult and student participants create that maximum profit?

chapter 2

Matrices

2.1 Basic Operations with Matrices

In this section, we are introducing a mathematical entity called a matrix. Our usual algebra that we think of includes rules about how to manipulate numbers and variables (that represent numbers) in expressions and equations. We know how to do arithmetic with numbers and variables, and we have multiplicative and additive identities, namely 1 and 0 respectively. We will now venture into a different type of algebra with similar rules as previously used, but our "entities" manipulated in this different algebra game are matrices and numbers. And, our variables will now be representing matrices, not just numbers. First, we need to describe the lay of the land for matrices.

VOCABULARY

Matrix—An array of numbers. We usually denote a matrix with a capital letter name, such as A. We would write A generically as

$$A = \begin{bmatrix} a_{11} & a_{12} & a_{13} & \cdot & \cdot & \cdot & a_{1n} \\ a_{21} & a_{22} & a_{23} & \cdot & \cdot & \cdot & a_{2n} \\ \cdot & \cdot & \cdot & \cdot & \cdot & \cdot & \cdot \\ \cdot & \cdot & \cdot & \cdot & \cdot & \cdot & \cdot \\ \cdot & \cdot & \cdot & \cdot & \cdot & \cdot & \cdot \\ a_{m1} & a_{m2} & a_{m3} & \cdot & \cdot & \cdot & a_{mn} \end{bmatrix}$$

Entry or Element—A single place or a number inside a matrix A, denoted as a_{ij}, where i is the row the element is in and j tells which column the element is in the matrix A.

Scalar—A real number. Previously, we would have called this a constant. In our algebra of matrices, we will now refer to constants as scalars.

Order (or size) of a Matrix—Usually denoted as $m \times n$ (read "m by n"). This means there are m rows of elements and n columns of elements in the matrix. The number of rows always comes first followed by the number of columns.

DEFINITIONS

Equal—Two matrices, A and B, are equal to each other if and only if every entry in A is exactly the same as the corresponding entry in B.

Transpose—A^T, read "A transpose," is the matrix formed by exchanging the rows of A with the columns of A. Thus, if A is size $m \times n$, then A^T is size $n \times m$.

Zero Matrix—A matrix filled entirely with zeroes in every entry.

Square Matrix—A matrix whose size is square, i.e., the number of its columns equals the number of its rows.

Column or Row Vector—A column vector is a matrix that has only one column and multiple rows. A row vector is a matrix that has only one row and multiple columns.

EXAMPLE 1

Given the matrix $A = \begin{bmatrix} 4 & 2 & -1 & 6 \\ 0 & 2 & -8 & 5 \\ 7 & 2 & 3 & -2 \end{bmatrix}$, answer the following questions.

(a) What is the size of A?

Since the matrix has 3 rows and 4 columns, its size is 3×4.

(b) What is a_{23}?

This is asking for the number that is in the place of the element in the second row and the third column, which is –8.

(c) What is A^T?

To find A^T, we need to exchange its rows with its columns. Thus, A^T will be size 4×3. This first row in A^T is the first column in A, etc.

$$A^T = \begin{bmatrix} 4 & 0 & 7 \\ 2 & 2 & 2 \\ -1 & -8 & 3 \\ 6 & 5 & -2 \end{bmatrix}$$

(d) Is *A* square?

No. Since *A* has more columns than rows, it is not a square matrix.

(e) Create the zero matrix that is the same size as *A*.

Basically, we want a 3×4 matrix with all zeros in its entries.

$$0 = \begin{bmatrix} 0 & 0 & 0 & 0 \\ 0 & 0 & 0 & 0 \\ 0 & 0 & 0 & 0 \end{bmatrix}$$

EXAMPLE 2

Construct the matrix *B* if *B* is a 2×2 matrix with entries given by the formula $b_{ij} = (-1)^{i+j}(2i + 3j)$.

For a 2×2 matrix, B will look like this, generically.

$$B = \begin{bmatrix} b_{11} & b_{12} \\ b_{21} & b_{22} \end{bmatrix}$$

We can determine each entry by the formula given, since we know the values of i and j for each entry.

When $i = 1$ and $j = 1$, we get:

$$b_{11} = (-1)^{1+1}(2(1) + 3(1)) = (-1)^2(2 + 3) = 1(5) = 5$$

When $i = 1$ and $j = 2$, we get:

$$b_{12} = (-1)^{1+2}(2(1) + 3(2)) = (-1)^3(2 + 6) = -1(8) = -8$$

When $i = 2$ and $j = 1$, we get:

$$b_{21} = (-1)^{2+1}(2(2) + 3(1)) = (-1)^3(4 + 3) = -1(7) = -7$$

When $i = 2$ and $j = 2$, we get:

$$b_{22} = (-1)^{2+2}(2(2) + 3(2)) = (-1)^4(4 + 6) = 1(10) = 10$$

So we put those entries into their places in the matrix, and we have

$$B = \begin{bmatrix} 5 & -8 \\ -7 & 10 \end{bmatrix}.$$

MATRIX ADDITION AND SCALAR MULTIPLICATION

Matrix Addition: We can only add (or subtract) matrices if they are exactly the same size. To add (or subtract) two matrices, add corresponding entries to create the result.

$$\text{For}\quad A = \begin{bmatrix} a_{11} & a_{12} & a_{13} & \cdots & a_{1n} \\ a_{21} & a_{22} & a_{23} & \cdots & a_{2n} \\ \cdot & \cdot & \cdot & \cdots & \cdot \\ \cdot & \cdot & \cdot & \cdots & \cdot \\ \cdot & \cdot & \cdot & \cdots & \cdot \\ a_{m1} & a_{m2} & a_{m3} & \cdots & a_{mn} \end{bmatrix} \text{ and } B = \begin{bmatrix} b_{11} & b_{12} & b_{13} & \cdots & b_{1n} \\ b_{21} & b_{22} & b_{23} & \cdots & b_{2n} \\ \cdot & \cdot & \cdot & \cdots & \cdot \\ \cdot & \cdot & \cdot & \cdots & \cdot \\ \cdot & \cdot & \cdot & \cdots & \cdot \\ b_{m1} & b_{m2} & b_{m3} & \cdots & b_{mn} \end{bmatrix},$$

$$A + B = \begin{bmatrix} a_{11}+b_{11} & a_{12}+b_{12} & a_{13}+b_{13} & \cdots & a_{1n}+b_{1n} \\ a_{21}+b_{21} & a_{22}+b_{22} & a_{23}+b_{23} & \cdots & a_{2n}+b_{2n} \\ \cdot & \cdot & \cdot & \cdots & \cdot \\ \cdot & \cdot & \cdot & \cdots & \cdot \\ \cdot & \cdot & \cdot & \cdots & \cdot \\ a_{m1}+b_{m1} & a_{m2}+b_{m2} & a_{m3}+b_{m3} & \cdots & a_{mn}+b_{mn} \end{bmatrix}.$$

Scalar Multiplication: We can multiply any matrix by a scalar by multiplying each entry of the matrix by that constant.

$$\text{For}\quad A = \begin{bmatrix} a_{11} & a_{12} & a_{13} & \cdots & a_{1n} \\ a_{21} & a_{22} & a_{23} & \cdots & a_{2n} \\ \cdot & \cdot & \cdot & \cdots & \cdot \\ \cdot & \cdot & \cdot & \cdots & \cdot \\ \cdot & \cdot & \cdot & \cdots & \cdot \\ a_{m1} & a_{m2} & a_{m3} & \cdots & a_{mn} \end{bmatrix} \text{ and } r \in \mathbb{R},$$

$$rA = \begin{bmatrix} ra_{11} & ra_{12} & ra_{13} & \cdots & ra_{1n} \\ ra_{21} & ra_{22} & ra_{23} & \cdots & ra_{2n} \\ \cdot & \cdot & \cdot & \cdots & \cdot \\ \cdot & \cdot & \cdot & \cdots & \cdot \\ \cdot & \cdot & \cdot & \cdots & \cdot \\ ra_{m1} & ra_{m2} & ra_{m3} & \cdots & ra_{mn} \end{bmatrix}.$$

EXAMPLE 3

Given $A = \begin{bmatrix} 1 & 5 & 4 & 0 \\ 4 & -2 & 3 & 1 \end{bmatrix}$, $B = \begin{bmatrix} -3 & 7 & 1 & 2 \\ 2 & -4 & 0 & 5 \end{bmatrix}$ and $C = \begin{bmatrix} 3 & -2 \\ 9 & 0 \\ 1 & -6 \\ 4 & 3 \end{bmatrix}$, answer the following questions.

(a) Compute $A + 3B$.

First, let's compute $3B$.

$$3B = \begin{bmatrix} 3(-3) & 3(7) & 3(1) & 3(2) \\ 3(2) & 3(-4) & 3(0) & 3(5) \end{bmatrix} = \begin{bmatrix} -9 & 21 & 3 & 6 \\ 6 & -12 & 0 & 15 \end{bmatrix}$$

Now, we can perform the addition.

$$A + 3B = \begin{bmatrix} 1 & 5 & 4 & 0 \\ 4 & -2 & 3 & 1 \end{bmatrix} + \begin{bmatrix} -9 & 21 & 3 & 6 \\ 6 & -12 & 0 & 15 \end{bmatrix} = \begin{bmatrix} 1+-9 & 5+21 & 4+3 & 0+6 \\ 4+6 & -2+-12 & 3+0 & 1+15 \end{bmatrix}$$

$$A + 3B = \begin{bmatrix} -8 & 26 & 7 & 6 \\ 10 & -14 & 3 & 16 \end{bmatrix}$$

(b) Compute $2A - C^T$.

First, we'll determine $2A$.

$$2A = \begin{bmatrix} 2(1) & 2(5) & 2(4) & 2(0) \\ 2(4) & 2(-2) & 2(3) & 2(1) \end{bmatrix} = \begin{bmatrix} 2 & 10 & 8 & 0 \\ 8 & -4 & 6 & 2 \end{bmatrix}$$

Next, we need to find C^T.

$$C^T = \begin{bmatrix} 3 & 9 & 1 & 4 \\ -2 & 0 & -6 & 3 \end{bmatrix}.$$

$$\Rightarrow \quad 2A - C^T = \begin{bmatrix} 2 & 10 & 8 & 0 \\ 8 & -4 & 6 & 2 \end{bmatrix} - \begin{bmatrix} 3 & 9 & 1 & 4 \\ -2 & 0 & -6 & 3 \end{bmatrix} = \begin{bmatrix} 2-3 & 10-9 & 8-1 & 0-4 \\ 8-(-2) & -4-0 & 6-(-6) & 2-3 \end{bmatrix}$$

$$2A - C^T = \begin{bmatrix} -1 & 1 & 7 & -4 \\ 10 & -4 & 12 & -1 \end{bmatrix}$$

(c) Compute $B + C$.

We actually can't allow this calculation because we can only add matrices of the same size, and B is a different size than C. B is size 2×4, and C is a 4×2 matrix!

2.1 Homework

Give the size of each of the following matrices:

1. $\begin{bmatrix} 1 & 2 \\ 3 & 4 \end{bmatrix}$

2. $\begin{bmatrix} 1 & 2 & 3 \\ 4 & 5 & 6 \\ 7 & 8 & 9 \end{bmatrix}$

3. $\begin{bmatrix} 10 & 7 & 6 & 1 \\ -5 & 4 & -3 & 2 \\ 1 & 2 & -10 & 12 \end{bmatrix}$

4. $\begin{bmatrix} 8 & 5 & 2 \\ 5 & -4 & 3 \\ 1 & -9 & 13 \\ 4 & 3 & -8 \end{bmatrix}$

5. $\begin{bmatrix} 1 & 7 & 5 & 1 \\ -10 & 6 & -5 & 2 \\ 7 & 2 & -10 & 14 \\ 5 & 5 & 2 & -7 \end{bmatrix}$

6. $\begin{bmatrix} -10 & 4 & 2 \\ 5 & -6 & 6 \\ 2 & -5 & 3 \end{bmatrix}$

7. $\begin{bmatrix} 6 & 2 & 2 & 1 & 2 & 3 \\ 6 & 3 & 4 & 5 & 7 & 8 \end{bmatrix}$

8. $\begin{bmatrix} 1 \\ 4 \end{bmatrix}$

9. $\begin{bmatrix} 3 & 2 \end{bmatrix}$

10. $\begin{bmatrix} -1 & 3 \\ 10 & -4 \\ 8 & 7 \\ 2 & 4 \\ 2 & 5 \\ 6 & -6 \end{bmatrix}$

Use these matrices to perform the following calculations:

$$A = \begin{bmatrix} 2 & 3 \\ 5 & 4 \end{bmatrix},\ B = \begin{bmatrix} -1 & 2 \\ 3 & -4 \end{bmatrix},\ C = \begin{bmatrix} 5 & -2 \\ -3 & 4 \end{bmatrix},\ D = \begin{bmatrix} -3 & -1 \\ -5 & 3 \end{bmatrix}$$

11. $3A$

12. $A + B$

13. $D - C$

14. $2B + 3C$

15. $3(A - D)$
16. $4D - 3B$
17. $A - 2(C + B)$
18. $4(A + B) - C$
19. $A - 2(B + C) + 4D$
20. $3(A + B) - 2(C - D)$

Use these matrices to perform the following calculations:

$$A = \begin{bmatrix} 4 & 2 & 3 \\ 5 & -6 & 4 \\ 2 & -3 & 4 \end{bmatrix},\ B = \begin{bmatrix} -2 & 1 & 3 \\ 5 & -3 & -4 \\ -3 & 2 & 4 \end{bmatrix},\ C = \begin{bmatrix} -5 & 4 & -3 \\ 4 & -3 & 2 \\ -1 & 2 & 10 \end{bmatrix},\ D = \begin{bmatrix} -5 & 3 & -2 \\ 4 & -2 & 3 \\ 1 & 3 & 2 \end{bmatrix}$$

21. $3A$
22. $A - D$
23. $D - C$
24. $3B + 2C$
25. $2(B - D)$
26. $3C - 2B$
27. $A - 3(B - C)$
28. $2(A + D) - B$
29. $B - 3(C + D) + 2A$
30. $2(D + B) - 3(C - A)$

Use these matrices to perform the following calculations, or indicate if the calculation cannot be performed:

$$A = \begin{bmatrix} 2 & 3 \\ 5 & 4 \\ -1 & -3 \end{bmatrix},\ B = \begin{bmatrix} 3 & -1 & 2 \\ -2 & 3 & -4 \end{bmatrix},\ C = \begin{bmatrix} -2 & 4 \\ -1 & -2 \\ -2 & 4 \end{bmatrix},\ D = \begin{bmatrix} -3 & 2 & -1 \\ -2 & 1 & -1 \end{bmatrix}$$

31. $A + C$
32. $A + B$
33. $B + C$
34. $B + D$
35. $B^T + C$
36. $B^T + A^T$
37. $C + 2B^T$
38. $B^T + D^T$
39. $B + (A - C)^T$
40. $(B + D)^T$

Use these matrices to perform the following calculations, or indicate if the calculation cannot be performed:

$$A = \begin{bmatrix} 3 \\ 5 \\ 2 \end{bmatrix}, \ B = \begin{bmatrix} 2 & 0 & 3 \end{bmatrix}, \ C = \begin{bmatrix} -3 & 5 & 2 \end{bmatrix}, \ D = \begin{bmatrix} 2 \\ -2 \\ -1 \end{bmatrix}$$

41. $A + B$

42. $A + C$

43. $A + D$

44. $B + C$

45. $B^T + C$

46. $B + A^T$

47. $C + 2D^T$

48. $B^T + D$

49. $B + (A - D)^T$

50. $(B + C)^T$

2.2 Matrix Multiplication

Remember that multiplication with numbers is commutative, meaning that the order doesn't matter. For example, 4(3) = 3(4) = 12. This is NOT true for multiplication of matrices. Order does matter!! In fact, we can't just multiply any two matrices together. We must meet one condition before multiplication of matrices is defined, namely, the number of columns of the first matrix must equal the number of rows of the second matrix in the product.

DEFINITIONS

Matrix Multiplication—For an $m \times n$ matrix A,

$$A = \begin{bmatrix} a_{11} & a_{12} & a_{13} & \cdot & \cdot & \cdot & a_{1n} \\ a_{21} & a_{22} & a_{23} & \cdot & \cdot & \cdot & a_{2n} \\ \cdot & \cdot & \cdot & \cdot & \cdot & \cdot & \cdot \\ \cdot & \cdot & \cdot & \cdot & \cdot & \cdot & \cdot \\ \cdot & \cdot & \cdot & \cdot & \cdot & \cdot & \cdot \\ a_{m1} & a_{m2} & a_{m3} & \cdot & \cdot & \cdot & a_{mn} \end{bmatrix}$$

and an $n \times p$ matrix B,

$$B = \begin{bmatrix} b_{11} & b_{12} & b_{13} & \cdot & \cdot & \cdot & b_{1p} \\ b_{21} & b_{22} & b_{23} & \cdot & \cdot & \cdot & b_{2p} \\ \cdot & \cdot & \cdot & \cdot & \cdot & \cdot & \cdot \\ \cdot & \cdot & \cdot & \cdot & \cdot & \cdot & \cdot \\ \cdot & \cdot & \cdot & \cdot & \cdot & \cdot & \cdot \\ b_{n1} & b_{n2} & b_{n3} & \cdot & \cdot & \cdot & b_{np} \end{bmatrix}$$

the product AB is an $m \times p$ matrix where the ij entry is given by $a_{i1}b_{1j} + a_{i2}b_{2j} + a_{i3}b_{3j} + \ldots + a_{in}b_{nj}$. Basically, the ij entry of the product AB is the sum of the products of the ith row elements of A with the corresponding jth column elements of B. (*Note:* $AB \neq BA$)

For multiplication with numbers, i.e., in our original algebra context, we have a multiplicative identity, namely 1. Any type of identity maintains the identity of the original entity when operated on with that specific operation. For example, notice that if we multiply ANY number by 1, we get back the original number we started with. The closest multiplicative identity we have for matrices is the identity matrix.

Identity Matrix—There are infinitely many identity matrices. Each identity matrix is square and denoted as I. I contains zeros in all non-diagonal entries and ones in all diagonal entries.

$$I = \begin{bmatrix} 1 & 0 & 0 & 0 & \cdot & \cdot & \cdot & 0 \\ 0 & 1 & 0 & 0 & \cdot & \cdot & \cdot & 0 \\ 0 & 0 & 1 & 0 & \cdot & \cdot & \cdot & 0 \\ \cdot & \cdot & \cdot & \cdot & \cdot & \cdot & \cdot & \cdot \\ \cdot & \cdot & \cdot & \cdot & \cdot & \cdot & \cdot & \cdot \\ \cdot & \cdot & \cdot & \cdot & \cdot & \cdot & \cdot & \cdot \\ 0 & 0 & 0 & 0 & 0 & 0 & 0 & 1 \end{bmatrix}$$

We choose the size of I based on the size of the matrix by which we will multiply it. For example, a 3x3 identity matrix would be

$$I = \begin{bmatrix} 1 & 0 & 0 \\ 0 & 1 & 0 \\ 0 & 0 & 1 \end{bmatrix}$$

If A is an $m \times n$ matrix, then $AI = A$, for I size $n \times n$ and $IA = A$, for I size $m \times m$. If A is a square matrix, size $m \times m$, then $AI = AI = A$, where I is size $m \times m$ also.

PROPERTIES OF MATRIX MULTIPLICATION

1. $(AB)C = A(BC)$	Associativity
2. $A(B + C) = AB + AC$	Right Distributivity through addition/subtraction
3. $(B + C)A = BA + CA$	Left Distributivity through addition/subtraction
4. $(AB)^T = B^TA^T$	

If A is a square matrix, size $n \times n$, and p is a positive integer, then we can define $A^P = A \cdot A \cdots A$ as the matrix A multiplied by itself p times. Since A is square, all the successive products are well-defined because an $n \times n$ matrix multiplied by an $n \times n$ matrix is another $n \times n$ matrix.

EXAMPLE 1

Given $A = \begin{bmatrix} 2 & -4 & 0 & 1 \\ 3 & 7 & -1 & 5 \end{bmatrix}$, $B = \begin{bmatrix} 1 \\ 2 \\ 5 \\ -4 \end{bmatrix}$ and $C = \begin{bmatrix} 0 & 2 & 6 \end{bmatrix}$, find the products if possible.

If not possible, state the reason.

(a) $AB = ?$

Let's first consider the sizes of the matrices A and B. A is 2×4 and B is 4×1. Since the number of columns of A matches the number of rows of B, the product AB is possible and well-defined. The product matrix will be a 2×1 matrix.

To get the first row element of AB, we need to multiply corresponding elements of the first row of A with the first (and only) column of B.

$$\Rightarrow \qquad AB = \begin{bmatrix} 2(1) + -4(2) + 0(5) + 1(-4) \\ ?? \end{bmatrix}$$

To finish the product and compute the second row, first column element of AB, we need to multiply corresponding elements of the second row of A with the first column of B.

$$\Rightarrow \qquad AB = \begin{bmatrix} 2(1) + -4(2) + 0(5) + 1(-4) \\ 3(1) + 7(2) + -1(5) + 5(-4) \end{bmatrix}$$

Now we only need to simplify the matrix to finish.

$$AB = \begin{bmatrix} 2(1)+-4(2)+0(5)+1(-4) \\ 3(1)+7(2)+-1(5)+5(-4) \end{bmatrix} = \begin{bmatrix} 2-8+0-4 \\ 3+14-5-20 \end{bmatrix} = \begin{bmatrix} -10 \\ -8 \end{bmatrix}$$

(b) $BA = ?$

Again, we first have to consider if the product is possible. We can do that by looking at the sizes of the matrices. B is 4×1 and A is 2×4. Since the number of columns of B does NOT match the number of rows of A (i.e., $1 \neq 2$), this product is not possible!

(c) $BC = ?$

B is size 4×1, and C is size 1×3. Therefore, the number of columns of B does indeed equal the number of rows of C, which means that the product BC is well-defined and possible. The resulting matrix will be size 4×3.

To get the first row/first column element of BC, we need to multiply corresponding elements of the first row of B with the first column of C. To get the first row/second column element of BC, we need to multiply corresponding elements of the first row of B with the second column of C. We continue in this fashion until we fill in all the 4×3 elements of BC.

$$\Rightarrow \qquad BC = \begin{bmatrix} 1(0) & 1(2) & 1(6) \\ 2(0) & 2(2) & 2(6) \\ 5(0) & 5(2) & 5(6) \\ -4(0) & -4(2) & -4(6) \end{bmatrix} = \begin{bmatrix} 0 & 2 & 6 \\ 0 & 4 & 12 \\ 0 & 10 & 30 \\ 0 & -8 & -24 \end{bmatrix}$$

(d) $A(BC) = ?$

To compute this product, we first need to compute BC since the parentheses indicate the need to perform that operation first. We already have the result for BC from part (c). A is size 2×4, and BC is size 4×3. So, since the number of columns of A equals the number of rows of BC ($4 = 4$), then we can perform the matrix multiplication. The resulting matrix for $A(BC)$ will be size 2×3.

$$A(BC) = \begin{bmatrix} 2 & -4 & 0 & 1 \\ 3 & 7 & -1 & 5 \end{bmatrix} \begin{bmatrix} 0 & 2 & 6 \\ 0 & 4 & 12 \\ 0 & 10 & 30 \\ 0 & -8 & -24 \end{bmatrix}$$

$$A(BC) = \begin{bmatrix} 2(0)+-4(0)+0(0)+1(0) & 2(2)+-4(4)+0(10)+1(-8) & 2(6)+-4(12)+0(30)+1(-24) \\ 3(0)+7(0)+-1(0)+5(0) & 3(2)+7(4)+-1(10)+5(-8) & 3(6)+7(12)+-1(30)+5(-24) \end{bmatrix}$$

$$A(BC) = \begin{bmatrix} 0+0+0+0 & 4-16+0-8 & 12-48+0-24 \\ 0+0+0+0 & 6+28-10-40 & 18+84-30-120 \end{bmatrix} = \begin{bmatrix} 0 & -20 & -60 \\ 0 & -16 & -48 \end{bmatrix}$$

(e) $AC = ?$

Matrix A is size 2×4, and C is 1×3. Since the number of columns of A does not match the number of rows of C, namely $4 \neq 1$, we cannot multiply A by C. This product is not possible.

EXAMPLE 2

Given $A = \begin{bmatrix} 2 & 1 & 0 \\ 3 & 4 & 5 \end{bmatrix}$, find A^TA and AA^T.

Let's first explicitly write A^T. Then we'll perform the multiplication.

$$A^T = \begin{bmatrix} 2 & 3 \\ 1 & 4 \\ 0 & 5 \end{bmatrix}$$

A^T is size 3×2, and A is 2×3, which means that A^TA will be size 3×3, since the number of columns of A^T is equal to the number of rows of A.

$$\Rightarrow \quad A^TA = \begin{bmatrix} 2 & 3 \\ 1 & 4 \\ 0 & 5 \end{bmatrix}\begin{bmatrix} 2 & 1 & 0 \\ 3 & 4 & 5 \end{bmatrix} = \begin{bmatrix} 2(2)+3(3) & 2(1)+3(4) & 2(0)+3(5) \\ 1(2)+4(3) & 1(1)+4(4) & 1(0)+4(5) \\ 0(2)+5(3) & 0(1)+5(4) & 0(0)+5(5) \end{bmatrix}$$

$$A^TA = \begin{bmatrix} 4+9 & 2+12 & 0+15 \\ 2+12 & 1+16 & 0+20 \\ 0+15 & 0+20 & 0+25 \end{bmatrix} = \begin{bmatrix} 13 & 14 & 15 \\ 14 & 17 & 20 \\ 15 & 20 & 25 \end{bmatrix}$$

To compute AA^T, notice that A is size 2×3 and A^T is size 3×2, which means that AA^T, will be size 2×2. This is a great example to remind us again that multiplication of matrices is not commutative; in other words, the order of the multiplication does indeed matter.

$$\Rightarrow \quad AA^T = \begin{bmatrix} 2 & 1 & 0 \\ 3 & 4 & 5 \end{bmatrix}\begin{bmatrix} 2 & 3 \\ 1 & 4 \\ 0 & 5 \end{bmatrix} = \begin{bmatrix} 2(2)+1(1)+0(0) & 2(3)+1(4)+0(5) \\ 3(2)+4(1)+5(0) & 3(3)+4(4)+5(5) \end{bmatrix}$$

$$AA^T = \begin{bmatrix} 5 & 10 \\ 10 & 50 \end{bmatrix}$$

EXAMPLE 3

For $C = \begin{bmatrix} 0 & 1 & -1 \\ 2 & 0 & 3 \\ 0 & 0 & 1 \end{bmatrix}$ compute C^3.

For this product, we need to multiply the square 3 × 3 matrix C by itself three times. First, we'll start by computing C^2.

$$\Rightarrow \quad C^2 = \begin{bmatrix} 0 & 1 & -1 \\ 2 & 0 & 3 \\ 0 & 0 & 1 \end{bmatrix}\begin{bmatrix} 0 & 1 & -1 \\ 2 & 0 & 3 \\ 0 & 0 & 1 \end{bmatrix} = \begin{bmatrix} 0(0)+1(2)+-1(0) & 0(1)+1(0)+-1(0) & 0(-1)+1(3)+-1(1) \\ 2(0)+0(2)+3(0) & 2(1)+0(0)+3(0) & 2(-1)+0(3)+3(1) \\ 0(0)+0(2)+1(0) & 0(1)+0(0)+1(0) & 0(-1)+0(3)+1(1) \end{bmatrix}$$

$$C^2 = \begin{bmatrix} 0+2+0 & 0+0+0 & 0+3-1 \\ 0+0+0 & 2+0+0 & -2+0+3 \\ 0+0+0 & 0+0+0 & 0+0+1 \end{bmatrix} = \begin{bmatrix} 2 & 0 & 2 \\ 0 & 2 & 1 \\ 0 & 0 & 1 \end{bmatrix}$$

We now need to multiply our last result by C once more.

$$\Rightarrow \quad C^3 = \begin{bmatrix} 2 & 0 & 2 \\ 0 & 2 & 1 \\ 0 & 0 & 1 \end{bmatrix}\begin{bmatrix} 0 & 1 & -1 \\ 2 & 0 & 3 \\ 0 & 0 & 1 \end{bmatrix} = \begin{bmatrix} 2(0)+0(2)+2(0) & 2(1)+0(0)+2(0) & 2(-1)+0(3)+2(1) \\ 0(0)+2(2)+1(0) & 0(1)+2(0)+1(0) & 0(-1)+2(3)+1(1) \\ 0(0)+0(2)+1(0) & 0(1)+0(0)+1(0) & 0(-1)+0(3)+1(1) \end{bmatrix}$$

$$C^3 = \begin{bmatrix} 0+0+0 & 2+0+0 & -2+0+2 \\ 0+4+0 & 0+0+0 & 0+6+1 \\ 0+0+0 & 0+0+0 & 0+0+1 \end{bmatrix} = \begin{bmatrix} 0 & 2 & 0 \\ 4 & 0 & 7 \\ 0 & 0 & 1 \end{bmatrix}$$

EXAMPLE 4

Solve this matrix equation for x.

$$4\begin{bmatrix} 1 \\ 2x \end{bmatrix} - \left(\begin{bmatrix} 15 \\ 9 \end{bmatrix} - 3\begin{bmatrix} 1 \\ 4 \end{bmatrix}\right) = \begin{bmatrix} -8 \\ 19 \end{bmatrix}$$

To solve for x, we first need to perform the matrix subtraction and simplify the left side of the equation. By performing the scalar multiplication, the equation becomes

$$\begin{bmatrix} 4 \\ 8x \end{bmatrix} - \left(\begin{bmatrix} 15 \\ 9 \end{bmatrix} - \begin{bmatrix} 3 \\ 12 \end{bmatrix} \right) = \begin{bmatrix} -8 \\ 19 \end{bmatrix}$$ Perform the subtraction inside the parentheses on the left.

$$\begin{bmatrix} 4 \\ 8x \end{bmatrix} - \begin{bmatrix} 12 \\ -3 \end{bmatrix} = \begin{bmatrix} -8 \\ 19 \end{bmatrix}$$ Perform the final matrix subtraction on the left side of the equation.

$$\begin{bmatrix} -8 \\ 8x + 3 \end{bmatrix} = \begin{bmatrix} -8 \\ 19 \end{bmatrix}$$ Now equate corresponding elements of the matrices.

$\Leftrightarrow \; -8 = -8$ and $8x + 3 = 19$ It is true that -8 is equal to itself. The second equation is simply a linear equation that we need to solve in order to get the value for x.

$$8x + 3 = 19 \;\Leftrightarrow\; 8x = 16 \;\Leftrightarrow\; x = 2$$

So, the solution is $x = 2$.

EXAMPLE 5

Solve this matrix equation for X.

$$\begin{bmatrix} 1 & -2 \\ 3 & 5 \end{bmatrix} X + \begin{bmatrix} 2 & -7 \\ 10 & 1 \end{bmatrix} = \begin{bmatrix} 0 & -19 \\ 15 & 20 \end{bmatrix}$$

Notice that the only way we can have a solution here is if the matrix X is size 2×2, since a 2×2 matrix multiplied by a 2×2 yields another 2×2 matrix which can then be added to another 2×2 matrix. Let's write X as $X = \begin{bmatrix} a & b \\ c & d \end{bmatrix}$ where we don't know the values of a, b, c, and d, so they're the single-value variables. Now we can rewrite the equation and solve for a, b, c, and d (and thus X).

$$\begin{bmatrix} 1 & -2 \\ 3 & 5 \end{bmatrix} \begin{bmatrix} a & b \\ c & d \end{bmatrix} + \begin{bmatrix} 2 & -7 \\ 10 & 1 \end{bmatrix} = \begin{bmatrix} 0 & -19 \\ 15 & 20 \end{bmatrix}$$ Perform the matrix multiplication on the left.

$$\begin{bmatrix} a - 2c & b - 2d \\ 3a + 5c & 3b + 5d \end{bmatrix} + \begin{bmatrix} 2 & -7 \\ 10 & 1 \end{bmatrix} = \begin{bmatrix} 0 & -19 \\ 15 & 20 \end{bmatrix}$$ Add the matrices on the left side of the equation.

$$\begin{bmatrix} a - 2c + 2 & b - 2d - 7 \\ 3a + 5c + 10 & 3b + 5d + 1 \end{bmatrix} = \begin{bmatrix} 0 & -19 \\ 15 & 20 \end{bmatrix}$$ Now we can equate corresponding elements of the matrices on both sides of the equation. We'll get four linear equations to solve.

$$a - 2c + 2 = 0, \quad b - 2d - 7 = -19, \quad 3a + 5c + 10 = 15, \quad 3b + 5d + 1 = 20$$
$$\Leftrightarrow \quad a - 2c = -2, \quad b - 2d = -12, \quad 3a + 5c = 5, \quad 3b + 5d = 19$$

Let's group these into two groups according to the equations that have the same variables in them.

$$\text{(1a)} \quad \begin{matrix} a - 2c = -2 \\ 3a + 5c = 5 \end{matrix} \quad \text{and} \quad \text{(2a)} \quad \begin{matrix} b - 2d = -12 \\ 3b + 5d = 19 \end{matrix}$$

For equations in (1a), multiply the top equation by -3 so we can eliminate a. For equations in (2a), multiply the top equation by -3 so we can eliminate b.

(1b) $\begin{aligned} -3a + 6c &= 6 \\ 3a + 5c &= 5 \end{aligned}$ and (2b) $\begin{aligned} -3b + 6d &= 36 \\ 3b + 5d &= 19 \end{aligned}$ For both sets of equations, add straight down.

We get: (1b) $11c = 11$, which means $c = 1$ and (2b) $11d = 55$, which means $d = 5$.
Substitute those values in for c and d into the top equations from (1a) and (2a).

(1a) becomes $a - 2c = -2 \iff a - 2(1) = -2 \iff a - 2 = -2 \iff a = 0$

(2a) becomes $b - 2d = -12 \iff b - 2(5) = -12 \iff b - 10 = -12 \iff b = -2$

We have all the values for a, b, c, and d now. This means we have solved for the unknown matrix X.

$$X = \begin{bmatrix} 0 & -2 \\ 1 & 5 \end{bmatrix}$$

2.2 Homework

Using these four matrices, find the indicated products, or state that the multiplication is not possible:

$$A = \begin{bmatrix} 2 & -6 \\ -2 & 3 \end{bmatrix},\ B = \begin{bmatrix} 6 \\ 3 \end{bmatrix},\ C = \begin{bmatrix} 4 & 7 \end{bmatrix},\ D = \begin{bmatrix} -8 & 6 & 1 \\ 5 & 7 & 7 \end{bmatrix}$$

1. AB
2. BA
3. BC
4. CB
5. CD
6. DC
7. AD
8. DA

Using these four matrices, find the indicated products, or state that the multiplication is not possible:

$$W = \begin{bmatrix} 7 & -7 \\ 0 & -6 \end{bmatrix},\ X = \begin{bmatrix} 6 \\ -2 \end{bmatrix},\ Y = \begin{bmatrix} 6 & 4 \end{bmatrix},\ Z = \begin{bmatrix} -9 & -1 \\ -8 & -4 \\ 0 & 5 \end{bmatrix}$$

9. Y^TY
10. ZY^T
11. YY
12. ZW
13. XY
14. W^TZ^T
15. W^TX^T
16. YW
17. XZ^T
18. WW
19. XX^T
20. XX

Using these four matrices, find the indicated products, or state that the multiplication is not possible:

$$P = \begin{bmatrix} 2 & 1 \\ 5 & 3 \end{bmatrix}, \; Q = \begin{bmatrix} 3 & -1 \\ -5 & 2 \end{bmatrix}, \; R = \begin{bmatrix} -5 & 9 \\ 1 & 2 \\ 1 & 3 \end{bmatrix}, \; S = \begin{bmatrix} 5 & 5 & 0 \\ -6 & -4 & 2 \\ 4 & -1 & 5 \end{bmatrix}$$

21. PQ

22. QP

23. RQ

24. QR

25. SS

26. PR

27. RP

28. S^TS

29. RR

30. SR

Solve for x and y.

31. $\begin{bmatrix} -7 & -10 \\ 6 & 9 \end{bmatrix}\begin{bmatrix} -18 & x \\ 16 & -10 \end{bmatrix} = \begin{bmatrix} -34 & 23 \\ y & -24 \end{bmatrix}$

Solve the following matrix equations.

32. $\begin{bmatrix} -8 & -2 \\ -6 & 2 \end{bmatrix}X = \begin{bmatrix} -50 & 2 & -8 \\ -6 & -30 & 22 \end{bmatrix}$

33. $\begin{bmatrix} -8 & -4 \\ 0 & -12 \end{bmatrix}X + \begin{bmatrix} -8 & -2 \\ 11 & -6 \end{bmatrix} = \begin{bmatrix} -36 & 30 \\ 143 & -150 \end{bmatrix}$

34. $\begin{bmatrix} 4 & 10 \\ -9 & 1 \end{bmatrix}X = \begin{bmatrix} -34 & -4 \\ -88 & -38 \end{bmatrix}$

35. $\begin{bmatrix} -4 & -1 \\ 8 & -5 \end{bmatrix}X + \begin{bmatrix} 2 & 6 \\ -7 & 5 \end{bmatrix} = \begin{bmatrix} 20 & 22 \\ -29 & -111 \end{bmatrix}$

36. $8\begin{bmatrix} -1 \\ -3x \end{bmatrix} - \left(\begin{bmatrix} -12 \\ -17 \end{bmatrix} - 2\begin{bmatrix} -3 \\ 11 \end{bmatrix}\right) = \begin{bmatrix} -2 \\ -81 \end{bmatrix}$

37. $\begin{bmatrix} -2x & 19 \\ -6 & 4 \end{bmatrix} + \left(\begin{bmatrix} -12 & -16 \\ 16 & -18 \end{bmatrix} + \begin{bmatrix} -17 & -9 \\ -15 & 1 \end{bmatrix} \right) = \begin{bmatrix} -19 & -6 \\ -5 & -13 \end{bmatrix}$

38. $\left(\begin{bmatrix} -5 \\ 10 \\ -19 \end{bmatrix} + \begin{bmatrix} -2x \\ 13 \\ -8 \end{bmatrix} \right) + \begin{bmatrix} -8 \\ -4 \\ -7 \end{bmatrix} = \begin{bmatrix} -5 \\ 19 \\ -34 \end{bmatrix}$

39. $\left(\begin{bmatrix} -4 \\ -19 \\ 17 \\ 2x \end{bmatrix} - \begin{bmatrix} 20 \\ 4 \\ 0 \\ -15 \end{bmatrix} \right) + 5\begin{bmatrix} -16 \\ 15 \\ -12 \\ 5 \end{bmatrix} = \begin{bmatrix} -104 \\ 52 \\ -43 \\ 26 \end{bmatrix}$

2.3 Gauss-Jordan Elimination

Now that we know how to do basic manipulations with our new entities (matrices), we can move on to an application that is very useful with matrices. Using Gauss-Jordan Elimination, we can use matrices to help us more efficiently solve a system of linear equations.

VOCABULARY

Augmented Matrix—A matrix that represents a system of linear equations. If we first set up the system of linear equations such that each equation has all the variables on the left side of the equation and the constant alone on the right side of the equation, then each row of the matrix will represent the coefficients of the variables in one of the linear equations, as well as the constant term at the end of the row.

Elementary Row Operations—The three operations allowed in manipulating a matrix. Each of these operations produces an equivalent system of equations for an augmented matrix.

Gauss-Jordan Elimination—The process of manipulating a matrix using the elementary row operations in order to get the matrix to look as close as possible to the identity matrix.

ELEMENTARY ROW OPERATIONS

1. *Switch two rows.* Since each row represents an entire equation, switching two rows is the same as simply rewriting the equation in a different location.
2. *Multiply any row by a nonzero constant.* Again, each row represents an equation and multiplying a row by a nonzero constant is equivalent to multiplying both sides of an equation by the same number.
3. *Replace one row with the result of adding it with a multiple of a different row.* This operation is truly the elimination piece. For this operation, we temporarily multiply one row by a nonzero constant and add it to another row in order to eliminate one of the entries. We replace one of the rows (choosing either of the rows we used) with the result of the addition.

Note: Keep in mind that these elementary row operations were exactly the steps used when doing the method of elimination to solve a system of two linear equations (with two variables). We are now just formalizing them to allow us to solve systems of linear equations with more than two variables.

EXAMPLE 1

(a) Write the augmented matrix for this system of equations:

$$x - 3y = -2$$
$$3y = -4x + 37$$

(b) Solve the system of equations using Gauss-Jordan Elimination.

(a) First, we must make sure that both equations are in the same order. In other words, it would be best to get all of the variable terms on the left side of the equations and leave the constants on the right side of the equations.

$x - 3y = -2$

$4x + 3y = 37$ We added $4x$ to both sides of the bottom equation. Now everything is aligned, i.e., in the same order for both equations.

Now, to put this information into an augmented matrix, the first (top) equation will correspond to the first row of the matrix and the second (bottom) equation will correspond to the second row of the matrix. The first column of the matrix will be the coefficients of x, the second column will be the coefficients of y, and the last column of the matrix will be the constants. We'll put a dashed line segment between the second and third columns in the augmented matrix to designate the location of the equals signs.

$$\textbf{Augmented matrix: } \left[\begin{array}{cc:c} 1 & -3 & -2 \\ 4 & 3 & 37 \end{array}\right]$$

(b) To solve the system of equations, we will use elementary row operations on the augmented matrix, until the left side of the dotted line looks like the identity matrix. Thus, we want to have a 1 in the diagonal entries and a 0 in the non-diagonal entries (where 4 and –3 are right now).

My first goal is to make the 4 turn into a 0. We can accomplish that intermediate goal by adding –4 to 4 to get 0. To do that, using elementary row operations, we can temporarily multiply the top row by –4 and add down to the third row, replacing the third row with the result of that addition. (Thus, we are using row operations #3.)

$$\begin{matrix}(-4)\\ \hookrightarrow\end{matrix}\left[\begin{array}{cc:c} 1 & -3 & -2 \\ 4 & 3 & 37 \end{array}\right]$$

$$= \left[\begin{array}{cc:c} 1 & -3 & -2 \\ -4+4 & -4(-3)+3 & -4(-2)+37 \end{array}\right] = \left[\begin{array}{cc:c} 1 & -3 & -2 \\ 0 & 15 & 45 \end{array}\right]$$

The next goal is to do a legal operation to turn the 15 into 1. We can use row operation #2 and multiply the second row by $\frac{1}{15}$.

$$= \begin{matrix}\\ \left(\frac{1}{15}\right)\end{matrix}\left[\begin{array}{cc:c} 1 & -3 & -2 \\ 0 & 15 & 45 \end{array}\right] = \left[\begin{array}{cc:c} 1 & -3 & -2 \\ 0 & 15\left(\frac{1}{15}\right) & 45\left(\frac{1}{15}\right) \end{array}\right] = \left[\begin{array}{cc:c} 1 & -3 & -2 \\ 0 & 1 & 3 \end{array}\right]$$

Now, the last goal is to change the –3 to be 0, in order to get the left side of the dotted line segment to look like the identity matrix. We can temporarily multiply the bottom row by 3 and add to the top row, replacing the top row with the result of the addition, thereby using row operation #3.

$$= \begin{matrix}\uparrow\\ (3)\end{matrix}\left[\begin{array}{cc:c} 1 & -3 & -2 \\ 0 & 1 & 3 \end{array}\right] = \left[\begin{array}{cc:c} 3(0)+1 & 3(1)+-3 & 3(3)+-2 \\ 0 & 1 & 3 \end{array}\right] = \left[\begin{array}{cc:c} 1 & 0 & 7 \\ 0 & 1 & 3 \end{array}\right]$$

We have reached our goal of getting the left side of the dotted line segment to be the identity matrix. Let's remember that each row represents an equation in our system. Rewrite the rows as their corresponding equations, keeping the same order of variables and constants as we started with.

$$\Rightarrow \begin{matrix} 1x + 0y = 7 \\ 0x + y = 3 \end{matrix} \Leftrightarrow \begin{matrix} x = 7 \\ y = 3 \end{matrix}$$

Thus, our solution is the point (7, 3).

Notice that in order to get the left side of the augmented matrix to look like the identity matrix, it's most efficient to go in some order with the intermediate goal-setting. First, we get the entries in the first column to be 1 followed by all zeros. Next, we work on getting the below-diagonal entries of the second column to be zeros. We continue in this fashion until all the below-diagonal entries of the left side of the matrix are zeros, with the diagonal entries all equal to 1. Following that goal, we focus on getting the above-diagonal entries to be zeros, moving from right to left. Let's do more examples.

EXAMPLE 2

Set up the augmented matrix for the following system of equations. Then use elementary row operations to solve the system of equations.

$$\begin{aligned} x + 2y - z &= 1 \\ 2x + 5y - 2z &= 5 \\ -x + y + 5z &= 24 \end{aligned}$$

We can set up the augmented matrix right away since every equation is in the same order; namely, the x term, then the y term, then the z term equals the constant term.

$$\textbf{Augmented matrix:} \quad \left[\begin{array}{ccc:c} 1 & 2 & -1 & 1 \\ 2 & 5 & -2 & 5 \\ -1 & 1 & 5 & 24 \end{array}\right]$$

We work first on getting the below-diagonal elements to be zero, using elementary row operations.

$$\left[\begin{array}{ccc:c} 1 & 2 & -1 & 1 \\ 2 & 5 & -2 & 5 \\ (-1) & 1 & 5 & 24 \end{array}\right]$$

Add the top row to the bottom row, replacing the bottom row with the result (row operation #3).

$$\left[\begin{array}{cccc:c} 1 & 2 & -1 & & 1 \\ 2 & 5 & -2 & & 5 \\ 1+-1 & 2+1 & -1+5 & & 1+24 \end{array}\right] = \begin{matrix}(-2)\\ \hookrightarrow \\ \\ \end{matrix}\left[\begin{array}{ccc:c} 1 & 2 & -1 & 1 \\ (2) & 5 & -2 & 5 \\ 0 & 3 & 4 & 25 \end{array}\right]$$

Temporarily multiply the top row by –2 and add to the middle row, replacing the middle row with the result (row operation #3).

$$\left[\begin{array}{ccc:c} 1 & 2 & -1 & 1 \\ -2(1)+2 & -2(2)+5 & -2(-1)+-2 & -2(1)+5 \\ 0 & 3 & 4 & 25 \end{array}\right] = \begin{matrix} \\ (-3) \\ \hookrightarrow \end{matrix}\left[\begin{array}{ccc:c} 1 & 2 & -1 & 1 \\ 0 & 1 & 0 & 3 \\ 0 & (3) & 4 & 25 \end{array}\right]$$

Notice that with the last step, we got the middle row, first column element to be zero, and we also ended up with the middle row diagonal element to be one. Now we need to multiply the middle row by –3, add down to the bottom row and replace the bottom row with the result (row operation #3).

$$\left[\begin{array}{ccc|c} 1 & 2 & -1 & 1 \\ 0 & 1 & 0 & 3 \\ -3(0)+0 & -3(1)+3 & -3(0)+4 & -3(3)+25 \end{array}\right] = \begin{array}{c} \\ \\ \left(\frac{1}{4}\right) \end{array}\left[\begin{array}{ccc|c} 1 & 2 & -1 & 1 \\ 0 & 1 & 0 & 3 \\ 0 & 0 & \textcircled{4} & 16 \end{array}\right]$$

Now that we have all the below-diagonal entries as zero, we next need to make sure the diagonal elements are all 1. To complete that goal, multiply the last row by ¼ (row operation #2).

$$\left[\begin{array}{ccc|c} 1 & 2 & -1 & 1 \\ 0 & 1 & 0 & 3 \\ 0\left(\frac{1}{4}\right) & 0\left(\frac{1}{4}\right) & 4\left(\frac{1}{4}\right) & 16\left(\frac{1}{4}\right) \end{array}\right] = \left[\begin{array}{ccc|c} 1 & 2 & -1 & 1 \\ 0 & 1 & 0 & 3 \\ 0 & 0 & 1 & 4 \end{array}\right]$$

The next intermediate goal is to manipulate the matrix in order to get the above-diagonal entries to be zero as well. Notice that, for this example, row 2 and row 3 already look like the identity matrix at this point. So we only need to turn the 2 and –1 entries (in the first row above the diagonal) into zeros.

$$\left[\begin{array}{ccc|c} 1 & 2 & \textcircled{-1} & 1 \\ 0 & 1 & 0 & 3 \\ 0 & 0 & 1 & 4 \end{array}\right]$$

Add the bottom row with the top row, replacing the top row with the result (row operation #3).

$$\left[\begin{array}{ccc|c} 1+0 & 2+0 & -1+1 & 1+4 \\ 0 & 1 & 0 & 3 \\ 0 & 0 & 1 & 4 \end{array}\right] = \begin{array}{c} \\ (-2) \\ \\ \end{array}\left[\begin{array}{ccc|c} 1 & \textcircled{2} & 0 & 5 \\ 0 & 1 & 0 & 3 \\ 0 & 0 & 1 & 4 \end{array}\right]$$

Next, temporarily multiply row 2 by –2 and add to the top row, replacing the top row with the result (row operation #3).

$$\left[\begin{array}{ccc|c} -2(0)+1 & -2(1)+2 & -2(0)+0 & -2(3)+5 \\ 0 & 1 & 0 & 3 \\ 0 & 0 & 1 & 4 \end{array}\right] = \left[\begin{array}{ccc|c} 1 & 0 & 0 & -1 \\ 0 & 1 & 0 & 3 \\ 0 & 0 & 1 & 4 \end{array}\right]$$

It's exciting to see that we now have the left side of the dotted line segment looking exactly like the identity matrix. So, we know we're done! If we convert this now into the corresponding system of equations, we'll have the following.

$$\begin{array}{c} 1x+0y+0z=-1 \\ 0x+1y+0z=3 \\ 0x+0y+1z=4 \end{array} \Leftrightarrow \begin{array}{c} x=-1 \\ y=3 \\ z=4 \end{array} \Leftrightarrow \text{The point of intersection is } (-1, 3, 4).$$

At this point, let's interject to get a visual idea of what we're doing when we solve a system of three linear equations, with three variables. Basically, we have three planes in 3-d space and we're looking for the one point all three planes have in common. However, we won't always have that nice case of one intersection point. These figures represent the different possibilities we have for the solution(s) of such a system.

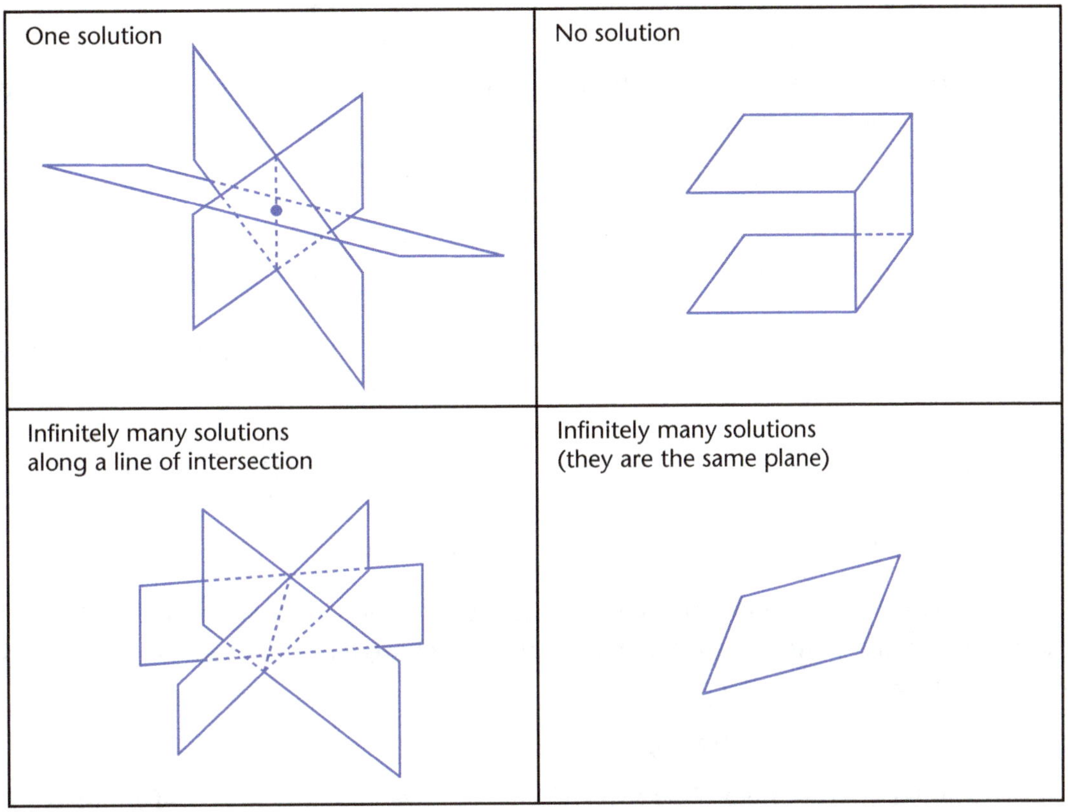

EXAMPLE 3

Solve the system of linear equations.

$$\begin{aligned} 2x + y - 3z &= 5 \\ x \quad\;\; + 5z &= 2 \\ 4x + 2y - 6z &= 8 \end{aligned}$$

Augmented matrix: $\left[\begin{array}{ccc|c} 2 & 1 & -3 & 5 \\ 1 & 0 & 5 & 2 \\ 4 & 2 & -6 & 8 \end{array}\right]$

For this example, we don't have a one in the first column/first row entry, i.e., on the top diagonal entry. Since we want that, we'll switch the first two rows (row operation #1).

$$(-2)\left[\begin{array}{ccc|c} 1 & 0 & 5 & 2 \\ 2 & 1 & -3 & 5 \\ 4 & 2 & -6 & 8 \end{array}\right]$$

To turn the 2 in the second row into a 0, we'll temporarily multiply the top row by –2 and add to the middle row, replacing the middle row with the result (row operation #3).

$$\left[\begin{array}{ccc|c} 1 & 0 & 5 & 2 \\ -2(1)+2 & -2(0)+1 & -2(5)+-3 & -2(2)+5 \\ 4 & 2 & -6 & 8 \end{array}\right] = \begin{array}{c}(-4)\\ \\ \end{array}\left[\begin{array}{ccc|c} 1 & 0 & 5 & 2 \\ 0 & 1 & -13 & 1 \\ \boxed{4} & 2 & -6 & 8 \end{array}\right]$$

To turn the 4 in the bottom row into a 0, we'll temporarily multiply the top row by –4 and add down to the bottom row, replacing the bottom row with the result (row operation #3).

$$\left[\begin{array}{ccc|c} 1 & 0 & 5 & 2 \\ 0 & 1 & -13 & 1 \\ -4(1)+4 & -4(0)+2 & -4(5)+-6 & -4(2)+8 \end{array}\right] = \begin{array}{c} \\ (-2)\\ \end{array}\left[\begin{array}{ccc|c} 1 & 0 & 5 & 2 \\ 0 & 1 & -13 & 1 \\ 0 & \boxed{2} & -26 & 0 \end{array}\right]$$

Next, we want to change the 2 in the bottom row into a 0, so temporarily multiply the middle row by –2 and add to the bottom row, replacing the bottom row with the result (row operation #3).

$$\left[\begin{array}{ccc|c} 1 & 0 & 5 & 2 \\ 0 & 1 & -13 & 1 \\ -2(0)+0 & -2(1)+2 & -2(-13)+-26 & -2(1)+0 \end{array}\right] = \left[\begin{array}{ccc|c} 1 & 0 & 5 & 2 \\ 0 & 1 & -13 & 1 \\ 0 & 0 & 0 & -2 \end{array}\right]$$

This is now a very interesting moment. Notice that the last step not only changed the 2 in the bottom row into 0, but it also changed the –26 into 0 as well. If we exchange the bottom row for its corresponding equation in our system right now, it will be

$$0x + 0y + 0z = -2 \quad \Leftrightarrow \quad 0 = -2$$

Is it ever true that 0 equals –2? No. This means that we have *no solution (N.S.)* for our system of linear equations!

EXAMPLE 4

Solve the system of linear equations.

$$\begin{aligned} x - y + 4z &= 1 \\ 3x + y &= 6 \\ x + 2y - 17z &= 4 \end{aligned}$$

Let's set up our augmented matrix and get going with our elementary row operations.

$$\begin{array}{c}(-3)\\ \\ \\ \end{array}\left[\begin{array}{ccc|c} 1 & -1 & 4 & 1 \\ \boxed{3} & 1 & 0 & 6 \\ 1 & 2 & -17 & 4 \end{array}\right]$$

To change the 3 in the second row to 0, temporarily multiply the top row by –3 and add to the middle row, replacing the middle row with the result (row operation #3).

$$\left[\begin{array}{ccc|c} 1 & -1 & 4 & 1 \\ -3(1)+3 & -3(-1)+1 & -3(4)+0 & -3(1)+6 \\ 1 & 2 & -17 & 4 \end{array}\right] = \begin{array}{c} (-1) \\ \downarrow \\ \rightarrow \end{array}\left[\begin{array}{ccc|c} 1 & -1 & 4 & 1 \\ 0 & 4 & -12 & 3 \\ \textcircled{1} & 2 & -17 & 4 \end{array}\right]$$

To change the 1 in the bottom row to 0, temporarily multiply the top row by –1 and add to the third row, replacing the third row with the result (row operation #3).

$$\left[\begin{array}{ccc|c} 1 & -1 & 4 & 1 \\ 0 & 4 & -12 & 3 \\ -1(1)+1 & -1(-1)+2 & -1(4)+-17 & -1(1)+4 \end{array}\right] = \left[\begin{array}{ccc|c} 1 & -1 & 4 & 1 \\ 0 & 4 & -12 & 3 \\ 0 & 3 & -21 & 3 \end{array}\right]$$

Our next intermediate goal is to exchange the 4 in the middle row (on the diagonal) for a 1. We can accomplish this a few different ways.

One choice is to multiply the middle row by ¼ (row operation #2).

$$\Rightarrow \begin{array}{c} \\ (\tfrac{1}{4}) \\ \\ \end{array}\left[\begin{array}{ccc|c} 1 & -1 & 4 & 1 \\ 0 & \textcircled{4} & -12 & 3 \\ 0 & 3 & -21 & 3 \end{array}\right] \Leftrightarrow \left[\begin{array}{ccc|c} 1 & -1 & 4 & 1 \\ 0 & 1 & -3 & \frac{3}{4} \\ 0 & 3 & -21 & 3 \end{array}\right]$$

The drawback with this approach, for most students, is that we've now introduced fractions into our matrix and that will make the computations more taxing to deal with as we continue on with the Gauss-Jordan Elimination process.

Another choice is to find some other creative way to legally manipulate the matrix to get that second row/second column element to be 1. We can temporarily multiply the bottom row by –1, add that to the middle row and replace the middle row with the result (row operation #3).

$$\Rightarrow \begin{array}{c} \\ \uparrow \\ (-1) \end{array}\left[\begin{array}{ccc|c} 1 & -1 & 4 & 1 \\ 0 & 4 & -12 & 3 \\ 0 & 3 & -21 & 3 \end{array}\right] \Leftrightarrow \left[\begin{array}{ccc|c} 1 & -1 & 4 & 1 \\ -1(0)+0 & -1(3)+4 & -1(-21)+-12 & -1(3)+3 \\ 0 & 3 & -21 & 3 \end{array}\right] \Leftrightarrow \left[\begin{array}{ccc|c} 1 & -1 & 4 & 1 \\ 0 & 1 & 9 & 0 \\ 0 & 3 & -21 & 3 \end{array}\right]$$

This matrix is nicer to handle, so we'll continue on with this choice.

$$\begin{array}{c} \\ (-3) \\ \downarrow \end{array}\left[\begin{array}{ccc|c} 1 & -1 & 4 & 1 \\ 0 & 1 & 9 & 0 \\ 0 & \textcircled{3} & -21 & 3 \end{array}\right]$$

To get the 3 in the bottom row to be zero, we can temporarily multiply the middle row by –3 and add to the bottom row, replacing the bottom row with the result (row operation #3).

$$\left[\begin{array}{ccc|c} 1 & -1 & 4 & 1 \\ 0 & 1 & 9 & 0 \\ -3(0)+0 & -3(1)+3 & -3(9)+-21 & -3(0)+3 \end{array}\right] = \begin{array}{c} \\ \\ (\tfrac{-1}{48}) \end{array}\left[\begin{array}{ccc|c} 1 & -1 & 4 & 1 \\ 0 & 1 & 9 & 0 \\ 0 & 0 & -48 & 3 \end{array}\right]$$

To change the –48 to a 1 (since it's on the diagonal), multiply the bottom row by $\frac{-1}{48}$ (row operation #2).

$$\left[\begin{array}{cccc} 1 & -1 & 4 & 1 \\ 0 & 1 & 9 & 0 \\ 0\left(\frac{-1}{48}\right) & 0\left(\frac{-1}{48}\right) & -48\left(\frac{-1}{48}\right) & 3\left(\frac{-1}{48}\right) \end{array}\right] = \left[\begin{array}{ccc|c} 1 & -1 & 4 & 1 \\ 0 & 1 & 9 & 0 \\ 0 & 0 & 1 & -\frac{1}{16} \end{array}\right]$$

At this point, let's do something slightly different than continuing on with our elimination process. We can translate each row of our matrix back into its corresponding equation.

$$x - y + 4z = 1$$
$$y + 9z = 0$$
$$z = \frac{-1}{16}$$

So, we have a value for z. We can now back-substitute that value into the middle equation to get a value for y.

$$y + 9\left(\frac{-1}{16}\right) = 0 \Leftrightarrow y + \frac{-9}{16} = 0 \Leftrightarrow y = \frac{9}{16}$$

Use the top equation to solve for x.

$$x - \left(\frac{9}{16}\right) + 4\left(\frac{-1}{16}\right) = 1$$

Simplify.

$$x - \frac{9}{16} - \frac{4}{16} = 1 \Leftrightarrow x - \frac{13}{16} = 1$$

Add $\frac{13}{16}$ to both sides of the equation.

$$x = \frac{16}{16} + \frac{13}{16} = \frac{29}{16}$$

Thus the point of intersection between these three planes is $\left(\frac{29}{16}, \frac{9}{16}, \frac{-1}{16}\right)$.

EXAMPLE 5

Solve this system of equations.

$$x + y + z = 5$$
$$3x - y + 2z = -4$$
$$4x + 3z = 1$$

Set up the augmented matrix and use elementary row operations as we've been doing thus far.

$$\begin{array}{c} (-3) \\ \hookrightarrow \\ \end{array}\left[\begin{array}{ccc|c} 1 & 1 & 1 & 5 \\ \textcircled{3} & -1 & 2 & -4 \\ 4 & 0 & 3 & 1 \end{array}\right]$$

To change the 3 in the middle row to zero, temporarily multiply the top row by –3 and add to the middle row, replacing the middle row with the result (row operation #3).

$$(-4)\left[\begin{array}{ccc|c} 1 & 1 & 1 & 5 \\ 0 & -4 & -1 & -19 \\ 4 & 0 & 3 & 1 \end{array}\right]$$

To change the 4 in the bottom row into a zero, temporarily multiply the top row by –4 and add to the bottom row, replacing the bottom row with the result (row operation #3).

$$\left[\begin{array}{ccc|c} 1 & 1 & 1 & 5 \\ 0 & -4 & -1 & -19 \\ 0 & -4 & -1 & -19 \end{array}\right] (-1)$$

Notice that both the middle and the bottom rows are exactly the same right now. If we multiply the middle row by –1 and add to the bottom row, replacing the bottom row with the result, every entry will turn to zero.

$$\Rightarrow \left[\begin{array}{ccc|c} 1 & 1 & 1 & 5 \\ 0 & -4 & -1 & -19 \\ 0 & 0 & 0 & 0 \end{array}\right]$$

Let's look at what the bottom row will be as its corresponding equation.

$$0x + 0y + 0z = 0 \quad \Leftrightarrow \quad 0 = 0$$

So, we have no pertinent information from our bottom row in the last matrix, but we do know that it's at least a true statement. What this means is that our system of linear equations has infinitely many solutions. However, we cannot stop here because that's not enough information. Our infinitely many solutions take on the form of a line, and we need to know how to describe those points on the line in 3-d space.

We'll jump out of the matrix now and convert the first two rows into their corresponding equations.

$$\Rightarrow \begin{array}{r} x + y + z = 5 \\ -4y - z = -19 \end{array}$$

From here, we can use the bottom equation to get z by itself.

$-4y - z = -19$	Add $4y$ to both sides of the equation.
$-z = 4y - 19$	Multiply both sides of the equation by –1.
$z = -4y + 19$	Now we have solved for z in terms of y.

Substitute that expression for z into the top equation and then get x by itself.

$x + y + z = 5 \Leftrightarrow x + y + (-4y + 19) = 5$	Simplify.
$x - 3y + 19 = 5$	Add $3y$ and subtract 19 from both sides of the equation.
$x = 3y - 14$	Now we have x by itself, again in terms of y.

Thus, every point on the line of intersection can be described by these equations.

$$x = 3y - 14,\ y = y, \text{ and } z = -4y + 19$$

Or, we can write this information as an ordered triple:

$$(3y - 14,\ y,\ -4y + 19)$$

This says that we can choose any arbitrary value of y, then plug it into the above ordered triple of expressions and we'll have a point on our line of intersection. We can find all of the infinitely many points on that intersection line by choosing all of the different real number values for y and substituting them into our ordered triple.

2.3 Homework

Write the augmented matrix for the system of linear equations.

1. $\begin{aligned} -4x + 8y - 9z &= 61 \\ -5x + 9y + 4z &= 55 \\ -6x - 2y - z &= -5 \end{aligned}$

2. $\begin{aligned} 5x + y &= 37 \\ -x + 3y &= 49 \end{aligned}$

3. $\begin{aligned} x + y - 2z &= 3 \\ y + 5z &= 1 \\ x - z &= 5 \end{aligned}$

4. $\begin{aligned} 3x - y &= 4 \\ x + 2y &= 5 \end{aligned}$

Use Gauss-Jordan Elimination to solve the system of equations.

5. $\begin{aligned} x + 8y &= -20 \\ 4x + 2y &= 10 \end{aligned}$

6. $\begin{aligned} x + 10y &= 35 \\ 5x + 4y &= -9 \end{aligned}$

7. $\begin{aligned} x - 7y &= -40 \\ 10x + 4y &= 44 \end{aligned}$

8. $\begin{aligned} x + 10y &= 23 \\ 9x + 7y &= 41 \end{aligned}$

9. $\begin{aligned} x - 3y &= -9 \\ 3x - 3y &= 9 \end{aligned}$

10. $\begin{aligned} 3x - 8y &= 4 \\ 2x - 9y &= -1 \end{aligned}$

11. $\begin{aligned} 9x - 10y &= -48 \\ 10x - 3y &= -29 \end{aligned}$

12. $\begin{aligned} 9x - 4y &= 19 \\ 4x - 8y &= -4 \end{aligned}$

13. $\begin{aligned} 2x + 6y &= -3 \\ x + 3y &= 2 \end{aligned}$

14. $\begin{aligned} 11x + 4y &= -2 \\ 5x + 9y &= 35 \end{aligned}$

15. $\begin{aligned} 22x + 7y &= 5 \\ 7x - 5y &= -49 \end{aligned}$

16. $\begin{aligned} 22x + 7y &= 5 \\ 7y - 9x + 41 &= 0 \end{aligned}$

17. $\begin{aligned} 7x + 4y - 19 &= 0 \\ 6y - 4x + 44 &= 0 \end{aligned}$

18. $\begin{aligned} 5x + y &= 4 \\ 15x + 3y &= 21 \end{aligned}$

19. $\begin{aligned} -x + 7y &= -3 \\ -\frac{1}{2}x + y &= 2 \end{aligned}$

20. $\begin{aligned} 13x - 1 &= 9y \\ 6x &= 8y + 12 \end{aligned}$

21. $3x - 2y = 12$
$-3x + 2y = -12$

22. $x + 2y = 4$
$2x + 4y = 8$

23. $9x = 6 - y$
$4x + 2y + 5 = 0$

24. $7x + 2y - 10 = 0$
$3x + 1 = 2y$

25. $-2x + 3y = 2$
$4x - 6y = -7$

26. $-2x + y = 1$
$2x - y = 7$

27. $2x - y = 10$
$-6x + 3y = -30$

28. $5x + 3y = 1$
$10x + 6y = 2$

29. $2x - y = 0$
$-2x + y = 3$

30. $2x + 3y + 10z = 16$
$x + 2y + 8z = 13$
$7x - 3y - z = 11$

31. $5x + 6y - 4z = 88$
$4x + 5y + 4z = 79$
$-6x - 7y + 10z = -99$

32. $9x + y - z = 10$
$8x + y = 3$
$2x - 5y + 4z = 3$

33. $7x + 5y - 5z = 16$
$3x + 4y + 2z = 17$
$3x - 3y + 2z = 10$

34. $3x - 4y - z = 1$
$2x - 3y + z = 1$
$x - 2y + 3z = 2$

35. $x - y + z = -1$
$x - y + z = 1$
$-x + y - z = 1$

36. $10x + y = 6$
$3x + y + 2z = 3$
$2x - y - 2z = 2$

37. $3x - 2y - 3z + 13 = 0$
$20 - y = -2x - 4z$
$3x + 2y + z = -13$

38. $3x - 2y - 7z = 0$
$x - y - z = 1$
$-x + 2y - 3z = -4$

39. $x - 2y + 3z = 1$
$2x - 4y + 6z = 2$
$3x + y - 2z = 0$

40. $z + 2 = 2x + 5y$
$x + 3y - 5z = -13$
$4y - 5x + 5z = -14$

41. $19x - 3z - 28 = -22y$
$6x - 3z = 7 - 7y$
$3y - 2x + 4z - 22 = 0$

42. $2x + y + z = 10$
$3x + 3y = 9$
$5x + 4y + z = 19$

43. $2x + 4y - 6z = 10$
$3x + 6y - 9z = 15$
$-2x - 3y + 4z = -6$

44. $3x - y = 3$
$x + z = 3$
$2x - y - z = 3$

45. $2x - 2y = 0$
$3x + y = 4$
$5x - y = 5$

46. $2x - 5y - 3z = 7$
$6x - 15y - z = -19$
$-4x + 10y + 2z = 6$

47. $x - y + z = 2$
$2x + 3y - z = 1$
$x - 2y + 3z = 0$

48. $x + y + z = 1$
$x - y - z = 1$
$-x + y - z = 1$

49. $x - 3y + 2z = 8$
$2x - y + z = 10$
$x + 2y - 3z = 10$

50. $4x - y + 2z = 3$
$8x - 2y + 9z = -1$
$-4x + y - 3z = -10$

51. $x - y - z = 1$
$2x + 3y + z = 2$
$3x + 2y = 0$

52. $2x + 3y + 12z = 4$
$4x - 6y + 6z = 1$
$x + y + z = 1$

53. $x - 2y + 3z = 1$
$2x - 3y - 2z = 3$
$x - y - 5z = 2$

54. $x - 3z = 6$
$y + 2z = 2$
$7x - 3y - 5z = 14$

55. $1 + y = 6x - 3z$
$3y - 8z - 2 = -5x$
$2x - 5y = -6z - 1$

56. $2z - 5x = -2y - 4$
$2y - 2x = 3z - 6$
$x = 5 + 3y - 2z$

2.4 Inverse Matrices

Look back to our algebra that we already know about with numbers and remember that we had a multiplicative identity, namely the number 1. This preserved the identity of any number by which we multiply 1. The next logical question to ask after establishing that we have an identity is whether or not a number has a multiplicative inverse. For example, if we start with 5, its multiplicative inverse is the number that we multiply by 5 to get back to the multiplicative identity of 1. In other words, we're asking $5 \cdot ? = 1$ and $? \cdot 5 = 1$. The multiplicative inverse for 5 is $\frac{1}{5}$.

Let's now see the relevance of inverses in our matrix algebra. We know that multiplying any matrix by the identity matrix, I, will definitely result in the original matrix, thereby making I some sort of multiplicative identity for matrix multiplication. (*Note:* It fails being a true identity because matrix multiplication is not commutative.) For any given matrix A, we want to know if there is an inverse matrix, written as A^{-1}, such that $A \cdot A^{-1} = A^{-1} \cdot A = I$. Let's analyze this by looking at the size of the matrices. We know the identity matrix is always square. If A is size $m \times n$, then A^{-1} would need to have n rows in order for the multiplication $A \cdot A^{-1}$ to be defined, and it would also need to have m columns in order to have the product $A \cdot A^{-1}$ turn into a square matrix. But, we need the product $A \cdot A^{-1}$ to turn into a square matrix I as well. If, according to our last statement, A^{-1} is size $n \times m$, then $A \cdot A^{-1}$ would result in an $n \times n$ matrix. Altogether, then, we'd have the product $A \cdot A^{-1}$ turn into an $m \times m$ matrix, but $A \cdot A^{-1}$ would be size $n \times n$. We need them to be the same identity matrix (meaning the same size). This means that the only way we can make them the same is to require $m = n$, i.e., we must have a square matrix A to start with.

INVERSE MATRIX

An **inverse matrix** for a square matrix A size $n \times n$, denoted by A^{-1} and read "A inverse," satisfies $A \cdot A^{-1} = A^{-1} \cdot A = I$, where A^{-1} is also size $n \times n$ and I is the $n \times n$ identity matrix.

(*Note:* $A^{-1} \neq \frac{1}{A}$. The -1 in the place where an exponent would be is not an exponent in this notation. It is simply referring to the inverse matrix. There is no such operation as matrix division!)

Now that we know what an inverse matrix is, let's learn the process to uncover that inverse matrix. Unlike our algebra with numbers where all real numbers except 0 have a multiplicative inverse, all square matrices do NOT have inverses.

TO FIND A^{-1}, GIVEN MATRIX A:

1. Check first to make sure A is square. If A is not square, then there cannot be an inverse matrix.
2. Augment A with the appropriately sized identity matrix. (Make sure to do this step *first* before doing any of the following steps in the process.)
3. Perform elementary row operations on the entire augmented matrix until the left side of the dotted vertical line segment is exactly the identity matrix. (If the left side ends up with a whole row of zeros, then the inverse matrix does not exist for A.)
4. The matrix made up of the exact entries on the right side of the dotted vertical line segment is the inverse matrix.

EXAMPLE 1

Find A^{-1}, if it exists.

(a) $A = \begin{bmatrix} 4 & -1 \\ 0 & 2 \\ 3 & 5 \end{bmatrix}$

Notice that this matrix is not square. This means that it can't possibly have an inverse matrix and there is nothing left to do.

(b) $A = \begin{bmatrix} 2 & 5 \\ 1 & 3 \end{bmatrix}$

If we follow the above process, we first check that this is indeed a square matrix. This means that we can continue on in our process.

Augment the matrix with the 2 × 2 identity matrix and perform elementary row operations, keeping the goal in mind—make the left side look like the identity matrix.

$$\left[\begin{array}{cc|cc} 2 & 5 & 1 & 0 \\ 1 & 3 & 0 & 1 \end{array}\right]$$

Switch the order of the two rows.

$$\begin{array}{c} (-2) \\ \end{array}\left[\begin{array}{cc|cc} 1 & 3 & 0 & 1 \\ 2 & 5 & 1 & 0 \end{array}\right]$$

Multiply the first row by –2 temporarily and add that to the bottom row, replacing the bottom row with the result.

$$\begin{array}{c} \\ (-1) \end{array}\left[\begin{array}{cc|cc} 1 & 3 & 0 & 1 \\ 0 & -1 & 1 & -2 \end{array}\right]$$

Multiply the bottom row by –1.

$$\begin{array}{c} \\ (-3) \end{array}\left[\begin{array}{cc|cc} 1 & 3 & 0 & 1 \\ 0 & 1 & -1 & 2 \end{array}\right]$$

Now the bottom left side is exactly as we need it. We only need to get the 3 to be a 0 in order for the left side to be the identity matrix. Multiply the bottom row by –3 temporarily and add to the top row, replacing the top row with the result.

$$\left[\begin{array}{cc|cc} 1 & 0 & 3 & -5 \\ 0 & 1 & -1 & 2 \end{array}\right]$$

We have reached our goal! The right side gives the inverse matrix.

$$\Rightarrow A^{-1} = \begin{bmatrix} 3 & -5 \\ -1 & 2 \end{bmatrix}$$

(c) $A = \begin{bmatrix} 4 & 1 \\ -8 & -2 \end{bmatrix}$

Again, this is a square matrix, so we can proceed with our outlined process to investigate the inverse matrix.

Augment the matrix with the 2 × 2 identity matrix and perform elementary row operations, attempting to get the left side to look like the identity matrix.

$$(2)\left[\begin{array}{cc|cc} 4 & 1 & 1 & 0 \\ -8 & -2 & 0 & 1 \end{array}\right]$$

To turn the –8 into 0, multiply the top row by 2 and add to the bottom row, replacing the bottom row with the result of that addition.

$$\left[\begin{array}{cc|cc} 4 & 1 & 1 & 0 \\ 8+(-8) & 2+(-2) & 2+0 & 0+1 \end{array}\right] = \left[\begin{array}{cc|cc} 4 & 1 & 1 & 0 \\ 0 & 0 & 2 & 1 \end{array}\right]$$

Since we got a row of zeros on the left side of our augmented matrix, we can halt the process here because we know that the inverse matrix does not exist for this matrix.

EXAMPLE 2

Find A–1 for the generic matrix $A = \begin{bmatrix} a & b \\ c & d \end{bmatrix}$.

If we can do this, then we'll have a formula for any 2 × 2 inverse matrix.

$$\left(\tfrac{1}{a}\right)\left[\begin{array}{cc|cc} a & b & 1 & 0 \\ c & d & 0 & 1 \end{array}\right]$$

Multiply the top row by $\frac{1}{a}$ to get the first row, first column term to 1.

$$(-c)\left[\begin{array}{cc|cc} 1 & \frac{b}{a} & \frac{1}{a} & 0 \\ c & d & 0 & 1 \end{array}\right]$$

Multiply the top row by $-c$ and add to the bottom row, replacing the bottom row with the result.

$$\left[\begin{array}{cc|cc} 1 & \frac{b}{a} & \frac{1}{a} & 0 \\ -c+c & \frac{-bc}{a}+d & \frac{-c}{a}+0 & 0+1 \end{array}\right] = \left[\begin{array}{cc|cc} 1 & \frac{b}{a} & \frac{1}{a} & 0 \\ 0 & \frac{ad-bc}{a} & \frac{-c}{a} & 1 \end{array}\right]\left(\tfrac{a}{ad-bc}\right)$$

Multiply the bottom row by $\frac{a}{ad-bc}$ to get the second row, second column entry to be 1.

$$\begin{array}{r} \\ \left(\frac{-b}{a}\right) \end{array}\left[\begin{array}{cc|cc} 1 & \frac{b}{a} & \frac{1}{a} & 0 \\ 0 & 1 & \frac{-c}{ad-bc} & \frac{a}{ad-bc} \end{array}\right]$$

Multiply the bottom row by $\frac{-b}{a}$ and add to the top row, replacing the top row with the sum.

$$\left[\begin{array}{cc|cc} 0+1 & \frac{-b}{a}+\frac{b}{a} & \frac{bc}{a(ad-bc)}+\frac{1}{a} & \frac{-b}{ad-bc}+0 \\ 0 & 1 & \frac{-c}{ad-bc} & \frac{a}{ad-bc} \end{array}\right] = \left[\begin{array}{cc|cc} 1 & 0 & \frac{bc+ad-bc}{a(ad-bc)} & \frac{-b}{ad-bc} \\ 0 & 1 & \frac{-c}{ad-bc} & \frac{a}{ad-bc} \end{array}\right]$$

$$= \left[\begin{array}{cc|cc} 1 & 0 & \frac{d}{ad-bc} & \frac{-b}{ad-bc} \\ 0 & 1 & \frac{-c}{ad-bc} & \frac{a}{ad-bc} \end{array}\right]$$

$$\Rightarrow A^{-1} = \begin{bmatrix} \frac{d}{ad-bc} & \frac{-b}{ad-bc} \\ \frac{-c}{ad-bc} & \frac{a}{ad-bc} \end{bmatrix} = \frac{1}{ad-bc}\begin{bmatrix} d & -b \\ -c & a \end{bmatrix}$$

Notice that if $ad - bc = 0$, then we have a problem because we can't divide by zero.

For $A = \begin{bmatrix} a & b \\ c & d \end{bmatrix}$, the inverse matrix is given by $A^{-1} = \frac{1}{ad-bc}\begin{bmatrix} d & -b \\ -c & a \end{bmatrix}$, assuming $ad - bc \neq 0$. If $ad - bc = 0$, then A^{-1} does not exist.

EXAMPLE 3

Given $A = \begin{bmatrix} 1 & 2 & 1 \\ 3 & 0 & 1 \\ 1 & -1 & 1 \end{bmatrix}$, find A^{-1}.

A is a square matrix, so we can go forward using our process to investigate whether this matrix has an inverse or not.

$$\begin{array}{r} (-1) \\ \\ \\ \end{array}\left[\begin{array}{ccc|ccc} 1 & 2 & 1 & 1 & 0 & 0 \\ 3 & 0 & 1 & 0 & 1 & 0 \\ 1 & -1 & 1 & 0 & 0 & 1 \end{array}\right]$$

Multiply the top row by –1, add to the bottom row and replace the bottom row.

$$(-3)\left[\begin{array}{ccc|ccc} 1 & 2 & 1 & 1 & 0 & 0 \\ 3 & 0 & 1 & 0 & 1 & 0 \\ 0 & -3 & 0 & -1 & 0 & 1 \end{array}\right]$$

Multiply the top row by –3, add to row two and replace row two.

$$(-2)\left[\begin{array}{ccc|ccc} 1 & 2 & 1 & 1 & 0 & 0 \\ 0 & -6 & -2 & -3 & 1 & 0 \\ 0 & -3 & 0 & -1 & 0 & 1 \end{array}\right]$$

Multiply the bottom row by –2, add to the middle row and replace row two.

$$\left(\tfrac{-1}{3}\right)\left[\begin{array}{ccc|ccc} 1 & 2 & 1 & 1 & 0 & 0 \\ 0 & 0 & -2 & -1 & 1 & -2 \\ 0 & -3 & 0 & -1 & 0 & 1 \end{array}\right]$$

Multiply the bottom row by $\frac{-1}{3}$.

$$\left[\begin{array}{ccc|ccc} 1 & 2 & 1 & 1 & 0 & 0 \\ 0 & 0 & -2 & -1 & 1 & -2 \\ 0 & 1 & 0 & \frac{1}{3} & 0 & \frac{-1}{3} \end{array}\right]$$

Switch the bottom and middle rows.

$$\left(\tfrac{-1}{2}\right)\left[\begin{array}{ccc|ccc} 1 & 2 & 1 & 1 & 0 & 0 \\ 0 & 1 & 0 & \frac{1}{3} & 0 & \frac{-1}{3} \\ 0 & 0 & -2 & -1 & 1 & -2 \end{array}\right]$$

Multiply the bottom row by $\frac{-1}{2}$.

$$(-2)\left[\begin{array}{ccc|ccc} 1 & 2 & 1 & 1 & 0 & 0 \\ 0 & 1 & 0 & \frac{1}{3} & 0 & \frac{-1}{3} \\ 0 & 0 & 1 & \frac{1}{2} & \frac{-1}{2} & 1 \end{array}\right]$$

Multiply row two by –2, add to the top row and replace the top row.

$$\left[\begin{array}{ccc|ccc} 1 & 0 & 1 & \frac{-2}{3}+1 & 0+0 & \frac{2}{3}+0 \\ 0 & 1 & 0 & \frac{1}{3} & 0 & \frac{-1}{3} \\ 0 & 0 & 1 & \frac{1}{2} & \frac{-1}{2} & 1 \end{array}\right] = \left[\begin{array}{ccc|ccc} 1 & 0 & 1 & \frac{1}{3} & 0 & \frac{2}{3} \\ 0 & 1 & 0 & \frac{1}{3} & 0 & \frac{-1}{3} \\ 0 & 0 & 1 & \frac{1}{2} & \frac{-1}{2} & 1 \end{array}\right](-1)$$

Multiply the bottom row by –1, add to the top row and replace the top row.

$$\left[\begin{array}{ccc|ccc} 1 & 0 & 0 & \frac{-1}{2}+\frac{1}{3} & \frac{1}{2}+0 & -1+\frac{2}{3} \\ 0 & 1 & 0 & \frac{1}{3} & 0 & \frac{-1}{3} \\ 0 & 0 & 1 & \frac{1}{2} & \frac{-1}{2} & 1 \end{array}\right] = \left[\begin{array}{ccc|ccc} 1 & 0 & 0 & \frac{-1}{6} & \frac{1}{2} & \frac{-1}{3} \\ 0 & 1 & 0 & \frac{1}{3} & 0 & \frac{-1}{3} \\ 0 & 0 & 1 & \frac{1}{2} & \frac{-1}{2} & 1 \end{array}\right]$$

$$\Rightarrow A^{-1} = \begin{bmatrix} \frac{-1}{6} & \frac{1}{2} & \frac{-1}{3} \\ \frac{1}{3} & 0 & \frac{-1}{3} \\ \frac{1}{2} & \frac{-1}{2} & 1 \end{bmatrix}$$

SOLVING A SYSTEM OF LINEAR EQUATIONS USING INVERSE MATRICES

Notice that we can write a system of linear equations using matrix notation.

$$\begin{aligned} x + 2y + z &= 4 \\ 3x \qquad + z &= 2 \\ x - \ y + z &= 1 \end{aligned}$$

$$\Leftrightarrow AX = B, \text{ where } A = \begin{bmatrix} 1 & 2 & 1 \\ 3 & 0 & 1 \\ 1 & -1 & 1 \end{bmatrix}, X = \begin{bmatrix} x \\ y \\ z \end{bmatrix} \text{ and } B = \begin{bmatrix} 4 \\ 2 \\ 1 \end{bmatrix}$$

Let's try it out.

$$AX = B \Leftrightarrow \begin{bmatrix} 1 & 2 & 1 \\ 3 & 0 & 1 \\ 1 & -1 & 1 \end{bmatrix}\begin{bmatrix} x \\ y \\ z \end{bmatrix} = \begin{bmatrix} 4 \\ 2 \\ 1 \end{bmatrix}$$

Perform the matrix multiplication, AX.

$$\Leftrightarrow \begin{bmatrix} 1x + 2y + 1z \\ 3x + 0y + 1z \\ 1x + -1y + 1z \end{bmatrix} = \begin{bmatrix} 4 \\ 2 \\ 1 \end{bmatrix} \Leftrightarrow \begin{bmatrix} x + 2y + z \\ 3x + z \\ x - y + z \end{bmatrix} = \begin{bmatrix} 4 \\ 2 \\ 1 \end{bmatrix}$$

And this is exactly the system of equations we started with.

If A is a square matrix, and its inverse exists, then we can multiply both sides of the matrix equation by A^{-1} (on the left).

$$\Rightarrow A^{-1}(AX) = A^{-1}(B)$$

$$\Leftrightarrow (A^{-1}A)X = A^{-1}B \Leftrightarrow IX = A^{-1}B$$

Since $A^{-1}A = I$

$$\Leftrightarrow X = A^{-1}B$$

In other words, if A^{-1} exists, then we can find the solution for our system of linear equations by multiplying the constant matrix, B, on the left by A^{-1}.

EXAMPLE 4

Solve this system of equations.

$$\begin{aligned} x + 2y + z &= 4 \\ 3x \qquad + z &= 2 \\ x - \ y + z &= 1 \end{aligned}$$

First, notice that this is the example we used above when describing this new method for solving a system of linear equations. We have

$$AX = B, \text{ where } A = \begin{bmatrix} 1 & 2 & 1 \\ 3 & 0 & 1 \\ 1 & -1 & 1 \end{bmatrix}, X = \begin{bmatrix} x \\ y \\ z \end{bmatrix} \text{ and } B = \begin{bmatrix} 4 \\ 2 \\ 1 \end{bmatrix}$$

Also, in Example 3 we already found A^{-1} for this matrix.

$$A^{-1} = \begin{bmatrix} \frac{-1}{6} & \frac{1}{2} & \frac{-1}{3} \\ \frac{1}{3} & 0 & \frac{-1}{3} \\ \frac{1}{2} & \frac{-1}{2} & 1 \end{bmatrix}$$

Now we know $X = A^{-1}B$.

$$\Rightarrow X = \begin{bmatrix} \frac{-1}{6} & \frac{1}{2} & \frac{-1}{3} \\ \frac{1}{3} & 0 & \frac{-1}{3} \\ \frac{1}{2} & \frac{-1}{2} & 1 \end{bmatrix} \begin{bmatrix} 4 \\ 2 \\ 1 \end{bmatrix} = \begin{bmatrix} \frac{-1}{6}(4) + \frac{1}{2}(2) + \frac{-1}{3}(1) \\ \frac{1}{3}(4) + 0(2) + \frac{-1}{3}(1) \\ \frac{1}{2}(4) + \frac{-1}{2}(2) + 1(1) \end{bmatrix} = \begin{bmatrix} \frac{-4+6-2}{6} \\ \frac{4+0-1}{3} \\ 2-1+1 \end{bmatrix} = \begin{bmatrix} 0 \\ 1 \\ 2 \end{bmatrix}$$

This means that $x = 0$, $y = 1$ and $z = 2$, or the point of intersection of these three planes is (0, 1, 2).

2.4 Homework

Find the inverse of each matrix, or state that no inverse exists:

1. $\begin{bmatrix} 3 & 6 \\ -4 & 9 \end{bmatrix}$

2. $\begin{bmatrix} 9 & 8 \\ 4 & 0 \end{bmatrix}$

3. $\begin{bmatrix} 4 & 6 \\ -2 & -3 \end{bmatrix}$

4. $\begin{bmatrix} -5 & 2 \\ 4 & -3 \end{bmatrix}$

5. $\begin{bmatrix} 7 & 1 \\ -7 & 8 \\ -1 & -6 \end{bmatrix}$

6. $\begin{bmatrix} -8 & 5 \\ -2 & 3 \end{bmatrix}$

7. $\begin{bmatrix} 7 & 4 \\ 5 & 3 \end{bmatrix}$

8. $\begin{bmatrix} 7 & 2 \\ 2 & 1 \end{bmatrix}$

9. $\begin{bmatrix} 1 & 5 \\ -4 & 7 \end{bmatrix}$

10. $\begin{bmatrix} 9 & -7 \\ -7 & 5 \end{bmatrix}$

11. $\begin{bmatrix} 9 & \frac{1}{2} \\ 6 & 2 \end{bmatrix}$

12. $\begin{bmatrix} 7 & 6 \\ \frac{2}{3} & 4 \end{bmatrix}$

13. $\begin{bmatrix} 5 & 0 \\ 4 & 2 \end{bmatrix}$

14. $\begin{bmatrix} 2 & \frac{3}{2} \\ \frac{2}{3} & \frac{1}{2} \end{bmatrix}$

15. $\begin{bmatrix} 3 & \frac{2}{3} \\ \frac{3}{4} & -\frac{1}{4} \end{bmatrix}$

16. $\begin{bmatrix} 1 & 1 & 1 \\ -3 & -2 & -2 \\ 4 & 13 & 14 \end{bmatrix}$

17. $\begin{bmatrix} 7 & -2 & -4 \\ -4 & 3 & -3 \\ 7 & -3 & -1 \end{bmatrix}$

18. $\begin{bmatrix} -10 & -9 & 3 \\ -8 & -10 & 1 \\ 6 & 11 & 1 \end{bmatrix}$

19. $\begin{bmatrix} 1 & 1 & 1 \\ 1 & 2 & 2 \\ 1 & 2 & 3 \end{bmatrix}$

20. $\begin{bmatrix} 2 & 3 & 1 \\ 2 & 5 & 3 \\ 2 & 5 & 4 \end{bmatrix}$

21. $\begin{bmatrix} 1 & 0 & 2 \\ 1 & 1 & 7 \\ 1 & 1 & 8 \end{bmatrix}$

22. $\begin{bmatrix} -2 & 16 & -13 \\ -1 & 5 & -4 \\ 0 & -1 & 1 \end{bmatrix}$

23. $\begin{bmatrix} 2 & -3 & -4 \\ 25 & 0 & 7 \\ 7 & 2 & 5 \end{bmatrix}$

24. $\begin{bmatrix} 5 & -1 & 0 \\ -11 & 4 & -1 \\ 7 & -3 & 1 \end{bmatrix}$

Solve each system of linear equations using an inverse matrix.

25. $\begin{aligned} 2x + 4y &= 5 \\ -x + 3y &= -2 \end{aligned}$

26. $\begin{aligned} 2x + y &= 5 \\ 5x - y &= 0 \end{aligned}$

27. $\begin{aligned} -x + 2y &= 30 \\ 8x - y &= 60 \end{aligned}$

28. $\begin{aligned} x + 2y + z &= 4 \\ 3x \qquad + z &= 2 \\ x - y + z &= 1 \end{aligned}$

29. $\begin{aligned} x + y + z &= 2 \\ x - y + z &= -2 \\ x - y - z &= 0 \end{aligned}$

30. $\begin{aligned} x + y \qquad &= 2 \\ 2x + 3y - z &= 0 \\ x \qquad + 2z &= 4 \end{aligned}$

2.5 Application Problems with Matrices

In this section, we'll tie together all the computational pieces from the previous four sections, using them to solve systems of equations that occur from real-life problems. Also notice that we've chosen to employ a variety of the strategies available to us for solving these systems of linear equations, rather than using the same technique every single time.

EXAMPLE 1

ENCRYPTION A very basic, rudimentary form of encryption uses matrix multiplication with some square matrix A to encode a message, then it uses matrix multiplication with A^{-1} to decode that message on the other end. We need to choose a matrix A that has an inverse (or we could say that A is invertible). We can use

$$A = \begin{bmatrix} 1 & 2 & 0 \\ 3 & 1 & 10 \\ -1 & 8 & -10 \end{bmatrix} \quad \text{whose inverse is} \quad A^{-1} = \frac{-1}{10}\begin{bmatrix} -18 & 4 & 4 \\ 4 & -2 & -2 \\ 5 & -2 & -2 \end{bmatrix}.$$

The person sending the message knows the matrix A, and the person receiving the message knows the inverse matrix. Let's say the message to encrypt is:

"Blackhawk has landed."

Assign each letter in the alphabet a number to represent its sequential place in the alphabet. For example, $a = 1$, $b = 2$, $c = 3$ and so on until $z = 26$. Let 0 = a space (denoted by a dash below) and 27 = a period. First, we need to break up the message into chunks of three letters.

Bla ckh awk –ha s–l and ed.

Next, each of these three-character chunks can be represented by a 3×1 matrix.

$$\text{Bla} \rightarrow \begin{bmatrix} 2 \\ 12 \\ 1 \end{bmatrix} \quad \text{ckh} \rightarrow \begin{bmatrix} 3 \\ 11 \\ 8 \end{bmatrix} \quad \text{awk} \rightarrow \begin{bmatrix} 1 \\ 23 \\ 11 \end{bmatrix} \quad \text{–ha} \rightarrow \begin{bmatrix} 0 \\ 8 \\ 1 \end{bmatrix}$$

$$\text{s–l} \rightarrow \begin{bmatrix} 19 \\ 0 \\ 12 \end{bmatrix} \quad \text{and} \rightarrow \begin{bmatrix} 1 \\ 14 \\ 4 \end{bmatrix} \quad \text{ed.} \rightarrow \begin{bmatrix} 5 \\ 4 \\ 27 \end{bmatrix}$$

To encode each three-character chunk, a.k.a. 3 × 1 matrix, we multiply each 3 × 1 matrix by A.

$$\text{Bla} \rightarrow \begin{bmatrix} 1 & 2 & 0 \\ 3 & 1 & 10 \\ -1 & 8 & -10 \end{bmatrix}\begin{bmatrix} 2 \\ 12 \\ 1 \end{bmatrix} = \begin{bmatrix} 2+24+0 \\ 6+12+10 \\ -2+96+-10 \end{bmatrix} = \begin{bmatrix} 26 \\ 28 \\ 84 \end{bmatrix}$$

$$\text{ckh} \rightarrow \begin{bmatrix} 1 & 2 & 0 \\ 3 & 1 & 10 \\ -1 & 8 & -10 \end{bmatrix}\begin{bmatrix} 3 \\ 11 \\ 8 \end{bmatrix} = \begin{bmatrix} 3+22+0 \\ 9+11+80 \\ -3+88+-80 \end{bmatrix} = \begin{bmatrix} 25 \\ 100 \\ 5 \end{bmatrix}$$

$$\text{awk} \rightarrow \begin{bmatrix} 1 & 2 & 0 \\ 3 & 1 & 10 \\ -1 & 8 & -10 \end{bmatrix}\begin{bmatrix} 1 \\ 23 \\ 11 \end{bmatrix} = \begin{bmatrix} 1+46+0 \\ 3+23+110 \\ -1+184+-110 \end{bmatrix} = \begin{bmatrix} 47 \\ 136 \\ 73 \end{bmatrix}$$

$$\text{-ha} \rightarrow \begin{bmatrix} 1 & 2 & 0 \\ 3 & 1 & 10 \\ -1 & 8 & -10 \end{bmatrix}\begin{bmatrix} 0 \\ 8 \\ 1 \end{bmatrix} = \begin{bmatrix} 0+16+0 \\ 0+8+10 \\ 0+64+-10 \end{bmatrix} = \begin{bmatrix} 16 \\ 18 \\ 54 \end{bmatrix}$$

$$\text{s-l} \rightarrow \begin{bmatrix} 1 & 2 & 0 \\ 3 & 1 & 10 \\ -1 & 8 & -10 \end{bmatrix}\begin{bmatrix} 19 \\ 0 \\ 12 \end{bmatrix} = \begin{bmatrix} 19+0+0 \\ 57+0+120 \\ -19+0+-120 \end{bmatrix} = \begin{bmatrix} 19 \\ 177 \\ -139 \end{bmatrix}$$

$$\text{and} \rightarrow \begin{bmatrix} 1 & 2 & 0 \\ 3 & 1 & 10 \\ -1 & 8 & -10 \end{bmatrix}\begin{bmatrix} 1 \\ 14 \\ 4 \end{bmatrix} = \begin{bmatrix} 1+28+0 \\ 3+14+40 \\ -1+112+-40 \end{bmatrix} = \begin{bmatrix} 29 \\ 57 \\ 71 \end{bmatrix}$$

$$\text{ed.} \rightarrow \begin{bmatrix} 1 & 2 & 0 \\ 3 & 1 & 10 \\ -1 & 8 & -10 \end{bmatrix}\begin{bmatrix} 5 \\ 4 \\ 27 \end{bmatrix} = \begin{bmatrix} 5+8+0 \\ 15+4+270 \\ -5+32+-270 \end{bmatrix} = \begin{bmatrix} 13 \\ 289 \\ -243 \end{bmatrix}$$

At this point, the sender would transmit these matrices in order.

$$\begin{bmatrix} 26 \\ 28 \\ 84 \end{bmatrix} \quad \begin{bmatrix} 25 \\ 100 \\ 5 \end{bmatrix} \quad \begin{bmatrix} 47 \\ 136 \\ 73 \end{bmatrix} \quad \begin{bmatrix} 16 \\ 18 \\ 54 \end{bmatrix} \quad \begin{bmatrix} 19 \\ 177 \\ -139 \end{bmatrix} \quad \begin{bmatrix} 29 \\ 57 \\ 71 \end{bmatrix} \quad \begin{bmatrix} 13 \\ 289 \\ -243 \end{bmatrix}$$

The receiver would then decode these matrices by multiplying each matrix by A^{-1}.

$$\Rightarrow \frac{-1}{10}\begin{bmatrix} -18 & 4 & 4 \\ 4 & -2 & -2 \\ 5 & -2 & -1 \end{bmatrix}\begin{bmatrix} 26 \\ 28 \\ 84 \end{bmatrix} = \frac{-1}{10}\begin{bmatrix} -468+112+336 \\ 104+-56+-168 \\ 130+-56+-84 \end{bmatrix} = \frac{-1}{10}\begin{bmatrix} -20 \\ -120 \\ -10 \end{bmatrix} = \begin{bmatrix} 2 \\ 12 \\ 1 \end{bmatrix} \Rightarrow \text{bla}$$

$$\frac{-1}{10}\begin{bmatrix} -18 & 4 & 4 \\ 4 & -2 & -2 \\ 5 & -2 & -1 \end{bmatrix}\begin{bmatrix} 25 \\ 100 \\ 5 \end{bmatrix} = \frac{-1}{10}\begin{bmatrix} -450+400+20 \\ 100+-200+-10 \\ 125+-200+-5 \end{bmatrix} = \frac{-1}{10}\begin{bmatrix} -30 \\ -110 \\ -80 \end{bmatrix} = \begin{bmatrix} 3 \\ 11 \\ 8 \end{bmatrix} \Rightarrow \text{ckh}$$

$$\frac{-1}{10}\begin{bmatrix} -18 & 4 & 4 \\ 4 & -2 & -2 \\ 5 & -2 & -1 \end{bmatrix}\begin{bmatrix} 47 \\ 136 \\ 73 \end{bmatrix} = \frac{-1}{10}\begin{bmatrix} -846+544+292 \\ 188+-272+-146 \\ 235+-272+-73 \end{bmatrix} = \frac{-1}{10}\begin{bmatrix} -10 \\ -230 \\ -110 \end{bmatrix} = \begin{bmatrix} 1 \\ 23 \\ 11 \end{bmatrix} \Rightarrow \text{awk}$$

We can continue on this way, but these three examples sufficiently show that we're able to decode the encoded message when performing matrix multiplication with A^{-1}!

EXAMPLE 2

TICKET SALES My niece, Kristin, was in a play at school to raise funds for a new playground. They sold a total of 120 tickets, for a total of \$960. The number of adult tickets sold was 20 more than the number of children's tickets and senior citizen tickets together. Each child's ticket cost \$4, an adult ticket was \$10, and each senior citizen ticket sold for \$7. For extra credit, Kristin needed to figure out how many of each of the three types of tickets sold.

Let's first set up a table that will help us organize the information and set up equations.

Let x = number of adult tickets sold
y = number of children's tickets sold
z = number of senior citizen tickets sold

	# TICKETS (tix)	COST PER TICKET (\$/tix)	TOTAL COST (\$)
Adult	x	\$10	$10x$
Child	y	\$4	$4y$
Senior Citizen	z	\$7	$7z$
Totals	120	—	\$960

Notice that each entry in the "Total Cost" column is the product of the number of tickets and the cost per ticket.

Now we can look down the "# Tickets" column and retrieve an equation from that column. In order to make an equation in a story problem, we need to find two different ways to represent the exact same value. For example, the total number of tickets given was 120. We also know only three types of tickets sold, so if we add the numbers of those three types of tickets together, that will also represent the total number of tickets sold.

$$\Rightarrow \quad x + y + z = 120$$

We can follow a similar process in the last column and get another equation.

$$\Rightarrow \quad 10x + 4y + 7z = 960$$

Since we have three variables (unknowns), we need three different equations in order to solve for all of the variables. We can get our last equation from the given problem—"The number of adult tickets sold was 20 more than the number of children's tickets and senior citizen tickets together."

$$\Rightarrow \quad x = 20 + y + z$$

$$\Leftrightarrow \quad x - y - z = 20$$

Basically, we need to solve a system of three linear equations:

$$x + y + z = 120$$
$$10x + 4y + 7z = 960$$
$$x - y - z = 20$$

We'll use Gauss-Jordan elimination with an augmented matrix.

$$\left[\begin{array}{ccc|c} 1 & 1 & 1 & 120 \\ 10 & 4 & 7 & 960 \\ 1 & -1 & -1 & 20 \end{array}\right]$$

This will seem out of order perhaps, but the result will be worthwhile. Add the top and bottom rows together and replace the bottom row with the result.

$$\left[\begin{array}{ccc|c} 1 & 1 & 1 & 120 \\ 10 & 4 & 7 & 960 \\ 1+1 & 1+-1 & 1+-1 & 120+20 \end{array}\right] = \left[\begin{array}{ccc|c} 1 & 1 & 1 & 120 \\ 10 & 4 & 7 & 960 \\ 2 & 0 & 0 & 140 \end{array}\right]$$

At this point, we could definitely keep going with our elementary row operations (or Gaussian elimination process), as we've done many times previously. However, it may be a more efficient process to jump out of the matrix and use simple substitution to finish this solution, mainly because we already know the value of one of the variables now (from the bottom row). If we convert these rows back into equations, we'll have the following.

(A): $x + y + z = 120$

(B): $10x + 4y + 7z = 960$

(C): $2x = 140$

$\Rightarrow$ $x = 70$ From equation (C).

Substitute 70 for x in equations (A) and (B).

$\Rightarrow$ (A): $70 + y + z = 120 \quad \Leftrightarrow \quad y + z = 50 \quad \Leftrightarrow \quad y = 50 - z$

(B): $10(70) + 4y + 7z = 960 \quad \Leftrightarrow \quad 4y + 7z = 260$

Substitute the last form of (A) into (B).

$\Rightarrow$ $4(50 - z) + 7z = 260$ Distribute the 4 through the parentheses.

$200 - 4z + 7z = 260$ Simplify the left side of the equation.

$200 + 3z = 260$ Subtract 200 from both sides of the equation.

$3z = 60$ To finish getting z alone, divide both sides of the equation by 3.

$z = 20$

$\Rightarrow$ $y = 50 - z = 50 - 20 = 30$

Now, we have $x = 70$, $y = 30$ and $z = 20$. Thus, 70 adult tickets, 30 child tickets and 20 senior citizen tickets sold for the play.

EXAMPLE 3

INVESTMENTS **Sam decided to invest his money in three different funds, each with varying levels of risk and return. Fund A has a 5% annual yield, fund B yields 8% annually and fund C yields 7.2% every year. He invests twice as much in fund B than fund A. His total investment was $84,000, and his total annual return was $6,000. How much did he invest in each fund?**

A table might be helpful to organize the information for this problem, like we used for the last problem.

Let x = $ invested in fund A
y = $ invested in fund B
z = $ invested in fund C

	$ IN FUND	ANNUAL YIELD (%)	RETURN ($)
A	x	5.00%	$0.05x$
B	y	8.00%	$0.08y$
C	z	7.20%	$0.072z$
Totals	$84,000	—	$6,000

From this information, we can see that we have the following equations.

1. $x + y + z = 84000$
2. $0.05x + 0.08y + 0.072z = 6000$

We need one more equation. We can get that from the problem, where it states: "He invests twice as much in fund B than fund A." The equation representing this information is:

3. $y = 2x$

With this last equation, we can substitute $2x$ for y in the first two equations and simplify them to have only two variables.

$\Rightarrow$ 1. $x + 2x + z = 84000 \quad \Rightarrow \quad 3x + z = 84000$

and 2. $0.05x + 0.08(2x) + 0.072z = 6000$

$\Leftrightarrow \quad 0.05x + 0.16x + 0.072z = 6000$ Simplify the left side of the equation.

$0.21x + 0.072z = 6000$ Multiply both sides of the equation by 1,000 to eliminate the decimal.

$$1000(0.21x + 0.072z) = (6000)1000 \quad \Leftrightarrow \quad 210x + 72z = 6{,}000{,}000$$

Divide both sides of the equation by 6 (which is a common factor to every term).

$$\frac{6(35x + 12z)}{6} = \frac{6(1{,}000{,}000)}{6} \quad \Leftrightarrow \quad 35x + 12z = 1{,}000{,}000$$

Now we have a system of two equations with two variables:

$$3x + z = 84000$$
$$35x + 12z = 1{,}000{,}000$$

Let's solve this system with a matrix inverse. Remember that we can represent a system of linear equations as $AX = B$, with A as the coefficient matrix, X as the variable matrix and B as the constant matrix.

$$A = \begin{bmatrix} 3 & 1 \\ 35 & 12 \end{bmatrix} \quad X = \begin{bmatrix} x \\ z \end{bmatrix} \quad \text{and} \quad B = \begin{bmatrix} 84000 \\ 1000000 \end{bmatrix}$$

If we can find A^{-1}, then our solution is $X = A^{-1}B$. For a 2×2 matrix, we have a formula to determine the inverse.

For $A = \begin{bmatrix} a & b \\ c & d \end{bmatrix}$, $A^{-1} = \dfrac{1}{ad - bc}\begin{bmatrix} d & -b \\ -c & a \end{bmatrix}$ (from section 2.4).

$$\Rightarrow A^{-1} = \frac{1}{3(12) - (35)}\begin{bmatrix} 12 & -1 \\ -35 & 3 \end{bmatrix} = \frac{1}{1}\begin{bmatrix} 12 & -1 \\ -35 & 3 \end{bmatrix} = \begin{bmatrix} 12 & -1 \\ -35 & 3 \end{bmatrix}$$

$$\Rightarrow X = A^{-1}B = \begin{bmatrix} 12 & -1 \\ -35 & 3 \end{bmatrix}\begin{bmatrix} 84000 \\ 1000000 \end{bmatrix} = \begin{bmatrix} 12(84000) + -1(1000000) \\ -35(84000) + 3(1000000) \end{bmatrix} = \begin{bmatrix} 8000 \\ 60000 \end{bmatrix}$$

This means that x = \$8,000 and z = \$60,000. And, from (3), we know that $y = 2x$, which means that y = \$16,000. Thus, \$8,000 was invested in fund A, \$16,000 in fund B and \$60,000 in fund C.

EXAMPLE 4

ROUTES/GRAPHS **According to the popular party game "Six Degrees of Kevin Bacon," every actor can link to Kevin Bacon within six actor connections. We can represent this phenomenon with a mathematical graph, connecting people. Let's use this hypothetical graph for someone named Karley and explore the connections she has with the people depicted here.**

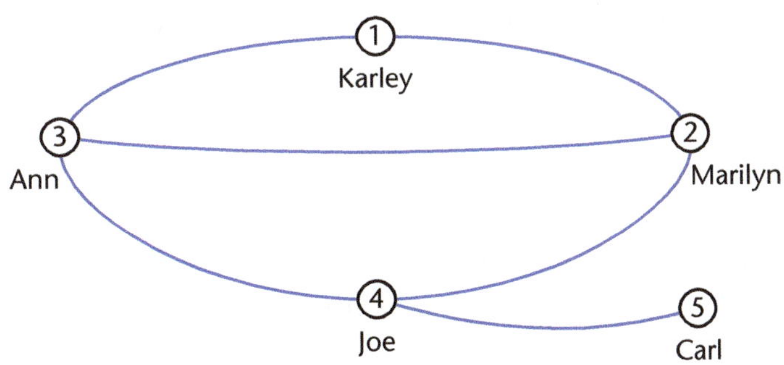

With this graph, we can use a matrix to represent the number of non-stop connections between each pair of people. Entry a_{ij} in matrix A represents the number of non-stop connections between person i and person j. For example, a_{13} is the number of non-stop connections between person 1 (Karley) and person 3 (Ann), which is only 1. Notice, then, that a_{ij} will be the same number as a_{ji} because we have no direction on the connections. So, $a_{13} = a_{31} = 1$. Also, $a_{ii} = 0$ because there is no connection from a person to him or herself. If we fill in the entire matrix A following this pattern, we'll get

$$A = \begin{bmatrix} 0 & 1 & 1 & 0 & 0 \\ 1 & 0 & 1 & 1 & 0 \\ 1 & 1 & 0 & 1 & 0 \\ 0 & 1 & 1 & 0 & 1 \\ 0 & 0 & 0 & 1 & 0 \end{bmatrix}$$

If we square A, then each entry will be the number of one-stop connections between one person to another.

$$A^2 = \begin{bmatrix} 0 & 1 & 1 & 0 & 0 \\ 1 & 0 & 1 & 1 & 0 \\ 1 & 1 & 0 & 1 & 0 \\ 0 & 1 & 1 & 0 & 1 \\ 0 & 0 & 0 & 1 & 0 \end{bmatrix} \begin{bmatrix} 0 & 1 & 1 & 0 & 0 \\ 1 & 0 & 1 & 1 & 0 \\ 1 & 1 & 0 & 1 & 0 \\ 0 & 1 & 1 & 0 & 1 \\ 0 & 0 & 0 & 1 & 0 \end{bmatrix} = \begin{bmatrix} 2 & 1 & 1 & 2 & 0 \\ 1 & 3 & 2 & 1 & 1 \\ 1 & 2 & 3 & 1 & 1 \\ 2 & 1 & 1 & 3 & 0 \\ 0 & 1 & 1 & 0 & 1 \end{bmatrix}$$

For example, the number of one-stop connections between person 1 (Karley) and person 5 (Carl) would still be zero since the shortest path between Karley and Carl is three connections. The number of one-stop connections between Karley (person 1) and Marilyn (person 2) will be represented by $a_{12} = 1$ because we can go from Karley to Ann and then to Marilyn, thereby making one stop from Karley to Marilyn. That's the only way to get from Karley to Marilyn with one stop. To get from Joe (person 4) to himself (person 4) with exactly one stop, it will be $a_{44} = 3$ ways because we can go (a) from Joe to Marilyn to Joe, (b) from Joe to Ann to Joe and (c) from Joe to Carl to Joe.

If we cube A, that is multiply A^2 by A again, then each entry a_{ij} in A^3 will represent the number of ways to get from person i to person j with exactly two stops.

$$A^3 = \begin{bmatrix} 2 & 1 & 1 & 2 & 0 \\ 1 & 3 & 2 & 1 & 1 \\ 1 & 2 & 3 & 1 & 1 \\ 2 & 1 & 1 & 3 & 0 \\ 0 & 1 & 1 & 0 & 1 \end{bmatrix} \begin{bmatrix} 0 & 1 & 1 & 0 & 0 \\ 1 & 0 & 1 & 1 & 0 \\ 1 & 1 & 0 & 1 & 0 \\ 0 & 1 & 1 & 0 & 1 \\ 0 & 0 & 0 & 1 & 0 \end{bmatrix} = \begin{bmatrix} 2 & 5 & 5 & 2 & 2 \\ 5 & 4 & 5 & 6 & 1 \\ 5 & 5 & 4 & 6 & 1 \\ 2 & 6 & 6 & 2 & 3 \\ 2 & 1 & 1 & 3 & 0 \end{bmatrix}$$

For example, the number of two-stop connections between person 1 (Karley) and person 3 (Ann) is represented in the first-row and third-column entry, namely 5. Here are the five ways to get from Karley to Ann with two stops in between: (a) Karley to Marilyn to Joe to Ann, (b) Karley to Ann to Karley to Ann, (c) Karley to Ann to Marilyn to Ann, (d) Karley to Ann to Joe to Ann and (e) Karley to Marilyn to Karley to Ann. Notice that we still have one zero entry in this matrix, in the fifth-row and fifth-column entry. This means that person 5 (Carl) has no way to get back to himself with exactly two stops. Since the rest of the matrix has non-zero entries, it means that within two stops every person connects to every other person in this scenario.

2.5 Homework

For problems 1 through 6, encrypt and decrypt messages using this table for the value of each letter.

(space)	A	B	C	D	E	F	G	H
0	1	2	3	4	5	6	7	8
I	J	K	L	M	N	O	P	Q
9	10	11	12	13	14	15	16	17
R	S	T	U	V	W	X	Y	Z
18	19	20	21	22	23	24	25	26

1. Encrypt the message "This is a secret" using the encryption matrix

$$A = \begin{bmatrix} 1 & -2 & 3 \\ -4 & 5 & -6 \\ 3 & -2 & 2 \end{bmatrix}.$$

(Pad the message with spaces to get the number of characters to be a multiple of 3.)

2. Decrypt the message "–26 59 –18 44 –101 52 –25 70 –30 –12 12 12 48 –84 25 2 –20 27" using the inverse of the encryption matrix

$$A = \begin{bmatrix} 1 & -2 & 3 \\ -4 & 5 & -6 \\ 3 & -2 & 2 \end{bmatrix}.$$

3. Encrypt the message "Ignore this message" using the encryption matrix

$$A = \begin{bmatrix} -1 & 1 & -2 \\ 3 & -3 & 1 \\ 1 & -2 & 2 \end{bmatrix}.$$

(Pad the message with spaces to get the number of characters to be a multiple of 3.)

4. Decrypt the message "–46 43 41 –32 61 31 –21 23 21 18 –54 –37 –7 –4 2 –52 61 52 –31 –7 17" using the inverse of the encryption matrix

$$A = \begin{bmatrix} -1 & 1 & -2 \\ 3 & -3 & 1 \\ 1 & -2 & 2 \end{bmatrix}.$$

5. Encrypt the message "Meet me at two" using the encryption matrix

$$A = \begin{bmatrix} -2 & -3 & 1 \\ 3 & 3 & -2 \\ 2 & -1 & -2 \end{bmatrix}.$$

(Pad the message with spaces to get the number of characters to be a multiple of 3.)

6. Decrypt the message "–73 87 13 –18 20 3 –1 –3 –8 –39 58 38 –20 20 0 –65 68 –12 –10 15 10" using the inverse of the encryption matrix

$$A = \begin{bmatrix} -2 & -3 & 1 \\ 3 & 3 & -2 \\ 2 & -1 & -2 \end{bmatrix}.$$

7. There are two types of tickets for a school play, student tickets and adult tickets. The student tickets cost \$2 and the adult tickets cost \$5. There were \$223 worth of tickets sold. There were 18 more adult tickets sold than student tickets. How many tickets of each type of ticket were sold?

8. A movie theater is showing two different movies. Tickets for the first movie cost \$4.50 per person while tickets for the second movie cost \$7.50. The total ticket sales for the day was \$10,002. The second movie is newer than the first, so there were 100 more tickets sold for the first movie than for the second movie. How many tickets of each type of ticket were sold?

9. A school was selling cakes, pies, and boxes of cookies at a bake sale. The cakes sold for \$3.50 apiece, the pies sold for \$5.00 apiece, and the boxes of cookies were priced at \$2.00. Altogether, 122 items were sold at the bake sale. The pies sold the best, selling 8 more pies than cakes and boxes of cookies combined. The total sales at the bake sale were \$505. How many of each type of item were sold?

10. A farmer has 1000 acres to be planted. He can plant beans, rice, or corn. He has 15,200,000 gallons of irrigation water available per day for the growing season. The bean crop will take 11,000 gallons of water per acre each day. The rice crop will take 17,000 gallons of water per acre each day. The corn crop will take 15,000 gallons of water per acre each day. The farmer wishes to plant 3 times as many acres in rice as in beans. How many acres of each should the farmer plant?

11. A company produces individual resistors and transistors as well as computer chips. Each set of resistors requires 2 units of copper, 2 units of zinc, and 1 unit of glass to manufacture. Each set of transistors requires 3 units of copper, 3 units of zinc and 2 units of glass. Each set of computer chips requires 2 units of copper, 1 unit of zinc, and 3 units of glass. If there are 160 units of copper, 120 units of zinc, and 170 units of glass available, how many sets of resistors, transistors, and computer chips should the company manufacture to use all of its available supplies or raw materials?

12. If solution A contains 10% of a chemical and solution B contains 50% of the same chemical, how much of each of these two solutions must be mixed together to make a mixture that is 10 gallons and contains 26% of this chemical?

13. A grocer is going to mix three kinds of nuts to make 50 pounds of a mixture that will be priced at $5.20 per pound. The three kinds of nuts are peanuts that are priced at $4.00 per pound, almonds that are $4.90 per pound, and pistachios that cost $7.20 per pound. The mixture will contain twice as much in almonds as pistachios by weight. How many pounds of each type of nut should the grocer use to make this mixture?

14. A grocer is going to mix three kinds of nuts to make 40 pounds of a mixture that will be priced at $5.95 per pound. The three kinds of nuts are peanuts that are priced at $4.00 per pound, cashews that are $6.60 per pound, and pistachios that cost $8.20 per pound. The mixture will contain twice as much in peanuts as cashews by weight. How many pounds of each type of nut should the grocer use to make this mixture?

15. A mixture of 20 gallons of chemical A, 10 gallons of chemical B, and 13 gallons of chemical C is required for a certain type of pesticide. There are three types of commercial sprays available. Spray X is a mixture of 4 parts of chemical A and 1 part of chemical C. Spray Y is simply chemical B. Spray Z is a mixture of 1 part of chemical B and 1 part of chemical C. How much of each of the commercial sprays X, Y, and Z should be mixed together to get the correct mixture for this pesticide?

16. For a particular plant, the optimal fertilizer should contain 5 pounds of fertilizer A, 4 pounds of fertilizer B, and 3 pounds of fertilizer C. Commercial brand X contains only fertilizer A. Commercial brand Y contains 2 parts fertilizer B and 3 parts fertilizer C. Commercial brand Z contains 1 part fertilizer A and 1 part fertilizer B. How many pounds of each commercial brand must be mixed together to get the optimal fertilizer for this plant?

17. A businessman has $10,000 to invest in the stock market. He wishes to invest this money and make an 8% profit over the next year. His broker has told him that the best two investments are company A, which should return about 7%, and company B, which should return about 11%. How much should the businessman invest in each company to make the 8% return on the $10,000?

18. An investor has $25,000 to invest in either bonds or stocks. He wishes to invest this money and make a 6.2% profit over the next year. An investment in bonds will return about 5% and investing in selected stocks should return about 11% over the next year. How much should the businessman invest in bonds or stocks to make the 6.2% return on the $25,000?

19. Fred has $100,000 to invest in three different mutual funds. He wishes to invest this money and make an 8% profit over the next year. Over the last year, the first fund has had a return of 5%, the second fund has had a return of 9%, while the third fund has had a return of 12%. Fred likes the second fund, so he wishes to invest $10,000 more in the second fund than in the third fund. How much should he invest in each company to make the 8% return on the $100,000?

20. An investor has $600,000 to invest. There are three ways to invest this money. Municipal bonds pay 8% interest every year. Blue chip stocks should return about 12% annually. Growth or speculative stocks should pay about 15% annually. The investor wants to get a 10% return overall on his investments. Also, he wishes to invest twice as much in blue chip stocks as in growth or speculative stocks. How much should he invest in each type of investment?

21. A company has $1,000,000 available for spending on three different development projects. The company has projected that it needs to make 10% on the $1,000,000 over the next year to meet its next year's budget. The first development project is projected to make 5.5% profit of the money invested in it. The second development project is projected to make 19% profit of the money invested in it. The third development project will not be finished in the next year, so any money invested in it will generate only a 1% return. The company's management decides to invest three times as much in the second project as in the third, because of the potential losses involved in the third project. How much should the company invest in each project to make the 10% return on the $1,000,000?

22. A company needs to borrow $150,000. For tax and related reasons, the company wants to pay 7.3% interest on this loan. There are three lenders for this money. The first charges 6%, the second charges 7%, and the third charges 10%. The company is going to borrow twice as much from the first lender as from the third. How much should the company borrow from each lender?

23. A company needs to borrow $1,000,000. The company wishes to pay 5% interest on this loan. There are three lenders for this money. The first charges 4%, the second charges 6%, and the third charges 8%. The company is going to borrow twice as much from the first lender as from the second. How much should the company borrow from each lender?

24. Victoria is five years younger than Danielle. Grace is 15 years younger than Danielle. Joseph is 4 years older than Grace. The sum of their ages is 69. How old is Grace?

25. Jen has three times as many nickels as dimes and six times as many quarters as dimes. Jen has a total of 50 nickels, dimes, and quarters. The total value of the coins is $8.75. How many of each coin does she have?

26. Toshi met his friends at the coffee house. He bought a total of 12 items. The number of drinks purchased was two more than the number of danishes. Each mocha latte cost $3.50, each cup of regular coffee cost $2, and the danishes were $1.50 each. He spent a total of $27.50 for the lattes, coffee, and danishes. How many of each item did he purchase?

27. This graph shows all the plane routes from Salt Lake City to Detroit for a certain airline company. How many two-stop connections are there from Salt Lake City to Detroit?

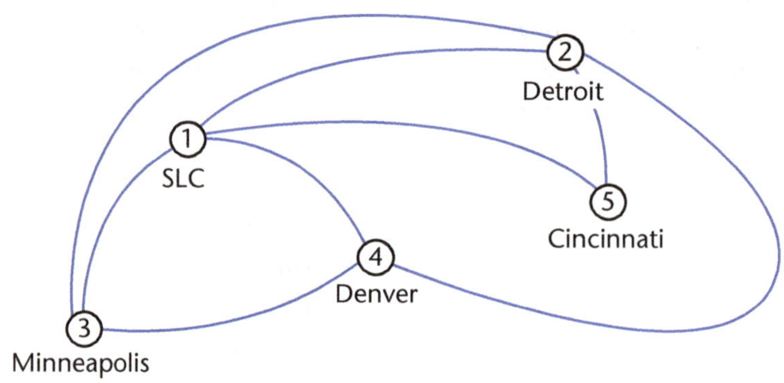

28. This graph depicts a sprinkler system, where each node of the graph represents a sprinkler head.

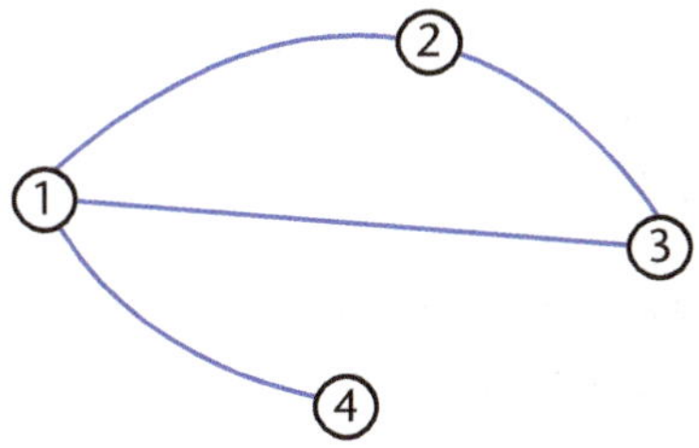

(a) Find the square matrix, A, that represents the number of non-stop connections between all 4 heads.
(b) What is the fewest number of stops made before water from head 4 returns to itself?

29. This graph represents the connections that Sue has with her cousins.

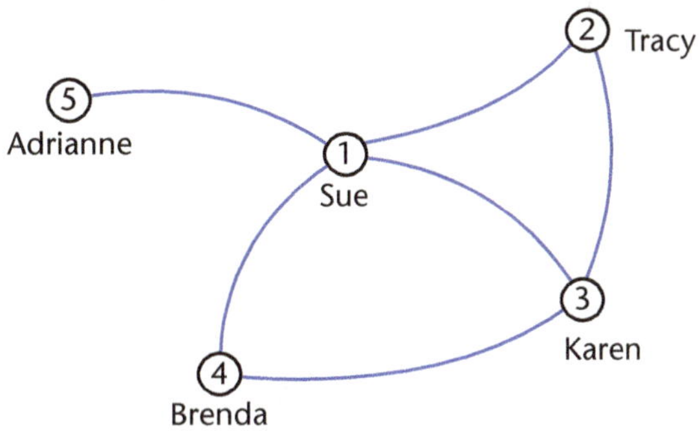

(a) Find the square matrix, A, that represents the number of non-stop connections between all five cousins.
(b) Find the square matrix that represents the number of ways to get from each person to another with two stops.
(c) How many one-stop connections are there from Karen to herself?

30. This graph represents the connections that Joey has with some people in his life.

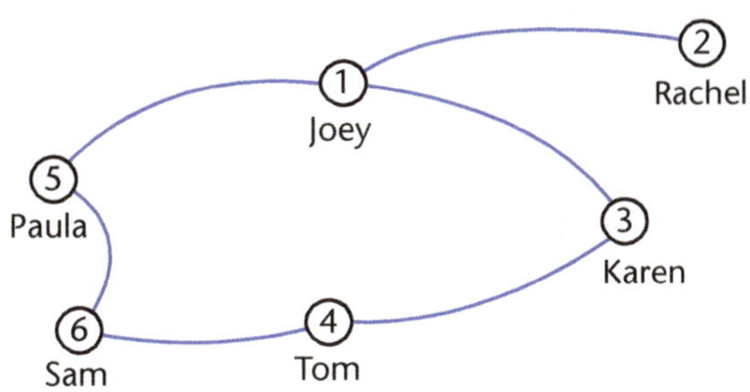

(a) Find the square matrix, *A*, that represents the number of non-stop connections between all six people.
(b) Find the square matrix that represents the number of ways to get from each person to another with two stops.
(c) How many one-stop connections are there between Karen and Sam?

31. A toxic waste storage facility has space for 11,000 tons of total waste. When it's full, the facility will be retired. They store three types (levels) of toxic waste, labeled L, M, and H. The cost per ton to store is \$4/ton for type L, \$5/ton for type M, and \$7/ton for type H. The total cost of storage is \$63,000. Profits need to be \$25,000. How much of each type of waste can be stored at this site when profit is \$1/ton for type L, \$2/ton for type M, and \$3/ton for type H?

32. Find three numbers such that their sum is 36, the sum of the smallest and largest is twice the middle number, and the sum of the two smallest numbers is eight more than the largest number.

33. Find three numbers such that their sum is 42, the sum of the two largest is two more than four times the smallest number, and the largest number less the smallest number is two more than the middle number.

34. A theater has 500 seats total, consisting of front seating, main section seating, and a balcony. If the theater sells all seats to a performance, its income is \$25,000. But, if no balcony seats are sold, the income is \$22,500. Front seats cost \$75 per ticket, main seats cost \$50 each, and balcony seats cost \$25 each. How many seats of each type are in the theater?

Chapter 2 Review

2.1

1. If $A = \begin{bmatrix} -9 & 9 \\ 5 & -9 \\ 3 & 4 \end{bmatrix}$, what is $4A$?

2. Given $A = \begin{bmatrix} 3 & 6 & 8 \\ 2 & 6 & 4 \end{bmatrix}$, and $B = \begin{bmatrix} -3 & 9 & 8 \\ 2 & -6 & 9 \end{bmatrix}$,

   compute the following.

   a. $A + B$

   b. $2A - 3B$

3. Let $A = \begin{bmatrix} 9 & 9 \\ 7 & 8 \end{bmatrix}$, $B = \begin{bmatrix} -9 & 7 \\ -3 & 8 \end{bmatrix}$, and $C = \begin{bmatrix} 4 & 3 \\ 7 & 5 \end{bmatrix}$.

   Find the following.

   a. A^T

   b. $A + B^T$

   c. $A - B + C^T$

4. For $A = \begin{bmatrix} 4 & 0 & -1 \\ 3 & 1 & 3 \\ 0 & -3 & 2 \end{bmatrix}$ and $B = \begin{bmatrix} 4 & 2 & -3 \\ 2 & 4 & -3 \\ 2 & -3 & -3 \end{bmatrix}$,

   find $2A + B^T$.

5. Solve for $x, y, z,$ and w.

$$\begin{bmatrix} x & 4 \\ 4y & w \end{bmatrix} - \begin{bmatrix} 4x & 2z \\ -3 & -2w \end{bmatrix} = \begin{bmatrix} 12 & 8 \\ y & 6 \end{bmatrix}$$

2.2

6. Let A be a 2×2 matrix, B is size 3×2 and C is a 3×3 matrix. Determine the size of the following matrices, if it's possible to perform the operation. Otherwise, state it's impossible.

   a. AB

   b. BA

   c. AB^T

   d. BC

   e. ABC

   f. CBA

   g. B^TC

   h. BC^T

7. Let $A = \begin{bmatrix} 9 & 1 & 7 \\ 7 & -1 & 8 \end{bmatrix}$ and $B = \begin{bmatrix} 5 & 0 \\ 8 & -8 \\ 6 & 6 \end{bmatrix}$.

   Compute the following, if possible.

   a. AB

   b. B^TA^T

8. Let $A = \begin{bmatrix} 5 & 5 \\ -5 & 5 \end{bmatrix}$ and $B = \begin{bmatrix} 3 & -1 \\ 2 & 4 \end{bmatrix}$.

   Compute the following, if possible.

   a. AB

   b. BA

   c. A^TB^T

9. Let $A = \begin{bmatrix} 1 & 4 & 9 \\ 0 & 1 & 5 \\ 0 & 0 & 2 \end{bmatrix}$.

Compute the following, if possible.

a. A^2

b. A^3

10. Let $A = \begin{bmatrix} -3 & -3 & -3 \\ 0 & -2 & -2 \\ 1 & 2 & 0 \end{bmatrix}$ and $B = \begin{bmatrix} 1 \\ 2 \\ 5 \end{bmatrix}$.

Compute the following, if possible.

a. AB

b. BA

c. B^TA

d. A^TB

2.3

Solve these systems of linear equations.

11. $\begin{aligned} 4x - 5y - 5z &= -6 \\ x + 6y - 6z &= 32 \\ 4x - 3y + 6z &= -46 \end{aligned}$

12. $\begin{aligned} 3x + 2y + z &= 0 \\ x + y + 2z &= 0 \\ 2x + y - z &= -1 \end{aligned}$

13. $\begin{aligned} 2x - 5y + z &= -9 \\ x + 4y - 6z &= 2 \\ 3x - 4y - 2z &= -10 \end{aligned}$

14. $\begin{aligned} 2x + y - z &= 9 \\ 3x - y &= 0 \\ x + 2y + 3z &= -5 \end{aligned}$

15. $\begin{aligned} x + 2y + 5z &= -5 \\ 4y - z &= 2 \\ 3x + y + z &= 13 \end{aligned}$

16. $\begin{aligned} 3x - y &= -1 \\ x + y + z &= 6 \\ 5x + y + 2z &= 11 \end{aligned}$

2.4

Find the inverse matrix, A^{-1}, for each matrix, if it exists. If it is not possible to find its inverse, state the reason why.

17. $A = \begin{bmatrix} 1 & -2 \\ 1 & 3 \end{bmatrix}$

18. $A = \begin{bmatrix} 3 & 0 & -3 \\ 1 & 1 & 2 \\ 1 & 2 & 1 \end{bmatrix}$

19. $A = \begin{bmatrix} 5 & 4 \\ -1 & 0 \end{bmatrix}$

20. $A = \begin{bmatrix} 2 & 3 & 1 \\ -1 & 0 & 5 \end{bmatrix}$

21. $A = \begin{bmatrix} 2 & 3 & 4 \\ 1 & 2 & 2 \\ 1 & 1 & 2 \end{bmatrix}$

22. $A = \begin{bmatrix} 2 & -5 & 1 \\ 1 & 4 & -6 \\ 3 & -4 & -2 \end{bmatrix}$

23. $A = \begin{bmatrix} 3 & 1 & 0 \\ 2 & -2 & 1 \\ 1 & 1 & 2 \end{bmatrix}$

24. Use $A^{-1} = \begin{bmatrix} 3 & 0 & -3 \\ 1 & 1 & 2 \\ 1 & 2 & 1 \end{bmatrix}$ to solve $AX = \begin{bmatrix} -2 \\ 5 \\ 4 \end{bmatrix}$ for X.

2.5

25. Encrypt the message "math is fun" using the encryption matrix

$$A = \begin{bmatrix} 1 & -2 & 3 \\ -4 & 5 & -6 \\ 3 & -2 & 2 \end{bmatrix}.$$

26. Decrypt the message

28 −55 22 58 −139 68 39 −69
20 −21 60 −26 68 −134 45

using the inverse of the encryption matrix from problem 25.

27. A brokerage house offers three stock portfolios. Portfolio 1 consists of two blocks of common stock and one municipal bond. Portfolio 2 consists of four blocks of common stock, two municipal bonds and three blocks of preferred stock. Portfolio 3 consists of seven blocks of common stock, three municipal bonds and three blocks of preferred stock. A customer wants 21 blocks of common stock, 10 municipal bonds and 9 blocks of preferred stock. How many of each Portfolio should be purchased for this customer?

28. A dietician is planning a meal that supplies 3225 mg of vitamin C, 2120 mg of calcium and 1255 mg of vitamin D. Three foods will be used. In food 1, there are 50 mg of vitamin C, 20 mg calcium and 10 mg of vitamin D. In food 2, there are 25 mg vitamin C, 20 mg calcium, and 15 mg vitamin D. Food 3 has 25 mg vitamin C, 30 mg calcium, and 15 mg vitamin D. What quantities of foods 1, 2 & 3 must be combined to form the desired mixture of these vitamins?

29. A grocer mixes peanuts, cashews, and almonds to obtain a 24-ounce package worth \$7.16. If peanuts, cashews and almonds cost \$0.19, \$0.40 and \$0.50 per ounce, respectively, and the amount of cashews in the mix is equal to the amount of almonds, how many ounces of each nut are in the package?

30. Marla's company needs to borrow \$80,000 for start-up costs. Three of her family members are willing to pitch in and loan her the money for a year. However, no one person has the total sum (or even half of it). Marla's sister is willing to loan money at 8% simple interest; her brother will charge 6.5% simple interest; and Marla's parents will charge 7.2% interest. Marla is only willing to pay a total of 7.1525% interest on the borrowed sum. Marla decides she will borrow twice as much from her parents than she borrows from her sister. How much should she borrow from each of her family members?

chapter 3

Functions and Graphs

3.1 Quadratic Equations in One Variable

QUADRATIC EQUATION IN ONE VARIABLE

An equation that can be written in the form:

$$ax^2 + bx + c = 0 \quad \text{where } a, b, c, \in \mathbb{R}, a \neq 0$$

ZERO PRODUCT PROPERTY

For $a, b \in \mathbb{R}$, $ab = 0 \quad \Leftrightarrow \quad a = 0$ or $b = 0$ or $a = b = 0$.

In other words, if we have two (or more) factors that multiply together to yield zero, then at least one of the factors must be zero. This property gives us leverage when solving quadratic equations, particularly when the quadratic expression factors.

There are basically four different methods or strategies for solving a quadratic equation.

STRATEGY	CONDITIONS FOR USE	EXAMPLE
Square Root Technique	Only works if $b = 0$, i.e., if quadratic expression is a perfect square. (*Note:* When taking the square root of both sides of an equation, we must consider the ± solutions.)	$4x^2 + 3 = 28 \Leftrightarrow {}^24x = 25$ $\Leftrightarrow x^2 = \frac{25}{4}$ $\Leftrightarrow \sqrt{x^2} = \pm\sqrt{\frac{25}{4}} \Leftrightarrow x = \frac{\pm 5}{2}$
Factoring Technique	Only works if you can factor the quadratic expression. (*Note:* This is where we use the Zero Product Property. But before we factor, we must have all quadratic terms on one side of the equation with only zero on the other side.)	$x^2 + 3x = 10$ $\Leftrightarrow x^2 + 3x - 10 = 0$ $\Leftrightarrow (x-2)(x+5) = 0$ $\Leftrightarrow x - 2 = 0$ or $x + 5 = 0$ $\Leftrightarrow x = 2$ or $x = -5$

STRATEGY	CONDITIONS FOR USE	EXAMPLE
Completing the Square	Always works!! (*Note:* The explanation for this process appears in more detail below.)	$x^2 + 6x + 2 = 0$ $\Leftrightarrow x^2 + 6x = -2$ $\Leftrightarrow x^2 + 6x + \left(\frac{6}{2}\right)^2 = -2 + \left(\frac{6}{2}\right)^2$ $\Leftrightarrow x^2 + 6x + 9 = 7$ $\Leftrightarrow (x+3)^2 = 7$ $\Leftrightarrow x + 3 = \pm\sqrt{7}$ $\Leftrightarrow x = -3 \pm \sqrt{7}$
Using the Quadratic Formula $x = \dfrac{-b \pm \sqrt{b^2 - 4ac}}{2a}$	Always works!! (*Note:* Make sure the quadratic equation is in the form stated above before substituting values in for *a*, *b* and *c* to the quadratic formula.)	$2x^2 + 3x - 4 = 0$ $\Leftrightarrow x = \dfrac{-3 \pm \sqrt{3^2 - 4(2)(-4)}}{2(2)}$ $\Leftrightarrow x = \dfrac{-3 \pm \sqrt{9 + 32}}{4}$ $\Leftrightarrow x = \dfrac{-3 \pm \sqrt{41}}{4}$

COMPLETING THE SQUARE PROCESS

You can view this process geometrically while simultaneously representing the steps algebraically.

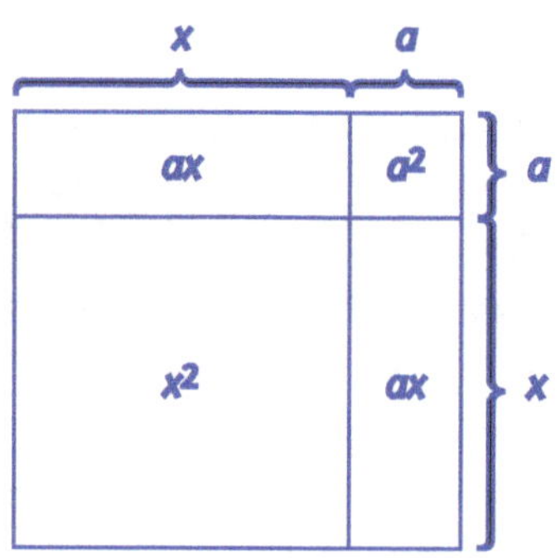

We can find the area of this square by squaring the side length.

$$\Rightarrow \quad A = (x + a)^2$$

A second way to find the area is to add up the areas of the individual rectangles and squares that make up the overall square.

$$\Rightarrow \quad A = x^2 + ax + ax + a^2 = x^2 + 2ax + a^2$$

When we "complete the square," it's as if we're literally given the bigger inside square and the two rectangular pieces. In order to finish making the overall square, we must add the area of the little corner square.

We know the little corner square has length a, which is one half of the coefficient of x. We need to square that length to "complete the square."

Given the quadratic equation $x^2 + bx + c = 0$, to complete the square,

1. Subtract c from both sides of the equation.

$$x^2 + bx = -c$$

2. Add $\left(\frac{b}{2}\right)^2$ to both sides of the equation.

$$x^2 + bx + \left(\frac{b}{2}\right)^2 = -c + \left(\frac{b}{2}\right)^2$$

3. Factor the quadratic expression into a perfect square.

$$\left(x + \frac{b}{2}\right)^2 = -c + \left(\frac{b}{2}\right)^2$$

4. Continue solving for x by taking the square root of both sides of the equation.

$$x + \frac{b}{2} = \pm\sqrt{-c + \left(\frac{b}{2}\right)^2}$$

5. Finally

$$x = \frac{-b}{2} \pm \sqrt{-c + \left(\frac{b}{2}\right)^2} = \frac{-b}{2} \pm \sqrt{-c\left(\frac{4}{4}\right) + \frac{b^2}{4}} = \frac{-b}{2} \pm \sqrt{\frac{-4c + b^2}{4}} = \frac{-b}{2} \pm \sqrt{\frac{b^2 - 4c}{4}} = \frac{-b \pm \sqrt{b^2 - 4c}}{2}$$

(*Note:* If we have a leading coefficient that is not 1, as in the above generic example, we must first factor out the leading coefficient from all terms containing x before we complete the square. In that case, we get the quadratic formula.)

EXAMPLE 1

Complete the square to solve this quadratic equation.

$$x^2 + 4x - 1 = 0$$

In general, if a quadratic equation is factorable, it's most efficient to use the factoring technique to solve the equation. However, there are many equations that we cannot factor and for those equations, completing the square is a very useful strategy.

$x^2 + 4x - 1 = 0$ — We cannot factor this. Add 1 to both sides of the equation in order to isolate the x-squared and x terms.

$x^2 + 4x = 1$ Add $\left(\frac{4}{2}\right)^2 = 2^2 = 4$ to both sides of the equation, to complete the square.

(We always take one half of the coefficient of x and square it to find the value to add.)

$x^2 + 4x + 4 = 1 + 4$ Simplify the right side of the equation and factor the left side of the equation.

$(x+2)^2 = 5$ Take the square root of both sides of the equation.

$x + 2 = \pm\sqrt{5}$

$x = -2 \pm \sqrt{5}$ Subtract 2 from both sides of the equation to finish getting x alone.

EXAMPLE 2

Solve the following quadratic equations, using a different strategy for each one.

(a) $49x^2 - 25 = 0$

We'll choose to use the "square root" strategy, since the x-term is missing, i.e., $b = 0$. This strategy is basically stripping off the layers of what happened to x in the opposite order in which it happened.

$49x^2 - 25 = 0$ Add 25 to both sides of the equation.

$49x^2 = 25$ Divide both sides of the equation by 49.

$x^2 = \frac{25}{49}$ Take the square root of both sides of the equation.

$$x = \pm\sqrt{\frac{25}{49}} = \frac{\pm 5}{7}$$

(b) $3x^2 - 5x = -7x^2$

Before we can decide which strategy to use here, it would be best if we have all terms on one side of the equation with zero on the other side.

$3x^2 - 5x = -7x^2$ Add $7x^2$ to both sides of the equation.

$10x^2 - 5x = 0$ We can factor the quadratic expression on the left side of the equation. Thus, we'll choose to use the "factor" strategy.

$5x(2x - 1) = 0$ Set each factor equal to zero separately.

$5x = 0$ or $2x - 1 = 0$ Solve each linear equation.

$x = 0$ or $2x = 1 \Leftrightarrow x = 0$ or $x = \frac{1}{2}$

(c) $x^2 + 3x - 1 = 0$

We cannot factor this quadratic expression on the left side of the equation. So we have to choose between the quadratic formula or completing the square techniques. We'll practice the quadratic formula.

We have $a = 1$, $b = 3$, $c = -1$.

$$\Rightarrow x = \frac{-3 \pm \sqrt{3^2 - 4(1)(-1)}}{2(1)} = \frac{-3 \pm \sqrt{9+4}}{2} = \frac{-3 \pm \sqrt{13}}{2}$$

(d) $2x^2 + 16x + 3 = 0$

The quadratic expression in this equation is also not factorable. This time we'll practice completing the square.

$2x^2 + 16x + 3 = 0$ Subtract 3 from both sides of the equation.

$2x^2 + 16x = -3$ Factor out the 2 (because it's the leading coefficient) on the left side of the equation before we complete the square.

$2(x^2 + 8x) = -3$ Divide both sides of the equation by 2.

$x^2 + 8x = \frac{-3}{2}$ Add $\left(\frac{8}{2}\right)^2 = 4^2 = 16$ to both sides of the equation.

$x^2 + 8x + 16 = \frac{-3}{2} + 16$ Simplify the right side of the equation and factor the left side.

$(x+4)^2 = \frac{29}{2}$ Take the square root of both sides of the equation.

$x + 4 = \pm\sqrt{\frac{29}{2}}$ Finally, subtract 4 from both sides of the equation.

$x = -4 \pm \sqrt{\frac{29}{2}}$

EXAMPLE 3

Solve this equation for y.

$$(y-5)^2 - 5(y-5) = 24$$

We have some good choices here to get us started. We can either pretend $y - 5 = x$ and solve this using one of the above-mentioned strategies. OR, we can multiply everything out, collect like terms and then use a quadratic equation solving strategy. We'll choose the first idea.

$(y-5)^2 - 5(y-5) = 24$ — Subtract 24 from both sides of the equation to get everything on one side.

$(y-5)^2 - 5(y-5) - 24 = 0$ — Factor the quadratic expression.

At this point, let's interject to make sure our next step is understandable. If we had $(\text{Bob})^2 - 5(\text{Bob}) - 24 = 0$, it would factor into $(\text{Bob} - 8)(\text{Bob} + 3) = 0$. Or if we had $p^2 - 5p - 24 = 0$, it would factor into $(p-8)(p+3) = 0$.

In other words, the name of our variable makes no difference. Going back to the real equation we're solving, then $y - 5$ is "Bob" or "p" in the above corollary examples.

So, now we have:

$((y-5)-8)((y-5)+3) = 0$ — Simplify each factor.

$(y-13)(y-2) = 0$ — Set each factor equal to zero.

$y - 13 = 0$ or $y - 2 = 0$ — Solve each equation separately.
$y = 13$ or $y = 2$

EXAMPLE 4

Solve the equation.

$$\frac{3}{w-2} + \frac{5}{w+1} = 1$$

This is a rational equation, not a quadratic equation (yet). We know that we can't allow w to be either 2 or -1 because that would make the denominator equal to zero, and we can't divide by zero. In general, to eliminate fractions in an equation, we can multiply both sides of the equation by the least common denominator (LCD).

$\frac{3}{w-2} + \frac{5}{w+1} = 1$ — Multiply both sides of the equation by $(w-2)$ and $(w+1)$.

$(w-2)(w+1)\left(\frac{3}{w-2}+\frac{5}{w+1}\right)=1(w-2)(w+1)$	Distribute the LCD to each term on the left side of the equation.
$\frac{3(w-2)(w+1)}{w-2}+\frac{5(w-2)(w+1)}{w+1}=(w-2)(w+1)$	Simplify each fraction.
$3(w+1)+5(w-2)=(w-2)(w+1)$	Multiply everything out using the Distributive Property.
$3w+3+5w-10=w^2-w-2$	Collect all terms on the right side of the equation.
$0=w^2-9w+5$	Finally, we have a quadratic equation. We cannot factor it, so we can choose to either complete the square or use the quadratic formula. We'll choose the quadratic formula.

$$w=\frac{9\pm\sqrt{(-9)^2-4(1)(5)}}{2(1)}=\frac{9\pm\sqrt{81-20}}{2}=\frac{9\pm\sqrt{61}}{2}$$

EXAMPLE 5

Solve the equation.

$$2x^2+2x+1=0$$

The left-hand side quadratic expression in this equation is not factorable. So we'll choose to use the quadratic formula to solve this equation.

We have $a=2$, $b=2$, $c=1$.

$$\Rightarrow \qquad x=\frac{-2\pm\sqrt{2^2-4(2)(1)}}{2(2)}=\frac{-2\pm\sqrt{-4}}{4}$$

But we run into a problem here since we cannot take the square root of a negative number. Thus our conclusion is that there are *no real solutions* to this equation.

3.1 Homework

Solve the equations in problems 1–16 using the square root technique.

1. $x^2 - 49 = 0$
2. $x^2 - 16 = 0$
3. $9x^2 - 36 = 0$
4. $x^2 - 9 = 16$
5. $2x^2 - 30 = 0$
6. $3x^2 - 32 = 16$
7. $5x^2 - 32 = 13$
8. $5x^2 - 32 = x^2 + 8$
9. $9x^2 - 16 = 0$
10. $(x - 1)^2 = 6$
11. $(x - 2)^2 = 25$
12. $(x + 1)^2 = 9$
13. $x^2 - 4 = 12$
14. $(x + 2)^2 = 49$
15. $(x - 13)^2 = 36$
16. $25x^2 - 4 = 0$

Solve the equations in problems 17–30 by factoring.

17. $x^2 - x - 12 = 0$
18. $x^2 + 7x + 10 = 0$
19. $x^2 - 11x + 24 = 0$
20. $x^2 - 2x = 63$
21. $2x^2 - 3x - 2 = 0$
22. $3x^2 + 7x = 20$
23. $6x^2 = 6 - 5x$
24. $4x^2 + 6x - 4 = 0$
25. $15x^2 + 24x - 12 = 0$
26. $2x^2 - 11x = 6$
27. $12x^2 + x - 6 = 0$
28. $x^2 - 24 = -5x$
29. $5x^2 - 12 = 7x - 5x^2$
30. $2x(5x + 6) = 16$

Solve the equations in problems 31–46 by completing the square.

31. $x^2 - 2x - 8 = 0$
32. $x^2 - 4x - 9 = 0$
33. $x^2 + 6x + 4 = 0$
34. $x^2 + x - 4 = 0$
35. $x^2 - 3x + 1 = 0$
36. $3x^2 - 3x - 5 = 0$
37. $2x^2 + 5x + 1 = 0$
38. $5x^2 - 3x - 1 = 0$
39. $x^2 - x - 2 = 0$
40. $x^2 + 9x = 8$
41. $x^2 + \frac{3}{2}x - 3 = 0$
42. $2y^2 - 22y = 11$
43. $x^2 + x - 1 = 0$
44. $t^2 + 10t = 4$
45. $y^2 - 8y = 15$
46. $p^2 + 6p = 16$

Solve the equations in problems 47–59 by using the quadratic formula.

47. $-x^2 - x + 10 = 0$
48. $6x^2 + 2x - 1 = 0$
49. $2x^2 + 3x - 1 = 0$
50. $-x^2 - 8x + 7 = 0$
51. $-3x^2 + 6x + 8 = 0$
52. $6x^2 - 2x - 2 = 0$
53. $2x^2 - 6x - 5 = 0$
54. $4x^2 - 10x - 1 = 0$
55. $3x^2 + 4x = 3$
56. $x^2 - 5 - 2x = 0$
57. $y + \frac{1}{y} = \frac{13}{6}$
58. $x^2 + 4 = 6x$
59. $14y^2 + 9y = 0$

Solve the rational equations in problems 60–67 by changing to a quadratic equation and then solving that quadratic equation.

60. $\frac{1}{x-5} - \frac{1}{x-2} = 1$
61. $\frac{1}{x-3} - \frac{1}{x+2} = \frac{1}{2}$
62. $x = \frac{1}{x+1}$
63. $2 + x = \frac{1}{x-5}$
64. $\frac{1}{x+4} + \frac{1}{x+3} = \frac{1}{3}$
65. $\frac{x}{(x-2)} = x + 1$
66. $\frac{2x}{x-6} = \frac{x+1}{x}$
67. $\frac{2x}{x-2} - \frac{x}{x+3} = 2$

Solve the equation in problems 68–80 by any method.

68. $6x^2 - 50 = x^2$

69. $x^2 + 8x + 10 = 0$

70. $x^2 = 7x - 7$

71. $x^2 = x + 56$

72. $3x^2 - x = 10$

73. $x^2 + 8x + 10 = 0$

74. $\frac{1}{x - 10} - \frac{1}{x - 9} = 1$

75. $6(x - 2)^2 = 48$

76. $2x^2 + 7x = 4$

77. $\frac{2x - 1}{x + 5} - \frac{x + 1}{x + 2} = 1$

78. $19y^2 + 8y = 0$

79. $4x + x(x - 3) = 0$

80. $25x = 3x^2 + 28$

3.2 Parabolas: Quadratic Equations in Two Variables

We can also call a quadratic equation in two variables a quadratic function. Typically, we call the input (independent) variable x and the output (dependent) variable y. In function notation, remember that $f(x)$ simply refers to the output, sometimes also known as y.

> **QUADRATIC EQUATION IN TWO VARIABLES**
>
> Any equation which can be written in the form
>
> $$f(x) = y = ax^2 + bx + c,$$
>
> where a, b, $c \in \mathbb{R}$ and $a \neq 0$. (This is called standard form.)

A quadratic function graphs into a parabola, which looks like these figures.

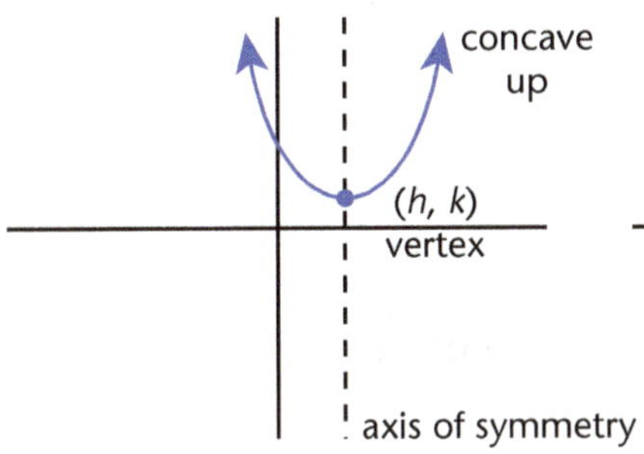

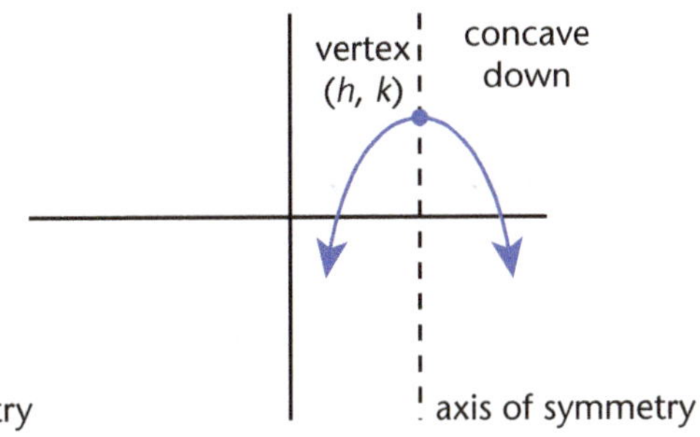

Note: We've previously referred to quadratic equations that we solved, in section 3.1. In those cases, we had equations with only one variable. Quadratic functions are essentially quadratic equations with TWO variables in the equations, namely x and y (usually). A quadratic equation in one variable has at most two solutions and they're x-values that would be graphed on a number line. A quadratic equation in two variables, aka a quadratic function, has infinitely many solutions and it graphs into a parabola, in two dimensions. This should appeal to your intuition, namely that an equation in one variable has solutions in 1-d and an equation in two variables has solutions in 2-d.

VOCABULARY

Vertex—The point on the parabola where it changes direction. The vertex is a minimum point (the lowest point possible on the graph) if the parabola is concave up and a maximum point (the highest point possible on the graph) if the parabola is concave down.

Axis of Symmetry—The vertical line about which the parabola is symmetric, i.e., the vertical line that behaves as a mirror and the graph of the parabola is reflected about that axis. Since the axis of symmetry is a vertical line, its equation will always be of the form $x = h$, where h = the x-coordinate of the vertex.

Concavity—A parabola is concave up or down, depending on the leading coefficient, a, of the quadratic function. If a is positive, the parabola is concave up. If a is negative, the parabola is concave down.

Zeros or Roots of the Function—The zeros or roots of any function are the input values, usually x-values, where the graph intersects the x-axis. Basically, they're the x-values of the x-intercepts.

We can also always rewrite the quadratic function in the form $f(x) = y = a(x - h)^2 + k$, which is more useful for graphing. In this form,

(a) the **vertex is at (*h*, *k*)**, and
(b) the **axis of symmetry is $x = h$.**
(c) The coefficient a tells us about how the graph is **stretched** (**if** $a > 1$) or **shrunk** (**if** $a < 1$).
(d) Also, if $a > 0$, the **parabola is concave up.** If $a < 0$, the **parabola is concave down.**

How can we get from the standard form of a quadratic function to this more useful form for graphing? Let's consider the parabola given by $y = ax^2 + bx + c$. When we substitute $x = 0$ into the standard form, we have

$$y = a(0^2) + b(0) + c = 0 + 0 + c = c,$$

which means the parabola goes through the point $(0, c)$.

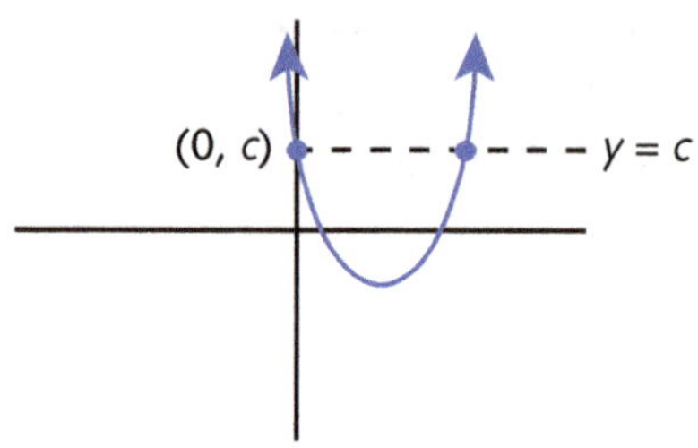

As we can see in the figure above, by symmetry of a parabola, there is another point whose y-coordinate is also c. Let's substitute c in for y (into the standard form of a quadratic function) and solve for x.

$\Rightarrow \quad c = ax^2 + bx + c$	Subtract c from both sides of the equation.
$0 = ax^2 + bx$	Factor x out of both terms on the right side of the equation.
$0 = x(ax + b)$	Set each factor equal to zero to solve for x.
$x = 0$ or $ax + b = 0$	Continue solving the right linear equation by subtracting b from both sides of the equation.
$x = 0$ or $ax = -b$	For the right equation, divide both sides of the equation by a (which is legal because we know a is not zero).
$x = 0$ or $x = \dfrac{-b}{a}$	

This means that two points on our parabola are $(0, c)$ and $\left(\dfrac{-b}{a}, c\right)$.

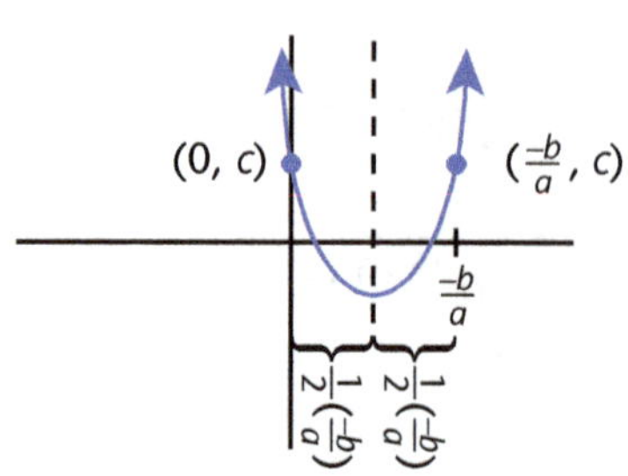

Since these two points have the same y-value, and we know the parabola is a mirror image about the line of symmetry, then the axis of symmetry must be halfway between those two mirrored points! That is, halfway between $x = 0$ and $x = \frac{-b}{a}$. Thus, the **axis of symmetry is at** $x = \frac{-b}{2a}$, and the **vertex is at the point** $(\frac{-b}{2a}, f(\frac{-b}{2a}))$, assuming these a and b values are from the quadratic function in standard form.

EXAMPLE 1

For $y = -3x^2 - 6x + 9$,

(a) Find the vertex

(b) Is the vertex a maximum or a minimum point?

(a) The parabola, as given, is in standard form with $a = -3$, $b = -6$ and $c = 9$. Thus, the x-coordinate of the vertex is at

$$x = \frac{-b}{2a} = \frac{-(-6)}{2(-3)} = \frac{6}{-6} = -1.$$

This also means that the axis of symmetry is $x = -1$. We can find the y-coordinate of the vertex by substituting $x = -1$ into the function.

$$\Rightarrow \quad y = -3(-1)^2 - 6(-1) + 9 = -3 + 6 + 9 = 12.$$

Thus, the vertex is at the point $(-1, 12)$.

(b) Because the leading coefficient is negative in standard form, namely $a = -3$, this means that the parabola is concave down. So the vertex is a maximum point.

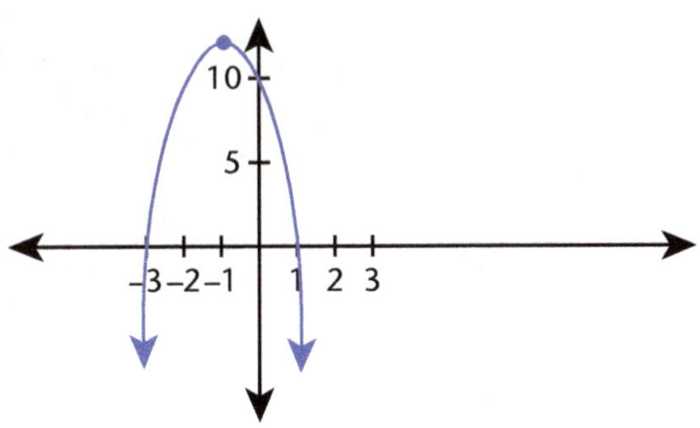

EXAMPLE 2

For the function $y = -2(x + 3)^2 + 8$,

(a) What is the vertex?

(b) Find the axis of symmetry.

(c) Is the parabola concave up or down?

(d) What are the x-intercepts of the parabola (or zeros of the function)?

(e) Sketch the graph.

(a) We can find the vertex immediately in this form of a quadratic function. The vertex is at (–3, 8).

(b) The axis of symmetry is the vertical line that goes through the vertex, namely $x = -3$.

(c) Since the leading coefficient, $a = -2$, is negative, the parabola is concave down. This means that the vertex is a maximum point for the graph of the parabola.

(d) We can find the zeros (roots) of the function by substituting $y = 0$.

$\Rightarrow \quad 0 = -2(x+3)^2 + 8$	Subtract 8 from both sides of the equation.
$-8 = -2(x+3)^2$	Divide both sides of the equation by –2.
$4 = (x+3)^2$	Take the square root of each side of the equation, remembering that we have to consider the plus/minus square roots.
$\pm\sqrt{4} = \sqrt{(x+3)^2} \Leftrightarrow \pm 2 = x + 3$	Subtract 3 from both sides of the equation.
$x = -3 \pm 2$	Separate these two values of x.
$x = -3 + 2 = -1$ or $x = -3 - 2 = -5$	

So, the zeros (roots) are $x = -1$ and -5. That's equivalent to saying that the x-intercepts are (–1, 0) and (–5, 0).

(e) To graph this parabola, we can plot the points (–3, 8) as the vertex and (–1, 0) and (–5, 0) as the x-intercepts. Connect the points with a parabola and we're done.

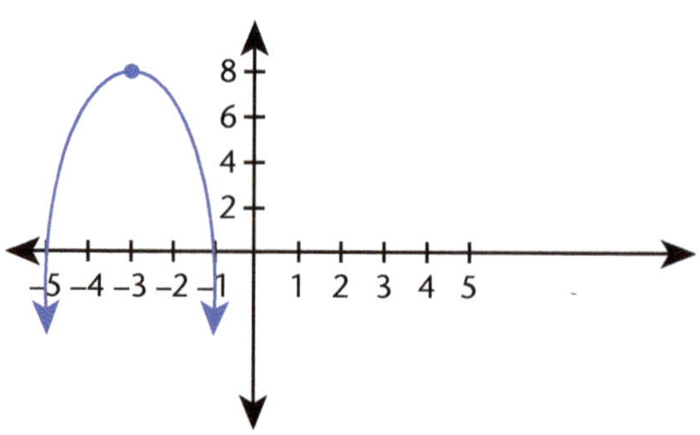

EXAMPLE 3

Given 200 feet of fencing to enclose a rectangular garden, what is the formula for the resulting area? What are the dimensions of the rectangle that give the maximum possible area?

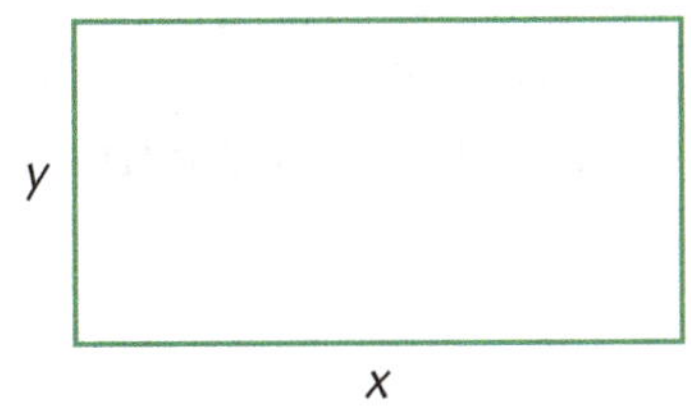

Area of a rectangle is given by length times width:

$A = xy$

We know that we have a limited amount of fence to work with. The perimeter of the rectangle gives the total fence required to create the rectangle.

$\Rightarrow P = 2x + 2y = 200$ Divide both sides of the last two parts of the equation by 2.

$\Leftrightarrow x + y = 100$

$y = 100 - x$

Substitute this expression for y into the area formula so that we can get everything in terms of only x.

$$\Rightarrow A = x(100 - x) = 100x - x^2 = -x^2 + 100x$$

We can explicitly write this in the standard form of a quadratic function.

$$A = -x^2 + 100x + 0 \text{ with } a = -1, b = 100, c = 0$$

Since a is negative, the graph of this parabola is concave down, which means the vertex is a maximum point for the area function.

$\Rightarrow$ x-coordinate of the vertex is $x = \dfrac{-b}{2a} = \dfrac{-100}{2(-1)} = \dfrac{-100}{-2} = 50$

$\Rightarrow$ y-coordinate of the vertex is $A(50) = -(50^2) + 100(50) = -2500 + 5000 = 2500$

This means that the maximum area is 2,500 square feet and the length, x, which yields that maximum area, is 50 ft. Thus the width, y, of the maximum area rectangle is $y = 100 - x = 100 - 50 = 50$ ft. In other words, the rectangle that yields the maximum area, given a fixed perimeter, is in fact a square!

EXAMPLE 4

Find the average rate of change $y = \frac{1}{3}x^2 + 5x - 2$ between $x = 3$ and $x = 0$.

The average rate of change is exactly the slope of the line connecting the two points on the graph of this parabola at the given x-values. In fact, whenever we see or hear the words "rate of change," "speed" or "velocity," we can automatically think of slope because they're all basically synonymous. The average rate of change is the average slope between those points, which will always be given by the slope of the line connecting the points.

First, we need to find the points at those x-values.

⇒ at $x = 3$, $$y(3) = \frac{1}{3}(3^2) + 5(3) - 2 = \frac{1}{3}(9) + 15 - 2 = 3 + 15 - 2 = 16$$

and at $x = 0$, $$y(0) = \frac{1}{3}(0^2) + 5(0) - 2 = 0 + 0 - 2 = -2$$

So, the parabola goes through the points (3, 16) and (0, –2). Now, we need to find the slope of the line going through those two points.

⇒ $$m = \frac{y_2 - y_1}{x_2 - x_1} = \frac{16 - (-2)}{3 - 0} = \frac{18}{3} = 6$$

The average rate of change between the two given points on the parabola is 6.

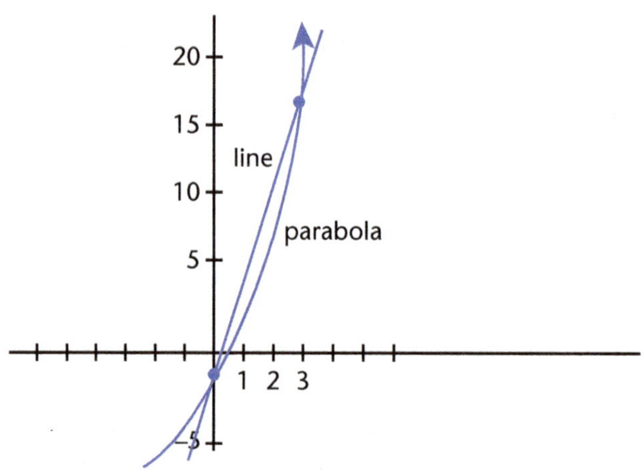

3.2 Homework

For problems 1–23, find the vertex of the parabola and then indicate whether the vertex is the maximum or minimum for the graph of the parabola.

1. $y = -7x^2 - 10$
2. $y = -2x^2 - 8x + 1$
3. $y = 7x^2 + 7x$
4. $y = -x^2 - 3x - 8$
5. $y = 8x^2 - 4x + 6$
6. $y = 9x^2 - 9x + 3$
7. $y = -7x^2 + 5x - 3$
8. $y = 8x^2 - 8x - 2$
9. $y = -x^2 - 6x - 18$
10. $y = -2x^2 + 8x - 7$
11. $y = x^2 + x + 14$
12. $y = 5x^2 + 10x + 12$
13. $y = 13x^2 - 17x + 5$
14. $y = -2x^2 - 8x + 1$
15. $y = 6x^2 - 13x + 1$
16. $y = 3x^2 + 2x - 1$
17. $y = -(x + 3)^2 - 2$
18. $y = -x^2 - 2x - 2$
19. $y = 3x^2 - 2x + 1$
20. $y = -x^2 + 3x + 12$
21. $y = x^2 - 2x + 4$
22. $y = -x^2 - 2x - 1$
23. $y = x^2 - 3x + 3$

For problems 24–43, find the vertex, the axis of symmetry, whether the parabola is concave up or down, and the x-intercepts of the parabola (or zeros of the function), if they exist.

24. $y = 9(x + 3)^2 + 6$
25. $y = (x + 9)^2 + 9$
26. $y = -9(x - 1)^2 + 8$
27. $y = (x - 1)^2$
28. $y = -(x - 4)^2 + 9$
29. $y = 5(x + 1)^2 - 8$

30. $y = 3(x + 8)^2 - 7$

31. $y = -5(x + 7)^2 - 3$

32. $y = 4(x - 17)^2 - 8$

33. $y = -17(x - 7)^2 - 16$

34. $y = 13(x - 16)^2 - 18$

35. $y = -16(x + 5)^2 + 15$

36. $y = (x - 11)^2 - 5$

37. $y = 5(x - 4)^2 + 7$

38. $y = -18(x - 12)^2 + 7$

39. $y = 2x^2 - 8x + 5$

40. $y = 8(x - 1)^2 + 2$

41. $y = -x^2 - 20x - 99$

42. $y = -3x^2 + 18x - 25$

43. $y = -4(x + 1)^2 - 4$

44. When an object is thrown upwards with a speed of 100 ft/sec, its height above the ground is given by the function $h = f(t) = -16t^2 + 100t$, where t is the time in seconds after it has been thrown. At what time will the object reach its highest point? How high will that point be in feet above the ground?

45. For this same object (as in problem 44) being thrown upwards and traveling according to the function $h = f(t) = -16t^2 + 100t$, what is its average speed for the first second of its travel?

46. When an object is thrown upwards with a speed of 100 ft/sec on the moon its height above the surface of the moon is given by the function $h = f(t) = -5.31t^2 + 100t$, where t is the time in seconds after it has been thrown. At what time will the object reach its highest point? How high will that point be in feet above the surface of the moon?

47. For this same object (as in problem 46) being thrown upwards on the moon and traveling according to the function $h = f(t) = -5.31t^2 + 100t$, what is its average speed for the first second of its travel?

48. A company has found that the annual profit in dollars from one of their products can be modeled by the function

$$P(x) = -\frac{1}{1000}x^2 + \frac{193}{20}x - 1425,$$

where x is the number of these products manufactured in a year. How many of these items should the company manufacture in a year to get the maximum profit? What is the profit?

49. An agency sells tickets for an opera. Based on previous experience, this agency has determined that the profit they can make on selling x tickets is given by the function $P(x) = 16x - 0.08x^2$. What is the maximum profit they can expect to make and how many tickets do they have to sell in order to earn this maximum profit?

50. A movie theater has found that for every 10 cent increase in ticket prices, 10 fewer people will buy tickets for that movie. When tickets are priced at $4.00, the theater can expect to sell 500 tickets per day. Given that total ticket sales equals the price of the ticket times the number of tickets sold, what is the price which will bring in the maximum amount of ticket sales and what is this maximum amount of sales?

51. A farmer wishes to build a rectangular corral for some livestock. He has 160 feet of fencing. The corral will use the side of a barn for one of its sides, the other three sides will be fencing. What dimensions should the corral have in order for it to have its maximum area?

52. A farmer wishes to build a set of rectangular corrals for some livestock. These corrals will actually be one large corral divided into two equal parts by an inner fence. One side of the outer wall of the corrals will be formed by a barn, the inner fence will be perpendicular to the barn wall and will divide the fenced area into two equal sized parts. The farmer has 270 feet of fencing. What dimensions should each of the smaller corrals have in order for them to have their maximum areas?

53. A rectangular playground is to be fenced in on all four sides. There is exactly 600 feet of fencing for this playground. What dimensions give the largest area for the playground?

54. Given a parabola with vertex at (0, 2) that passes through the point (1, 8), find the equation for the parabola.

3.3 Quadratic Business Applications

Since we've now examined quadratic functions and solving quadratic equations, let's look at some examples of applications that are useful in the business world. (*Note:* For this section, we will cover similar application problems to those covered previously in section 1.5. The difference now will be in the types of functions used to represent the quantities discussed. For vocabulary and process, refer back to section 1.5.)

PROFIT, REVENUE, COST: BREAK-EVEN POINTS AND OPTIMIZATION

Remember that Profit = Revenue – Cost. To find the break-even point(s) for profit, we need to find where Profit is zero, which is equivalent to finding where Revenue equals Cost.

EXAMPLE 1

Suppose a company has total costs given by $C(x) = 120x + 900$ and total revenue given by $R(x) = x(195 - x)$, where x = the number of products produced and sold.

(a) Find the break-even point(s).

(b) What number of products should it produce and sell to yield maximum profit?

(c) What is the maximum profit?

(a) We can find the break-even points by setting cost equal to revenue, i.e., $R(x) = C(x)$.

$\Rightarrow \quad x(195 - x) = 120x + 900$	Simplify the left side of the equation.
$195x - x^2 = 120x + 900$	Now we have a quadratic equation to solve. The strategy we'll use is to move all terms to the right side of the equation with zero remaining on the left side.
$0 = x^2 - 75x + 900$	Factor the right side of the equation, if possible.
$0 = (x - 60)(x - 15)$	Set each linear factor equal to zero and solve each of those equations. This works because the only way two factors multiply to become zero is if one of them is exactly zero.
$0 = x - 60$ or $0 = x - 15$	Solve each separate equation.
$x = 60$ or $x = 15$	

So, (60, 0) and (15, 0) give the break-even points for the profit. In other words, if the company produces and sells 15 units or 60 units, they'll earn zero profit.

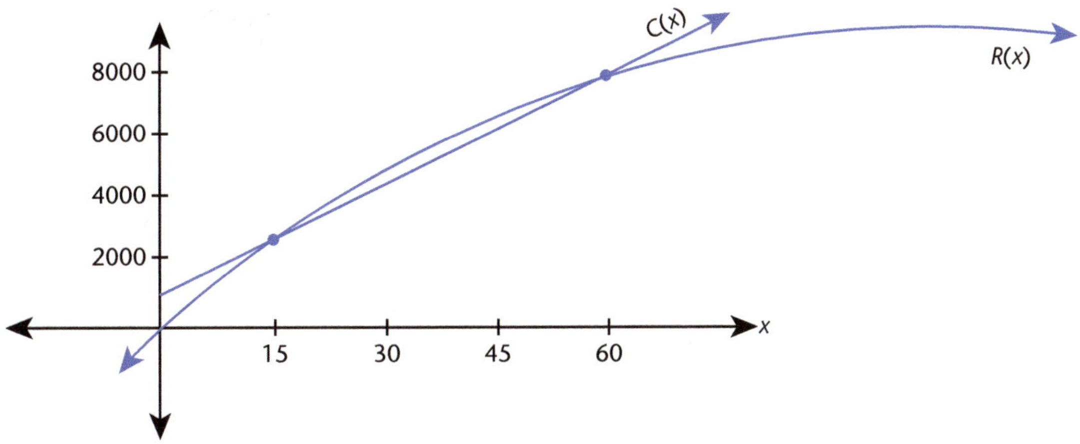

(b) First, we need to find the profit function. We know $P(x) = R(x) - C(x)$.

$$\Rightarrow \quad P(x) = x(195 - x) - (120x + 900) \qquad \text{Simplify.}$$

$$P(x) = 195x - x^2 - 120x - 900 = -x^2 + 75x - 900$$

Notice that the profit function is a quadratic function that will graph into a parabola. Since the leading coefficient, $a = -1$, is negative, the parabola is concave down. This means that the vertex is the maximum point on the graph, i.e., that there is no other point higher than the vertex on the graph of the profit function. We are looking for the value of x that produces the maximum profit, which is exactly the same as asking for the x-value of the vertex.

$$\Rightarrow \quad x = \frac{-b}{2a} = \frac{-75}{2(-1)} = \frac{-75}{-2} = 37.5$$

This says that for maximum profit, the company should produce and sell 37.5 units. Most likely, it's impossible to produce half of a unit. So, we'll need to look at producing either 37 or 38 units for maximum profit.

(c) We can find the maximum profit by substituting the x-value of the vertex into the profit function. However, remember that we really need to try 37 or 38, since we can't produce half a unit.

$$\Rightarrow \quad P(37) = -(37^2) + 75(37) - 900 = -1369 + 2775 - 900 = 506$$

and $\quad P(38) = -(38^2) + 75(38) - 900 = -1444 + 2850 - 900 = 506$

Notice that the profit at 37 and 38 units is the same amount. That's because the vertex is at $x = 37.5$, which is halfway between 37 and 38, and parabolas are symmetric about the vertical axis through the vertex. That means that the y-value (or profit value in this case) is the same at two x-values that are the same distance from the vertex on a parabola.

So, the maximum profit is \$506.

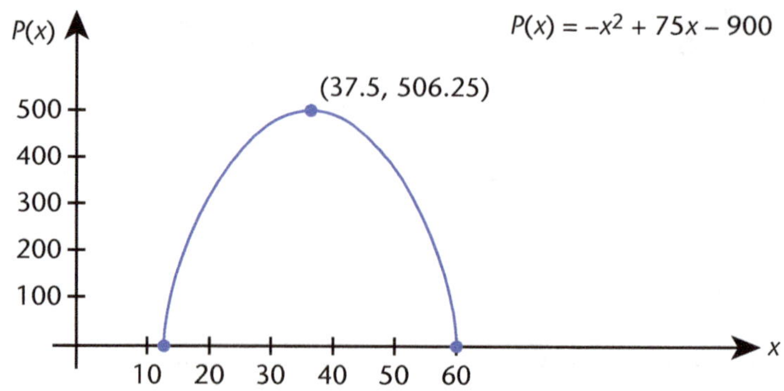

EXAMPLE 2

Jaclyn Enterprises sells personal computer services. Suppose that the company has monthly fixed costs of \$500 and variable costs of $\frac{2}{5}x + 835$ dollars per service, where x = the total number of service calls. Also, the selling price is $900 - \frac{1}{5}x$ dollars per service.

(a) How many monthly service calls does the company need to make in order to break even?

(b) How many monthly service calls will maximize revenue?

(c) How many monthly service calls will maximize profit?

(d) Should Jaclyn Enterprises focus on maximizing revenue or profit? Why?

(a) To break even, we need to either set $R(x) = C(x)$ or, equivalently, set $P(x) = 0$. The profit function, $P(x)$, is given by $R(x) - C(x)$, where $C(x)$ and $R(x)$ are given below.

$$C(x) = 500 + \left(\frac{2}{5}x + 835\right)x = 500 + \frac{2}{5}x^2 + 835x$$

This represents the total cost made up of the fixed costs, as is, plus the variable costs. The total variable costs is the "per unit" variable cost multiplied by the number of service calls, which is x.

$$\text{And, } R(x) = \left(900 - \frac{1}{5}x\right)x = 900x - \frac{1}{5}x^2.$$

This represents the total revenue, which is the "per unit" selling price times the number of service calls sold.

$$\Rightarrow \quad P(x) = R(x) - C(x) = 900x - \frac{1}{5}x^2 - \left(500 + \frac{2}{5}x^2 + 835x\right) \qquad \text{Simplify.}$$

$$P(x) = 900x - \frac{1}{5}x^2 - 500 - \frac{2}{5}x^2 - 835x = \frac{-3}{5}x^2 + 65x - 500$$

To find the number of service calls necessary to break even, we set this profit function equal to zero and solve for x.

$\frac{-3}{5}x^2 + 65x - 500 = 0$ — To eliminate the fraction, multiply both sides of the equation by –5.

$3x^2 - 325x + 2500 = 0$ — Factor the quadratic expression.

$(3x - 25)(x - 100) = 0$ — Set each factor equal to zero and solve for x.

$3x - 25 = 0 \text{ or } x - 100 = 0$

$x = \frac{25}{3} = 8.\overline{3} \text{ or } x = 100$

So, we have two break-even points, namely $(8.\overline{3}, 0)$ and $(100, 0)$. That means Jaclyn Enterprises needs to make at least 9 service calls to make a little profit; because they can't make a partial service call, we have to round the $8.\overline{3}$ up. Also, if they make 100 service calls per month, then the profit will decrease down to zero again, due to the costs overcoming the revenue.

(b) Notice that the revenue function $R(x) = 900x - \frac{1}{5}x^2$ graph is a concave down parabola. This means that the vertex is the point representing the maximum revenue.

$$\Rightarrow \text{vertex is at } x = \frac{-b}{2a} = \frac{-900}{2\left(\frac{-1}{5}\right)} = \frac{-900}{\frac{-2}{5}} = -900 \cdot \left(\frac{5}{-2}\right) = 450 \cdot 5 = 2250$$

This means that to maximize revenue, Jaclyn Enterprises needs to sell 2,250 service calls per month.

(c) We can find maximum profit similarly. Recall that $P(x) = \frac{-3}{5}x^2 + 65x - 500$ gives the profit function as computed in part (a). This is also a concave down parabola whose maximum point occurs at the vertex.

$$\Rightarrow \text{The vertex is at } x = \frac{-b}{2a} = \frac{-65}{2\left(\frac{-3}{5}\right)} = \frac{-65}{\frac{-6}{5}} = -65 \cdot \left(\frac{5}{-6}\right) = \frac{325}{6} = 54\frac{1}{6} = 54.1\overline{6}$$

Thus, the maximum profit occurs when Jaclyn Enterprises sells about 54 service calls per month.

(d) Jaclyn Enterprises will make the most money if they focus on maximizing profit rather than revenue. Remember that profit is how much money is left over after paying the costs. This is the category most companies care most about increasing. If they focus only on maximizing revenue, and ignore the associated costs, the company could have great losses instead of overall profit.

SUPPLY/DEMAND AND MARKET EQUILIBRIUM

EXAMPLE 3

If $p = q^2 + 5q + 18$ gives the supply equation for a commodity, and $p = 60 - 3q - q^2$ gives the demand equation for the same commodity, sketch both curves and find the equilibrium quantity and price.

First, let's sketch both curves on the same coordinate axes. Remember that the quantity (q) goes on the horizontal axis and price (p) goes on the vertical axis.

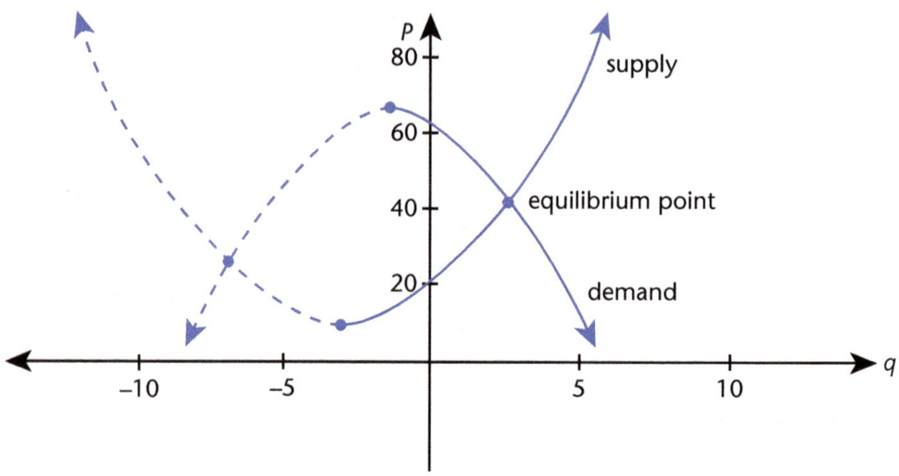

Second, we need to find the equilibrium q and p, which means the intersection point(s) of their graphs. Basically, we have a system of two equations with two variables to solve simultaneously.

1. $p = q^2 + 5q + 18$ and
2. $p = 60 - 3q - q^2$

We can substitute the expression for p in equation (1) into p in equation (2).

$\Rightarrow \quad q^2 + 5q + 18 = 60 - 3q - q^2$ — Since it's a quadratic equation, we need to move all terms to one side of the equation, leaving zero on the other side.

$2q^2 + 8q - 42 = 0$ — Factor out 2 from all terms on the left side of the equation.

$2(q^2 + 4q - 21) = 0$ — Factor the remaining quadratic expression.

$2(q+7)(q-3) = 0$ — Set each factor equal to zero and solve.

$2 = 0$ (this can't happen, so it yields no solution) or $q + 7 = 0$ or $q - 3 = 0$

$\Leftrightarrow \quad q = -7$ or $q = 3$

We can't let q be a negative quantity, so the only solution that makes sense is $q = 3$. Thus, the equilibrium quantity is 3 units of the commodity.

Next, we need to find the equilibrium price. We can do this by substituting $q = 3$ into either equation (1) or equation (2), i.e., either of the supply or demand equations.

$$\Rightarrow p = q^2 + 5q + 18 = 3^2 + 5(3) + 18 = 9 + 15 + 18 = \$42$$

The equilibrium price is $42.

EXAMPLE 4

Find the new equilibrium quantity and price for example 3 if the government places a $6 tax on the production of each commodity and that cost then gets passed through the supplier.

This tax will only change the supply equation. The consumers are still only willing to behave as the given demand equation states.

$$\Rightarrow \text{ new supply equation is } p = q^2 + 5q + 18 + 6 = q^2 + 5q + 24$$

Now, we need to solve the system of equations given by:

1. $p = q^2 + 5q + 24$
2. $p = 60 - 3q - q^2$

Again employing the same strategy as in the last example, substitute the expression in equation (1) for p into equation (2).

$q^2 + 5q + 24 = 60 - 3q - q^2$	Move all terms to the left side of equation, with zero on the right.
$2q^2 + 8q - 36 = 0$	Factor out the 2 from all terms on the left side of the equation.
$2(q^2 + 4q - 18) = 0$	Divide both sides of the equation by 2.
$q^2 + 4q - 18 = 0$	Since the quadratic expression doesn't factor, we can use the quadratic formula to solve this equation.

We have $a = 1$, $b = 4$, and $c = -18$.

$$q = \frac{-4 \pm \sqrt{4^2 - 4(1)(-18)}}{2(1)} = \frac{-4 \pm \sqrt{16 + 72}}{2} = \frac{-4 \pm \sqrt{88}}{2}$$

We know that $88 = 4(22) \Rightarrow \sqrt{88} = \sqrt{4}\sqrt{22} = 2\sqrt{22}$.

$$\Rightarrow q = \frac{-4 \pm \sqrt{88}}{2} = \frac{-4 \pm 2\sqrt{22}}{2} = \frac{2\left(-2 \pm \sqrt{22}\right)}{2} = -2 \pm \sqrt{22}$$

If we break this down into its two solutions, we'd have $q = -2 + \sqrt{22} \cong -2 + 4.69 = 2.69$ or $q = -2 - \sqrt{22} \cong -2 - 4.69 = -6.69$. However, we need to keep in mind that q represents the quantity sold, which means it must be non-negative. So we only keep the first solution, namely $q = 2.69$. The other fact to remember is that, in some circumstances, we can only produce and sell whole quantities of commodities, not a partial amount. If that is the case, then we'd need to round up to a quantity of 3. For this problem, we'll assume that a partial amount of the commodity is possible (for instance, we can sell partial amounts of stocks). Then we need to calculate the corresponding price with this quantity.

$$\Rightarrow q = q^2 + 5q + 24 \cong 2.69^2 + 5(2.69) + 24 = 44.6861$$

Since p represents a price, we need to round this answer to the nearest cent, which means the equilibrium price is \$44.69.

3.3 Homework

Find the number of units that need to be produced and sold to break even given the revenue and cost functions.

1. $R(x) = 900x - 0.1x^2$
   $C(x) = 800 - 10x + 0.25x^2$

2. $R(x) = 8x$
   $C(x) = \frac{22}{9}x + 5000$

3. $R(x) = 100\sqrt{x}$
   $C(x) = 2x + 100$

4. $R(y) = 0.1y^2 + 7y$
   $C(y) = 2y + 500$

5. $R(t) = 120t$
   $C(t) = t^2 + 40t + 1000$

6. $R(w) = -w^2 + 1600w$
   $C(w) = 1500w + 1600$

7. $R(x) = -x^2 + 1600x$
   $C(x) = 1500x + 1600$

8. $R(x) = 75x - 3x^2$
   $C(x) = 16x + 125$

Given the supply and demand equations, find the equilibrium price and quantity.

9. supply: $p = q^2 + 10$
   demand: $p = -3q^2 - 6q + 86$

10. supply: $p = q^2 + 8q + 10$
    demand: $p = -q^2 - 4q + 64$

11. supply: $p = q + 10$
    demand: $p = \frac{100}{q}$

12. supply: $p = \frac{q}{2} + 25$
    demand: $p = \frac{100}{q} + 20$

13. supply: $p = q^2 + 8q + 20$
    demand: $p = -3q^2 + 6q + 400$

14. Find the maximum profit if the revenue per watch sold is \$10 and cost is given by $C(x) = (x + 1)^2 - 499$.

15. If the demand equation is $p = 2000 - 6q$, find the maximum revenue. (*Hint:* Revenue is number of units sold times price.)

16. Given the revenue function $R(x) = 2000x - 60x^2$, a fixed cost of $4000, and variable costs of $500 per unit:

    (a) What is the profit function?
    (b) Find the maximum profit.

17. A company has daily production costs of $C(x) = 800 - 10x + 0.25x^2$. How many items produced each day yield a minimum cost?

18. A clothing company estimates their profits to behave according to $P(x) = 220 + 20x - \frac{1}{2}x^2$. If this is a correct prediction, what is the maximum profit?

19. The total costs for a company to produce and sell x units of a product are given by $C(x) = 96 + 5x + x^2$ (in dollars). The sale price for one item is $25.

    (a) Find the revenue function, $R(x)$.
    (b) Find the profit function, $P(x)$.
    (c) Find the break-even point(s).
    (d) Find the number of items sold to get the maximum profit.

20. The monthly revenue of a company is given by $R(p) = 1125p - 10p^2$ where p is the price in dollars of the product manufactured. At what price will the revenue be $25,000 if the price must be greater than $50?

21. The profit function for a firm making widgets is $P(x) = 88x - x^2 - 1200$.

    (a) Find the number of units at which maximum profit is achieved.
    (b) Find the maximum profit.

22. If the cost of production for a product is given by $C(x) = x^2 + 6x + 84$ and the revenue is given by $R(x) = 25x$,

    (a) Find the profit function $P(x)$.
    (b) Find the break-even point(s).

23. If 160 feet of fence is used to fence in a rectangular yard, then the resulting area is given by

    $$A(x) = x(80 - x)$$

    where x feet is the width of the rectangle and $(80 - x)$ feet is the length. Determine the width and length that give the maximum area.

24. Mr. Jones makes and sells bicycles. His cost function is given by $C(x) = 30x + 500$. His revenue function is described by $R(x) = -x^2 + 90x$.

    (a) How many bicycles should Mr. Jones make and sell to maximize profit?
    (b) What is Mr. Jones' maximum profit?
    (c) How many bicycles must he sell to break even?

25. If total costs are $C(x) = 1600 + 1500x$ and total revenue is given by $R(x) = 1600x - x^2$, how many units must be sold to break even?

26. If the demand function for a commodity is given by $p = 100 - 4q - q^2$ and the supply function is given by $p = q^2 + 8q + 20$, find the equilibrium quantity and price.

27. Given $C(x) = 2500 + 45x + 0.3x^2$ and $R(x) = 255x - 0.7x^2$, find the level of production that maximizes profit, knowing that x represents how many units are produced. What is the maximum profit?

3.4 Polynomial Functions

VOCABULARY

Degree—The degree of a polynomial is the highest exponent on the variable.

Descending Order—Typically we write polynomials in descending order, meaning that the terms are ordered from highest degree to lowest degree terms.

Coefficient—The number multiplied by the variable expression.

Leading Coefficient—The number multiplied by the highest degree term in the polynomial.

Zeros (or Roots)—The zeros or roots of a function are the input values (usually called x-values) that make the function zero.

> **POLYNOMIAL FUNCTION**
>
> A polynomial function is a function that can be written in the form of
>
> $$f(x) = a_n x^n + a_{n-1}x^{n-1} + a_{n-2}x^{n-2} + \dots + a_2x^2 + a_1x + a_0$$
>
> where $a_i \in \mathbb{R}$ for $i = 0, 1, 2, \dots, n$, $a_n \neq 0$, and $n \in \mathbb{Z}$, $n > 0$.
>
> (Literally a many-termed function.)

We have actually already discussed two particular types of polynomial functions, namely linear and quadratic polynomials, i.e., something like $f(x) = y = 4x + 7$ or $f(x) = y = -2x^2 + x - 5$ respectively.

Here are some general graphs of polynomial functions. We can see how the degree of the polynomial dominates its shape. The degree tells the maximum number of directions the graph can go. For example, a linear function, degree = 1, only has one direction. A quadratic function, degree = 2, graphs into a parabola that has two directions, namely it either goes up and then down or down and then up. With this logic, we expect a cubic polynomial, degree = 3, to have three directions, and so on. In general, that's true, but for polynomial functions of degree 3 or higher, we can choose the coefficients to "flatten" out the "humps" of the graph and thereby make the graph look like it has fewer directions.

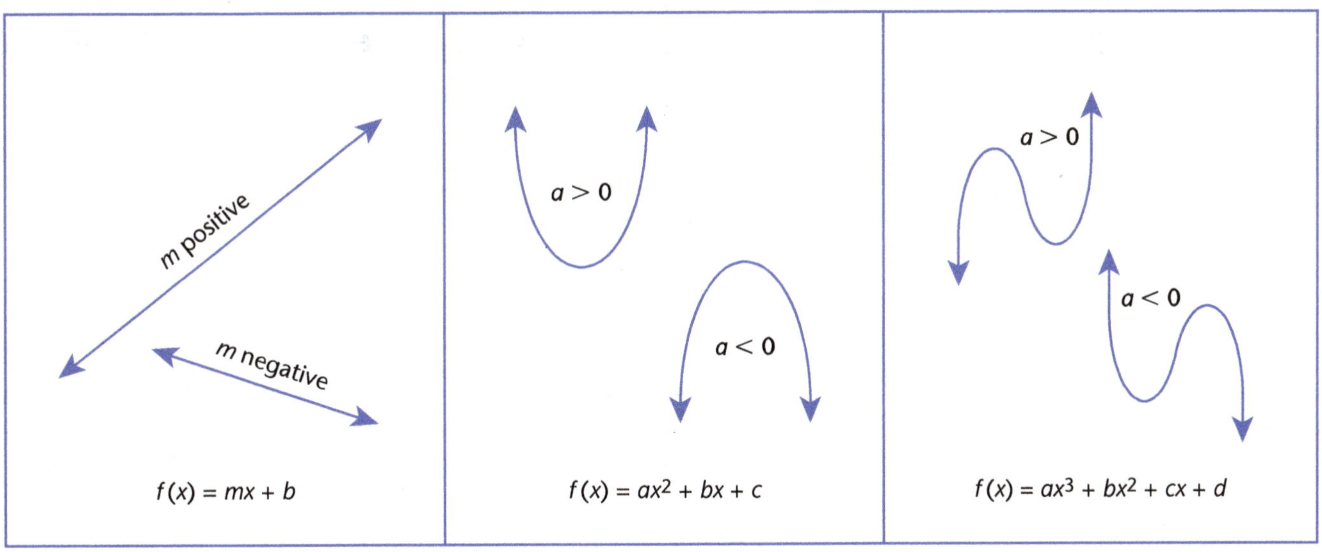

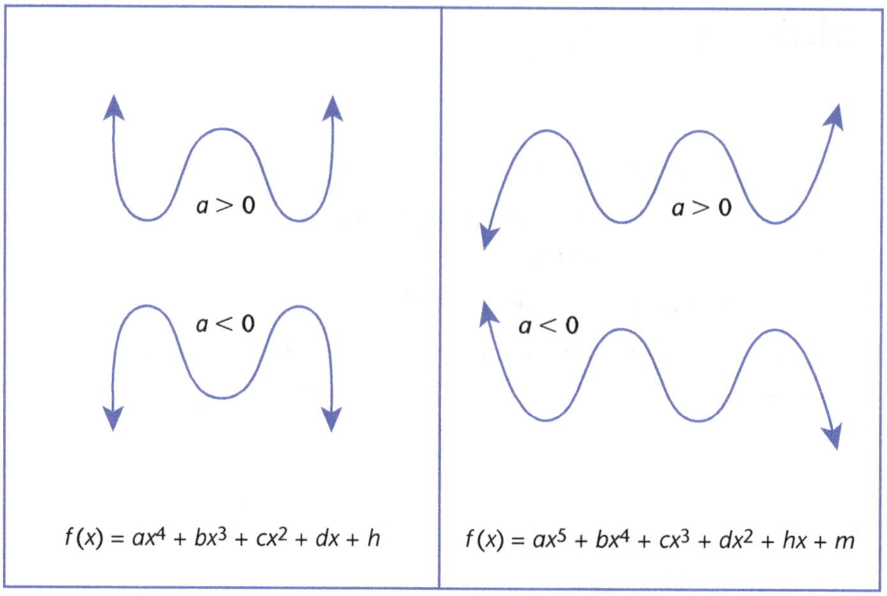

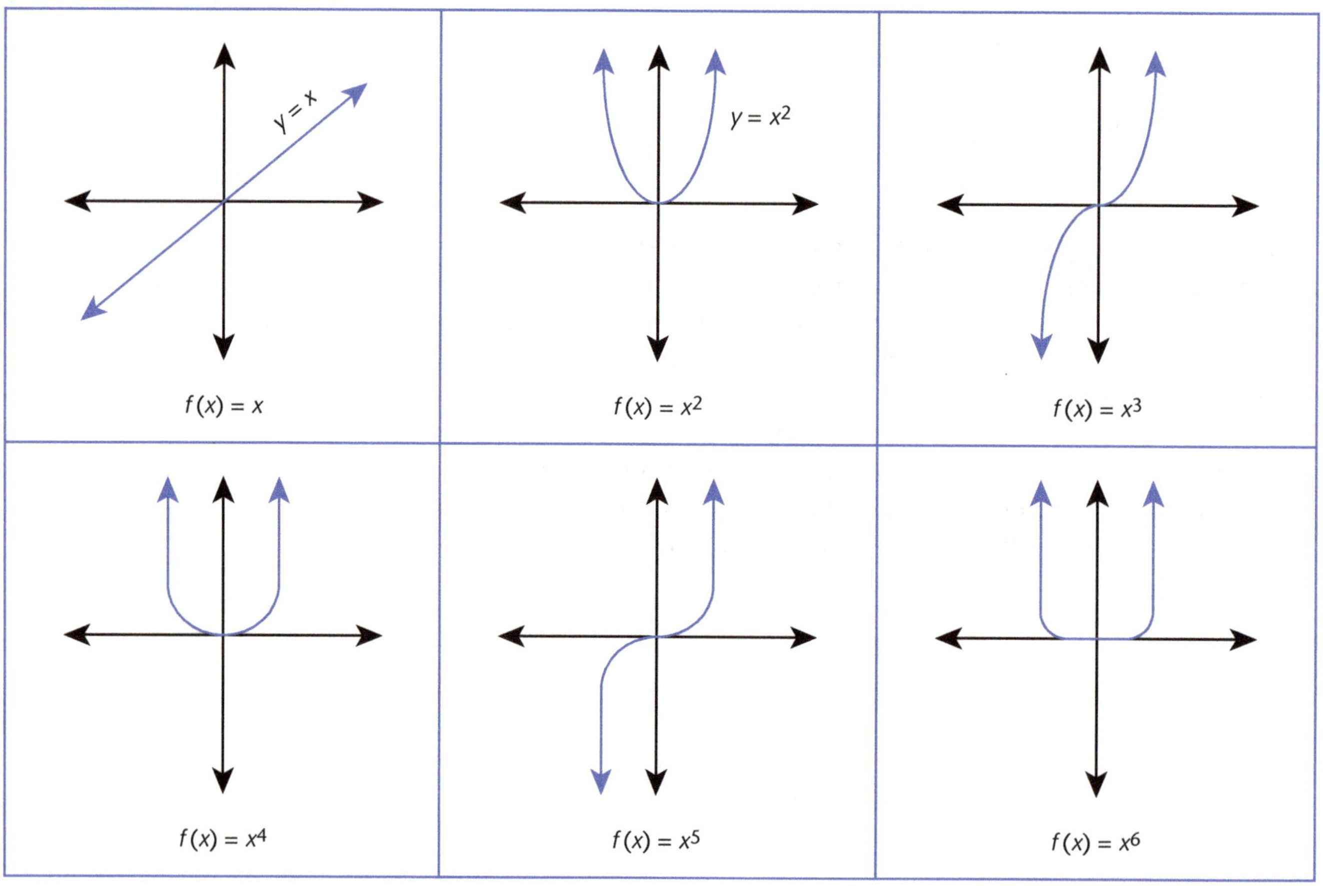

Notice that even degree polynomials eventually either go up or down forever on both sides. And odd degree polynomials eventually go up forever on one side and down forever on the other side of the graph. Both of these phenomena are dependent on the leading coefficient. This table describes the effects.

	POSITIVE LEADING COEFFICIENT	NEGATIVE LEADING COEFFICIENT
Odd degree polynomial	Goes up on right and down on left side of graph	Goes down on right and up on left side of graph
Even degree polynomial	Goes up on both sides of graph	Goes down on both sides of graph

EXAMPLE 1

Answer the following questions for the polynomial function $f(x) = -1 + 3x^3 + x - 6x^5 + 5x^2$.

(a) Rewrite the polynomial function in descending order.

(b) What is the degree?

(c) What is the leading coefficient?

(d) Does this graph go up or down on the right side?

(a) We need to rearrange the terms so that the highest degree term comes first, then the next highest degree term, etc.

$$\Rightarrow f(x) = -6x^5 + 3x^3 + 5x^2 + x - 1$$

(b) The highest exponent on x is 5, which means the degree of this polynomial is 5.

(c) The leading coefficient is the number being multiplied by the highest degree term, which in this case is –6.

(d) Because it's an odd degree polynomial with a negative leading coefficient, its graph will go down forever on the right side and up forever on the left side.

EXAMPLE 2

Answer the following questions for the polynomial function $f(x) = 2x^3 - 4x^2 - 6x$.

(a) What is the degree?

(b) What is the leading coefficient?

(c) Find all the zeros of the function.

(d) What are the x-intercepts, if any?

(e) What is the y-intercept?

(f) Sketch the graph.

(a) The degree is 3 since that is the highest exponent on x.

(b) The leading coefficient is 2.

(c) To find the zeros of the function, we need to substitute 0 in for the function value and solve for x.

$0 = 2x^3 - 4x^2 - 6x$ To solve this equation, we must factor the polynomial completely.

$0 = 2x(x^2 - 2x - 3)$

$0 = 2x(x - 3)(x + 1)$

Remember (from previously solving quadratic equations) that the only way to multiply several factors together to produce zero is if one of those factors is exactly zero. Thus, now we need to set each of these factors on the right side of the equation equal to zero separately and to solve for x.

$$2x = 0 \text{ or } x - 3 = 0 \text{ or } x + 1 = 0$$

$$x = 0 \text{ or } x = 3 \text{ or } x = -1$$

(d) Now that we have the zeros, or roots, of the polynomial function, we also know the x-intercepts. The roots, or zeros, of the function are the x-values where the function is equal to zero. Graphically, the zeros are the x-values of the x-intercepts for the graph of the function.

$\Rightarrow$ x-intercepts are (0, 0), (3, 0), and (–1, 0)

(e) To find the y-intercept, we need to substitute 0 in for x and simplify to get the y-value.

$\Rightarrow$ $f(0) = 2(0^3) - 4(0^2) - 6(0) = 0$

$\Rightarrow$ The y-intercept is the point (0, 0).

(f) Let's put all of this information together to sketch the graph of this polynomial function.

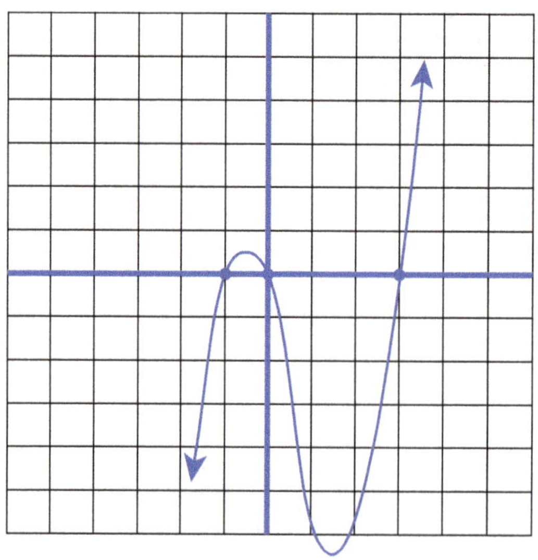

PIECEWISE FUNCTIONS

A piecewise function is a function that we define in pieces. That means that its graph may or may not be a smooth and/or connected curve. We define each piece of the graph for a certain interval of the input variable. For a piecewise function, we have to graph each piece separately.

EXAMPLE 3

Given the piecewise function below, answer the following questions.

$$f(x) = \begin{cases} x^2 + 4, \ x \geq 1 \\ -x + 3, \ -2 \leq x < 1 \\ -(x+3)^2, \ x < -2 \end{cases}$$

(a) Fill in the following table with function values.

x	$f(x)$
0	
1	
2	
−1	
−2	
−3	
−4	
3	

(b) Graph the function.

(a) We'll fill in the table of function (output) values. For each value of x, we need to choose which of the three pieces to use for evaluating the output value. The piecewise function definition will always tell you which piece to use. We just need to pay attention to the conditions on x for each piece.

x	$f(x)$	Which piece of the function do we choose?
0	$f(0) = -0 + 3 = 3$	Choose the middle piece since 0 is in between −2 and 1.
1	$f(1) = 1^2 + 4 = 1 + 4 = 5$	Choose the top piece of the function because we want to evaluate at $x = 1$, and that is the only piece that allows that.
2	$f(2) = 2^2 + 4 = 4 + 4 = 8$	Choose the top piece of the function again since $x = 2$ satisfies that condition; namely, it's greater than 1.

x	$f(x)$	Which piece of the function do we choose?
–1	$f(-1) = -(-1) + 3 = 4$	Choose the middle piece because $x = -1$ is in between –2 and 1.
–2	$f(-2) = -(-2) + 3 = 5$	Choose the middle piece because that's the only piece that allows x to be exactly –2.
–3	$f(-3) = -(-3 + 3)^2 = -0^2 = 0$	Choose the bottom piece because $x = -3$ meets the condition that x be less than –2.
–4	$f(-4) = -(-4 + 3)^2 = -(-1)^2 = -1$	Choose the bottom piece because –4 is definitely less than –2.
3	$f(3) = 3^2 + 4 = 9 + 4 = 13$	Choose the top piece because $x = 3$ is bigger than 1.

(b) We'll plot each of these points on our coordinate axes. Then we need to use the knowledge we have about the shape of each piece to fill in the rest of the graph.

$$f(x) = \begin{cases} x^2 + 4, \ x \geq 1 \\ -x + 3, \ -2 \leq x < 1 \\ -(x+3)^2, \ x < -2 \end{cases}$$

We know the top piece (for $x \geq 1$) is a concave up parabola, but we only graph the portion of the parabola that satisfies $x \geq 1$.

The middle piece (for $-2 \leq x < 1$) is a line with a slope of –1 and a y-intercept (0, 3). Again, we'll only graph the piece of the line where $-2 \leq x < 1$, including the endpoint at $x = -2$ and not including the endpoint at $x = 1$.

The bottom piece (for $x < -2$) is a concave down parabola. We'll only graph the portion of the parabola that satisfies $x < -2$, and we don't include the endpoint since x is not allowed to actually equal –2 for this piece of the curve.

As x gets close to 1 in "middle" piece, the line approaches (1, 2).

As x gets close to –2 in "bottom" piece, the parabola approaches (–2, –1).

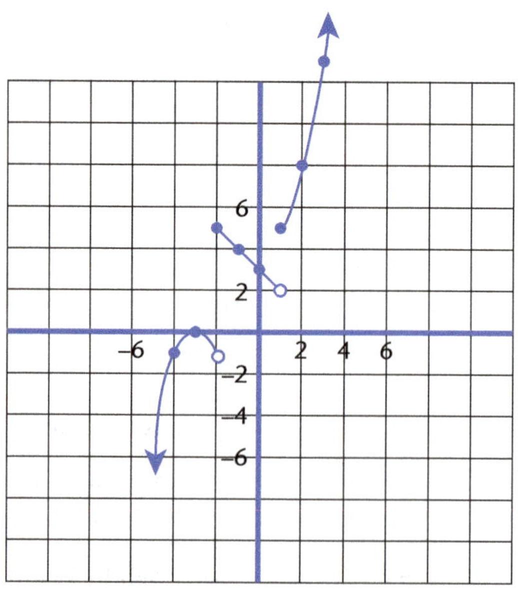

EXAMPLE 4

For this piecewise function, fill in the table of points and sketch the graph.

$$f(x) = \begin{cases} 5x - 6, \ x > 3 \\ x^2, \ 0 < x \leq 3 \\ 5, \ x \leq 0 \end{cases}$$

x	$f(x)$
0	$f(0) = 5$
1	$f(1) = 1^2 = 1$
3	$f(3) = 3^2 = 9$
4	$f(4) = 5(4) - 6 = 20 - 6 = 14$
2.9	$f(2.9) = 2.9^2 = 8.41$
–1	$f(-1) = 5$
–2	$f(-2) = 5$
0.01	$f(0.01) = 0.01^2 = 0.0001$

We know the top piece of the graph (for $x > 3$) is a line with slope 5 and y-intercept (0, –6). But we'll only graph the portion of the line where $x > 3$.

The middle piece of the graph is the basic parabola, but again, we only want to graph the piece of that parabola that satisfies the condition $0 < x \leq 3$. We'll include the endpoint at $x = 3$ but not include the endpoint at $x = 0$.

The bottom piece of the curve is the line $y = 5$, which is a horizontal line. We only want the piece of the line when $x \leq 0$.

As x gets close to 3 in the "top" piece of the function, the line approaches the point (3, 9). So the line and the parabola meet there.

As x gets close to 0 in the "middle" piece, the parabola approaches the origin, (0, 0).

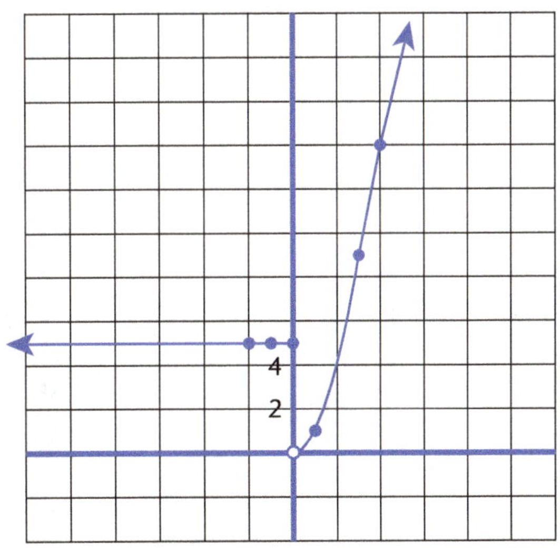

3.4 Homework

For each polynomial function, answer the following questions:

(a) *Rewrite it in standard (descending) order.*
(b) *Find its degree.*
(c) *What is the leading coefficient?*
(d) *Does its graph go up or down on the right?*

1. $f(x) = x^3 - 3x^2 + 1 + x$
2. $f(x) = 40 - 2x^2 + 12x$
3. $f(p) = -4p^2 + 11 - p^3 + p$
4. $f(y) = 12y^2 - 15y - 3y^3 + 12$
5. $f(x) = -4x^2 + 12 - 3x + 12x^3$
6. $f(x) = 4x - 12 - 2x^3 - x^2$
7. $f(x) = x^7 - 3x^9 + x^4 - 3x^2 + 11$
8. $f(y) = -27y + y^2 - 31y^5$
9. $f(x) = 14x^4 - 19x^3 + 20x^6 - 31$
10. $f(x) = 3x^7 - 14x + 3x^2 - 4x^4 - 5$

For each polynomial function, find the following:

(a) *Its degree.*
(b) *All the zeros (or roots).*
(c) *The y-intercept.*
(d) *The x-intercept(s).*
(e) *Sketch the graph.*

11. $f(x) = 2x^2 - 7x - 4$
12. $g(x) = 2x^2 + 5x - 3$
13. $h(x) = x^3 - 3x^2 - 10x$
14. $y(x) = 3x^2 - 3x - 6$
15. $f(x) = x^3 - x$
16. $f(x) = x^4 - 8x^2 + 16$
17. $f(x) = 2x^3 - 2x^2 - 4x$
18. $f(x) = 2(x - 1)^4$
19. $g(x) = x(x + 3)(x - 2)$
20. $y(x) = 5x^2 - 80$

For each piecewise function, fill in the table of points and sketch the graph.

21. $f(x) = \begin{cases} 2x+1, & x \geq 0 \\ x^2, & x < 0 \end{cases}$

x	0	1	−1	2	−2
y					

22. $g(x) = \begin{cases} x+5, & x \geq 1 \\ -2x+8, & x < 1 \end{cases}$

x	1	0	−1	2	3
y					

23. $h(x) = \begin{cases} x+1, & x \geq 2 \\ x^2-1, & 0 \leq x < 2 \\ -x+5, & x < 0 \end{cases}$

x	−1	0	1	2	3	4
y						

24. $y(x) = \begin{cases} x^2, & x \geq 2 \\ -x^2+x+6, & -1 \leq x < 2 \\ -x-1, & x < -1 \end{cases}$

x	−2	−1	0	1	2	3
y						

25. $p(x) = \begin{cases} 2x-1, & x \geq 0 \\ x+5, & -3 \leq x < 0 \\ -x, & x < -3 \end{cases}$

x	−4	−3	−1	0	1	2
y						

26. $f(x) = \begin{cases} -2x, & x \geq 1 \\ x^2, & -1 \leq x < 1 \\ x+3, & x < -1 \end{cases}$

x	−2	−1	0	1	2	3
y						

27. $f(x) = \begin{cases} \sqrt{x-2}, & x \geq 2 \\ -x, & 0 \leq x < 2 \\ 2x+1, & x < 0 \end{cases}$

x	−1	0	1	2	3	6
y						

28. $h(x) = \begin{cases} 4, & x \geq 3 \\ |x|, & -3 \leq x < 3 \\ -1, & x < -3 \end{cases}$

x	−4	−3	0	1	3	4
y						

29. $r(x) = \begin{cases} (x-1)^2-1, & x \geq 1 \\ -x, & -1 \leq x < 1 \\ -(x+1)^2+1, & x < -1 \end{cases}$

x	−2	−1	0	1	2	3
y						

3.5 Rational Functions

RATIONAL FUNCTION

A **rational function** is a ratio (quotient) of two polynomials. That is, if $f(x)$ is a rational function, it can be written as

$$f(x) = \frac{n(x)}{d(x)}$$

where $n(x)$ and $d(x)$ are both polynomials.

VOCABULARY

Domain—The set of allowable inputs. For a rational function, the domain will be all real numbers except for all input values that would make the denominator equal to zero.

Asymptotes for a Rational Function

VERTICAL ASYMPTOTES	HORIZONTAL ASYMPTOTES
▪ The line(s) $x = a$, such that $d(a) = 0$. ▪ The graph can NEVER touch or cross the vertical asymptotes. ▪ The graph will always approach (get extremely close to) the vertical asymptotes. ▪ We find the vertical asymptotes by looking for the disallowed values of x (from the domain). ▪ Vertical asymptotes are RESTRICTIONS, telling us what values of x are not allowed.	▪ The line $y = b$ such that b is the limiting value of f as x gets outrageously huge (in either direction). ▪ The graph can touch or cross the horizontal asymptote many times. ▪ The graph will always approach, but not touch, the horizontal asymptotes as x (the input variable) gets very large in either direction. ▪ We find the horizontal asymptotes by looking at the left and right end behavior of the rational function. ▪ Horizontal asymptotes are DESCRIPTIONS of end behavior.

TO FIND THE HORIZONTAL ASYMPTOTES OF A RATIONAL FUNCTION

1. If the degree of $n(x) >$ degree of $d(x)$, then there is no horizontal asymptote.
2. If the degree of $n(x) <$ degree of $d(x)$, then the horizontal asymptote is the line $y = 0$.
3. If the degree of $n(x) =$ degree of $d(x)$, then the horizontal asymptote is the line $y = \frac{a}{b}$, where $a =$ leading coefficient of $n(x)$ and $b =$ leading coefficient of $d(x)$.

Remember that the horizontal asymptote describes the behavior of the rational function's graph as x gets outrageously large, either positively or negatively. Since the numerator and denominator are both polynomials, then as x gets large, the highest degree term dominates the value of both polynomials. Thus, if the degree of the numerator is larger than the degree of the denominator polynomial, then it means that as x gets large, the numerator grows larger much faster than the denominator, which means that the quotient of the two polynomials grows infinitely large eventually. Thus, in that case, there would be no horizontal asymptote because the output, y, is not approaching anything finite. We will use this same logic to find the horizontal asymptotes for all rational functions.

EXAMPLE 1

Find the vertical and horizontal asymptotes for the following rational functions.

(a) $f(x) = \dfrac{2x^2 - x + 5}{x + 1}$

The vertical asymptote (VA) will be the line representing the x-value that is not allowed. For this rational function, the only x-value not allowed is -1 because if $x = -1$, the denominator would be zero, and it's impossible to divide by zero.

$$\Rightarrow \text{VA: } x = -1$$

The horizontal asymptote (HA) describes the behavior of the function value (or y-value) as x gets really huge. For this function, the dominating term in the numerator is $2x^2$ because it's the highest degree term for that polynomial. The reason we need to look only at the highest degree term is because, as x gets large, the highest degree term grows the fastest. The other terms eventually become inconsequential compared to the highest degree term. We can liken it to money—if you have \$1,000,000, and then you add or subtract \$50 to or from that, it doesn't make a big dent in the overall value. In the denominator, the dominating term is x (the highest degree term in that polynomial).

Thus the function, as x gets huge, looks and behaves like

$$\frac{2x^2}{x} = 2x \to \infty \text{ as } x \to \infty$$

That is, the function goes off to infinity eventually, as x gets really huge. This means there is no horizontal asymptote because y is not approaching a finite number.

$$\Rightarrow \text{HA: none}$$

(b) $f(x) = \dfrac{3x - 1}{x + 4}$

The vertical asymptote in this case will be $x = -4$ because that is the value of x we cannot allow in the denominator as it would make the denominator equal zero, which is never allowed.

$$\Rightarrow \text{VA: } x = -4$$

Let's examine the behavior of the function as x gets huge in order to determine the horizontal asymptote. As x gets ridiculously large, the dominating term in the numerator polynomial is $3x$ and the dominating term in the denominator polynomial is x. Thus, as $x \to \infty$, the function behaves like

$$y = f(x) = \frac{3x}{x} = 3.$$

In other words, the horizontal asymptote will be the line y = the ratio of the numerator leading coefficient to the denominator leading coefficient, because the degree of both polynomials is the same in the numerator and denominator.

$$\Rightarrow \text{HA: } y = 3$$

(c) $f(x) = \dfrac{2x + 3}{x^2 - 16}$

To find the vertical asymptotes for this function, we need to factor the denominator to see exactly what x-values would make it go to zero.

$$f(x) = \frac{2x + 3}{x^2 - 16} = \frac{2x + 3}{(x - 4)(x + 4)}$$

So, the disallowed x-values are 4 and –4.

$$\Rightarrow \text{VA: } x = 4 \text{ and } x = -4$$

To find the horizontal asymptote, we look at what happens to the function as x gets very large.

As $x \to \infty$ the function behaves like $y = f(x) = \dfrac{2x}{x^2}$ which reduces to $y = \dfrac{2}{x} \to 0$.

Notice that in this case, the degree of the numerator polynomial is smaller than the degree of the denominator polynomial. This means that the denominator will grow huge much faster than the numerator, as x continues to get larger and larger. As a result, the fraction continues to get smaller and smaller, until it goes to zero.

$$\Rightarrow \text{HA: } y = 0$$

TO ANALYZE AND GRAPH A RATIONAL FUNCTION

1. Find the domain and any vertical and horizontal asymptotes.
2. Find the y-intercept.
3. Find the x-intercept(s).
4. Plot all the intercepts and at least one point on both sides of the vertical asymptote(s).
5. Fill in the graph with smooth curves that definitely approach all the asymptotes.

EXAMPLE 2

Analyze and graph the function $f(x) = \dfrac{x^2 + 2x - 15}{x^2 - 4}$ by answering the following questions.

(a) Find the domain.

(b) Find the vertical asymptote(s), if it exists.

(c) Find the horizontal asymptote, if it exists.

(d) Find the *y*-intercept.

(e) Find the *x*-intercept(s).

(f) Sketch the graph.

Notice first that we can factor out the numerator and denominator of the function.

$$f(x) = \frac{x^2 + 2x - 15}{x^2 - 4} = \frac{(x+5)(x-3)}{(x-2)(x+2)}$$

(a) The domain is the set of all allowable inputs, or x-values. In this case, we can let x be any number except whatever x-value will make the denominator equal zero. From the factored form of the function, we can see that the denominator will be zero if $x = 2$ or if $x = -2$. Thus, the domain of this function is $x \in \mathbb{R}$, $x \neq \pm 2$ (i.e., x can be any real number except 2 or –2).

(b) The vertical asymptotes of a rational function are easy to find once we've established the domain. It's exactly the vertical lines $x = 2$ and $x = -2$, that is, the values of x we had to exclude in the domain.

(c) We find the horizontal asymptotes by considering what happens to the function values as we allow x to get outrageously large (either positively or negatively). In this case, as x approaches infinity, the dominating term in the numerator is x^2 and the dominating term in the denominator is also x^2. So, we have as $x \to \pm\infty$, $y = f(x)$ behaves like $\frac{x^2}{x^2} = 1$. This means that the horizontal asymptote is the line $y = 1$.

(d) The y-intercept is the point where the graph crosses the y-axis, which means the x-value is zero. To find the y-intercept, we need to substitute 0 in for x in the function.

$$y = f(0) = \frac{0^2 + 2(0) - 15}{0^2 - 4} = \frac{-15}{-4} = \frac{15}{4} \text{ or } 3\frac{3}{4}$$

$$\Rightarrow \text{The } y\text{-intercept is the point } \left(0, \frac{15}{4}\right)$$

(e) The x-intercepts are those points where the graph crosses the x-axis, which means the y-value is zero. To find the x-intercepts, we need to substitute 0 in for y this time.

$$0 = \frac{x^2 + 2x - 15}{x^2 - 4} = \frac{(x+5)(x-3)}{(x-2)(x+2)}$$

To solve for x, use the factored form. Multiply both sides of the equation by the denominator on the right side.

$$(x-2)(x+2)0 = \frac{(x+5)(x-3)}{(x-2)(x+2)}(x-2)(x+2)$$

Simplify.

$$0 = (x+5)(x-3)$$

Set each factor equal to zero and solve for x.

$$0 = (x+5) \text{ or } 0 = (x-3)$$

$$x = -5\ ,\ 3$$

⇒ The x-intercepts are the points (–5, 0) and (3, 0).

(f) Now we get to put all the information together into the graph of the function. We've plotted the x-intercepts and y-intercept. We draw the vertical and horizontal asymptotes as dotted lines. Notice that the graph approaches the vertical asymptotes on either side, and also the horizontal asymptote, as x gets huge. We need another couple of points in the middle region between the two vertical asymptotes to see whether the graph is going up, down, or up on one side and down on the other side of the y-axis. The regions to the left and the right of the vertical asymptotes don't really need any more points. We know the graph has to approach the vertical and horizontal asymptotes, so we draw in smooth curves to represent that fact.

Another couple of points:

x	$y = f(x)$
1	$y = f(1) = \frac{1^2 + 2(1) - 15}{1^2 - 4} = \frac{-12}{-3} = 4$
–1	$y = f(-1) = \frac{(-1)^2 + 2(-1) - 15}{(-1)^2 - 4} = \frac{-16}{-3} = \frac{16}{3} \text{ or } 5\frac{1}{3}$

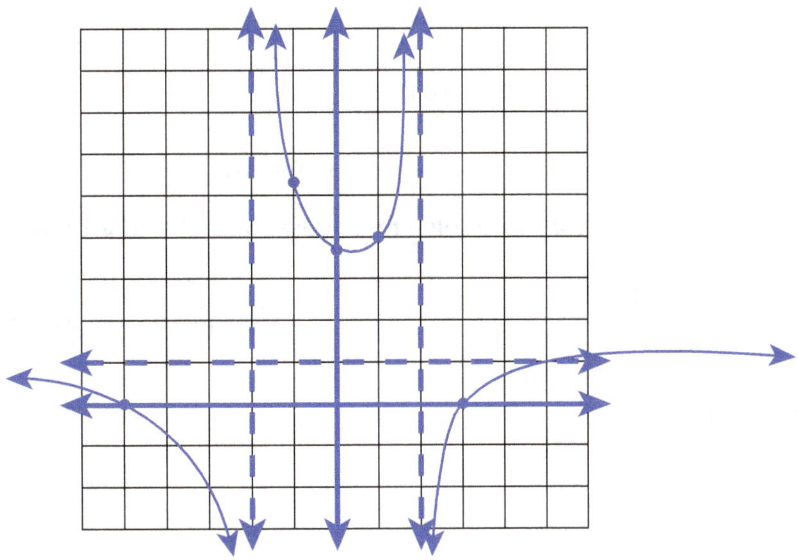

EXAMPLE 3

Analyze and graph the function $g(x) = \dfrac{x^2 + x - 6}{x - 3}$.

Let's first write the function in its factored form as well. For some questions, the original form of the function, as given, is better to work with, and for other questions, the factored form is much faster to use. Thus, having both forms is useful.

$$g(x) = \frac{x^2 + x - 6}{x - 3} = \frac{(x-2)(x+3)}{(x-3)}$$

Domain: All the allowable x-values. The only x-value not allowed is 3 because it would make the denominator equal to zero.

$\Rightarrow$ Domain: $x \in \mathbb{R}, x \neq 3$ (Read "x can be any real number except 3.")

VA: Now that we have the domain, the vertical asymptote is quick to find.

$\Rightarrow$ VA: $x = 3$

HA: As x gets huge, the dominating term in the numerator is x^2, and the dominating term in the denominator is x. So, the end behavior of the function is

$$g(x) = \frac{x^2}{x} = x \text{ which goes to } \pm\infty \text{ as } x \to \pm\infty$$

$\Rightarrow$ HA: none

y-intercept: Substitute zero in for x.

$$y = g(0) = \frac{0^2 + 0 - 6}{0 - 3} = \frac{-6}{-3} = 2$$

$\Rightarrow$ *y*-intercept: (0, 2)

x-intercept(s): Substitute zero in for y and solve for x. (Use the factored form of the function for this question.)

$$0 = \frac{(x-2)(x+3)}{(x-3)}$$ Multiply both sides of the equation by $(x - 3)$.

$$0 = (x-2)(x+3)$$ Set each linear factor equal to zero and solve for x.

$$0 = x - 2 \text{ or } 0 = x + 3 \Leftrightarrow x = 2 \text{ or } x = -3$$

$\Rightarrow$ *x*-intercepts: (2, 0) and (–3, 0)

Another couple of points: So far, we have three points, namely the *x*-intercepts and the *y*-intercept that all lie to the left of the vertical asymptote of $x = 3$. We need another couple of points on the right side of the vertical asymptote.

x	y
4	$y = g(4) = \frac{4^2 + 4 - 6}{4 - 3} = \frac{16 + 4 - 6}{1} = 14$
5	$y = g(5) = \frac{5^2 + 5 - 6}{5 - 3} = \frac{25 + 5 - 6}{2} = \frac{24}{2} = 12$
6	$y = g(6) = \frac{6^2 + 6 - 6}{6 - 3} = \frac{36 + 6 - 6}{3} = \frac{36}{3} = 12$

$\Rightarrow$ Some other points are (4, 14), (5, 12) and (6, 12).

Sketch graph: We plot all the points we know are on the graph, along with the vertical asymptote. We know as $x \to -\infty$, $y \to -\infty$, and as $x \to \infty$, $y \to \infty$, which we discovered when looking for the (nonexistent) horizontal asymptote. In other words, on the far left side of the graph the *y*-value will continue to decrease (go down) forever, and on the far right side of the graph, the *y*-value will continue to increase (go up) forever. With all this information, we have a fairly accurate idea of the shape of the graph.

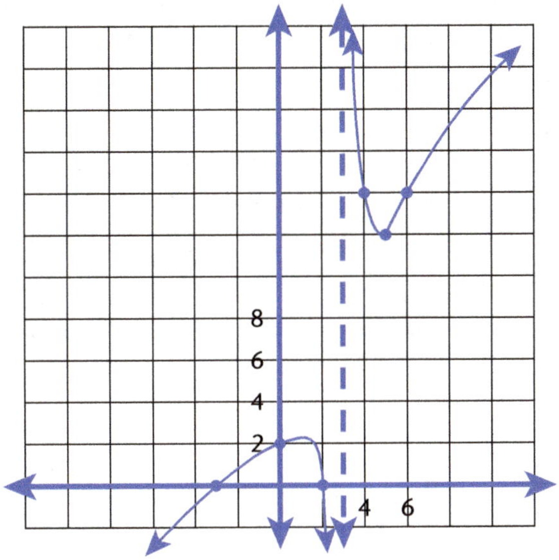

EXAMPLE 4

Analyze and graph the function $h(x) = \dfrac{x+5}{x^2 - 3x - 10}$.

Factored form of the function: $h(x) = \dfrac{x+5}{x^2-3x-10} = \dfrac{x+5}{(x-5)(x+2)}$

Domain: $x \in \mathbb{R}, x \neq 5, -2$

VA: $x = 5$ and $x = -2$

HA: As $x \to \pm\infty, y = h(x) \to \dfrac{x}{x^2} = \dfrac{1}{x} \to 0 \Rightarrow y = 0$ is the horizontal asymptote.

y-intercept: $y = h(0) = \dfrac{0+5}{0^2 - 3(0) - 10} = \dfrac{5}{-10} = \dfrac{-1}{2} \Rightarrow$ y-intercept is $\left(0, -\dfrac{1}{2}\right)$

x-intercept(s):

$$0 = \frac{x+5}{(x-5)(x+2)}$$

Multiply by the right-side denominator on both sides of the equation.

$$0 = x + 5 \Leftrightarrow x = -5$$

$\Rightarrow$ x-intercept is $(-5, 0)$

More points: So far we have vertical asymptotes at $x = 5$ and $x = -2$. We have the x-intercept, which is a point in the left-most region of the plane. We have the y-intercept, which lands in the middle region. We need a couple more points in each of the regions.

x	y
–3	$y = h(-3) = \dfrac{-3+5}{(-3)^2 - 3(-3) - 10} = \dfrac{2}{8} = \dfrac{1}{4}$
–6	$y = h(-6) = \dfrac{-6+5}{(-6)^2 - 3(-6) - 10} = \dfrac{-1}{44}$
–1	$y = h(-1) = \dfrac{-1+5}{(-1)^2 - 3(-1) - 10} = \dfrac{4}{-6} = \dfrac{-2}{3}$
4	$y = h(4) = \dfrac{4+5}{4^2 - 3(4) - 10} = \dfrac{9}{-6} = \dfrac{-3}{2}$
6	$y = h(6) = \dfrac{6+5}{6^2 - 3(6) - 10} = \dfrac{11}{8}$
7	$y = h(7) = \dfrac{7+5}{7^2 - 3(7) - 10} = \dfrac{12}{18} = \dfrac{2}{3}$

Sketch the graph:

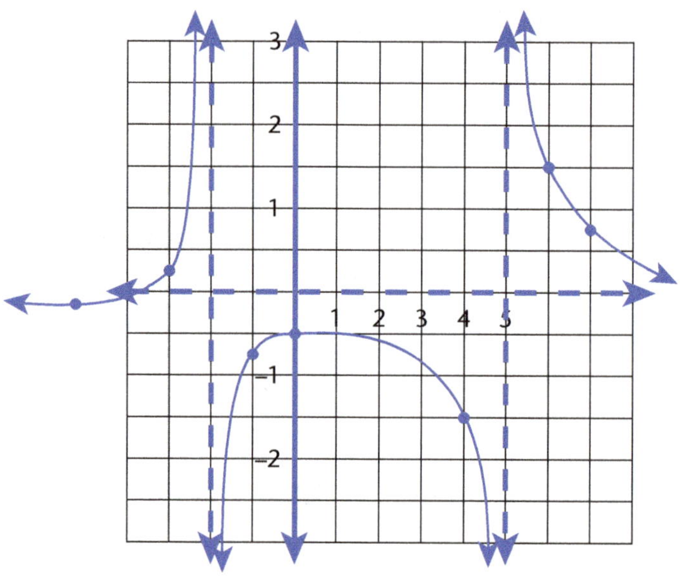

3.5 Homework

Find the vertical and horizontal asymptotes for these functions.

1. $f(x) = \dfrac{x}{x^2 - x - 2}$

2. $f(x) = \dfrac{2x - 1}{x}$

3. $f(x) = \dfrac{3}{x - 2}$

4. $g(x) = \dfrac{x^2 - 4}{x^2 + 1}$

5. $h(x) = \dfrac{2(x^2 - 9)}{x^2 - 4}$

6. $f(x) = \dfrac{1 - 3x}{3x + 1}$

7. $g(x) = \dfrac{x^2}{x^2 + 9}$

8. $h(x) = \dfrac{x + 5}{x^2 + 4x + 4}$

9. $g(x) = \dfrac{3x - 12}{x^2 - 1}$

10. $f(x) = \dfrac{2 + x}{1 - x}$

Analyze each graph by finding all the following information and then sketch the graph:

(a) *Domain*
(b) *Vertical asymptote(s), if any exist*
(c) *Horizontal asymptote, if any exist*
(d) *y-intercept, if any exist*
(e) *x-intercept(s), if any exist*

11. $f(x) = \dfrac{2}{(x + 2)^2}$

12. $f(x) = \dfrac{1}{x - 1}$

13. $g(x) = \dfrac{3x + 2}{2x - 1}$

14. $f(x) = \dfrac{10}{x^2 + 2}$

15. $f(x) = \dfrac{2x - 1}{x^2 + 4x}$

16. $g(x) = \dfrac{4 + 2x}{x}$

17. $f(x) = \dfrac{x^2 - 4}{x^2 + 6x + 9}$

18. $f(x) = \dfrac{2x - 6}{x + 4}$

19. $f(x) = \dfrac{2x}{x^2 - 4}$

20. $h(x) = \dfrac{2x + 4}{2x^2 - 5x - 3}$

Analyze each graph by finding all the following information. (Don't bother sketching the graph.)

(a) *Domain*
(b) *Vertical asymptote(s), if any exist*
(c) *Horizontal asymptote, if any exist*
(d) *y-intercept, if any exist*
(e) *x-intercept(s), if any exist*

21. $g(x) = \dfrac{x - 5}{x(x + 1)(x - 2)}$

22. $h(x) = \dfrac{(x + 1)(x - 2)}{2x(x - 3)}$

23. $k(x) = \dfrac{2(x - 4)(x + 7)}{(x + 5)^2(x + 1)}$

24. $l(x) = \dfrac{x^2(x - 6)}{(x^2 - 1)(2x + 3)}$

25. $f(x) = \dfrac{(3x - 1)(x + 5)}{(3x + 1)(x - 5)}$

3.6 Transformations of Graphs

TYPES OF TRANSFORMATIONS TO $y = f(x)$

(Assume c is a constant such that $c \in \mathbb{R}$, $c > 0$.)

(For all examples in the last column of this table, we'll use $y = f(x) = x^2$ as the base or parent function graph.)

1. **Shift:** $h(x) = f(x) \pm c$	Shifts graph up or down by c units (if we add c, shift up; if we subtract c, shift down)	$y = x^2 + 2$ shifts graph 2 units up
$h(x) = f(x \pm c)$	Shifts graph left or right by c units (if we add c, shift left; if we subtract c, shift right)	$y = (x - 3)^2$ shifts graph 3 units right
2. **Reflection:** $g(x) = -f(x)$	Reflects graph vertically (across x-axis)	$y = -x^2$ reflects graph vertically
$g(x) = f(-x)$	Reflects graph horizontally (across y-axis)	$y = (-x)^2$ reflects graph horizontally
3. **Stretch/Shrink:** $k(x) = cf(x)$	Stretches/shrinks graph vertically (if $c > 1$, it's a stretch; if $0 < c < 1$, it's a shrink)	$y = 5x^2$ stretches graph vertically by factor of 5
$k(x) = f(cx)$	Stretches/shrinks graph horizontally (if $c > 1$, it's a shrink; if $0 < c < 1$, it's a stretch)	$y = (4x)^2$ shrinks graph horizontally by one quarter

Note: ALL the vertical effects or changes to the graph appear "outside" the function, that is, outside the base or parent function that defines the overall shape. ALL the horizontal effects or changes to the graph appear "inside" the function, that is, before we perform the essence of the function. In the above examples, the main shape of the graph is the parabola given by $y = f(x) = x^2$. So, any algebraic change that happens before we square anything is "inside" the function, and any change that happens after the square is "outside" the function.

Note also that all the vertical shifts, stretches and shrinks are "intuitive," meaning that they're as expected. Adding two outside the function, for example, shifts the graph up, and we would expect a positive vertical shift to be up. Also, the horizontal shifts, stretches and shrinks are all "counter-intuitive," meaning that they're the opposite of what we'd expect. For example, adding three inside the function shifts the graph to the left by three units, which is perhaps the opposite of what one would expect from a positive horizontal change. Adding three inside the function means a smaller x-value is needed to produce the same y-value as before the shift. Because a smaller x-value is needed, we shift left instead of right.

Warning: Make sure to do all the reflections and shrink/stretch transformations FIRST before the shifts, when graphing these transformed functions. If you do the shifts first, you can get the wrong graph very easily.

Keep in mind that we can always use the default, back-up strategy of plotting lots of points and connecting the dots to graph any function. This method of understanding the transformations of graphs, along

TRANSFORMATION	EFFECT	EXAMPLE
$f(x)$	Base graph	$f(x)$ 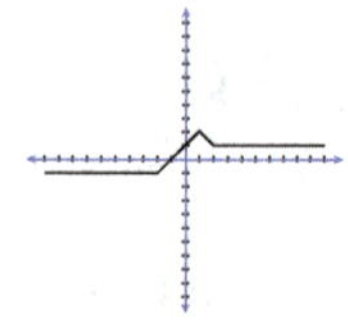
$h(x) = f(x) + c$	Shifts graph *up* by c units (down if c is negative)	$h(x) = f(x) + 2$ 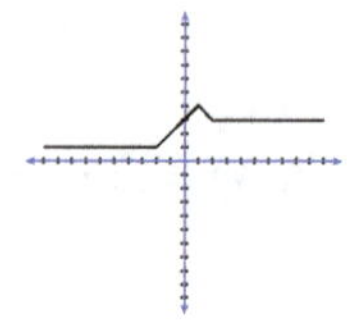
$h(x) = f(x + c)$	Shifts graph *left* by c units (right if c is negative)	$h(x) = f(x - 2)$ 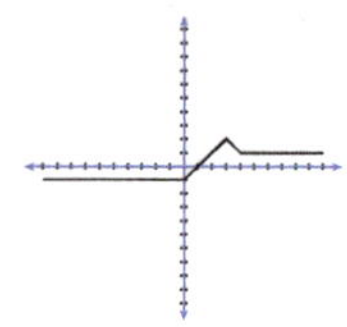
$h(x) = -f(x)$	Reflects graph vertically	$h(x) = -f(x)$ 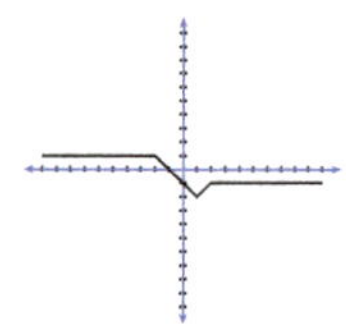
$h(x) = f(-x)$	Reflects graph horizontally	$h(x) = f(-x)$ 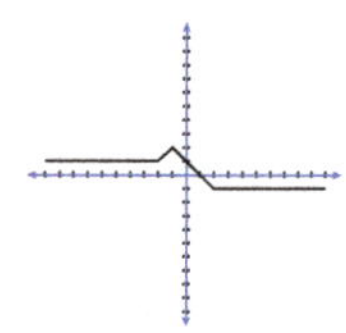
$h(x) = c \cdot f(x)$	Stretches graph vertically by a factor of c	$h(x) = 2f(x)$ 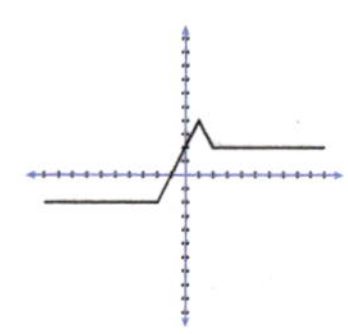
	(shrinks if $0 < c < 1$)	$h(x) = \frac{1}{2}f(x)$ 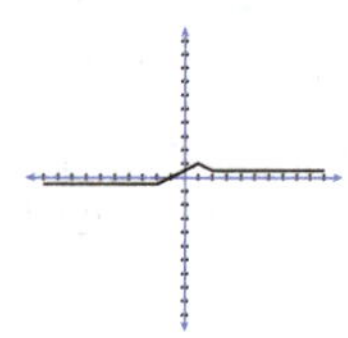
$h(x) = f(c \cdot x)$	Shrinks graph horizontally by a factor of c	$h(x) = f(2x)$ 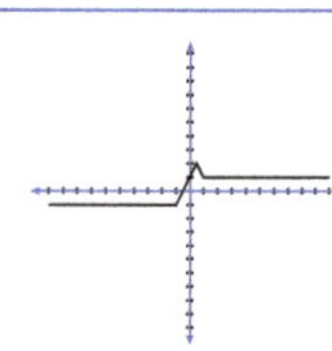
	(stretches if $0 < c < 1$)	$h(x) = f\left(\frac{1}{2}(x)\right)$ 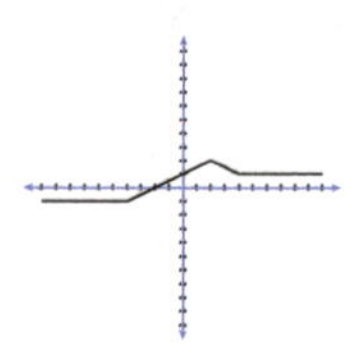

with the knowledge of the basic function graph shapes, is simply a more powerful and sophisticated method of graphing than plotting points.

Let's do several more examples to practice.

EXAMPLE 1

Graph these functions, using knowledge of transformations and the shapes of the base graphs.

(a) $y = -x^2 + 3$

(b) $y = |x - 2| + 1$

(c) $y = \sqrt{-x}$

(a) $y = -x^2 + 3$

The base function here is $y = x^2$, which is the basic parabola whose vertex is at (0, 0) and it's concave up. The graph we want has two changes to that base graph: (i) Multiplying the outside of the function by –1 means that graph will be vertically reflected, and (ii) adding three "outside" the square means the graph will be shifted up three units. This means the actual graph reflects in a vertical motion, or across the x-axis and is then shifted up by three.

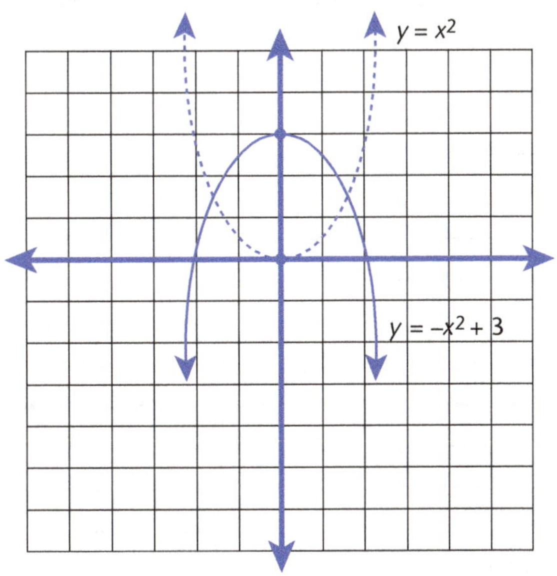

(b) $y = |x - 2| + 1$

The base graph here is the v-shaped absolute value function $y = |x|$ whose vertex is at the point (0, 0). The transformations for this function's graph, compared to the base function's graph, are two shifts: (i) Adding 1 outside the function shifts the graph up by 1, and (ii) subtracting 2 inside the function shifts the graph two to the right.

When we graph this, we can graph the base graph as a dotted graph. Then basically just shift all the points 1 unit up and 2 units right to get the solid curved graph of the function we're given. In practice, we can simply shift the vertex, namely (0, 0), and then draw in the shape of the graph from there.

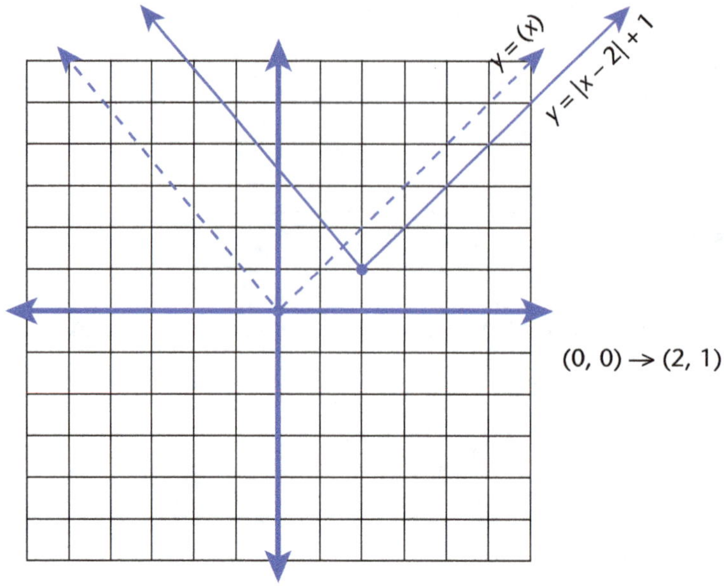

(c) $y = \sqrt{-x}$

The base function, or graph, for this transformed function is $y = \sqrt{x}$. The only change we've made to the base graph is to multiply by –1 inside the function which means it's a horizontal reflection. This means that the graph's reflection is a horizontal motion; in other words, we reflect the original curve across the *y*-axis.

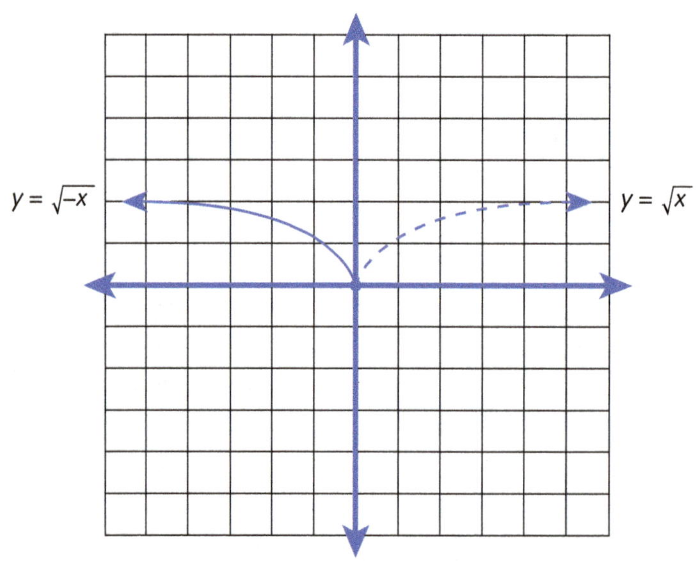

EXAMPLE 2

Write an equation for these graphs, knowing they are simply transformations of some base graph.

(a)

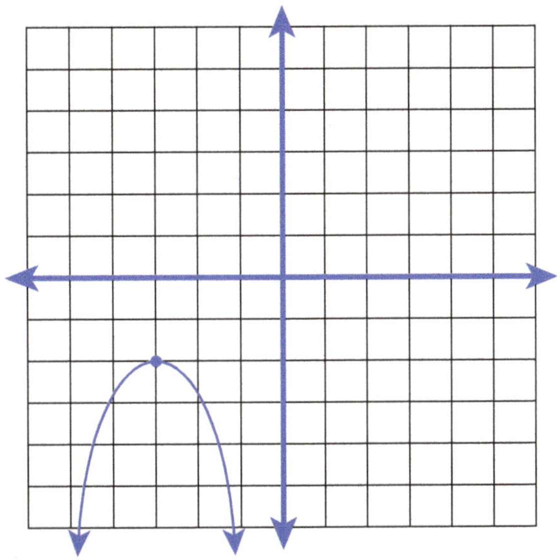

Notice that the shape of this graph looks like the base function is $y = x^2$. With that base graph, this has been moved to the left 3 units and down 2 units. We can tell that by looking at the location of the graph's vertex. The vertex here is at the point (–3, –2), compared to the base graph's vertex at (0, 0). This means the function for this graph will be $y = (x + 3)^2 - 2$.

(b)

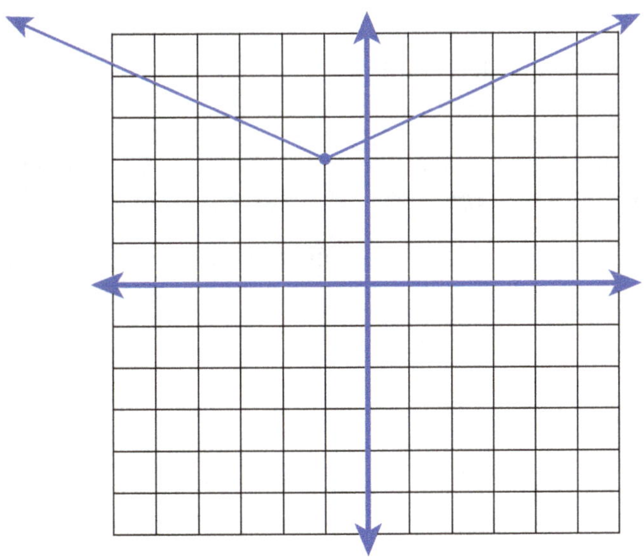

The base function for this graph is $y = |x|$, as we can tell by the v-shaped curve. It looks like the graph got squished vertically a bit, which suggests vertical shrinkage, i.e., multiplication by a proper fraction outside the function. The vertex has also shifted to the left 1 and up 3. So far, then, we have some function that will look like $y = c|x + 1| + 3$ where c will be the shrinkage factor. How can we figure out what the shrinkage is? Normally, from the vertex of an absolute value graph, we could go to the right one and up one to get to another point on the graph. In this case, we go right 2 to get up the same 1 unit for another point on the curve, which means it is growing vertically half as fast. This indicates that the shrinkage factor is one-half. So, $y = \frac{1}{2}|x + 1| + 3$ can represent our function.

EXAMPLE 3

Describe the transformations compared to the base graphs.

(a) $g(x) = 2(x + 1)^5 - 9$

(b) $h(x) = -|x - 4| - 3$

(c) $k(x) = -\sqrt{-\frac{1}{3}x} + 2$

(d) $m(x) = 3(-x)^4 - 1$

(a) $g(x) = 2(x + 1)^5 - 9$

The base graph is $y = x^5$. This function has three changes or transformations occurring.

1. Multiplying by 2 outside the function means the graph has been vertically stretched by a factor of 2. In other words, it grows vertically 2 times as fast as the base graph.
2. Adding 1 inside the function means the graph has shifted to the left by 1 unit.
3. Subtracting 9 outside the function means the graph has shifted down by 9 units.

(b) $h(x) = -|x - 4| - 3$

The base graph here is the absolute value graph $y = |x|$ that looks like a v, with the vertex at (0, 0). There are three changes for this graph, compared to the base graph.

1. Multiplying by –1 outside the function is a vertical reflection of the graph. This means the v-shape will be opening downward instead of upward like the base graph.
2. Subtracting 4 inside the function shifts the graph to the right by 4 units.
3. Subtracting 3 outside the function shifts the graph down by 3 units.

(c) $k(x) = -\sqrt{-\frac{1}{3}x} + 2$

The base function here is $y = \sqrt{x}$ that is half of a sideways parabola with the endpoint at the origin, (0, 0). This function has four different transformations to that base graph.

1. Multiplying by $\frac{1}{3}$ inside the function stretches the graph horizontally by a factor of 3 (the reciprocal of $\frac{1}{3}$).
2. Multiplying by –1 inside the function is a horizontal reflection, meaning that the motion of the reflection is horizontal (or across the y-axis).

3. Multiplying by –1 outside the function is a vertical reflection, meaning that the motion of the reflection is vertical (or across the x-axis).

4. Adding 2 outside the function shifts the graph up by 2 units.

(d) $m(x) = 3(-x)^4 - 1$

The base graph here is $y = x^4$, which looks like a concave up parabola that's flatter at the bottom than the real parabola, with the vertex at (0, 0). There are three changes to the base graph in this function.

1. Multiplying by –1 inside the function causes a horizontal reflection, meaning that the curve is reflected horizontally across the y-axis.

2. Multiplying by 3 outside the function is a vertical stretch (since $3 > 1$ it's a stretch, not a shrink). This means that the graph gets tall 3 times faster than the base graph.

3. Subtracting 1 outside the function shifts the graph down by 1 unit.

EXAMPLE 4

Given this graph for some $f(x)$, sketch the graph of the indicated transformed functions.

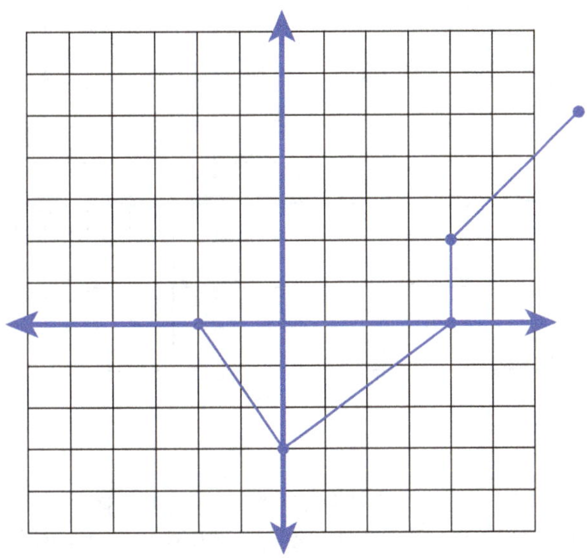

(a) $f(-x)$

(b) $f(x - 1) + 2$

(c) $-2f(x)$

When we transform this graph, we'll basically apply the transformations to the endpoints of each segment of this graph. Then, we'll connect those points with line segments, just as this graph has done.

(a) $f(-x)$

For this transformation, we want to change the graph by doing a horizontal reflection. That means we can take every endpoint on the original graph and reflect it horizontally. That means that every x-value of the points will be the opposite of its original value. Here's a table representing the points and their transformed points.

ORIGINAL (x,y)	TRANSFORMED (x,y)
(–2, 0)	(2, 0)
(0, –3)	(0, –3)
(4, 0)	(–4, 0)
(4, 2)	(–4, 2)
(7, 5)	(–7, 5)

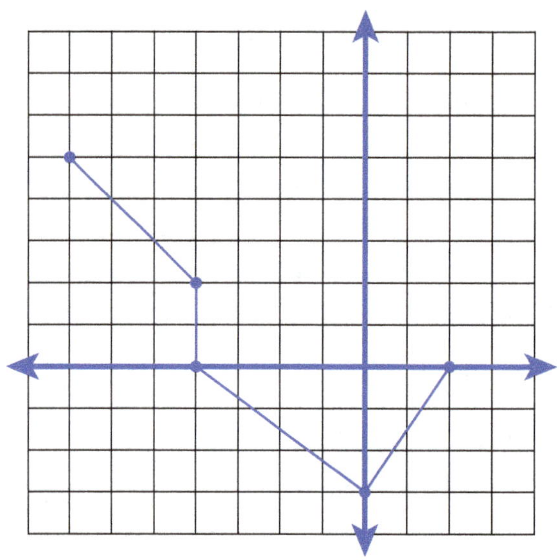

(b) $f(x-1)+2$

For this transformation, we want to (i) shift the graph's points up 2 units and (ii) shift the graph to the right by 1 unit. That means for each point on the graph we want to add 2 to the y-value and add 1 to the x-value to cause that movement.

ORIGINAL (x,y)	TRANSFORMED (x,y)
(–2, 0)	(–1, 2)
(0, –3)	(1, –1)
(4, 0)	(5, 2)
(4, 2)	(5, 4)
(7, 5)	(8, 7)

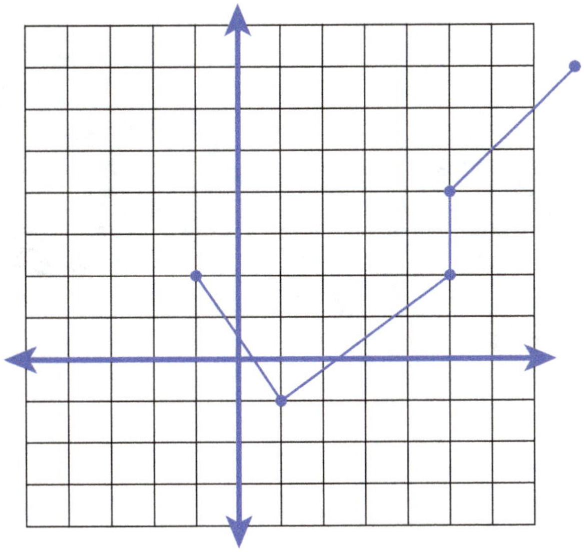

(c) $-2f(x)$

There are two changes made to the base graph for this function. (i) Multiplying by 2 outside the function causes a vertical stretch by a factor of two. This means we need to multiply all y-values of the points on the original graph by 2. (ii) Multiplying by –1 outside the function causes a vertical reflection across the x-axis, which means that all the y-values get multiplied by –1. Overall, these two changes combined together mean we simply need to multiply all the y-values of the original points by –2.

ORIGINAL (x,y)	TRANSFORMED (x,y)
(–2, 0)	(–2, 0)
(0, –3)	(0, 6)
(4, 0)	(4, 0)
(4, 2)	(4, –4)
(7, 5)	(7, –10)

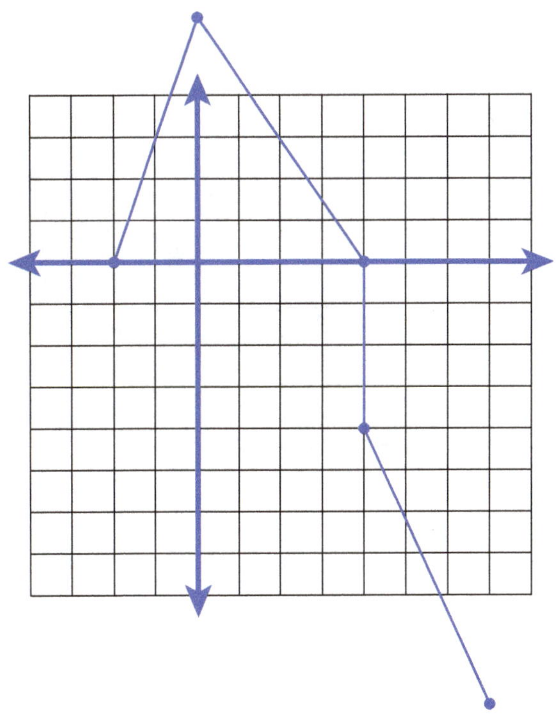

3.6 Homework

For each function, answer the following questions, compared to the base function graph:

(a) *What is the vertical shift?*
(b) *What is the horizontal shift?*
(c) *Is the graph stretched/shrunk vertically? If so, by what factor?*
(d) *Is the graph stretched/shrunk horizontally? If so, by what factor?*
(e) *Is the graph reflected? If so, is it a vertical or horizontal reflection?*
(f) *Sketch the graph of the function.*

1. $f(x) = 2x^2$
2. $f(x) = 2x^2 - 3$
3. $f(x) = -(x+1)^2 - 2$
4. $h(x) = 3|x-2| + 5$
5. $y(x) = -4|x-1| - 3$
6. $y(x) = \sqrt{-(x+3)} - 4$
7. $f(x) = \sqrt{-(x-2)} + 1$
8. $h(x) = \sqrt{\frac{1}{2}(x+3)} - 3$
9. $p(x) = -\sqrt{3(x-2)} + 4$
10. $q(x) = -2|x-2| + 1$
11. $f(x) = x^2 - 4x + 7$
12. $f(x) = -3x^2 - 6x + 1$
13. $g(x) = -x^2 - 4x + 5$
14. $p(x) = \frac{1}{2}\sqrt{x-3} + 5$
15. $h(x) = -4(x+1)^2 - \frac{1}{2}$

Given the graph of $f(x)$, sketch the graph of the indicated transformed graph.

(a) $f(-x) + 1$
(b) $-f(x) + 1$
(c) $f(x + 1)$
(d) $f(-(x + 1))$

16.

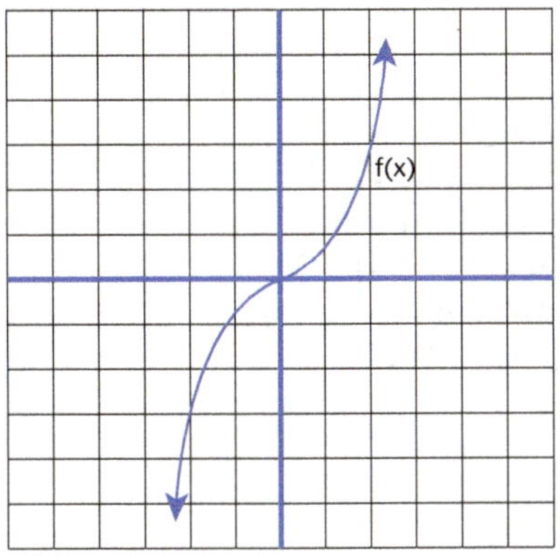

19.

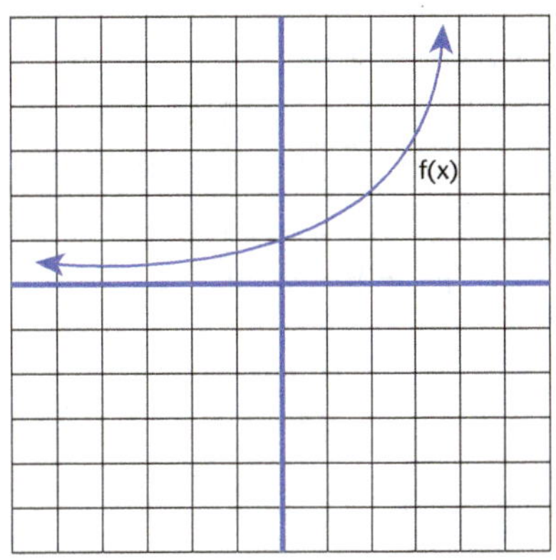

17.

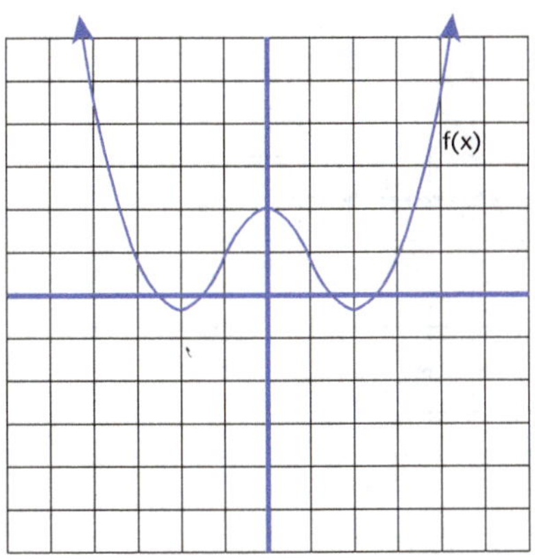

20.

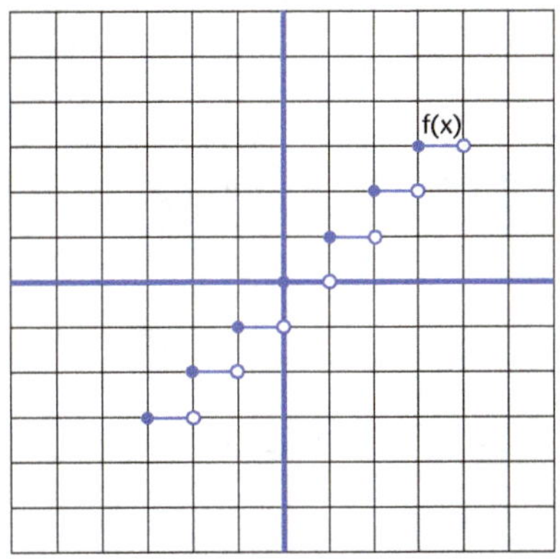

18.

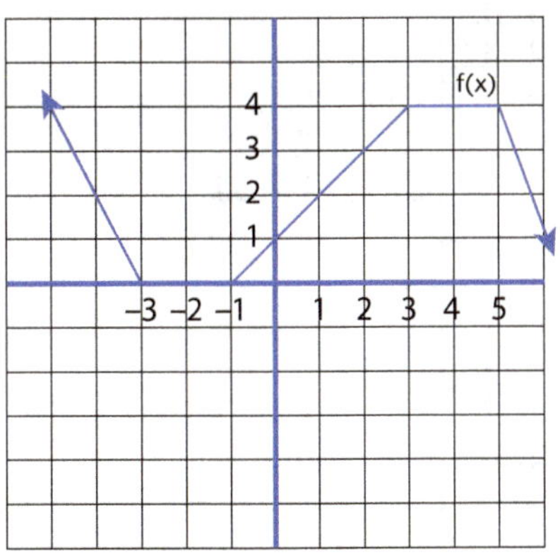

Write an equation to represent the function that would produce these graphs.

21.

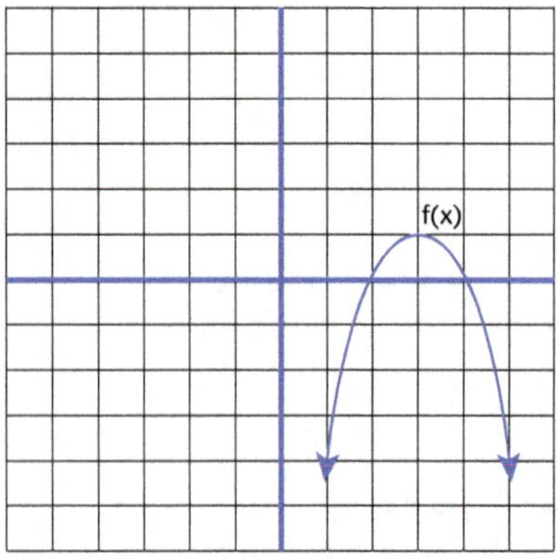

24.

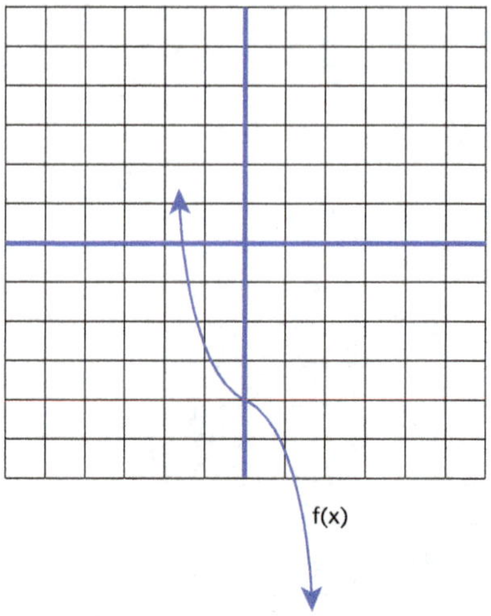

22.

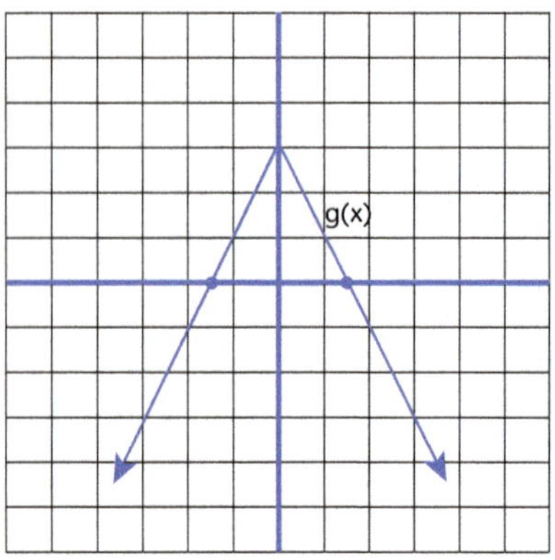

25.

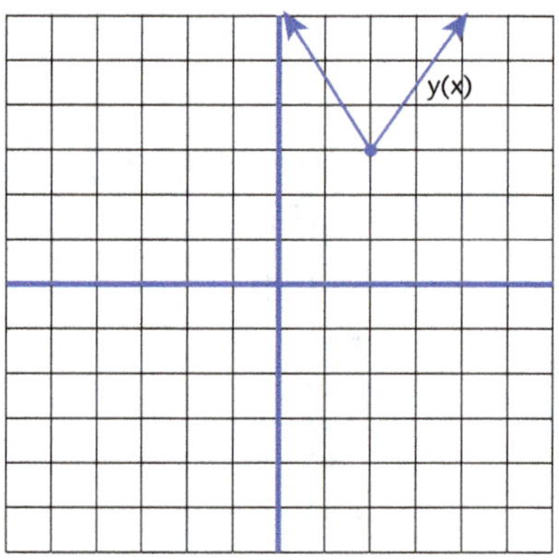

23.

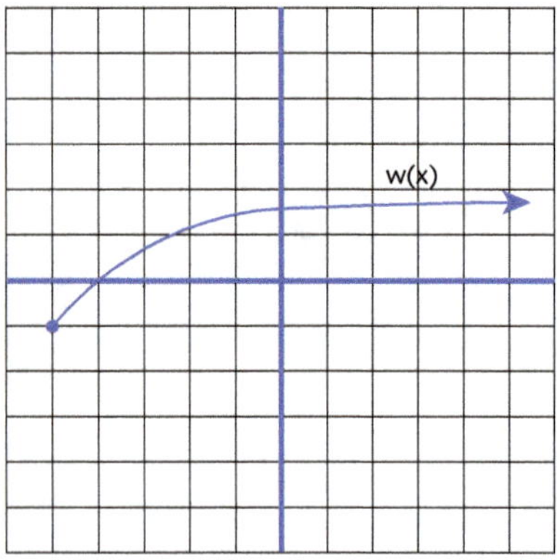

26.

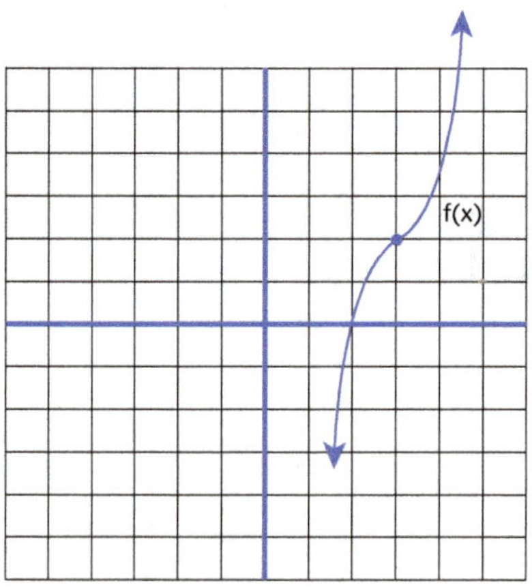

27. 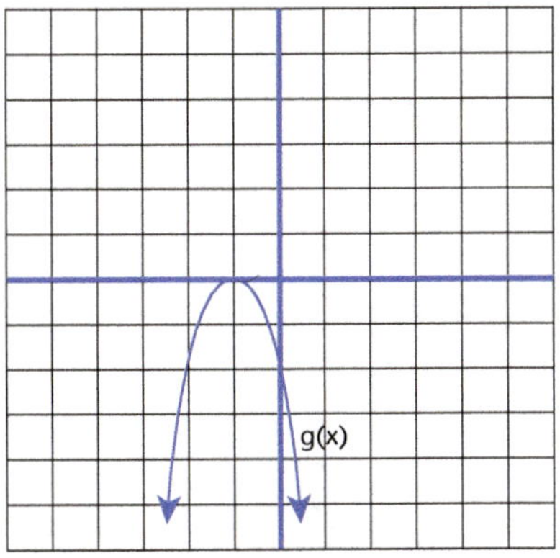

28. 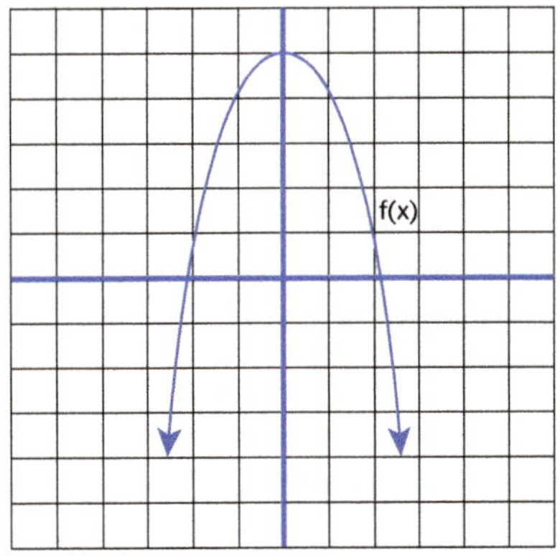

3.7 Combinations of Functions

ARITHMETIC OPERATIONS WITH FUNCTIONS

Addition	$(f+g)(x) = f(x) + g(x)$
Subtraction	$(f-g)(x) = f(x) - g(x)$
Multiplication	$(fg)(x) = f(x)\,g(x)$
Division	$\left(\frac{f}{g}\right)(x) = \frac{f(x)}{g(x)},\ g(x) \neq 0$

In other words, doing arithmetic with functions is exactly as we'd hope. The "sum function" is the sum of the individual functions; the "difference function" is the difference of the individual functions; the "product function" is the product of the individual functions; and the "quotient function" is the quotient of the individual functions.

COMPOSITION FUNCTION

The composition function of f with g is denoted by $(f \circ g)(x)$ and defined as $(f \circ g)(x) = f(g(x))$.

We can think of the composition function as a "nested" function. The idea here is that we apply g to x first and then apply the function, f, to the result of $g(x)$. This idea of nested operations occurs every day in our lives. For instance, if I'm getting ready to go skiing, I will first put on my long underwear. Then I put my fleece pants on. Finally, I will put on my waterproof ski pants as the final layer. The order that I choose to put my pants on for skiing definitely makes a difference. I wouldn't want my waterproof pants on first and then my long underwear on the outside. This idea of having to do one thing first before the next step is exactly the concept involved in "nested" composition functions. From this example we can also see that there may be several layers to the composition, not just two layers.

EXAMPLE 1

Given $f(x) = 2x^2 + x - 5$ and $g(x) = 3x - 1$, perform the following operations.

(a) $(f+g)(x)$,

(b) $(fg)(x)$,

(c) $(g-f)(x)$,

(d) $\left(\frac{f}{g}\right)(x)$,

(e) $(f \circ g)(x)$, and

(f) $(g \circ f)(x)$.

(a) $(f+g)(x) = f(x) + g(x)$	Add the functions together.
$= (2x^2 + x - 5) + (3x - 1)$	Simplify by collecting like terms.
$= 2x^2 + x + 3x - 5 - 1$	
$= 2x^2 + 4x - 6$	

(b) $(fg)(x) = f(x)\,g(x)$	Multiply the functions together.
$= (2x^2 + x - 5)(3x - 1)$	Use the distributive property to multiply the polynomials.
$= 2x^2(3x) + 2x^2(-1) + x(3x) + x(-1) - 5(3x) - 5(-1)$	Simplify.
$= 6x^3 - 2x^2 + 3x^2 - x - 15x + 5$	
$= 6x^3 + x^2 - 16x + 5$	

(c) $(g-f)(x) = g(x) - f(x)$	Subtract $f(x)$ from $g(x)$.
$= (3x - 1) - (2x^2 + x - 5)$	Distribute the subtraction through the second set of parentheses.
$= 3x - 1 - 2x^2 - x + 5$	Simplify by combining like terms.
$= -2x^2 + 2x + 4$	

(d) $\left(\frac{f}{g}\right)(x) = \frac{f(x)}{g(x)}$	Find the quotient of $f(x)$ and $g(x)$.
$= \frac{2x^2 + x - 5}{3x - 1}$	This is already completely simplified since there are no common factors in the numerator and denominator.

(e) $(f \circ g)(x) = f(g(x))$	Compose $f(x)$ with $g(x)$.
$= f(3x - 1)$	With composition functions, we always work from the "inside out." In this case, that means we find $g(x)$ first and then apply f to that.
$= 2(3x - 1)^2 + (3x - 1) - 5$	So, we treat $(3x - 1)$ as the input into $f(x)$.
$= 2(3x - 1)(3x - 1) + 3x - 1 - 5$	Distribute and simplify.
$= 2(9x^2 - 3x - 3x + 1) + 3x - 1 - 5$	
$= 18x^2 - 6x - 6x + 2 + 3x - 1 - 5$	
$= 18x^2 - 9x - 4$	

(f) $(g \circ f)(x) = g(f(x))$ Compose $g(x)$ with $f(x)$.

$= g(2x^2 + x - 5)$ We need to use the entire expression, $(2x^2 + x - 5)$, as the input in $g(x)$.

$= 3(2x^2 + x - 5) - 1$ Distribute and simplify.

$= 6x^2 + 3x - 15 - 1$

$= 6x^2 + 3x - 16$

EXAMPLE 2

Given $f(x) = \sqrt{x - 4}$, $g(x) = x^3 + 1$, and $h(x) = 2x - 9$, perform the following operations.

(a) $(f \circ g)(x)$,

(b) $g(h(x))$,

(c) $(h \circ g)(x)$,

(d) $f(h(g(x)))$,

(e) $(f + g)(5)$,

(f) $(hf)(6)$, and

(g) $(g - h)(1)$.

(a) $(f \circ g)(x) = f(g(x))$ Compose $f(x)$ with $g(x)$.

$= f(x^3 + 1)$ Substitute the entire expression for $g(x)$ into $f(x)$ wherever x originally was.

$= \sqrt{(x^3 + 1) - 4}$ Simplify.

$= \sqrt{(x^3 - 3)}$

(b) $g(h(x))$ Compose $g(x)$ with $h(x)$.

$= g(2x - 9)$ Substitute the $h(x)$ expression into $g(x)$, wherever x originally was in the definition for $g(x)$.

$= (2x - 9)^3 + 1$ We can choose to multiply this out and collect like terms, or leave it as it is. Since it is easier to leave it alone, we'll choose that option and not risk any possible errors.

(c) $(h \circ g)(x) = h(g(x))$ — Compose $h(x)$ with $g(x)$.

$= h(x^3 + 1)$ — Substitute the $g(x)$ expression into $h(x)$ for the input.

$= 2(x^3 + 1) - 9$ — Distribute and simplify.

$= 2x^3 + 2 - 9 = 2x^3 - 7$

(d) $f(h(g(x)))$ — Compose $f(x)$ with the composition of $h(x)$ with $g(x)$. Basically, here we have two different layers of composition.

$= f(h(x^3 + 1))$ — We already know from part (c) that $h(x^3 + 1) = 2x^3 - 7$.

$= f(2x^3 - 7)$ — Substitute that expression for $h(g(x))$ into f for its input.

$= \sqrt{(2x^3 - 7) - 4}$ — Simplify.

$= \sqrt{2x^3 - 11}$

(e) $(f + g)(5) = f(5) + g(5)$ — We want to know the sum of $f(x)$ and $g(x)$ evaluated at $x = 5$. We can do this in one of two ways: (1) Find the sum function and then evaluate at 5, or (2) evaluate $f(5)$ and $g(5)$ and then find the sum. We'll demonstrate both techniques for this problem.

1. $(f + g)(x) = f(x) + g(x)$ — Find the sum function.

$= \sqrt{x - 4} + x^3 + 1$ — Now, evaluate at $x = 5$.

$(f + g)(5) = \sqrt{5 - 4} + 5^3 + 1 = \sqrt{1} + 125 + 1 = 127$

2. $f(5) = \sqrt{5 - 4} = \sqrt{1} = 1$ — Evaluate both $f(5)$ and $g(5)$ separately.

$g(5) = 5^3 + 1 = 125 + 1 = 126$ — Add the two values together to get the sum function evaluated at 5.

$(f + g)(5) = f(5) + g(5) = 1 + 126 = 127$

(f) $(hf)(6) = h(6)f(6)$ — Again, we have two choices how to proceed. We can either (1) find the product function $h(x)f(x)$ and then evaluate at $x = 6$, or (2) find $h(6)$ and $f(6)$ separately and then find their product. In essence, we'll choose method (2) this time.

$h(6)f(6) = (2(6) - 9)(\sqrt{6 - 4})$ — Simplify.

$= (12 - 9)\sqrt{2} = 3\sqrt{2}$

(g) $(g-h)(1) = g(1) - h(1)$ — Substitute $x = 1$ into $g(x)$ and $h(x)$ and find their difference.

$= (1^3 + 1) - (2(1) - 9)$ — Simplify.

$= 2 - (-7) = 9$

EXAMPLE 3

If $f(x) = x^2 + x + 3$, what is $(f \circ f)(x)$ and $f^2(x)$?

This is an excellent question because we need to see what the difference is, if there is any, between composing a function with itself and the square of a function. Remember that squaring a function simply means to multiply a function by itself. So, we could write $f^2(x)$ as $f(x)f(x)$. Also, $(f \circ f)(x)$ is the same as $f(f(x))$. Thus, the overall question boils down to: Is $f(x)f(x)$ equal to $f(f(x))$? Even though the notation looks similar, the operations are completely different.

$f^2(x) = f(x)f(x)$	
$= (x^2 + x + 3)^2$	Multiply out.
$= (x^2 + x + 3)(x^2 + x + 3)$	Distribute.
$= x^4 + x^3 + 3x^2 + x^3 + x^2 + 3x + 3x^2 + 3x + 9$	Simplify.
$= x^4 + 2x^3 + 7x^2 + 6x + 9$	
$(f \circ f)(x) = f(f(x))$	
$= f(x^2 + x + 3)$	Substitute $f(x)$ expression into $f(x)$ as the input.
$= (x^2 + x + 3)^2 + (x^2 + x + 3) + 3$	Multiply out using the fact (from the left side of this table) that we already know how $(x^2 + x + 3)^2$ simplifies.
$= x^4 + 2x^3 + 7x^2 + 6x + 9 + x^2 + x + 3 + 3$	Simplify.
$= x^4 + 2x^3 + 8x^2 + 7x + 15$	

EXAMPLE 4

Find functions $f(x)$ and $g(x)$ such that $f(g(x)) = \dfrac{2}{3x^4 + 1}$.

For this problem, we're looking for two different functions that when nested together would create the result given. There will be many correct answers here. We'll find two such correct answers as an example.

1. For the first example of creating this function, I'm thinking that I can separate the two operations of "two over something" from the "polynomial in the denominator." If $f(*) = \frac{2}{*}$ ("two over something") and $g(x) = 3x^4 + 1$ (the "polynomial in the denominator"), then $f(g(x)) = f(3x^4 + 1) = \dfrac{2}{3x^4 + 1} = \dfrac{2}{3x^4 + 1}$.

2. For the next example of creating the same function, I'm going to keep more of the original function intact when nesting the two together. So, let's separate the operations of "two over a polynomial" from "raising the input of the polynomial to the fourth power." If $f(*) = \frac{2}{3* + 1}$ ("two over the polynomial") and $g(x) = x^4$ ("raising the input of the previous denominator's polynomial to the fourth power"), then $f(g(x)) = f(x^4) = \frac{2}{3(x^4) + 1} = \frac{2}{3x^4 + 1}$.

3.7 Homework

For each set of given functions, find the following:

(a) $(f+g)(x)$
(b) $(fg)(x)$
(c) $(g-f)(x)$
(d) $\left(\dfrac{f}{g}\right)(x)$
(e) $(f \circ g)(x)$
(f) $(g \circ f)(x)$

1. $f(x) = x^2 - 2, \quad g(x) = x + 1$

2. $f(x) = x + 3, \quad g(x) = x + 5$

3. $f(x) = x^2 + 1, \quad g(x) = x^2 - 4$

4. $f(x) = 2x - 1, \quad g(x) = x^2 - 3$

5. $f(x) = \sqrt{x+1}, \quad g(x) = \dfrac{2}{x}$

6. $f(x) = x^3, \quad g(x) = \dfrac{2}{x^2+1}$

7. $f(x) = \sqrt{x+1}, \quad g(x) = x + 5$

8. $f(x) = x^3, \quad g(x) = x - 1$

9. $f(x) = -x, \quad g(x) = \dfrac{x-4}{x-2}$

10. $f(x) = 2x^2, \quad g(x) = 1 - 3x^2$

Perform the following operations for each set of given functions:

(a) $(f \circ g)(x)$
(b) $g(h(x))$
(c) $(h \circ g)(x)$
(d) $f(h(g(x)))$
(e) $(f+g)(2)$
(f) $(hf)(3)$
(g) $(g-h)(1)$

11. $f(x) = 2x - 3, \quad g(x) = x + 5, \quad h(x) = x^2$

12. $f(x) = x^2 - 1, \quad g(x) = \dfrac{x}{2}, \quad h(x) = \sqrt{x-1}$

13. $f(x) = x^3, \quad g(x) = \sqrt{x^2 + 1}, \quad h(x) = \frac{1}{x}$

14. $f(x) = 2x + 1, \quad g(x) = 3x - 2, \quad h(x) = x^2$

15. $f(x) = \sqrt{x + 1}, \quad g(x) = \frac{2}{x}, \quad h(x) = x^2 + 3$

16. $f(x) = 2x - 1, \quad g(x) = \frac{1}{x + 5}, \quad h(x) = 3|x + 2|$

17. $f(x) = x^2 - 1, \quad g(x) = \sqrt{x - 1}, \quad h(x) = x$

18. $f(x) = 2\sqrt{x}, \quad g(x) = x^3, \quad h(x) = \frac{3}{x}$

19. If $f(x) = x^2 - 3x + 4$, find $(f \circ f)(x)$.

20. If $f(x) = x^2 - 3x + 4$, find $(f^2)(x)$. (*Hint:* $(f^2)(x) = f(x)f(x)$.)

21. If $g(x) = \frac{x^2 + 1}{x - 2}$, find $(g \circ g)(x)$.

22. If $f(x) = x^2 - 3$, find $\frac{f(x + h) - f(x)}{h}$.

23. If $g(x) = 3x - 1$, find $\frac{g(x + h) - g(x)}{h}$.

24. If $p(x) = 2x^2 + 1$, find $\frac{p(x + h) - p(x)}{h}$.

25. Find an $f(x)$ and $g(x)$ such that $(f \circ g)(x) = x^2 + \frac{1}{x^2}$. (*Note:* There are many correct answers.)

26. Find a $g(x)$ and $h(x)$ such that $h(g(x)) = -2x - 4$. (*Note:* There are many correct answers.)

27. Find $f(x)$, $g(x)$, and $h(x)$ such that $h(f(g(x))) = \frac{1}{x^2 + 4} - 3$. (*Note:* There are many correct answers.)

Chapter 3 Review

3.1

Solve these equations.

1. $4x^2 - 25 = 0$

2. $2x^2 + 5x = 12$

3. $x^2 + 4x - 7 = 0$

4. $6x^2 - 5 = 13x$

5. $2(w - 1)^2 - 8 = 0$

6. $x + 5 = \dfrac{24}{x}$

7. $\dfrac{2}{x-5} - \dfrac{1}{x+1} = \dfrac{1}{4}$

8. $2(y + 1)^2 - 14 = y^2 + 5y + 30$

3.2

For each parabola, find the following:

(a) its vertex,
(b) the axis of symmetry,
(c) whether the parabola is concave up or down,
(d) whether the vertex is a maximum or minimum point,
(e) the x-intercept(s), if any exist,
(f) sketch the graph.

9. $y = 2(x + 3)^2 - 4$

10. $y = -(x - 5)^2 + 2$

11. $y = (x - 2)^2 + 1$

12. $y = -\dfrac{1}{2}x^2 + \dfrac{3}{2}x + 5$

13. $y = -4(x + 1)^2 - 1$

14. $y = x^2 - 6x + 5$

15. $y = \dfrac{2}{9}x^2 - 2$

16. A farmer wants to fence a rectangular field and then divide it in half with a fence down the middle parallel to one side. If 1488 feet of fence is to be used, what is the maximum area of the overall lot?

17. If a profit function is given by $P(x) = -2x^2 + 2320x - 1000$, where x represents the number of motor bikes made and sold, how many motor bikes need to be sold for maximum profit? What is the maximum profit?

3.3

Find the number of units that need to be produced and sold to break even, given the revenue and cost functions.

18. revenue: $R(x) = 1600x - x^2$
    cost: $C(x) = 1.6(x - 500)^2 + 200000$

19. revenue: $R(x) = -0.4x^2 + 53.6x$
    cost: $C(x) = 0.4x^2 + 12.8x + 496$

20. revenue: $R(x) = -0.9x^2 + 73x$
    cost: $C(x) = 0.6x^2 + 10x + 588$

Given the supply and demand equations, find the equilibrium quantity and price.

21. supply: $p = q^2 + 8q + 20$
demand: $p = 100 - 4q - q^2$

22. supply: $p = 2(q + 5)^2 + 6$
demand: $p = -7(q + 3)^2 + 994$

23. supply: $5p - q = 18$
demand: $pq = 3540 + 19q$

24. The supply and demand functions for Margo's design products are given by $\frac{p}{q} = 2$ and $pq = 10(20 + 4)$ respectively. If Margo is taxed \$4 per product and passes that tax on to the consumer as a price increase, find the market equilibrium point.

25. The cost of producing Bill's handiwork is given by $C(x) = 0.4x^2 + 12.8x + 496$ and he's found the demand for his product obeys the equation $p = -0.4x + 53.6$. If he can only produce fewer than 30 of his handiwork projects per week, how many must he make and sell to break even? How much profit will he make if he sells all 30 units?

3.4

For each polynomial function, find/do the following:

(a) *its degree,*
(b) *all the zeros (roots),*
(c) *the y-intercept,*
(d) *the x-intercept(s), if any exist,*
(e) *sketch the graph.*

26. $y = \frac{1}{5}(x^3 + 2x^2 - 8x)$

27. $y = \frac{1}{8}(x^2 - 2x - 3)^2$

28. $y = -\frac{1}{16}(x + 4)(x^2 - 16)$

29. $y = -\frac{1}{24}(x - 2)^2\left((x - 2)^2 - 16\right)$

30. $y = x^5 - 2x^3 + x$

For each piecewise function, fill in the table of points and then sketch the graph.

31. $f(x) = \begin{cases} 3x - 1, & x \ge 0 \\ -x^2 + 2, & x < 0 \end{cases}$

x	−2	−1	0	1	2	3
y						

32. $y(x) = \begin{cases} \frac{1}{2}x^2, & x \ge 2 \\ x, & -1 \le x < 2 \\ -2x - 2, & x < -1 \end{cases}$

x	−2	−1	0	1	2	3
y						

33. $h(x) = \begin{cases} 8, & x \ge 2 \\ x^3, & -2 \le x < 2 \\ -1, & x < -2 \end{cases}$

x	−3	−2	0	1	2	3
y						

3.5

Analyze each rational function by answering the following questions.

(a) *What is the domain?*
(b) *Find the vertical asymptote(s), if there are any.*
(c) *Find the horizontal asymptote, if it exists.*
(d) *Find the y-intercept, if it exists.*
(e) *Find the x-intercept(s), if any exist.*
(f) *Sketch the graph.*

34. $f(x) = \frac{3}{x - 4}$

35. $y = \frac{2x + 1}{x}$

36. $y = \dfrac{x^2 - 4}{x^2 - x - 2}$

37. $y = \dfrac{x}{(x+2)(x-2)}$

38. $y = \dfrac{4}{x^2 - 9}$

3.6

For each function, answer the following questions, compared to the base function graph.

(a) *What is the vertical shift?*
(b) *What is the horizontal shift?*
(c) *Is the graph stretched/shrunk vertically? If so, by what factor?*
(d) *Is the graph stretched/shrunk horizontally? If so, by what factor?*
(e) *Is the graph reflected? If so, is it a vertical or horizontal reflection?*
(f) *Sketch the graph.*

39. $y = -|x+1| - 3$

40. $f(x) = \dfrac{1}{2}(x+2)^2 + 5$

41. $h(x) = \sqrt{-4(x-1)} + 2$

42. $g(x) = -x^2 - 6x + 1$

43. $y = 2(x-4)^3$

3.7

For each pair of given functions, find the following:

(a) $(f+g)(x)$
(b) $(fg)(1)$
(c) $(f \circ g)(x)$
(d) $g(f(x))$
(e) $(g-f)(2)$
(f) $\left(\dfrac{f}{g}\right)(3)$
(g) $(f \circ f)(x)$

44. $f(x) = \sqrt{2x+1}$
$g(x) = -x + 5$

45. $f(x) = x^2 + 3$
$g(x) = |x| - 1$

46. $f(x) = x^3 + x$
$g(x) = \dfrac{2}{x+1}$

47. $f(x) = \dfrac{x+5}{x-3}$
$g(x) = x^2$

48. $f(x) = 2x - 6$
$g(x) = \dfrac{x}{x^2 - 25}$

49. $f(x) = \sqrt{x}$
$g(x) = x^3 - 1$

50. $f(x) = -4x$
$g(x) = x^2 + 3x$

chapter 4

Exponential and Logarithmic Functions

4.1 Inverse Functions

We can only talk about an inverse function if we have a function (not a relation) to begin with. If you remember from section 3.2, a function has a unique output for every input. Graphically, that means a function's graph will pass the vertical line test (every vertical line passes through the graph in at most one point). A reasonable follow-up question to ask is whether or not the function is invertible, meaning "can we undo everything the function just did to x (the input)?" This is what an inverse function is, if it exists.

DEFINITION
If $f(g(x)) = g(f(x)) = x$, then $g(x)$ is the inverse function for $f(x)$ and its notation is $f^{-1}(x)$.
($f^{-1}(x)$ is read as "f inverse of x.")

Basically, the function composed with its inverse function gets us back to the original input x. The inverse function basically just undoes what the function did to x in the first place. We already know of many inverse functions. For example, if we add 3 to x, we know that to undo that action, we simply need to subtract the 3 and we'll get back to where we started. To undo multiplying by 8, we divide by 8.

Please notice that this notation for the inverse function is exactly the same as the inverse notation used for the matrices that were covered in a previous chapter. The inverse notation used here will be the same for many mathematical operators. When the inverse function undoes the original function, leaving only the input, this is basically saying it returns the identity of the original input.

VOCABULARY

One-to-One—A function is one-to-one if for every output there's only one input that yields that output. This means that given the output, we can definitely get back to the input. (Being one-to-one is equivalent to saying the function is **invertible,** meaning that an inverse function exists.)

EXAMPLE 1

Are the functions, as defined, one-to-one (or invertible)?

(a) A function defined by {(2, 5), (1, 3), (3, 5)}.

(b) A function defined by {(1, 2), (–1, 3), (0, 4), (8, 10), (3, 0)}.

For a function to be one-to-one, we need to make sure that there is only one output assigned to any input AND the output is only assigned once.

(a) For this set of ordered pairs (or points), we see that the inputs are 2, 1 and 3. The outputs are 5, 3 and 5. This means that two different inputs got assigned to the same output. So, it's a function, but it's NOT invertible.

(b) This set of ordered pairs has five distinct and different inputs and also five distinct outputs. Thus, it is a function and it's also invertible.

EXAMPLE 2

For the function $f(x) = x^2$: (a) Is it a function? (b) Is it one-to-one?

(a) To decide if it is a function, we need to determine whether or not every input yields exactly one output. This function does indeed fit that definition because for any input I have only one way to calculate the output, namely to square the input. The other aid in making this decision is to graph the parabola and determine if it passes the vertical line test, which it does.

(b) For the function to be one-to-one (or invertible), we need to ensure that every output can uniquely determine the input. For $f(x) = x^2$, we see that $f(5) = f(-5) = 25$, which means that two different inputs give the same output. So, if we have 25 as the output and ask for the input (or x-value) that gives that result for this function, we'd have two different answers of 5 and –5, instead of one answer. This means the function is NOT one-to-one (or invertible). In a sense, to explore the idea of an inverse function, we must go backwards from the output (answer) to the input (question), something like the way the popular game "Jeopardy" does.

HORIZONTAL LINE TEST

If every horizontal line passes through the graph of a function $f(x)$ at most one time, then the function is one-to-one, or we can say that the function is invertible. In other words, if the graph of $f(x)$ passes the horizontal line test, then $f^{-1}(x)$ exists.

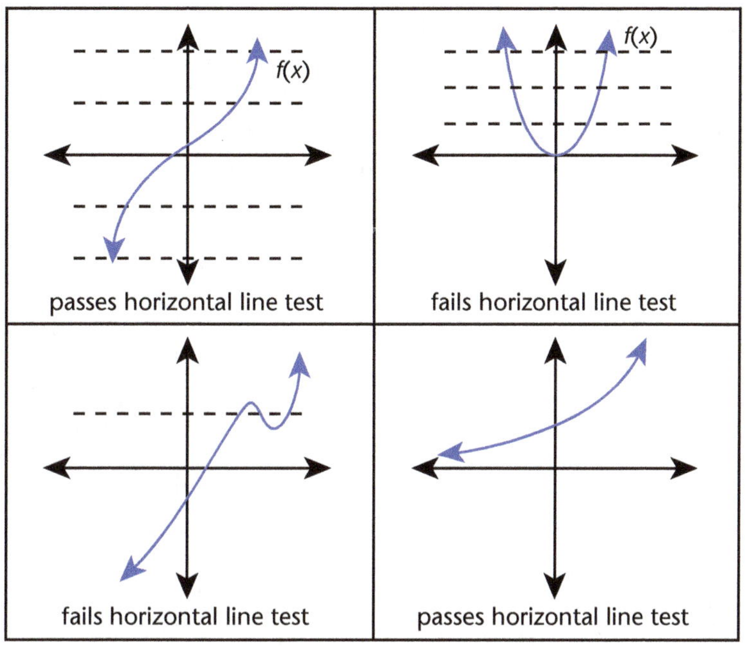

When we've established that a function is invertible, how then do we find the inverse function? Remember that the inverse function "undoes" everything that the function originally did. In essence, the inverse function's input is the original function's output and vice versa. Graphically, this means that if the point (x, y) is on the graph of $f(x)$, then the point (y, x) is the corresponding point on the graph of $f^{-1}(x)$. Overall, this means that the graph of $f^{-1}(x)$ will be the mirror image of $f(x)$, mirrored across the line $y = x$. Here are some examples.

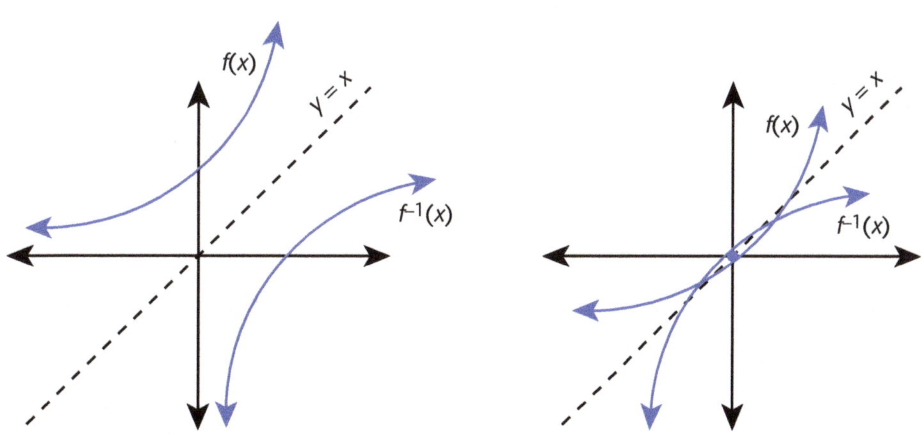

EXAMPLE 3

Let's come back to the function from Example 2, $f(x) = x^2$. Can we find an inverse function $f^{-1}(x)$?

We already established that this function has no inverse. That is, if we square a number, we cannot undo that action uniquely. However, in our reality, we would normally agree that taking the square root "undoes" the square, which suggests that we can undo a square. The unsaid hidden assumption in our last statement is that *if* we meet certain conditions first, the function actually does have an inverse. In this case, we need to restrict the domain of $f(x)$ to be $x \geq 0$. Then, we DO have an inverse! If we restrict the domain of this parabola as stated, then we basically only have the right half of the parabola, which does pass the vertical and horizontal line tests, suggesting that it is an invertible function. And, we can graph the inverse function by reflecting the half-parabola across the line $y = x$.

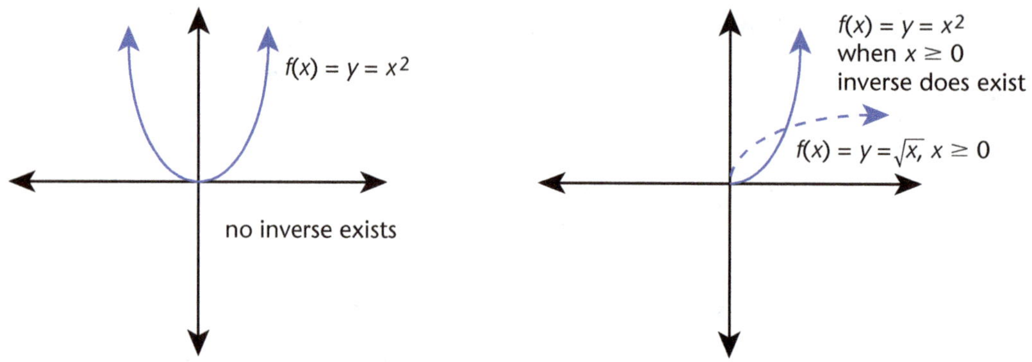

Thus, $f^{-1}(x) = \sqrt{x}$ if $f(x) = x^2$, $x \geq 0$. We can show this algebraically by using the definition of an inverse.

$$\Rightarrow f^{-1}(f(x)) = f^{-1}(x^2) = \sqrt{x^2} = x$$

and

$$f(f^{-1}(x)) = f(\sqrt{x}) = (\sqrt{x})^2 = x$$

In both instances, we definitely get back x since x was non-negative to begin with.

EXAMPLE 4

Are $f(x) = 2x + 7$ and $g(x) = \frac{x-7}{2}$ inverse functions?

We need to see if both of the composition functions get back to x and thereby satisfy the definition of the inverse function.

1. $f(g(x)) = f\left(\frac{x-7}{2}\right) = 2\left(\frac{x-7}{2}\right) + 7$ Simplify by dividing out the 2.

$= x - 7 + 7$ Continue simplifying.

$= x$

2. $g(f(x)) = g(2x+7) = \frac{2x+7-7}{2}$ Simplify the numerator by adding like terms.

$= \frac{2x}{2}$ Continue simplifying.

$= x$

Voila! We have $f(g(x)) = g(f(x)) = x$, which means that $f(x)$ and $g(x)$are indeed inverse functions.

FINDING AN INVERSE FUNCTION

We will offer two techniques for finding the inverse function. The first technique, which we'll nickname the "pants" technique, is only useful in limited settings. The second algebraic technique is useful in many more instances.

1. **The "Pants" Technique**—We can liken the "function" to preparing for a snow activity, like skiing. First you put on your long underwear, then fleece pants and finally the waterproof snow pants. At the end of the day, to undo the function, you would remove the outermost layer first. In other words, you'd remove the snow pants first, then the fleece pants and finally the long underwear. You would not attempt to remove the pants in the order you put them on because that would be impossible. Finding the inverse function in an algebraic setting is remarkably similar. (*Note:* We can only use this technique when there are just single layers of operations happening to the input. Visibly, we can say that if x appears in the function definition only once, then we can use this technique.)
2. **Formal Algebraic Technique**—(a) First, exchange the variables x and y in the original function. (b) Then use legal algebraic operations in that equation to move things around, getting y by itself on one side of the equation. (c) The final function is now the inverse function. (*Note:* This method is very useful in all the settings you'll encounter in this course. If you see more than one instance of x in the original function, you can still use this technique to unravel what happened to x in the function.)

EXAMPLE 5

Find the inverse function for the following functions.

(a) $f(x) = 4x + 6$

(b) $f(x) = 2x^3 + 5$

(c) $f(x) = 7\sqrt[5]{x-1} - 2$

First, notice that all of these functions have only one instance of x showing up in the function definition. This means we can practice the "pants" technique for finding the inverse function.

(a) For this function, we need to first list the operations that happened to x in the order in which they happened.

1. We multiplied by 4.
2. We added 6.

Now, undo those operations in the opposite order (like removing the pants). Remember here that we are still starting with x and undoing those operations on x.

0. Start with x.
1. Undo adding 6, by subtracting 6. $x - 6$
2. Undo multiplying by 4, by dividing by 4. $\frac{x-6}{4}$

$\Rightarrow \quad f^{-1}(x) = \frac{x-6}{4}$

Notice that every step in the process acted on the result from the previous step.

(b) We did the operations in this order for $f(x) = 2x^3 + 5$:

1. Cubed x.
2. Multiplied by 2.
3. Added 5.

For the inverse function, start with x and undo the above steps in the reverse order.

0. Start with x.
1. Undo adding 5 by subtracting 5. $x - 5$
2. Undo multiplying by 2 by dividing by 2. $\frac{x-5}{2}$
3. Undo cubing by taking the cube root. $\sqrt[3]{\frac{x-5}{2}}$

$\Rightarrow \quad f^{-1}(x) = \sqrt[3]{\frac{x-5}{2}}$

***For this example, let's take the extra time to prove these are truly inverse functions, so you're convinced this technique works.

$$f\left(f^{-1}(x)\right) = f\left(\sqrt[3]{\frac{x-5}{2}}\right) = 2\left(\sqrt[3]{\frac{x-5}{2}}\right)^3 + 5 = 2\left(\frac{x-5}{2}\right) + 5 = x - 5 + 5 = x$$

and

$$f^{-1}\left(f(x)\right) = f^{-1}\left(2x^3 + 5\right) = \sqrt[3]{\frac{2x^3 + 5 - 5}{2}} = \sqrt[3]{\frac{2x^3}{2}} = \sqrt[3]{x^3} = x$$

(c) We did the operations to x in the following order for $f(x) = 7\sqrt[5]{x-1} - 2$:

1. Subtracted one.
2. Took the fifth root.
3. Multiplied by 7.
4. Subtracted 2.

Again, for the inverse function we need to start with x and undo these above operations in the reverse order.

0. Start with x.
1. Undo subtracting 2 by adding 2. $x + 2$
2. Undo multiplying by 7 by dividing by 7. $\frac{x+2}{}$
3. Undo taking the fifth root, by raising it to the fifth power. $\left(\frac{x+2}{7}\right)^5$
4. Undo subtracting 1 by adding 1. $\left(\frac{x+2}{7}\right)^5 + 1$

EXAMPLE 6

Find the inverse function for the following functions.

(a) $f(x) = \dfrac{2x-1}{2x+5}$

(b) $h(x) = \left(\dfrac{x-2}{x-4}\right)^3$

There is more than one appearance of x for both of these functions, namely once in the numerator and once again in the denominator. So for this case, we cannot use the fun "pants" technique because there is more than one layer of operations happening to x simultaneously. We must use the formal algebraic technique.

(a) We have $y = \dfrac{2x-1}{2x-5}$.

1. Exchange x and y. $x = \dfrac{2y-1}{2y+5}$

2. Solve for y.

$x = \dfrac{2y-1}{2y+5}$	Multiply both sides of the equation by $(2y + 5)$.
$x(2y+5) = 2y-1$	Distribute x through the parentheses on the left side of the equation.
$2xy + 5x = 2y - 1$	Subtract $5x$ and $2y$ from both sides of the equation.
$2xy - 2y = -5x - 1$	Factor out the y on the left side of the equation.
$y(2x-2) = -5x-1$	Divide by $(2x - 2)$ on both sides of the equation.
$y = \dfrac{-5x-1}{2x-2}$	And, we finally have y by itself.

3. $f^{-1}(x) = \dfrac{-5x-1}{2x-2}$

(b) To start, we have the function $y = \left(\dfrac{x-2}{x-4}\right)^3$.

1. Exchange x and y. $x = \left(\dfrac{y-2}{y-4}\right)^3$

2. Solve for y.

$x = \left(\dfrac{y-2}{y-4}\right)^3$	Take the cube root of both sides of the equation.
$\sqrt[3]{x} = \dfrac{y-2}{y-4}$	Multiply both sides of the equation by $(y - 4)$.
$\sqrt[3]{x}(y-4) = y-2$	Distribute the cube root of x through the parentheses on the left side.
$y\sqrt[3]{x} - 4\sqrt[3]{x} = y - 2$	Subtract y and add $4\sqrt[3]{x}$ to both sides of the equation.
$y\sqrt[3]{x} - y = 4\sqrt[3]{x} - 2$	Factor out the y on the left side of the equation.
$y\left(\sqrt[3]{x} - 1\right) = 4\sqrt[3]{x} - 2$	Divide both sides by $\sqrt[3]{x} - 1$ to get y by itself.
$y = \dfrac{4\sqrt[3]{x}-2}{\sqrt[3]{x}-1}$	

3. $f^{-1}(x) = \dfrac{4\sqrt[3]{x}-2}{\sqrt[3]{x}-1}$

4.1 Homework

Is the relation, as given, (a) a function, and (b) is it one-to-one (or invertible)?

1. $\{(1, 3), (2, 4), (3, -1), (4, 3), (5, 0), (6, 2)\}$

2. $\{(a, b), (b, p), (c, p), (a, r), (b, s)\}$

3. A relation defined by the set of inputs = {students in my Math 1010 class}, outputs = {the students' final letter grades for the class}

4. $f(x) = x^3$

5. $g(x) = \sqrt{x} + 1$

6. A relation defined by: inputs = {citizens of U.S.}, outputs = {Social Security numbers}

Determine whether the given function has an inverse or not. If it doesn't have an inverse, state how to restrict the domain so that the function does have an inverse.

7. $f(x) = (x - 2)^2$

8. $f(x) = -4x + 5$

9. $f(x) = -4x^2$

10. $g(x) = x^2 + 9$

11. $h(x) = \dfrac{-3}{x}$

12. $p(x) = \dfrac{4}{x + 2}$

13. $f(x) = x^3 - 1$

14. $f(x) = \dfrac{4}{\sqrt[3]{x}}$

15. $y(x) = |3x - 2|$

16. $k(x) = 2x^4 - 1$

17. $f(x) = 3(x - 2)^2$

18. $m(x) = 3(x - 3)^3$

Show that these functions are inverses of each other. That is, show $f(g(x)) = g(f(x)) = x$.

19. $f(x) = \dfrac{3}{\sqrt[3]{x}}, \quad g(x) = \dfrac{27}{x^3}$

20. $f(x) = 3 - 2x, \quad g(x) = \dfrac{-1}{2}(x - 3)$

21. $f(x) = 2\sqrt[3]{x - 1}, \quad g(x) = \dfrac{1}{8}x^3 + 1$

22. $f(x) = \dfrac{-3}{x}, \quad g(x) = \dfrac{-3}{x}$

23. $f(x) = \dfrac{4}{x + 2}, \quad g(x) = \dfrac{4 - 2x}{x}$

For each function, find its inverse function, $f^{-1}(x)$.

24. $f(x) = 2x - 3$

25. $f(x) = 4\sqrt[3]{x + 2}$

26. $f(x) = \sqrt[5]{3x + 1}$

27. $g(x) = \dfrac{5}{x} - 1$

28. $p(x) = \dfrac{\sqrt[3]{2x - 3}}{5}$

29. $f(x) = 2x + 4$

30. $y(x) = \sqrt[3]{x + 1} + 2$

31. $h(x) = \dfrac{1}{3}x - \dfrac{1}{2}$

32. $k(x) = \dfrac{-3x - 4}{x - 2}$

33. $f(x) = \dfrac{4}{x + 2}$

34. $h(x) = \dfrac{3x + 1}{x}$

35. $m(x) = \dfrac{x^3 - 5}{x^3 + 1}$

36. $g(x) = \dfrac{4x + 2}{5x - 1}$

37. $w(x) = \sqrt[5]{\dfrac{x + 2}{x - 3}}$

38. $f(x) = 5x^3 - 2$

39. $y(x) = \dfrac{\sqrt[5]{2x - 1}}{3}$

40. $g(x) = \sqrt[3]{\dfrac{x + 1}{2x + 3}}$

41. $h(x) = 4 - 5x$

42. $k(x) = \dfrac{-1}{x - 3}$

43. $n(x) = \dfrac{4x}{2x + 7}$

44. For $f(x) = \sqrt{x} + 4$, can we find an inverse function? If so, what is it?

45. Are $f(x) = 2\sqrt{x} - 1$ and $g(x) = \frac{1}{4}(x + 1)^2$ inverse functions? If not, explain why.

46. For $f(x) = (x - 2)^2 + 3$, find the inverse, if it exists.

47. Are $f(x) = x^4$ and $g(x) = \sqrt[4]{x}$ inverse functions? If not, explain why.

48. Find the inverse for $p(x) = 3(x - 1)^2 + 2$ given $x \geq 1$.

49. Find the inverse for $g(x) = \sqrt{\frac{x-1}{4}} + 5$.

For each graph, determine if the funtion graphed passes the horizontal line test. If it does pass the horizontal line test, sketch its inverse function graph.

50.

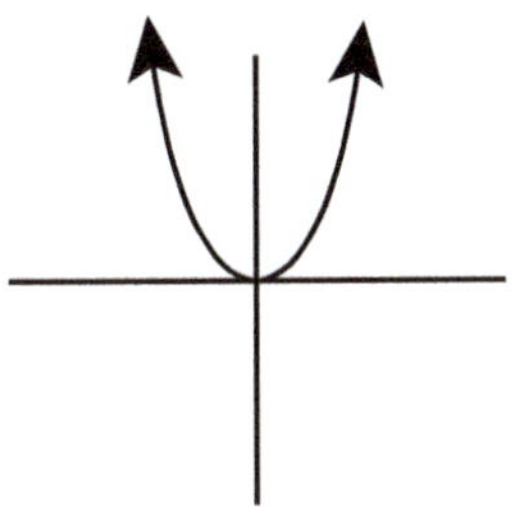

51.

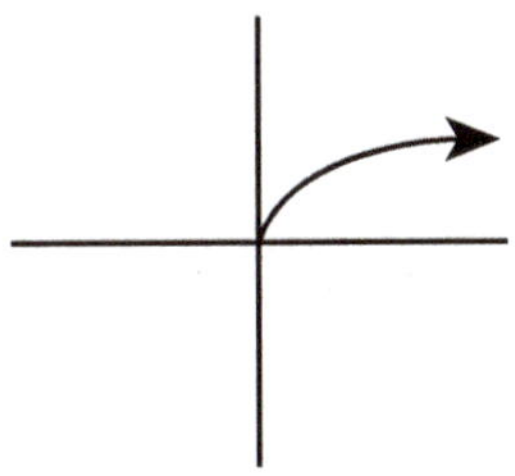

52.

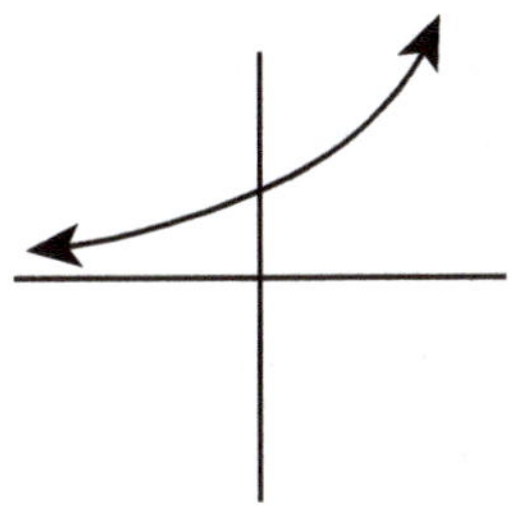

53.

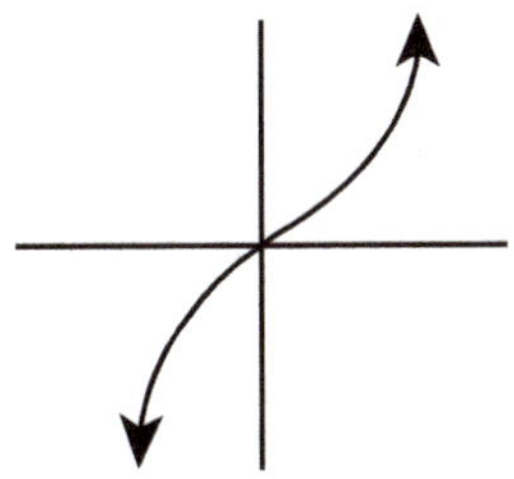

4.2 Exponential Functions

We will now explore a new type of function, one where the input variable, usually called x, appears in the exponent. Up until now, if there were exponents in the function, they were constants and the input variable only appeared in the base of the power, not in the exponent. Exponential functions will have the constant in the base and the variable in the exponent.

VOCABULARY

Exponential Expression/Equation/Function—An expression/equation/function where the variable is in the exponent, not in the base.

e—This is an irrational number (a real number that, when expressed in decimal form, never ends and never repeats). It is approximately equal to 2.7182818. This number has its own special name of e (similar to π) as its use and appearance in applications is incredibly frequent and it warrants its own name.

DEFINITION

If $a \in \mathbb{R}$, $a > 0$ and $a \neq 1$, then $y = f(x) = a^x$ is an **exponential function** with base a.

Notice that we disallow a from being 1. That's because 1 to any power is still 1, so $y = f(x) = 1^x = 1$ would just be a horizontal line when graphed. That is, the output would always be one regardless of the input value of x. Although that is a reasonable function, it does not fit the nature of the exponential function.

This graph displays the basic shape of an exponential function.

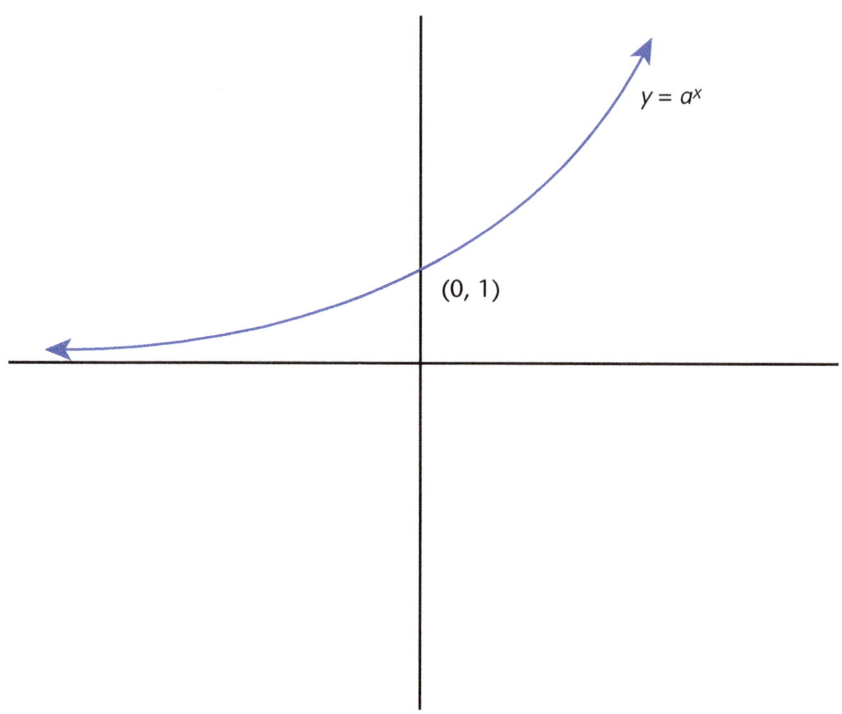

Let's use this example to compare an exponential function to something that we already know: For a linear function, if we look at the outputs for inputs of 0, 1, 2, 3, 4, . . . , we would see that the outputs are computed by repeatedly adding the same value (the slope) to the y-values. For an exponential function, the outputs are computed by repeatedly multiplying the same value (the base of the exponential) by the y-values, instead of by adding.

The following table demonstrates this.

x	$y = 3x + 5$	$y = 5(3^x)$
0	$y = 3(0) + 5 = 5$	$y = 5(3^0) = 5$
1	$y = 3(1) + 5 = 8$	$y = 5(3^1) = 15$
2	$y = 3(2) + 5 = 11$	$y = 5(3^2) = 45$
3	$y = 3(3) + 5 = 14$	$y = 5(3^3) = 135$
4	$y = 3(4) + 5 = 17$	$y = 5(3^4) = 405$

You can see that both the linear and the exponential functions start with a y-value of 5. From there, the linear function grows by adding the slope, 3, over and over again. The exponential function grows by multiplying the base, 3, over and over again. Repeatedly adding the same value is also known as arithmetic growth and repeatedly multiplying the same value is known as geometric growth.

GRAPHS OF EXPONENTIAL FUNCTIONS

Here are some examples of exponential function graphs.

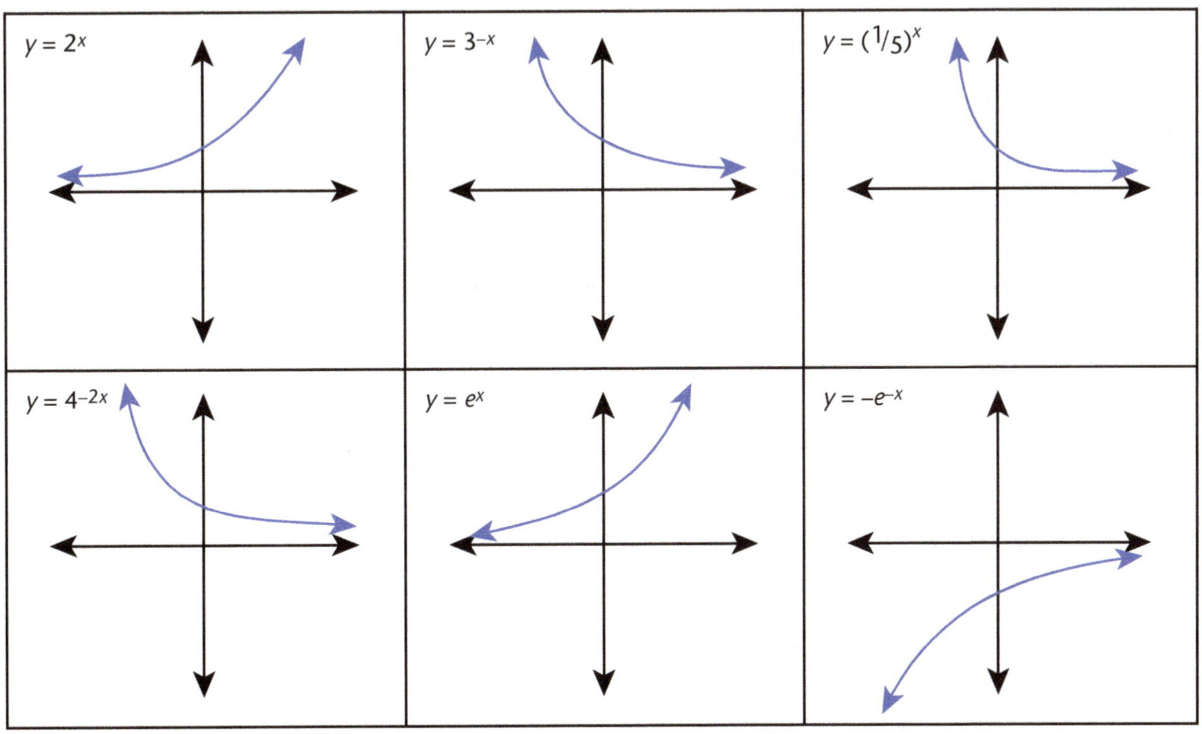

EXAMPLE 1

Graph these functions by plotting a few points and using the knowledge of the basic shape of an exponential function graph.

(a) $y = 5^x$

(b) $y = 2^{x-1} + 3$

(a) Let's first fill in a table of points for this function.

x	y
0	$y = 5^0 = 1$
1	$y = 5^1 = 5$
–1	$y = 5^{-1} = \frac{1}{5}$
2	$y = 5^2 = 25$

We know the basic shape of an exponential function's graph, and there have been no transformations of the graph for this function. So, we can simply plot these points and connect the points with a smooth curve that fits the basic shape.

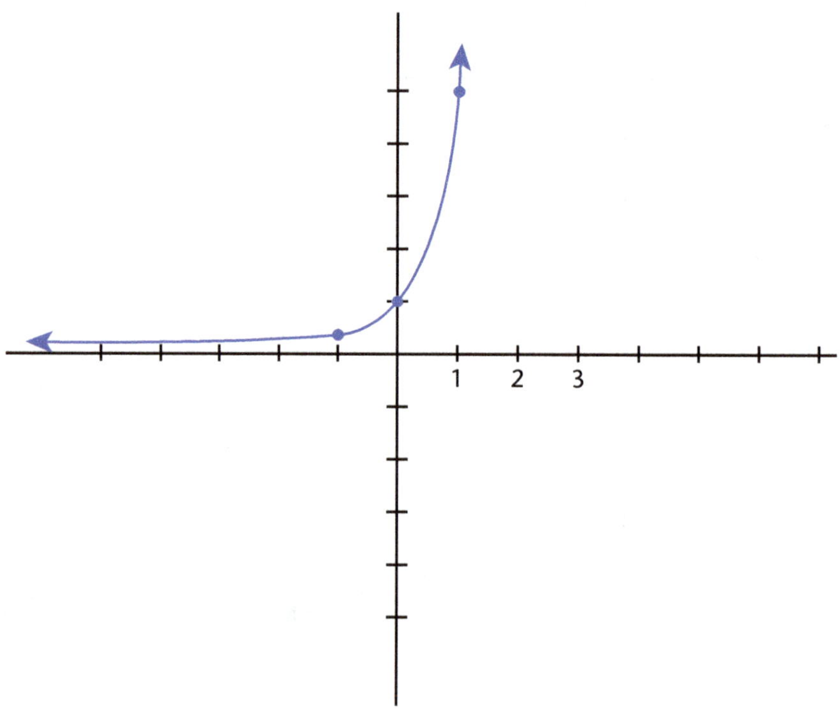

(b) $y = 2^{x-1} + 3$

This function is basically the regular exponential graph, with base of 2. So, the base function (or parent function) will be $y = 2^x$. It's also been shifted, as we can see from the adding/subtracting that happens in the exponent ("inside the function") and after the exponential takes place ("outside the function"). Subtracting 1 in the exponent shifts the graph horizontally right by 1 unit. And adding 3 outside the exponential shifts the graph vertically up 3 units.

We can plot points to see where this graph lives, or we can simply graph the base function and then shift it accordingly, as pointed out above. For this example, we'll choose to plot the base function as a dotted curve and then do the shifting. Here is a table of points to show explicitly what is happening.

(x, y) of base graph $y = 2^x$	(x, y) of shifted graph $y = 2^{x-1} + 3$
(0, 1)	(1, 4)
(1, 2)	(2, 5)
(2, 4)	(3, 7)

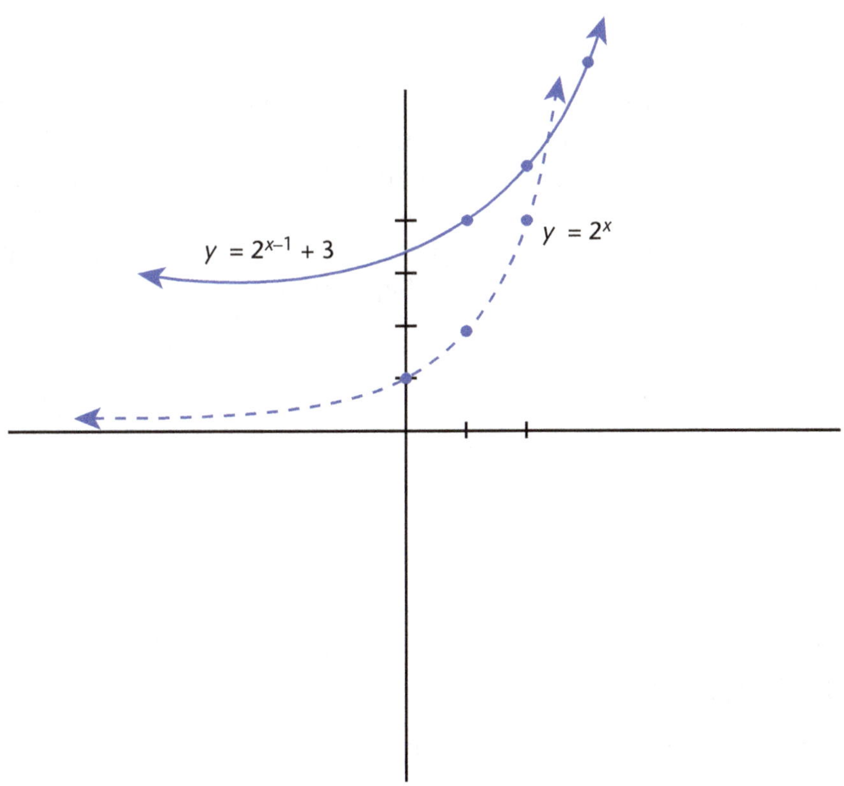

EXAMPLE 2

Match the function with the graph for these graphs.

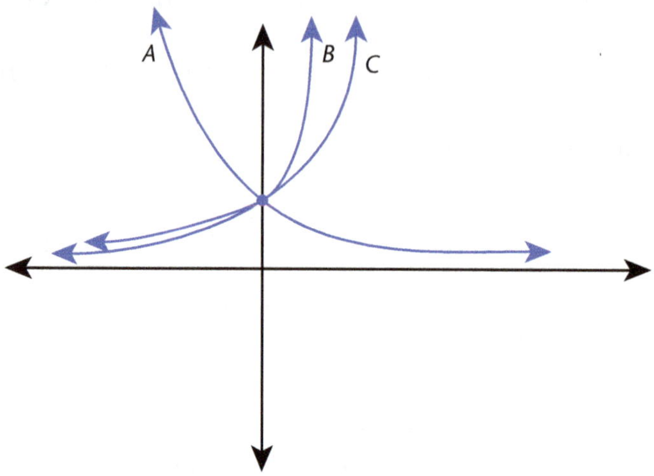

(a) $y = 6^x$

(b) $y = e^x$

(c) $y = 0.4^x$

Looking at the graphs, we can see that two of them grow bigger, meaning that they go up forever as you look at the graph from left to right. One of the graphs decreases, meaning that it goes down when you look from left to right. For the growth graphs, that means the base of the exponential is bigger than 1. For the decaying graph, the base of the exponential must be smaller than 1. Thus, we already have narrowed down that graph A is matched with the exponential function in (c) because that's the only exponential function with a base less than 1, namely 0.4. That leaves graphs B and C. We can see that graph B grows faster than graph C. That must mean that the base of B's exponential function is bigger than C's exponential function. Thus, graph B goes with (a) since $6 > e$. That leaves graph C to be matched with (b).

$\Rightarrow$ A → (c)
B → (a)
C → (b)

EXAMPLE 3

If a person invests \$5,000 for t years at 8% interest compounded continuously, the equation $A = 5000e^{0.08t}$ gives the value (A) after t years. What will the value be after 10 years?

For this problem, we need to substitute 10 in for t and use our calculator to evaluate A.

$\Rightarrow$ $A = 5000e^{0.08(10)}$ Simplify exponent.
$A = 5000e^{0.8}$ Use a calculator to evaluate the right side expression.
$A = \$11,127.70$

EXAMPLE 4

The population of Bricktown has been modeled to grow according to the function $P = 135{,}000(1.012)^t$, where t represents the number of years after 2002 and P is the population.

(a) What was the population in 2005?

(b) What will the population be in the year 2012?

(a) To compute the population in 2005, we need to convert the year to a value of t. In 2005, $t = 3$, since t measures the number of years after 2002. Substitute that value of t into the given equation and compute P.

$$\Rightarrow \quad P = 135{,}000(1.012)^3$$

Remember to round to the nearest whole number since this represents the number of people, and we can't have a portion of a person.

$$P = 139{,}919$$

(b) In 2012, $t = 10$.

$$\Rightarrow \quad P = 135{,}000(1.012)^{10} = 152{,}103$$

4.2 Homework

Use a calculator to evaluate the expression, rounding to 3 decimal places.

1. $(4.3)^{5.6}$
2. 2^{10}
3. e^{2}
4. e^{-5}
5. π^{3}
6. π^{π}
7. $e^{\sqrt{2}}$
8. e^{π}
9. π^{e}
10. 10^{e}

Match the function with the graph.

11. $y = 2^{x}$
12. $y = 4^{x}$
13. $y = \left(\frac{1}{2}\right)^{x}$
14. $y = 10^{x}$

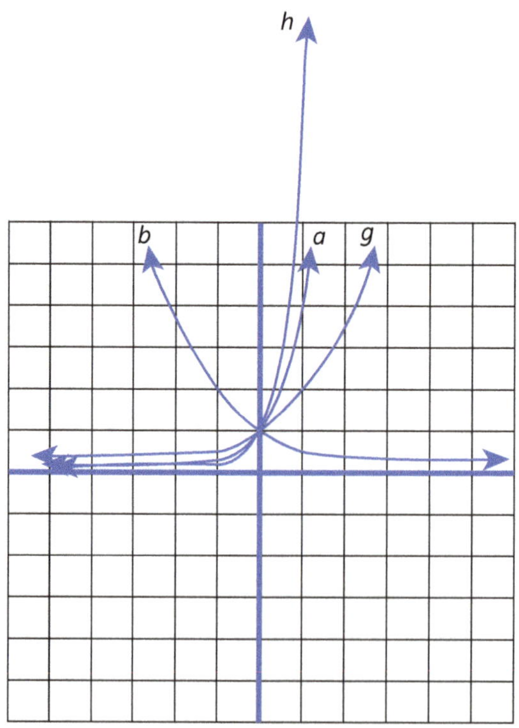

Simplify each expression.

15. $\dfrac{e^{2x}}{e^{1-x}}$

16. $\dfrac{\pi^{x-5}}{\pi^{x+2}}$

17. $(2^{4x})^{(x-1)}$

18. $\dfrac{5^{3-x}}{5^{x+3}}$

19. $\left(\dfrac{3}{2}\right)^{-x}$

20. $(2e^{-1.2})^x$

Sketch the graph of each function. Be sure to label the y-intercept and the horizontal asymptote.

21. $y = 2^x$

22. $y = e^{-x}$

23. $y = -2^x$

24. $y = 10e^{0.02x}$

25. $f(x) = e^x - 2$

26. $f(x) = 2^{x-2}$

27. $g(x) = 3^{\frac{x}{100}}$

28. $h(x) = 1 + e^{-x}$

29. $(1.14)^x = y$

30. $\pi^x = y$

31. The demand function is given by $p(x) = 500 - 0.2e^{0.004x}$, where x is the number of units demanded. Find the price, p, when demand is 50.

32. Radium has a half-life of 1620 years. The amount, y, of radium after t years is $y = 25\left(\frac{1}{2}\right)^{\frac{t}{1620}}$, measured in pounds.

    (a) How much radium is left after 75 years?
    (b) What is the initial amount of radium?

33. Carbon-14 has a half-life of 5730 years. The amount $y(t)$ of carbon-14 present after t years is given by $y = y_0\left(\frac{1}{2}\right)^{\frac{t}{5730}}$, where y_0 is the amount present at $t = 0$. How much is left after 18,000 years if the initial amount was 10 grams?

34. A car costs \$20,000 new. Its depreciating value is given by $v(t) = 20{,}000\left(\frac{3}{4}\right)^t$, where t is time after purchase, measured in years. What is the car worth after 7 years?

35. A colony of bacteria grows according to the exponential function $N(t) = 100e^{0.2197t}$, where N is the number of bacteria at time t, measured in hours. How many bacteria are present after 11 hours?

36. A town in 2009 increases in population according to $P(t) = 25{,}000e^{0.0293t}$, where P is the population at time t given in years, i.e. t measures the number of years since the year 2009. What is the town's population in 2012?

37. A company's advertising program hopes to reach 2 million people. The number of people actually aware of the advertised product is modeled by $N(t) = 2(1 - e^{-0.037t})$, where N is the number of people, in millions, aware of the advertised product t days after the ads start. How many people are aware of the product after 3 days?

38. World population growth, in billions, is modeled by $P(t) = 6.5e^{0.0114t}$, where P is the population t years after 2009. What is the population in (a) 2015 and (b) 2030?

39. Referring to problem 39, interpret the factor 0.0114 in $P(t) = 6.5e^{0.0114t}$.

4.3 Logarithmic Functions

Now that we've explored both exponential functions and inverse functions, we will continue on to discuss logarithmic functions. First, let's consider the graph of the basic exponential function $f(x) = y = a^x$, where $a > 0$, $a \neq 1$. Notice that it passes both the vertical line test (which means it is a function) and the horizontal line test (which means it is invertible, that is it has an inverse function). Remember that an inverse function's graph is the reflection of the function's graph across the line $y = x$. So we already have enough information to graph the inverse exponential function.

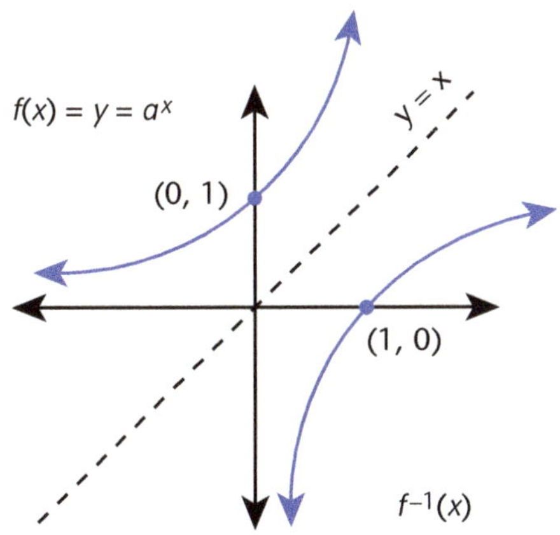

This is the fun part—now that we've established that the exponential function has an inverse, what do we name this inverse function? We can choose or create any symbol for this new function. . . . Well, mathematicians already made this decision long ago, so we'll stay with the accepted choice.

> **DEFINITION**
> The **logarithmic function,** notated as $y = \log_a x$, is equivalent to the exponential statement $a^y = x$, if $a > 0$, $a \neq 1$.

The domain of the logarithmic function is $x > 0$. This is due to the fact that x is the result of raising a to some power. If a is positive, then any power of a will also be positive, which means x must be positive.

VOCABULARY

log *x*—This logarithmic expression that has no explicit base indicated is "log base 10."

ln *x*—This is a logarithmic expression of base e, called the natural logarithm.

Keep in mind that the logarithmic function is simply the name of the inverse function for the exponential function. It's similar to saying that subtraction is the inverse function for addition. The logarithm "undoes" the exponential and vice versa. Remember that $f(f^{-1}(x)) = f^{-1}(f(x)) = x$ by definition of an inverse function. For these exponential and logarithmic functions, this means for

$$f(x) = a^x \text{ and } f^{-1}(x) = \log_a x,$$

$$\Rightarrow \quad f(f^{-1}(x)) = f(\log_a x) = a^{\log_a x} = x \text{ and } f^{-1}(f(x)) = f^{-1}(a^x) = \log_a a^x = x.$$

Remember from elementary school, when you first learned division (which is the inverse function for multiplication), you probably knew your multiplication facts reasonably well. So when you encountered the question $15 \div 3 = ?$, you may have converted it into the more familiar question $3 \cdot ? = 15$. This works perfectly well exactly because multiplication and division are inverse operators, which means these two questions are ultimately the same question. Logarithms are no different than this. Since it's a bit unfamiliar to evaluate a logarithmic expression, it will be best to convert the logarithmic statement into its equivalent exponential statement. Let's do some examples.

EXAMPLE 1

Rewrite these statements using the definition of the logarithm.

(a) $5^2 = 25$

(b) $\log_4 (\frac{1}{16}) = -2$

(c) $\log_{12} 144 = 2$

(d) $3^5 = 243$

(a) Using the definition of the logarithm for this example, we have $a = 5$, $y = 2$, $x = 25$. Use those values to rewrite the original exponential statement into its equivalent logarithmic statement.

$$\log_5 25 = 2$$

(b) According to the definition, for this logarithmic statement $a = 4$, $y = -2$, $x = \frac{1}{16}$. Then, the equivalent exponential statement is $4^{-2} = \frac{1}{16}$.

(c) Using the definition, we have $\log_{12} 144 = 2 \Leftrightarrow 12^2 = 144$.

(d) Again, using the definition, we have $3^5 = 243 \Leftrightarrow \log_3 243 = 5$.

EXAMPLE 2

Evaluate these logarithmic expressions.

(a) $\log_6 36$

(b) $\log_2 64$

(c) $\log 1000$

(d) $\ln 1$

(a) $\log_6 36 = ?$ is equivalent to asking $6^? = 36$, i.e., "6 raised to what power gives back 36?" We know $6^2 = 36$, which means $\log_6 36 = 2$.

(b) $\log_2 64$ is equivalent to asking $2^? = 64$, or "2 raised to what power gives back 64?" We can figure out that $2^6 = 64$, which means that $\log_2 64 = 6$.

(c) $\log 1000$ is equivalent to asking $10^? = 1000$, or "10 raised to what power gives back 1,000?" We know that $10^3 = 1000$, which means that $\log 1000 = 3$.

(d) $\ln 1$ is equivalent to asking $e^? = 1$, or "e raised to what power gives back 1?" We know that any nonzero base raised to the zero power gives back one, that is $e^0 = 1$. Thus, $\ln 1 = 0$.

EXAMPLE 3

Match the function with the graph for these graphs.

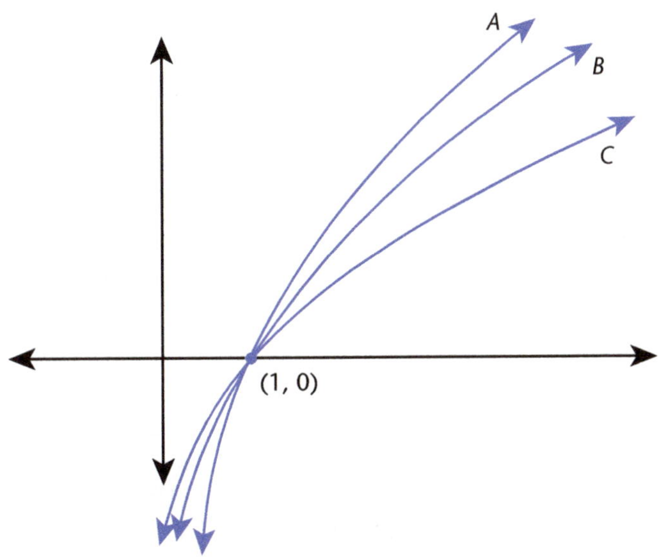

(a) $y = \log_5 x$

(b) $y = \ln x$

(c) $y = \log x$

First of all, we can see that the shapes of these logarithmic function graphs are all very similar. Basically they're the same shape, except that they increase vertically at different rates, reading the graphs from left to right. They all go through the point (1, 0). The main difference in the graphs comes from the different bases of the logarithms. Notice from the graphs that graph A increases (goes up) faster than graphs B or C. This means that graph A goes with the logarithmic function with the smallest base. We can think of this by looking at where the graphs have a y-value of 1.

(a) goes through the point (5, 1)
(b) goes through the point (e, 1) (and e is between 2 and 3)
(c) goes through the point (10, 1)

Thus, the smallest *x*-value of these functions that yields a *y*-value of 1 would be *e*, for the graph of (b). This is the graph that increases vertically the most rapidly. The next most rapid vertical increase comes from (a) and then finally (c).

$\Rightarrow$ A → (b)
B → (a)
C → (c)

EXAMPLE 4

Graph these logarithmic functions.

(a) $f(x) = y = \log x$

(b) $f(x) = y = \log_3 x + 2$

(c) $f(x) = y = -\ln x$

(d) $f(x) = y = \log_5 (x + 1) - 3$

(a) We know the basic shape (as given earlier in this section). We really need to just make sure to plot a few points that the function goes through and then connect the dots with the shape we know for a logarithmic function's graph. For this table, we've chosen good values for *x* and then substituted those values into the function to find their corresponding *y*-values.

x	$f(x) = y = \log x$
1	$y = \log 1 \Leftrightarrow 10^y = 1 \Rightarrow y = 0$
10	$y = \log 10 \Leftrightarrow 10^y = 10 \Rightarrow y = 1$
$\frac{1}{10}$	$y = \log (\frac{1}{10}) \Leftrightarrow 10^y = \frac{1}{10} \Rightarrow y = -1$

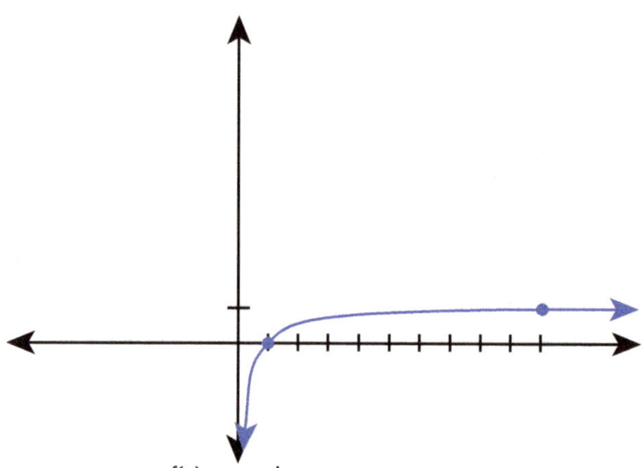

$f(x) = y = \log x$

(b) Here's a table with a few points for this function. One thing to notice first is that adding 2 "outside" the logarithmic function is simply a vertical shift up by 2, compared to the base function's graph for $y = \log_3 x$.

x	$f(x) = y = \log_3 x + 2$
1	$? = \log_3 1 \Leftrightarrow 3^? = 1 \Rightarrow ? = 0 \Rightarrow y = \log_3 1 + 2 = 0 + 2 = 2$
3	$? = \log_3 3 \Leftrightarrow 3^? = 3 \Rightarrow ? = 1 \Rightarrow y = \log_3 3 + 2 = 1 + 2 = 3$
$\frac{1}{3}$	$? = \log_3 (\frac{1}{3}) \Leftrightarrow 3^? = \frac{1}{3} \Rightarrow ? = -1 \Rightarrow y = \log_3 (\frac{1}{3}) + 2 = -1 + 2 = 1$

Note: $\log_a x + c = (\log_a x) + c \neq \log_a(x + c)$

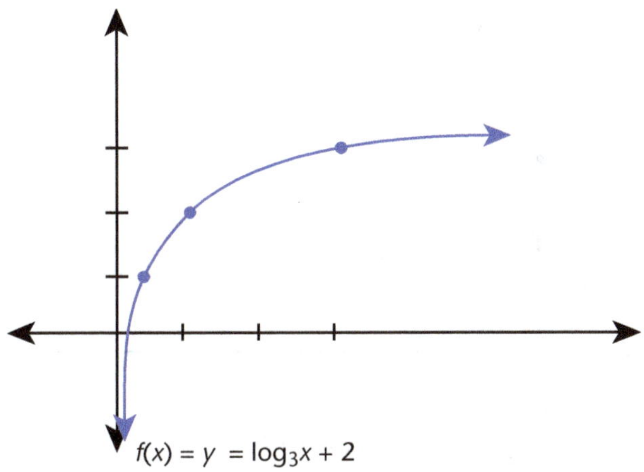

(c) For this logarithmic function, the base is e. The only difference between this function and the base function of $y = \ln x$ is that we multiplied by -1 "outside" the logarithmic function. Graphically, this means the graph of this function will be vertically reflected (across the x-axis) compared to the base function graph.

x	$f(x) = y = -\ln x$
1	$y = -\ln 1 = -(0) = 0$
e	$y = -\ln e = -(1) = -1$
e^2	$y = -\ln e^2 = -(2) = -2$

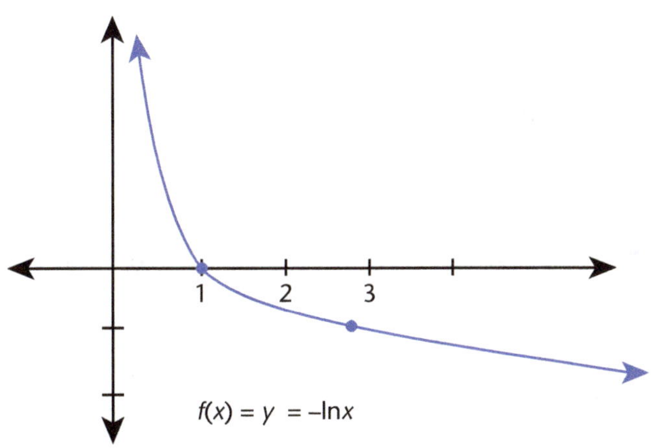

(d) Notice that subtracting 3 "outside" the logarithmic function shifts the base graph of $y = \log_5 x$ down by 3 units for this function. Also, adding 1 "inside" the logarithmic function shifts the graph left by 1 unit. We'll fill in a table of values for x and y from the base graph and then simply shift those points according to the last two sentences to get the points on our function in question. That is, from the base graph point (x, y), we can get the point on our function's graph by subtracting 1 from the x-value and subtracting 3 from the y-value.

x	$y = \log_5 x$ (base function)	Point (x, y) on $y = \log_5(x + 1) - 3$
1	$y = \log_5 1 = 0$	$(1 - 1, 0 - 3) = (0, -3)$
5	$y = \log_5 5 = 1$	$(5 - 1, 1 - 3) = (4, -2)$
$\frac{1}{5}$	$y = \log_5 (\frac{1}{5}) = -1$	$(\frac{1}{5} - 1, -1 - 3) = (\frac{-4}{5}, -4)$

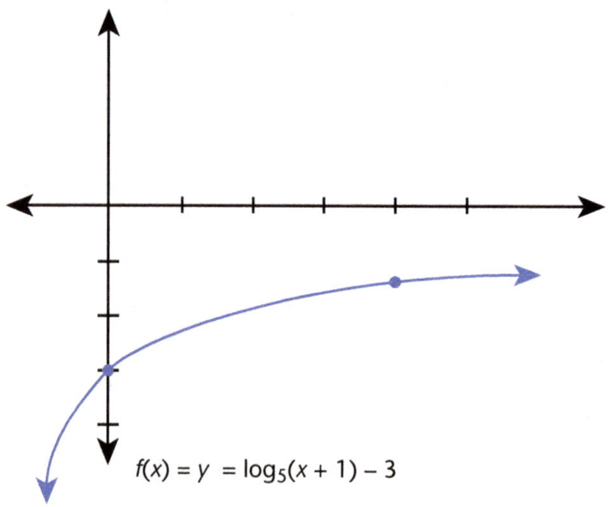

EXAMPLE 5

The formula for loudness of sound is

$$L = 10\log\left(\frac{I}{I_0}\right)$$

where loudness (L) is measured in decibels (dB) and intensity (I) is measured in watts per meter squared $\left(\frac{W}{m^2}\right)$.

I_0 is equal to $10^{-12}\left(\frac{W}{m^2}\right)$, which is the threshold of human hearing. This means the formula is

$$L = 10\log\left(\frac{I}{10^{-12}}\right)$$

(a) If you're at a party and the stereo is playing music with intensity $3 \times 10^{-5}\left(\frac{W}{m^2}\right)$, what is the loudness of sound?

(b) At the same party, someone decides to turn on a second stereo (at the same time) with the same intensity. How much louder is the sound from both stereos compared to the sound from only one stereo?

(a) To evaluate the loudness, we need to substitute the given intensity in for I in the given formula.

$\Rightarrow \quad L = 10\log\left(\frac{3(10^{-5})}{10^{-12}}\right)$ Simplify the number inside the logarithmic expression.

$$L = 10\log\left(3(10^{7})\right)$$

At this point, we can use a calculator to get an approximate value of this numerical expression. Remember that this is using log base 10, which our calculator will have as a button.

$$L = 74.77 \text{ dB}$$

(b) Basically, we need to compare two loudness values, one for the original intensity from one stereo, and another for twice that intensity, or $6 \times 10^{-5}\frac{W}{m^2}$. In part (a), we already calculated the loudness from one stereo. We can now calculate the loudness from both stereos playing.

$\Rightarrow \quad L = 10\log\left(\frac{6(10^{-5})}{10^{-12}}\right)$ Simplify the number inside the logarithmic expression.

$$L = 10\log\left(6(10^{7})\right)$$

With the help of a calculator, this evaluates to about $L = 77.78$ dB.

This means that doubling the intensity only affects the loudness by about $77.78 - 74.77 = 3.01$ dB.

4.3 Homework

Rewrite the logarithmic statement as its equivalent exponential statement using the definition of logarithm.

1. $\log_3 27 = 3$
2. $\log_2 32 = 5$
3. $\ln 1 = 0$
4. $\log_4 8 = \frac{3}{2}$
5. $\log 100 = 2$
6. $\log 1 = 0$
7. $\log_4 64 = 3$
8. $\log_7 \frac{1}{49} = -2$
9. $\log_{16} 8 = \frac{3}{4}$
10. $\log_3 81 = 4$

Use the logarithmic definition to rewrite these exponential statements into their equivalent logarithmic statements.

11. $10^{-3} = 0.001$
12. $8^2 = 64$
13. $9^{\frac{3}{2}} = 27$
14. $81^{\frac{1}{4}} = 3$
15. $e^x = 2$
16. $49 = 7^2$
17. $a = e^x$
18. $36 = 6^2$
19. $e^{-x} = 5$
20. $2^7 = 128$

Simplify these expressions, without a calculator.

21. $\log_2 4$
22. $\log_{10} 10^{-4}$
23. $\log_7 1$
24. $\log 10$
25. $\log_{16} 4$
26. $\ln e^3$
27. $\log_a a^2$
28. $\log_3 81$
29. $\log_{27} 9$
30. $\log_{25} \left(\frac{1}{5}\right)$

31. What is $\log_a 1$, for any base a?

32. Sound intensity is given by the formula $\beta = 10 \log \frac{I}{I_0}$, where I_0 is the reference intensity, just below the threshold of hearing; $I_0 = 10^{-16}$ watts per square centimeter. Beta is called the decibel rating. What decibel rating is a sound of intensity $I = 10^{-13}$ W/cm^2 (a whisper)?

33. Find β for $I = 10^{-8}$ W/cm^2 (heavy traffic). (Use the formula given in problem 32.)

34. The time t for any quantity to double, if it is increasing at the rate of r, is given by $t = \frac{\ln 2}{r}$. Complete the table (round to the nearest integer).

r	0.005	0.01	0.02	0.03	0.04
t					

For each function, answer the following:

(a) *What is the domain?*
(b) *What is the x-intercept?*
(c) *Sketch its graph. Be sure to draw and label the vertical asymptote.*

35. $f(x) = \log x$
36. $g(x) = \log_5 (x + 1)$
37. $h(x) = \log (x - 3)$
38. $f(x) = \ln (3 - x)$
39. $y(x) = \ln (-x) + 5$
40. $p(x) = -\log (x - 1)$

4.4 Properties of Logarithms

PROPERTIES OF LOGARITHMS	
For all of these properties, we're assuming that $x \in \mathbb{R}$, $a > 0$, $a \neq 1$, $m > 0$, $n > 0$.	**Proofs:**
1. $\log_a 1 = 0$	**1.** By the definition of logarithms, $a^0 = 1$ (which is true since $a \neq 0$) $\Leftrightarrow \log_a 1 = 0$.
2. $\log_a a = 1$	**2.** Again, using the logarithm definition, we have $a^1 = a \Leftrightarrow \log_a a = 1$.
3. $\log_a a^x = x$	**3.** This will also use the definition of logarithm. $a^x = (a^x) \Leftrightarrow \log_a (a^x) = x$. We can claim this is true as well by basically remembering that we defined the logarithm specifically to be the inverse of the exponential, which means the logarithm "undoes" the exponential function of the same base.
4. $a^{\log_a x} = x$, for all $x > 0$	**4.** This is also taking advantage of the definition of the logarithm being the inverse function of the exponential of the same base. We have to restrict x to be positive here because the logarithmic function is only defined for a domain of positive x. By definition then, $\log_a x = (\log_a x) \Leftrightarrow a^{\log_a x} = x$.
5. $\log_a(mn) = \log_a m + \log_a n$	**5.** Let $u = \log_a m$ and $w = \log_a n$. Then, by the definition of the logarithm, we also have the equivalent statements: $a^u = m$ and $a^w = n$. $\Rightarrow \log_a(mn) = \log_a(a^u a^w)$ Using substitution. $= \log_a(a^{u+w})$ By rules of exponents. $= u + w$ By property (3) above. $= \log_a m + \log_a n$ Using substitution again.
6. $\log_a(m^n) = n(\log_a m)$	**6.** Let $u = n(\log_a m)$. $\Leftrightarrow \frac{u}{n} = \log_a m$ Dividing the above equation by n. $\Leftrightarrow a^{\frac{u}{n}} = m$ By the logarithm definition. So, $\log_a(m^n) = \log_a((a^{\frac{u}{n}})^n)$ By substitution. $= \log_a(a^u)$ Using rules of exponents. $= u$ Using property (3) above. $= n(\log_a m)$
7. $\log_a\left(\frac{m}{n}\right) = \log_a m - \log_a n$	**7.** Now that we've proven properties (5) and (6), we can use those to prove this property. $\log_a\left(\frac{m}{n}\right) = \log_a(mn^{-1})$ Rewriting in power form. $= \log_a m + \log_a(n^{-1})$ Using property (5). $= \log_a m + (-1)\log_a n$ Using property (6). $= \log_a m - \log_a n$

NOTE OF WARNING:

(i) The logarithm function does NOT distribute, as multiplication distributes through addition and/or subtraction. In other words,

$$\log(a \pm b) \neq \log a \pm \log b.$$

Be very careful to avoid this commonly made mistake!!!

(ii) Another commonly made mistake to avoid is regarding property (6):

$$(\log a)^m \neq m \log a.$$

Looking at property (6), we see that to apply that rule, we must have the exponent only on the expression inside the logarithm, NOT if the exponent occurs outside the whole logarithmic expression.

We can use these logarithmic properties to expand or condense logarithmic expressions. We'll need these skills in order to solve logarithmic equations!

EXAMPLE 1

Use logarithmic properties to expand these expressions into as many terms as possible:

(a) $\log_2\left(\frac{3x}{x+4}\right)$

(b) $\log\left(y^4\sqrt{y-3}\right)$

(c) $\ln((2x+1)(x-5))$

(d) $\log_3\left(\frac{x^2y^3}{w^4}\right)$

(a) $\log_2\left(\frac{3x}{x+4}\right) = \log_2(3x) - \log_2(x+4)$ — By property (7).

$= \log_2 3 + \log_2 x - \log_2(x+4)$ — By property (5).

(b) $\log(y^4\sqrt{y-3}) = \log y^4 + \log\sqrt{y-3}$ — By property (5).

$= \log y^4 + \log(y-3)^{\frac{1}{2}}$ — Rewriting the square root as a power.

$= 4\log y + \frac{1}{2}\log(y-3)$ — By property (6).

(c) $\ln((2x+1)(x-5)) = \ln(2x+1) + \ln(x-5)$ — By property (5).

There is nothing else to do here using the logarithmic properties. Remember that we cannot "distribute" the logarithmic function through addition. The logarithm of a product is the sum of the logarithms, but that is quite different than distributivity!

(d) $\log_3\left(\dfrac{x^2y^3}{w^4}\right) = \log_3(x^2y^3) - \log_3(w^4)$ By property (7).

$= \log_3(x^2) + \log_3(y^3) - \log_3(w^4)$ By property (5)

$= 2\log_3 x + 3\log_3 y - 4\log_3 w$ By property (6).

EXAMPLE 2

Use logarithmic properties to condense these expressions into one logarithmic term.

(a) $\log_5 8 - \frac{1}{3}\log_5 2$

(b) $\log_4(3x+1) + 5\log_4(x-2)$

(c) $3(\ln 6 + 4\ln 5 - \ln 2)$

(d) $\log 5x + 2(\log x + \log(x+2))$

(a) $\log_5 8 - \frac{1}{3}\log_5 2 = \log_5 8 - \log_5(2^{\frac{1}{3}})$ By property (6).

$= \log_5\left(\dfrac{8}{2^{\frac{1}{3}}}\right)$ By property (7).

$= \log_5\left(\dfrac{2^3}{2^{\frac{1}{3}}}\right)$ Rewriting 8 as 2^3.

$= \log_5 2^{\frac{8}{3}}$ Using the rule of exponents $\left(\dfrac{a^m}{a^n} = a^{m-n}\right)$ to finish simplifying.

(b) $\log_4(3x+1) + 5\log_4(x-2) = \log_4(3x+1) + \log_4(x-2)^5$ By property (6).

$= \log_4((3x+1)(x-2)^5)$ By property (5).

(c) $3(\ln 6 + 4\ln 5 - \ln 2) = 3(\ln 6 + \ln 5^4 - \ln 2)$ By property (6).

$= 3(\ln(6(5^4)) - \ln 2)$ By property (5).

$= 3\ln\left(\dfrac{6(5^4)}{2}\right)$ By property (7).

$= 3\ln(3(5^4))$ Simplifying the fraction, by dividing out 2 on top and bottom.

$= \ln(3(5^4))^3$ By property (6).

$= \ln(3^3 5^{12})$ Using the rule of exponents $((a^m b^n)^p = a^{mp} b^{np})$ to finish simplifying.

For this problem, let's point out another equally efficient path to condense this expression. Keep in mind that there is *always* more than one way to answer any mathematical question. For most of our examples we choose one efficient and correct path to show you because it is the most efficient, but that is not to imply it is the only way to proceed. Correctness is the first priority and efficiency is secondary.

$3(\ln 6 + 4\ln 5 - \ln 2) = 3\ln 6 + 12\ln 5 - 3\ln 2$ Using distributivity (because multiplication distributes through addition and subtraction).

$= \ln 6^3 + \ln 5^{12} - \ln 2^3$ By property (6).

$= \ln 6^3 + \ln\left(\frac{5^{12}}{2^3}\right)$ By property (7).

$= \ln\left(\frac{5^{12}6^3}{2^3}\right)$ By property (5).

$= \ln(5^{12}3^3)$ Using rules of exponents to simplify $\frac{6^3}{2^3} = \left(\frac{6}{2}\right)^3 = 3^3$.

(d) $\log(5x) + 2(\log x + \log(x+2)) = \log(5x) + 2\log(x(x+2))$ By property (5).

$= \log(5x) + \log(x(x+2))^2$ By property (7).

$= \log((5x)(x(x+2))^2)$ By property (5).

EXAMPLE 3

Evaluate these logarithmic expressions exactly (without the use of a calculator).

(a) $\log_2 \sqrt{2} + \log_3 \sqrt[3]{3} - \log_4 \sqrt[4]{4}$

(b) $\log_7 \left(7\sqrt{7}\right)^3$

(c) $\ln e^{6.23} - \log 100 + \log_5\left(\frac{1}{125}\right)$

(a) $\log_2 \sqrt{2} + \log_3 \sqrt[3]{3} - \log_4 \sqrt[4]{4} = \log_2 2^{\frac{1}{2}} + \log_3 3^{\frac{1}{3}} - \log_4 4^{\frac{1}{4}}$ Rewriting the roots as powers.

$= \frac{1}{2} + \frac{1}{3} - \frac{1}{4}$ By property (3), since the base of each logarithm matches the base of the power.

$= \frac{6}{12} + \frac{4}{12} - \frac{3}{12} = \frac{7}{12}$

(b) $\log_7 (7\sqrt{7})^3 = \log_7 \left(7 \cdot 7^{\frac{1}{2}}\right)^3$ Rewriting the square root as a power.

$= \log_7 \left(7^{\frac{3}{2}}\right)^3$ Using the rule of exponents ($a^m a^n = a^{m+n}$) to simplify.

$= \log_7 \left(7^{\frac{9}{2}}\right)$ Using the rule of exponents ($(a^m)^n = a^{mn}$) to simplify further.

$= \frac{9}{2}$ By property (3).

(c) $\ln e^{6.23} - \log 100 + \log_5 \left(\frac{1}{125}\right) = \ln e^{6.23} - \log 10^2 + \log_5 5^{-3}$ Rewriting each number as a power, with a base that matches the base of its logarithm.

$= 6.23 - 2 + (-3)$ By property (3).

$= 6.23 - 5 = 1.23$

EXAMPLE 4

Given $\log_b u = 3.21$, $\log_b v = 1.75$, and $\log_b w = -0.69$, evaluate the following expressions.

(a) $\log_b \left(\frac{uv}{w}\right)$

(b) $\log_b (vw)^2$

(a) $\log_b \left(\frac{uv}{w}\right) = \log_b (uv) - \log_b w$ By property (7).

$= \log_b u + \log_b v - \log_b w$ By property (5).

$= 3.21 + 1.75 - (-0.69)$ Substituting the values given.

$= 3.21 + 1.75 + 0.69$

$= 3.90 + 1.75 = 5.65$

(b) $\log_b (vw)^2 = 2\log_b (vw)$ By property (6).

$= 2(\log_b v + \log_b w)$ By property (5).

$= 2(1.75 + -0.69)$ Substituting the values given.

$= 2(1.06) = 2.12$

4.4 Homework

Use the properties of logarithms to expand these expressions completely.

1. $\log\left[x^2\left(x^4 - 3x^3\right)^{\frac{1}{2}}\right]$

2. $\log\left[\dfrac{(x+2)^2}{x-1}\right]$

3. $\ln\left(e^{\ln x}\right)$

4. $\ln\left(\dfrac{e^x}{x^e}\right)$

5. $\log_5\left[\dfrac{\left(x^2 - 3x + 2\right)^{\frac{3}{4}}}{\left(x^2 - 1\right)^{\frac{3}{4}}}\right]$

6. $\log_3\left[\dfrac{x^7}{\left(x^2 - 1\right)^{\frac{1}{2}}}\right]$

7. $\ln\left(\dfrac{e^x - e^{-x}}{e^{2x}}\right)$

8. $\log\left[x^2\sqrt{\dfrac{x+3}{x-1}}\right]$

9. $\ln\left[\dfrac{(x-1)^3}{x^2 - 1}\right]^{\frac{1}{2}}$

10. $\ln\left(\dfrac{x^4\sqrt{2x-3}}{\sqrt[3]{4x+2}}\right)^{\frac{2}{3}}$

Use logarithm properties to write these expressions in condensed form.

11. $\log(2x-1) - \frac{3}{4}\log(2x-1) + \frac{1}{3}\log(x^4 - 3)$

12. $\ln x^2 + 2\ln x + \ln x^2$

13. $\log_5 x^3 - \log_5 y + \log_5(xy) - \log_5\left(\dfrac{x}{y}\right)$

14. $\log_2 2 + \log_2 x^2 + 3\log_2 y + \frac{1}{2}\log_2 7 - \log_2 3$

15. $\ln 6 - \ln 3 + \ln(2x)$

16. $\log x^4 + \log(x+2)^{\frac{1}{2}} - \frac{1}{2}\log(x+2)$

17. $\ln(2x) + \ln(3x) + \ln(4x) - \ln(5x)$

18. $\ln x^2 + \ln x^3 + \ln x^4 + \ln x^5$

19. $\log_3 4 + \log_3 x - \log_3 4x$

20. $\log_2(2x) + \log_2(3y) - \log_2(xy)$

Evaluate these logarithmic expressions without a calculator.

21. $\log_2 2 + \log 10 + \log_4 4$

22. $\ln e + \ln e^2 + \ln e^3$

23. $\log_7 7^3 - \log_7 \sqrt[3]{7}$

24. $\ln e^2 - \ln e^{-2}$

25. $\ln e^{10} + \log 10^e$

26. $\log_2 8 - \log 100$

27. $\log_5\left(\frac{100(15)}{12(125)}\right)$

28. $\log_4\left(\frac{1}{64}\right) + \ln(e^5)$

29. $\log_3\left(\frac{1}{81}\right) - \log(300) + \log 3$

30. $\log_5(125) + \log_2(64)$

Given $\log_b x = 1.2$, $\log_b y = 3.1$, and $\log_b z = 11.1$, evaluate the expressions.

31. $\log_b\left(\frac{xz}{y}\right)$

32. $\log_b(xzy)^{\frac{1}{2}}$

33. $\log_b x - \log_b\left(\frac{yz}{x}\right)$

34. $\log_b(xz) - \log_b(xy) - \log_b(x^2)$

35. $\log_b(xz)^{\log_b y}$

36. $(\log_b y)(\log_b x)(\log_b z)$

37. $\log_b\left(\frac{xy}{z^3}\right) - \log_b(x^2 z)$

38. $\log_b\left(\frac{x}{y}\right) - \log_b\left(\frac{y}{x}\right)$

4.5 Logarithmic and Exponential Equations

We can classify an equation as logarithmic if the variable is inside the logarithmic operator. Likewise, an equation is exponential if the variable is in the exponent. These types of equations require the following strategies to solve with the most efficiency. (Remember that there are always many ways to solve an equation. These strategies are by no means the "only" way to get to the answer. However, this book offers them as the most efficient strategies.)

STRATEGY TO SOLVE A LOGARITHMIC EQUATION	STRATEGY TO SOLVE AN EXPONENTIAL EQUATION
1. Use addition/subtraction to move all the logarithmic expressions on one side of the equation. 2. Use logarithmic properties to condense the logarithmic expressions into one term on one side of the equation, so that the logarithmic expression is totally isolated. 3. Either (a) Use the definition of logarithm to rewrite the statement as its exponential equivalence. OR (b) Exponentiate both sides of the equation, using the same base as the logarithm. (This is to "undo" the logarithm since exponentiation is the inverse of the logarithm, of the same base.) 4. Continue solving the equation with whatever strategy is most useful and efficient for that type of equation. (Typically, at this point, the equation will be linear or quadratic.) 5. Check all solutions to ensure they are in the domain of the original equation. *(Remember, for a logarithmic function we can only take the logarithm of a positive number! And we always must find the domain from the original equation before we do any simplification steps.)*	1. Isolate the exponential expression on one side of the equation. 2. Either (a) Use the definition of logarithm to rewrite the statement as its logarithmic equivalence. OR (b) Take the logarithm of both sides of the equation, using either the same base as the logarithm or using the natural logarithm. (This is to "undo" the exponential since logarithm is the inverse operator of a logarithm of the same base.) 3. Continue solving the equation with whatever strategy is most useful and efficient for that type of equation. (Again, the resulting equation at this point will usually be linear or quadratic.) * For exponential equation solutions, you are welcome to check your answers. However, it's not completely required to ensure correctness because the domain of an exponential expression is unrestricted, unlike a logarithmic expression.

Please notice how similar the strategies are. Basically, we need to isolate the logarithmic/exponential expression, rewrite the statement using the logarithm definition and then continue solving. Let's try it now.

EXAMPLE 1

Solve the following logarithmic equations.

(a) $3 \log_2 x = 12$

(b) $\ln (x + 3)^7 = 4$

(c) $\ln (4 - x) + \ln 2 = 2 \ln x$

(d) $\log_3 x + \log_3(x - 8) = 2$

(e) $\log_5(7x) - \log_5(x + 1) = 1$

(f) $\log_5(x^2) = (\log_5 x)^2$

(a) $3 \log_2 x = 12$ — To isolate the logarithmic expression, we need to divide both sides of the equation by 3.

$\log_2 x = 4$ — Rewrite this statement using the definition of logarithms.**

$2^4 = x$ — Now, we have x alone. We simply need to calculate the left side value.

$x = 16$ — Check to make sure this value is allowed in the original equation.

Since our original domain was $x > 0$, we can keep this solution because it meets that condition.

**At this point, we could have exponentiated both sides of the equation instead, with base 2.

$\Rightarrow 2^{\log_2 x} = 2^4 \Leftrightarrow x = 2^4$, since the exponential with base 2 and the logarithm with base 2 are inverse functions of one another. This takes us back to the same next step in the above process.

(b) $\ln(x + 3)^7 = 4$ — It will be most efficient here to use the logarithmic property to pull the 7 exponent down in front of the logarithmic expression.

$7 \ln(x + 3) = 4$ — To isolate the logarithmic expression, divide both sides of the equation by 7.

$\ln(x + 3) = \frac{4}{7}$ — Use the logarithm definition to rewrite this as an exponential statement.**

$e^{\frac{1}{2}} = x + 3$ — Finish solving for x by subtracting 3 from both sides of the equation.

$e^{\frac{1}{2}} - 3 = x$ — We can rewrite this as $x = \sqrt[7]{e^4} - 3$, if that's easier to think about.

We need to make sure this solution is in the domain of the original equation. Notice that the original logarithmic statement is of something to the 8th power. Anything to the 8th power is positive, so the only value of x that must be excluded is -3, because then we would be trying to take the logarithm of zero, which is not allowed. So, our solution of $x = \sqrt[7]{e^4} - 3$ is completely fine.

**Again, we could have exponentiated both sides of the equation at this point using e as the base.

$\Rightarrow e^{\ln(x+3)} = e^{\frac{1}{2}} \Leftrightarrow x + 3 = e^{\frac{1}{2}}$ (because the exponential with base e is the inverse function of the natural logarithm), which takes us back to the same next step as above.

(c) $\ln(4 - x) + \ln 2 = 2 \ln x$ — Use logarithmic properties to condense the logarithmic expressions on each side of the equation.

$\ln(2(4 - x)) = \ln x^2$ — Subtract $\ln x^2$ from both sides of the equation to get all logarithmic expressions on one side of the equation.

$\ln(2(4 - x)) - \ln x^2 = 0$ — Use logarithmic properties to condense the logarithmic expressions.

$\ln\left(\frac{(2(4 - x))}{x^2}\right) = 0$ — Rewrite the logarithmic statement into its equivalent exponential statement using the logarithm definition.**

$\frac{(2(4 - x))}{x^2} = e^0$ — Remember that $e^0 = 1$. Multiply both sides of the equation by x^2.

$2(4 - x) = x^2$ — Distribute the 2 through the parentheses on the left side of the equation.

$8 - 2x = x^2$ — Since it's now a quadratic equation, get all terms on one side of the equation.

$0 = x^2 + 2x - 8$ — Factor the quadratic expression.

$0 = (x + 4)(x - 2)$ — Set each linear factor equal to zero and solve for x.

$x + 4 = 0$ or $x - 2 = 0 \quad \Leftrightarrow \quad x = -4$ or 2

Do we get to keep both of these solutions? From our original equation, the first logarithmic expression needs $4 - x > 0 \Leftrightarrow x < 4$, and the logarithmic expression on the right side needs $x > 0$. Altogether, that means we can only allow x-values that fit this condition: $0 < x < 4$. The solution $x = 2$ fits this condition, but $x = -4$ does not, so we must throw that away as a solution. This means the only solution for this equation is $x = 2$.

**If we had exponentiated both sides of the equation with base e here, we would have the following next step.

$e^{\ln\left(\frac{(2(4-x))}{x^2}\right)} = e^0 \quad \Leftrightarrow \quad \frac{2(4 - x)}{x^2} = 1$ — This is exactly the next step in our above process.

(d) $\log_3 x + \log_3(x - 8) = 2$ — Use logarithmic properties to condense the logarithmic expressions on the left side of the equation.

$\log_3(x(x - 8)) = 2$ — Rewrite this logarithmic statement as its equivalent exponential statement using the definition of logarithm.**

$3^2 = x(x - 8)$ — Simplify each side of the equation.

$9 = x^2 - 8x$ — Subtract 9 from both sides of the equation to get all terms on one side with zero on the other side (our best strategy for solving a quadratic equation).

$0 = x^2 - 8x - 9$ — Factor the quadratic expression.

$0 = (x - 9)(x + 1)$ — Set each factor equal to zero and solve for x.

$0 = x - 9$ or $0 = x + 1 \quad \Leftrightarrow \quad x = 9$ or -1

Can we keep both solutions? If we look back to our original equation, the first logarithmic expression needs $x > 0$. The second logarithmic expression needs $x > 8$. The intersection (or overlap) of those restrictions is $x > 8$. Thus, we can only keep the solution $x = 9$.

**Using the other technique of exponentiating both sides of the equation with base 3, we would have $3^{\log_3(x(x-8))} = 3^2 \Leftrightarrow x(x-8) = 3^2$ which is identical to the following step above.

(e) $\log_5(7x) - \log_5(x+1) = 1$ — Use logarithmic properties to condense into one term on the left side of the equation.

$\log_5\left(\frac{7x}{x+1}\right) = 1$ — Use the logarithm definition to rewrite in its equivalent exponential form.**

$5^1 = \frac{7x}{x+1}$ — Multiply both sides of the equation by $(x + 1)$.

$5(x+1) = 7x$ — Distribute and simplify the left side of the equation.

$5x + 5 = 7x$ — Subtract $5x$ from both sides of the equation.

$5 = 2x$ — Divide both sides of the equation by 2 to get x by itself.

$x = \frac{5}{2}$

Let's check this solution with the domain from the original equation. For the first logarithmic term, we know $7x > 0$ must be true, which is the same as the condition $x > 0$. For the second logarithmic term, we need $(x+1) > 0 \Leftrightarrow x > -1$. The stricter condition between those two (or the intersection of the two restrictions) is $x > 0$. Our solution fits that condition, which means we can keep the solution.

**Using the other technique of exponentiating both sides of the equation with base 5, we have.

$$5^{\log_5\left(\frac{7x}{x+1}\right)} = 5^1 \quad \Leftrightarrow \quad \frac{7x}{x+1} = 5^1.$$

(f) $\log_5(x^2) = (\log_5 x)^2 \quad \Leftrightarrow \quad 2\log_5 x = (\log_5 x)^2$

This one is a bit tricky. Notice that, on the left side, we can apply one of the logarithmic properties and bring down the 2 power in front of the logarithm. However, we cannot do that on the right side because the square is outside the entire logarithmic expression, not just on x. Thus, we must use a more clever approach here.

$2\log_5 x = (\log_5 x)^2$ — Subtract $2\log_5 x$ from both sides of the equation.

$0 = (\log_5 x)^2 - 2(\log_5 x)$ — Factor out a $\log_5 x$ from both terms on the right side of the equation.

$0 = \log_5 x(\log_5 x - 2)$ — Now, set each factor equal to zero and solve those equations for x.

$\log_5 x = 0$ or $\log_5 x - 2 = 0$ — Isolate the logarithmic expressions in each equation.

$\log_5 x = 0$ or $\log_5 x = 2$ — Use the definition of the logarithm to rewrite these statements as their equivalent exponential statements.

$5^0 = x$ or $5^2 = x$ — Simplify!

$1 = x$ or $25 = x$

Since the only restriction on x from the original equation is $x > 0$, and both of these solutions satisfy that condition, then we have found two valid solutions for this logarithmic equation.

EXAMPLE 2

Solve the following exponential equations. Leave solutions in exact and simplified form.

(a) $(3^{(2x-5)})^2 = 3^4$

(b) $4(5^x - 1) = 20$

(c) $\frac{7}{2^x} = 14$

(d) $10^{x+1} + 3(10^x) = 39$

(e) $e^{2x} - 9e^x = -20$

(a) $(3^{(2x-5)})^2 = 3^4$ For this equation, it would be best to use the rule of exponents (that allows us to multiply the exponents together when a power is raised to a power) to simplify the left side of the equation.

$3^{(2x-5)2} = 3^4$ Now that the exponential expressions on both sides of the equation are simply 3 to some power, it must be true that the powers are the same since they have the same base.

$(2x - 5)2 = 4$ Distribute and simplify the left side of the equation.

$4x - 10 = 4$ Add 10 to both sides of the equation and then divide everything by 4 to solve for x.

$4x = 14 \quad \Leftrightarrow \quad x = \frac{14}{4} \quad \Leftrightarrow \quad x = \frac{7}{2}$

(b) $4(5^x - 1) = 20$ For this equation, we want to get the exponential term by itself. So divide both sides of the equation by 4.

$5^x - 1 = 5$ Add 1 to both sides of the equation to finish isolating the exponential term.

$5^x = 6$ Rewrite the statement as its equivalent logarithmic statement, using the definition of logarithm.**

$\log_5 6 = x$ This is our final answer since we cannot simplify it any further.

**If we wanted to get an approximate value for this, we would need to do something a bit different, since our calculator typically has only log base 10 and the natural log function. In this case, we can instead take the log base 10 of both sides of the equation.

$\Rightarrow \log 5^x = \log 6$ Use one of the logarithmic properties to bring the x down in front of the log expression on the left side of the equation.

$x(\log 5) = \log 6$ Divide both sides of the equation by log 5 to get x by itself.

$x = \frac{\log 6}{\log 5}$

This is now a value that we can plug into the calculator and get an approximate answer back.

(c) $\dfrac{7}{2^x} = 14$ Multiply both sides of the equation by 2^x to get it out of the denominator.

$7 = 14(2^x)$ Divide both sides of the equation by 14 to isolate the exponential expression.

$\dfrac{1}{2} = 2^x$ Remember that $\dfrac{1}{2}$ you can write as 2^{-1}.

$2^{-1} = 2^x$ So, now we have 2 to some power equals 2 to some power, which means the powers must be the same since they have the same base.

$\Rightarrow x = -1$

It would be great to take this opportunity to show a completely different way to solve this equation. Remember that we could fill twice as many pages in this book by showing multiple ways to do each example problem. In general, one way is good enough. However, occasionally it's better to remind ourselves that there are always multiple paths to find a solution.

$\dfrac{7}{2^x} = 14 \Leftrightarrow 7(2^{-x}) = 14$ Since $\dfrac{1}{2^x} = 2^{-x}$. Divide both sides of the equation by 7.

$2^{-x} = 2$ Use the definition of logarithm to rewrite this in its equivalent logarithmic statement.

$\log_2 2 = -x$ Simplify the left side of the equation.

$1 = -x \Leftrightarrow -1 = x$

(d) $10^{x+1} + 3(10^x) = 39$ This is a tricky equation to solve. We can tell because it has two instances of an exponential expression. This means we can't just "isolate the exponential term" because there are two of them. So we need to try something else first. In this case, let's rewrite 10^{x+1} as $10(10^x)$.

$10(10^x) + 3(10^x) = 39$ Now we see a next step we can take. Factor 10^x out of both terms on the left side of the equation.

$10^x(10 + 3) = 39$ Simplify.

$10^x(13) = 39$ At this point, we have only one exponential term. To isolate it, divide both sides of the equation by 13.

$10^x = 3$ Use the logarithm definition to rewrite this statement in its equivalent logarithmic form.**

$\log 3 = x$ This is our final solution.

**We also could have taken the logarithm of both sides of the equation.

$\Rightarrow \log 10^x = \log 3 \Leftrightarrow x(\log 10) = \log 3 \Leftrightarrow x = \log 3$ because $\log 10 = 1$

(e) $e^{2x} - 9e^x = -20$ Again, we need a clever technique for this example because we have more than one exponential term, so we can't just isolate the exponential term to start with. First, let's rewrite e^{2x} as $(e^x)^2$.

$(e^x)^2 - 9(e^x) = -20$ The way it's written now, it looks similar to a quadratic equation. We'll treat it as a quadratic equation and try to solve using techniques we know work for quadratic equations. Add 20 to both sides of the equation.

$(e^x)^2 - 9(e^x) + 20 = 0$ Factor, if possible.

$(e^x - 4)(e^x - 5) = 0$ Now set each factor equal to zero and solve for x.

$e^x - 4 = 0$ or $e^x - 5 = 0$ Add 4 or 5 to isolate the exponential in each equation.

$e^x = 4$ or $e^x = 5$ Use the definition of the logarithm to rewrite these statements.**

$\ln 4 = x$ or $\ln 5 = x$

**If we had taken the natural logarithm of both sides of the equation, the rest of the solution process would have been:

$\ln e^x = \ln 4$ or $\ln e^x = \ln 5$ $\Leftrightarrow$ $x(\ln e) = \ln 4$ or $x(\ln e) = \ln 5$ $\Leftrightarrow$ $x = \ln 4$ or $x = \ln 5$.

CHANGE OF BASE FORMULA

For $a > 0$, $a \neq 1$, $b > 0$, $b \neq 1$, we can change the base of the logarithm according to

$$\log_b x = \frac{\log_a x}{\log_a b} = \frac{\ln x}{\ln b} = \frac{\log x}{\log b}.$$

Proof:

Let's start with $x = b^y \Leftrightarrow y = \log_b x$. We can treat the first equation as an equation that we're trying to solve for y. Take the logarithm base a of both sides of the equation.

$\Leftrightarrow \log_a x = \log_a b^y$ Use logarithmic properties to bring the y down.

$\Leftrightarrow \log_a x = y(\log_a b)$ Now divide both sides of the equation by the coefficient of y to get y by itself.

$$\Leftrightarrow y = \frac{\log_a x}{\log_a b}$$

Thus, we now have two equal expressions for y, namely $\log_b x = \dfrac{\log_a x}{\log_a b}$.

The change of base formula is extremely useful when we need to get an approximate value for any final logarithmic numerical expression. In that case, where we need the use of a calculator, we need to either be using log base 10 or the natural log, since those are typically the only logarithmic bases offered on a calculator.

EXAMPLE 3

Evaluate these expressions as approximate values, by changing the base.

(a) $\log_6 18$

(b) $\log_4 23.5$

(a) $\log_6 18$ For this example,, let's change the base to 10.

$$\Rightarrow \quad \log_6 18 = \frac{\log 18}{\log 6}$$

If we plug this value into a calculator, we get about 1.613147.

(*Note:* If we had changed the base to e instead, we would still get the same answer.)

(b) $\log_4 23.5$ For this example, let's change the base to the natural logarithmic base e.

$$\Rightarrow \quad \log_4 23.5 = \frac{\ln 23.5}{\ln 4}$$

On a calculator, this evaluates to about 2.277294.

4.5 Homework

Solve each equation.

1. $3e^{2x} - 7 = 11$

2. $\left(\frac{3}{7}\right)^x = 4$

3. $\left(\frac{1}{4}\right)^x = \frac{1}{4}$

4. $3^{2x-1} = 2$

5. $3e^{2x} + 4 = 2e^{2x} + 11$

6. $2^{\pi x} = 3$

7. $5e^{x^2+2x-3} = 5$

8. $2\left(31^{2x}\right) = 8$

9. $10^x = 10^{x^2}$

10. $\pi^x = 4$

11. $\log x = 2$

12. $\log\left(x^2 - x\right) + \log 2 - \log x = 1$

13. $\log\left(x^2\right) + 1 = \log(3x + 1)$

14. $\log_\pi x = 2$

15. $\ln\left(x^2\right) = 4$

16. $\log_7(2x + 3) = \log_7 x - \log_7 2$

17. $\log_5\left(5^x\right) = 2x - 1$

18. $\log_2(3x - 2) - \log_2\left(x^2\right) = 0$

19. $\log_4\left(x^2 - 9\right) - \log_4(x + 3) = 3$

20. $e^{x^2} = e^{4x}e^{12}$

21. $\log_2\left(\log_x 16\right) = 2$

22. $3\left(4^{3-x}\right) + 2 = 7$

23. $\ln\left(x^2 - 4\right) + \log_3\left(x^2 - 4\right) = 1$

24. $12e^{x^2-2} = 17$

25. $\ln x^2 - 15 = \log x^2$

26. $11^4 = 121^{x^2+x}$

27. $\ln e^x + e^{\ln x} + \ln e = 4$

28. $3^{4-x} = 15$

29. $\log x + \log_5 x = 5$

30. $\log x^2 - \log\left(x^2 - 4\right) + \log x = \log(x - 1)$

31. $16^{x+1} = 4^{4x}$

32. $x^2 e^x - 5xe^x = 0$

33. $5^{2x} - 2\left(5^x\right) = 3$

34. $\log\left[(x + 5)(x - 1)\right] - \log(x + 3) = \log(x - 1)$

35. $3e^{2x} - 30 = 9e^x$

36. $\log_5(x - 2) - \log_5(x - 3) = \log_5(x + 4) - \log_5(x + 1)$

37. $2^{4x} - 81 = 0$

38. $\ln x^3 = (\ln x)^3$

Use the change of base formula and a calculator to evaluate each logarithm. (Give your answer rounded to three decimal places.)

39. $\log_2(5)$

40. $\log_3(12)$

41. $\log_5(2)$

42. $\log_{10}(e)$

43. $\log_2(2048)$

44. $\log_4(9)$

45. $\log_7(10)$

4.6 Logarithmic and Exponential Business Applications

In this section, we'll put everything we know about solving exponential and logarithmic equations together to use in real-life applications.

EXAMPLE 1

COMPOUND INTEREST **An investment compounded n times per year with an interest rate of r will double according to the formula $2 = (1 + \frac{r}{n})^{nt}$, where t is the number of years it takes to double. If the interest rate is 8% and the investment is compounded quarterly, how long will it take for the investment to double?**

An interest rate of 8% means $r = 0.08$. Quarterly compounding means that the account compounds four times per year, so $n = 4$. We need to substitute those values in for n and r (into the given formula) and then solve for t.

$$\Rightarrow \quad 2 = \left(1 + \frac{0.08}{4}\right)^{4t}$$

We have an exponential equation here, since our variable is in the exponent. First, simplify inside the parentheses.

$$2 = (1.02)^{4t}$$

Now that we have isolated the exponential term, we can either use the definition of the logarithm to change this statement into its equivalent logarithmic form or we can take the logarithm (either base e or base 10) of both sides. In this case, we'll choose to take the logarithm of both sides of the equation because the other choice would leave us with a logarithm in base 1.02, which isn't calculator-friendly.

$$\ln 2 = \ln(1.02)^{4t}$$

By logarithmic properties, we can bring the exponent down on the right side of the equation.

$$\ln 2 = 4t(\ln 1.02)$$

Solve for t by dividing both sides of the equation by the coefficient of t.

$$\frac{\ln 2}{4(\ln 1.02)} = t$$

Plug this value into a calculator to get an approximate value for t.

$$t \simeq 8.75 \text{ years}$$

Thus, for an interest rate of 8%, compounded quarterly, an investment takes just under 9 years to double its value.

EXAMPLE 2

NATIONAL DEBT **According to statistics from the U.S. Department of the Treasury, the national debt in 1980 was about \$930 billion. By 2007, it had reached approximately \$9,008 billion. Assuming that the growth is exponential (with the natural base *e*), find the function that represents this increase. Then use that function to predict the answers to the following questions.**

(a) In what year will the debt reach \$25,000 billion?

(b) What is the predicted national debt for the year 2050?

First, we need to take the two given data points and find the exponential function that goes through those points. We're looking for some function of the form $y = a(e^{bx})$, since we were told that the base of the exponential is e.

We'll use the two data points to find the values of a and b that will best describe this growth.

Let x = the time passed after 1980, measured in years.

(Since 1980 was the first year listed, we can choose to let that be basically time zero when we start measuring the years for the change in national debt.)

Then, y = the national debt, measured in billions of dollars (\$).

Thus, our points are (0, 930), since $x = 0$ in the year 1980, and (27, 9008), since $x = 2007 - 1980 = 27$ in the year 2007.

Now, let's substitute the first point into our generic equation.

$\Rightarrow \quad 930 = a(e^{b(0)})$ Simplify the exponential factor, to solve for a.

$930 = a(1)$ $\Leftrightarrow 930 = a$

Our function thus far is $y = 930e^{bx}$. At this point, we can substitute in our other point to find the value of b for this example.

$\Rightarrow \quad 9008 = 930e^{b(27)}$ Divide both sides of the equation by 930 to isolate the exponential term.

$\frac{9008}{930} = e^{b(27)}$ Take the natural logarithm of both sides of the equation.

$\ln\left(\frac{9008}{930}\right) = \ln(e^{27b})$ Use a logarithmic property to bring down the exponent on the right side of the equation.

$\ln\left(\frac{9008}{930}\right) = 27b\ln e$ Simplify the right side of the equation, remember that $\ln(e) = 1$.

$\ln\left(\frac{9008}{930}\right) = 27b$ Divide both sides of the equation by 27 to solve for b.

$\frac{1}{27}\ln\left(\frac{9008}{930}\right) = b$ Plug this computation into a calculator to get an approximate value.

$b \simeq 0.0841$ (We rounded to the nearest ten-thousandth place value.)

So far, we know the exponential function that will interpolate between the two data points given:

$$y = 930e^{0.0841x}$$

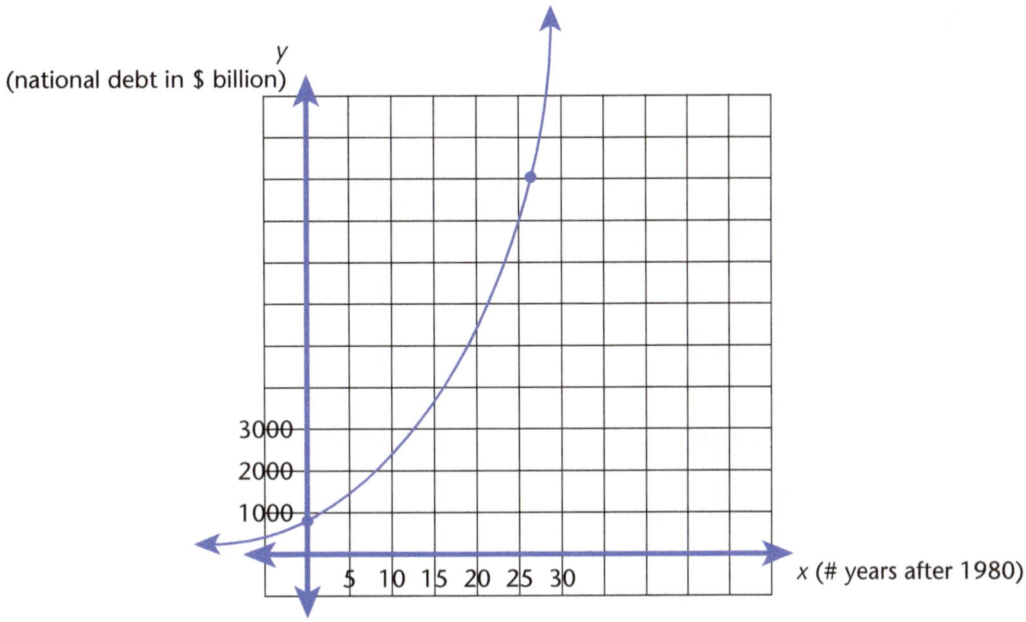

We can use this function to predict the growth of the national debt in future years. This is assuming, of course, that it's reasonable to model the growth with an exponential function.

(a) To determine the year that the debt will grow $25,000 billion means we need to use our function to find x when $y = 25000$.

$\Rightarrow$ $25000 = 930e^{0.0841x}$ Divide both sides of the equation by 930.

$\frac{2500}{93} = e^{0.0841x}$ Use the logarithm definition to rewrite this statement into its equivalent logarithmic statement. (Another equally good choice is to take the natural logarithm of both sides of the equation. We'll end up in the same place either way.)

$\ln\left(\frac{2500}{93}\right) = 0.0841x$ Divide both sides of the equation by 0.0841 to solve for x.

$\frac{1}{0.0841}\ln\left(\frac{2500}{93}\right) = x$ Use a calculator to compute an approximate value for x.

$x \simeq 39.14$ years

Thus, it will take a little more than 39 years for the national debt to grow to $25,000 billion. Remember that $x = 0$ corresponds to the year 1980. So, $x \simeq 39.14$ corresponds to $1980 + 39.14 = 2019.14$. Then, according to our model, we'll reach that debt in the year 2019.

(b) In the year 2050, x will be $2050 - 1980 = 70$. Plug that value of x into our function and solve for y.

$$\Rightarrow \quad y = 930e^{0.0841(70)} = 930e^{5.887} \simeq 335100.1047$$

This tells us that in the year 2050, according to our function's prediction, the national debt will be \$335,100 billion.

EXAMPLE 3

RICHTER SCALE **The Richter scale typically measures an earthquake's intensity. If I is the intensity, or the energy released from the earthquake (measured in joules), then**

$$M = \frac{2}{3}\log\left(\frac{I}{I_0}\right)$$

gives the magnitude of the earthquake. I_0 is the minimum intensity that one can feel (or a zero-level earthquake).

(a) What is the intensity for an earthquake that has a magnitude of 4.8?

(b) If an earthquake is twice as intense as that in part (a), what is its magnitude?

(c) An earthquake in San Francisco measured 7.1 in magnitude. How many times more intense was that earthquake than the earthquake in part (a)?

(a) We know M is 4.8 and we want to determine I. So, substitute our value for M into the above equation and solve for I.

$$4.8 = \frac{2}{3}\log\left(\frac{I}{I_0}\right)$$ Multiply both sides of the equation by $\frac{3}{2}$ to isolate the logarithmic term.

$$\frac{3}{2}(4.8) = \log\left(\frac{I}{I_0}\right)$$ Use the definition of logarithm to rewrite this statement into its equivalent exponential statement.

$$10^{1.5(4.8)} = \frac{I}{I_0}$$ Multiply both sides of the equation by I_0 to solve for I.

$$I_0(10^{7.2}) = I$$ Compute an approximate value for $10^{7.2}$.

$$I \simeq 15{,}848{,}931.9\, I_0$$

In other words, such an earthquake would be over 15 million times more intense than a zero-level earthquake.

(b) For this question, we're trying to solve for M, given that the intensity, I, is two times more than what we found in part (a), or $I \simeq 2(15848931.9\, I_0) = 31697863.85\, I_0$.

$$\Rightarrow \quad M \simeq \frac{2}{3}\log\left(\frac{31697863.85(I_0)}{I_0}\right)$$ Divide out the common factor I_0 in the fraction.

$M = \frac{2}{3}\log(31697863.85)$ Compute an approximate value for M with the aid of a calculator.

$M \simeq 5.00068$

Notice the difference in magnitudes for the supposed earthquakes in parts (a) and (b). An earthquake whose intensity feels twice as powerful equates to a small magnitude change from 4.8 to about 5.0!

(c) To determine how much more intense an earthquake with $M = 7.1$ is compared to an earthquake with $M = 4.8$, we need to first know the intensity of the 7.1-magnitude earthquake. So let's calculate that.

$7.1 = \frac{2}{3}\log\left(\frac{I}{I_0}\right)$ Multiply both sides of the equation by $\frac{3}{2}$.

$\frac{3}{2}(7.1) = \log\left(\frac{I}{I_0}\right)$ Rewrite into its equivalent exponential statement, using the logarithm definition.

$10^{1.5(7.1)} = \frac{I}{I_0}$ Multiply both sides of the equation by I_0 to solve for I.

$I_0(10^{10.65}) = I$

We want to know how this intensity compares to the intensity for the 4.8-magnitude earthquake. Specifically, we want to know how many times more intense it is. Then we need to take the ratio of the two intensities, with the 7.1-magnitude earthquake's intensity in the numerator.

$\Rightarrow \frac{\text{7.1-magnitude earthquake's intensity}}{\text{4.8-magnitude earthquake's intensity}} = \frac{10^{10.65} I_0}{10^{7.2} I_0}$ Divide out the common factor of I_0.

$= \frac{10^{10.65}}{10^{7.2}} = 10^{10.65-7.2} = 10^{3.45} \simeq 2818.38$

This tells us that the 7.1-magnitude earthquake is about 2,818 times more intense than the 4.8-magnitude earthquake.

EXAMPLE 4

MEDICINE DECAY Suppose that a patient has a painkiller given intravenously with an initial concentration of 400 mg/ml. (Assume the bloodstream completely absorbs the painkiller immediately.) After 2.5 hours, the patient's blood contains the painkiller in a concentration of only 150 mg/ml. Assuming that the concentration of the painkiller decays exponentially (with base e), find the function that represents this decay.

(a) When will the patient's blood have a concentration of only 10 mg/ml?

(b) What is the concentration of painkiller after 6 hours?

(c) What is the half-life of the medicine?

Before we can answer the specific questions posed, we must determine the exponential function that describes this medicinal decay in the body. The independent variable will be time, and the dependent variable will be the concentration of painkiller.

⇒ Let x = the time the medicine is in the body, measured in hours and
y = the concentration of medicine in the body, measured in mg/ml.

Then, we have been given two data points, namely (0, 400) and (2.5, 150). We also know that we are to model this decay with an exponential function, with base e, which means the function looks like $y = ae^{bx}$. We need to use the two points to solve for a and b here.

⇒ Plug in (0, 400). $400 = ae^{b(0)}$ Simplify.

$400 = ae^0 \Leftrightarrow 400 = a$

Now we know our function is $y = 400e^{bx}$, and we can plug in the other point to find b.

⇒ $150 = 400e^{b(2.5)}$ Divide both sides of the equation by 400 to isolate the exponential expression.

$\frac{3}{8} = e^{2.5b}$ Rewrite this statement into its equivalent logarithmic statement using the logarithm definition.

$\ln\left(\frac{3}{8}\right) = 2.5b$ Divide both sides of the equation by 2.5 to get b by itself.

$\left(\frac{1}{2.5}\right)\ln\left(\frac{3}{8}\right) = b$ Finally, do this computation with a calculator to get an approximate value of b.

$b \simeq -0.39233$

Notice that b is negative, which is as we expected. That indicates exponential decay rather than exponential growth, which is appropriate for this example.

The final form for our predictive function is $y = 400e^{-0.39233x}$. Now we can use that function to answer the questions.

(a) We're asked for the value of x when y = 10 mg/ml. Substitute $y = 10$ into our function and solve for x.

⇒ $10 \simeq 400e^{-0.39233x}$ Divide both sides of the equation by 400 to isolate the exponential expression.

$\frac{1}{40} \simeq e^{-0.39233x}$ Use the logarithmic definition to rewrite this statement.

$\ln\left(\frac{1}{40}\right) \simeq -0.39233x$ Divide both sides of the equation by the coefficient of x.

$\left(\frac{1}{-0.39233}\right)\ln\left(\frac{1}{40}\right) \simeq x$ Get an approximate value for x, using a calculator.

$x \simeq 9.402$

So, x is about 9.4 hours, which means it will take almost 9.5 hours for the medicine to decay to a point of only 10 mg/ml in the patient's blood.

(b) For a value of 6 hours, we use $x = 6$ in our function to determine the y-value, or the medicinal concentration in the bloodstream.

$$\Rightarrow \quad y \simeq 400e^{-0.39233(6)} \simeq 37.966$$

This means that the medicine concentration is about 38 mg/ml after 6 hours.

(c) The half-life means the time it takes for half of the medicine to dissipate or decay in the body. In other words, we are looking for the time, or x-value, when y is half of its original amount.

$\Rightarrow \quad \frac{1}{2}(400) \simeq 400e^{-0.39233x}$ — Divide both sides of the equation by 400.

$0.5 \simeq e^{-0.39233x}$ — Rewrite this exponential statement as its equivalent logarithmic statement using the definition of logarithm.

$\ln(0.5) \simeq -0.39233x$ — Divide both sides of the equation by the coefficient of x to solve for x.

$\left(\frac{1}{-0.39233}\right)\ln(0.5) \simeq x$ — Get an approximate value for x by plugging this into a calculator.

$x \simeq 1.766745$

This indicates that the half-life is a little less than 2 hours. Thus, half of the medicine dissipates in the body after about 1.76 hours.

EXAMPLE 5

LOGISTIC GROWTH Human population growth typically grows exponentially for some period of time before it levels off. This is also often true for the spread of infectious diseases. Suppose a high school student becomes infected with a disease on an international trip. He returns to his high school of 500 students and is initially the only one infected with this disease. If the function

$$P = \frac{500}{1 + 499e^{-0.86t}}$$

can model the spread of the disease, where t represents the number of days after the infected student's return to the high school, and P is the number of students infected at that time, how long will it take for 100 students at the high school to become infected?

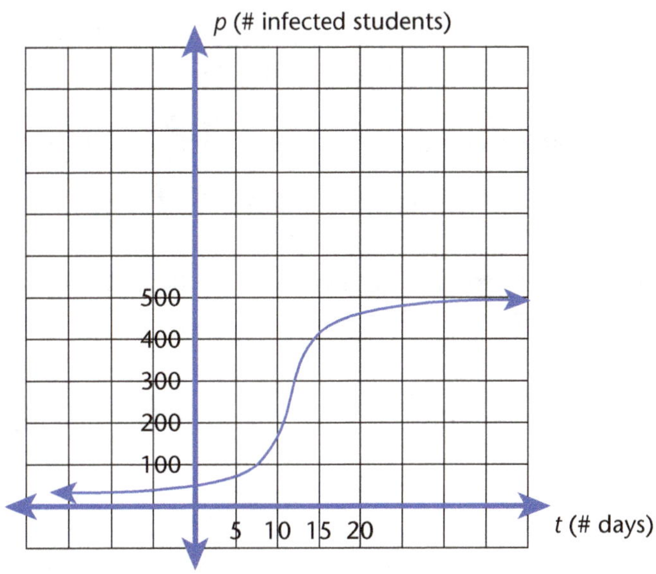

We're looking for the t-value when P is 100. Plug in 100 for P and solve for t.

$\Rightarrow \quad 100 = \dfrac{500}{1 + 499e^{-0.86t}}$	Multiply both sides of the equation by the right-side denominator.
$100(1 + 499e^{-0.86t}) = 500$	Divide both sides of the equation by 100 to start stripping off the layers of operations acting on t. Basically, we need to work toward isolating the exponential term.
$1 + 499e^{-0.86t} = 5$	Subtract 1 from both sides of the equation.
$499e^{-0.86t} = 4$	Divide both sides of the equation by 499.
$e^{-0.86t} = \dfrac{4}{499}$	Rewrite this statement into its equivalent logarithmic statement using the logarithm definition.
$-0.86t = \ln\left(\dfrac{4}{499}\right)$	Divide both sides of the equation by -0.86.
$t = \left(\dfrac{1}{-0.86}\right)\ln\left(\dfrac{4}{499}\right)$	Approximate the value of t by using a calculator to evaluate this.
$t \simeq 5.61199$ days	

So, after about 5.6 days, 100 students are infected.

EXAMPLE 6

DEMAND/REVENUE Suppose the demand for a particular product behaves according to the function $p = \frac{250}{\ln(q+1)}$ where q is the quantity demanded and p is its corresponding price.

(a) What is the price when consumers are demanding a quantity of 80 products?

(b) What is the corresponding revenue?

(a) We want to know the value of p when q is 80, so plug in $q = 80$ and find the value of p.

$$\Rightarrow \quad p = \frac{250}{\ln(80+1)} = \frac{250}{\ln 81}$$

Use a calculator to get an approximate value for p, rounding to the nearest hundredth since we measure p in dollars.

$$\Rightarrow \quad p \simeq \$56.89$$

This means that the market price is \$56.89 when consumers are demanding 80 products.

(b) Remember that the revenue function is the product of the demanded price and the number of products purchased at that price.

$$\Rightarrow \quad R(q) = \left(\frac{250}{\ln(q+1)}\right)q$$

We want to know the revenue for 80 products.

$$\Rightarrow \quad R(80) = \left(\frac{250}{\ln(80+1)}\right)80 \simeq (56.89)80 \simeq \$4551.20$$

4.6 Homework

Note: Most of these problems will be dealing with exponential growth/decay. Such problems can all be modeled with $y = y_0e^{kt}$ where y is measuring whatever quantity is growing or decaying, y_0 is the original amount of what's being measured, t is time, and k is the growth/decay rate (per the time period that t is being measured in), which is often given as a percentage. (See examples 2 and 4.)

1. If oil is being consumed at the rate of 2.3% per year, how long does it take to consume one half of all present oil reserves? How long will it take to deplete 90% of all the reserves? (*Hint:* Use $y = y_0e^{kt}$.)

2. Radioactive Iodine-131, a daughter product of uranium decay and nuclear reactions, has a half-life of 8 days. How long does it take to reduce an initial amount of Iodine-131 to 1% of the initial amount?

3. Radioactive Strontium-90 is a decayed product from some nuclear reactions and has a half-life of 28.9 days. If there are 100 grams now, how much is left after 180 days?

4. Uranium-238 is the most commonly occurring isotope of uranium. Its half-life is 4.47 billion years. How much (as a percentage) of an initial amount of uranium-238 is left after 100 years? What percent is left after 100,000 years?

5. Carbon-14 has a half-life of 5740 years. An archeaological site found charcoal fragments from a fire pit. The charcoal had only 25% as much carbon-14 as a present sample would have. How old was the fire pit?

6. Radioactive contamination from the Chernobyl nuclear power plant accident in 1984 was carried by winds to surrounding lands. The hay crop in neighboring Austria was contaminated by iodine-131, a daughter product from uranium decay, which has a half-life of 8 days. A rule of thumb is that after 10 half-lives have passed, such contamination is reduced to safe levels. What percent of an initial amount of iodine-131 contamination is safe using this guideline?

7. Population of mosquitoes grows exponentially at a rate of 58.8% per year. How long does it take for the mosquito population to double in size?

8. Refer to the information in problem 7. How long would it take a population of 1,000 mosquitoes to reach 1 million? How long would it take for the mosquito population to reach 100 million?

9. Refer back to problem 7 and 8. How long would it take for the weight of the mosquitoes to equal the mass of the earth, if 1000 mosquitoes weighs 2.5 grams and the earth weighs 5.9736×10^{24} kg? Is this answer realistic? Why aren't we overtaken by these mosquitoes?

10. Fill in the following table with values from exponential decay, where y represents the amount present at any time t and y_0 is the initial amount. This means that y_1 represents the amount of the indicated substance left at time t_1 and y_2 represents the amount of the indicated substance left at time t_2. Half-life is the time it takes for half of the initial amount of the substance to dissipate. What do the values for y_2 suggest? (*Hint:* Use $y = y_0e^{kt}$.)

ISOTOPE	HALF-LIFE	y_0	t_1	y_1	t_2 (10 HALF-LIVES)	y_2
Iodine-131	8 days	100 g	30 days		80 days	
Strontium-90	28.9 days	10 g	90 days		289 days	
Radium-226		1000 g	100 years	959.82 g	16,900 years	
Potassium	1.3 billion years		1000 years		13 billion years	0.0009765625 g

11. If the population of metropolitan Salt Lake City in 2009 is 300,000, and it was 250,000 in 1995, what is the growth rate for that time period, assuming exponential growth? Discuss your answer.

12. A thermometer reads 70 degrees Fahrenheit and is placed in a refrigerator that has a constant temperature of 0 degrees Fahrenheit. If the thermometer reads 46° F, after 5 minutes in the refrigerator, what will it read after 10 minutes? Use the modeling function given by $T = T_1 + (T_0 - T_1)e^{kt}$, where T is the temperature at any time t, T_0 is the initial temperature of the thermometer and T_1 is the constant temperature of the surrounding medium.

13. A thermometer reads 10 degrees Celsius and is put in a room with a constant temperature of 30 degrees Celsius. If the thermometer reads 17° C after 6 minutes in the room, what does the thermometer read after 11 minutes in that room? (Use the formula given in problem 12.)

14. A certain insect population grows exponentially at a rate of 2% per year. The initial population was 500 insects. How long does it take for the population to triple in size?

15. The city of Kenosha, Wisconsin, had a population of 78,000 in the year 2000. If this city grows exponentially at a rate of 1.1% per year, how long will it take for the population to double?

16. A pictograph in southern Utah was painted using pigments used from native plants and human urine. A recent test on a small sample of the pigment showed that it contained 86% of the carbon-14 to be expected from a modern sample. How long ago was this pictograph painted on the rock?

17. A population of 100 bacteria grows to 500 bacteria in 7 days. How long will it take to grow to 700 bacteria?

18. When table salt is dissolved in water, it dissociates into sodium and chlorine ions, according to an exponential growth/decay model. If 25 pounds of salt is placed in a container filled with water, and after 10 hours, 15 pounds is undissolved, how long will it take to reduce the undissolved amount of salt to 2 pounds?

19. Sound intensity is measured in decibels, according to $\beta = 10 \log\left(\frac{I}{I_0}\right)$, where I is the intensity of sound measured in watts per square meter. The threshold intensity for human hearing, I_0, equal to 10^{-12} watts per square meter, is the least intense sound a human ear can detect. What is the intensity of a sound measured at 130 decibels, the threshold of pain in humans?

20. A jet plane at full throttle can produce a level of sound intensity of 0.15 watts per square meter. What does this measure in decibels? (Use the formula given in problem 19.)

21. A relatively recent earthquake in Salt Lake City, Utah, measured 5.1 on the Richter scale. A large earthquake hit Salt Lake City measuring 7.5 on the Richter scale in the distant past. How much greater was the intensity of the past earthquake compared to the more recent one? Use $M = \log\left(\frac{I}{I_0}\right)$, where M is the magnitude of the earthquake, I is the intensity of the earthquake and $I_0 = 10^{-3}$ is the zero-level earthquake, or the minimum intensity that can be felt.

22. The earthquake that caused the tsunami of 2004 killed over 200,000 people and was measured at $M = 9.1$ on the Richter scale. How much larger was the tsunami quake compared to the recent Salt Lake City earthquake from problem 21?

23. An investment compounded n times per year with an interest rate of r will double according to the formula $2 = (1 + \frac{r}{n})^{nt}$, where t is the number of years it takes to double and n is the number of compoundings per year. If the interest rate is 5% and the investment is compounded monthly, how long will it take to double?

24. Refer to the formula given in problem 23. For what interest rate will your money double if it's invested for 20 years with quarterly compounding?

25. Suppose the demand for high-quality staplers behaves according to the formula given by $5\log(q)\, p = 50$, where q is the quantity demanded at a price of p dollars. What is the price consumers are demanding for a quantity of 100? What is the revenue when 100 staplers are sold?

Chapter 4 Review

4.1

For each function, find its inverse.

1. $f(x) = 2(x+3)^3 - 4$
2. $h(x) = \dfrac{\sqrt[5]{2x-1}}{3}$
3. $f(w) = \dfrac{3w}{4w+9}$
4. $p(y) = 6y^{\frac{-1}{3}} + 2$
5. $g(x) = \dfrac{x^3+1}{5-x^3}$
6. $y(x) = \sqrt[3]{\dfrac{x+7}{2x-11}}$
7. $f(p) = \dfrac{-2}{p+1}$
8. For each function, restrict the domain so it has an inverse.Then, find the inverse function.
   a. $f(x) = 5(x-1)^2 + 3$
   b. $y(x) = \sqrt{\dfrac{x}{2}} - 1$
   c. $h(x) = -2x^4 + 7$

4.2

Sketch the graph of each function. Label the y-intercept.

9. $y = e^x + 2$
10. $y = 5^x - 1$
11. $f(x) = 3^{-x}$
12. $g(x) = -2^x + 3$
13. $y = 4^{\frac{x}{2}}$

Simplify each expression.

14. $(3^{4-x})^{3x}$
15. $\dfrac{4^{x+1}}{4^{3-x}}$
16. $\dfrac{\pi^{5x+3}}{\pi^2}$
17. $(-6^2 e)^x$
18. $\dfrac{\pi^{2x}e^{3y}}{e^{2x}\pi^{3y}}$

4.3

Rewrite the logarithmic statement as its equivalent exponential statement using the definition of logarithm.

19. $\log_5 25 = 2$
20. $\log_4\left(\dfrac{1}{4}\right) = -1$

21. $\log 1 = 0$

22. $\ln\left(\frac{1}{\sqrt{e}}\right) = -\frac{1}{2}$

Simplify each expression, without using a calculator.

23. $\log_8 1$

24. $\log_2 16$

25. $\ln e^{-5}$

26. $\log_3\left(\frac{1}{81}\right)$

For each function, find the domain and sketch the graph. Label the x-intercept.

27. $f(x) = \log_3(-x)$

28. $y = \log(x-2)$

29. $g(x) = \ln(x) + 3$

30. $h(x) = 2\log_5(x) - 1$

31. $y = -\log_2(x+1)$

4.4

Use properties of logarithms to expand each expression completely.

32. $\log\left(\frac{x^2+5}{x^3}\right)^4$

33. $\ln\left(x^2\sqrt[3]{\frac{(x-1)^4}{x+9}}\right)$

34. $\log_5\left(\frac{(x+1)^2(x-2)^3(x+3)^4}{x}\right)$

Use properties of logarithms to condense each expression completely.

35. $3\log_2 x - 4\log_2(x+5) + \log_2 9$

36. $\frac{1}{2}\log_4(x-3) + \frac{1}{4}\log\ x - \log_4(\sqrt{x})$

37. $3\ln(e^x) - \ln x^2 + \ln(5x-2)$

38. $\log(4x) + \log(5y) - \log(xy)$

39. $\log_5(x^2-5) + 3\log_5 x - \log_5\left(\frac{x}{2}\right)$

Evaluate each expression, without using a calculator.

40. $3\ln e - \ln 1 + \ln\left(\frac{1}{\sqrt{e}}\right) + \ln e^5$

41. $\log_4\left(\frac{120(24)}{18(40)}\right) - \log_5 625 + \log 1000$

42. $\log_2(2^\pi) + \log_\pi(\pi^2) - \ln(e^3)$

4.5

Solve each equation.

43. $2e^{5x} - 3 = 9$

44. $5^{x^2}5^{3x} = 5^{10}$

45. $3^{x+1} + 2 = 4(3^{x+1}) - 7$

46. $2^{3w+5} + 1 = 2^0 + \frac{1}{16}$

47. $2(e^{2x} - 10) = -3e^x$

48. $\log_4 5 + \log_4 x = \log_4(3x+10)$

49. $\log_2 x^2 - \log_2(x+5) = 2$

50. $\log x + \log(x+5) = \log 66$

51. $e^{\ln(x^2+x)} = 90$

52. $\ln(x+2) = \ln(x+1) + 3$

53. $3 = \log_5(x+2) + \log_5\left(\frac{1}{x}\right)$

54. $2x10^x = x^2 10^x$

4.6

55. Suppose the number of rabbits on a small island is given by $N = 400(0.1^{0.2^t})$, where t is the number of years after 2005.

 a. How many rabbits are on the island in 2005?

 b. How many rabbits are on the island in 2006?

 c. In the year 2020, what will the rabbit population be?

56. The number N of people in a community who are reached by a particular rumor at time t (in days) is given by $N(t) = \dfrac{680}{1+169e^{-0.4t}}$.

 a. Find $N(0)$, the number of people who initially know the rumor.

 b. How long will it take 340 people to know the rumor?

57. The demand function for a product is given by $p = 180(4^{\frac{-q}{15}})$.

 a. At what price will there be 2 units demanded?

 b. If the unit price is \$110, how many units will be demanded?

58. The half-life of radioactive radium is 1620 years. What percent of a present amount of radioactive radium will remain after 870 years?

59. The number of units of a product sold after t years is given by $n(t) = 150(0.2^{0.03t})$. After how many years will there be 125 units sold?

60. The quantity, measured in milligrams, of a radioactive substance present after t years is given by $q = 180e^{-0.04t}$. After how many years will there be 14 mg present?

61. The population of Mathville, with initial population of 14,000 in the year 2000, grows at a rate of 3% per year.

 a. What is the population function?(Let P = population and t = number of years past year 2000.)

 b. What is the population in 2002?

 c. In what year will the population be 35,000?

62. Melida saved \$5,000 from her weekly cash over the last two years. She wants to invest her money in an account that grows according to the formula $A(t) = P(1.05^{2t})$, where P is the original amount invested, t is the number of years she leaves her money in that account and A is the value of that account at time t.

 a. How much will the account be worth after two years?

 b. After how many years will she have \$8,500?

chapter 5

Mathematics of Finance

5.1 Arithmetic and Geometric Sequences

VOCABULARY

Sequence—An ordered list of numbers that establish a pattern. We can also think of a sequence as a function whose domain is the set of natural (or whole) numbers. We typically denote a sequence as $\{a_n\}$, where $n \in \mathbb{N}$ and a_1 = the first term in the sequence, a_2 = the second term in the sequence, etc. We can call each number in the sequence a term or an element of the sequence.

Recursive—A recursive formula for the next term in a sequence depends on the previous term (or possibly several previous terms). A recursive formula is implicit, meaning that we cannot simply calculate the nth term in the sequence without knowing the previous $(n - 1)$ terms.

Iterative—An iterative formula for the next term in a sequence does not depend on any previous term. It is a formula given explicitly, meaning that we can simply substitute a value of n into the formula and get back the corresponding nth term of the sequence.

ARITHMETIC SEQUENCE

A sequence where the difference between any two consecutive terms is constant. We can write each term of such a sequence as

$$a_n = a_{n-1} + d,\ n = 2, 3, 4, \ldots$$

given a_1 (the starting value).

GEOMETRIC SEQUENCE

A sequence where the quotient between any two consecutive terms is constant. We can write each term of a geometric sequence as

$$a_n = da_{n-1},\ n = 2, 3, 4, \ldots$$

given a_1 (the starting value).

In other words, in an arithmetic sequence we get each next term by adding the same constant number, repeatedly. In a geometric sequence, we get each next term by multiplying by the same constant number, repeatedly. Graphically, for an arithmetic sequence, every time we go to the right one unit, we go up the same amount, over and over, exactly like the graph of a line (if we were to connect the dots). For a geometric sequence, its graph would be an exponential function (if we connected the dots) because we'd keep multiplying by the same number every time we went one unit to the right.

An example of an arithmetic sequence is

$$4, 7, 10, 13, 16, \ldots$$

Notice that the first term is 4, and every term from there on is the previous term plus 3. So, the pattern here is that we continue to add 3 to each successive term.

An example of a geometric sequence is

$$4, 12, 36, 108, 324, \ldots$$

Again, notice that we also started at 4, but then we continued to multiply by 3 to get each successive term in the sequence.

It's pretty easy to see with the above examples which sequence can grow faster. The geometric sequence grows fast very quickly compared to the arithmetic sequence with the same starting place and the same difference or quotient. If $d < 1$, then the geometric sequence actually decays, where the arithmetic sequence grows, although rather slowly.

Notice that in both boxes defining the arithmetic and geometric sequences, the formulas for the next term in the sequence are given recursively. Can we determine the iterative formulas?

ARITHMETIC SEQUENCE	GEOMETRIC SEQUENCE
$a_n = a_{n-1} + d,\ n = 2, 3, 4, \ldots, d \neq 0$	$a_n = da_{n-1},\ n = 2, 3, 4, \ldots, d \neq 0, 1$
a_1 given $a_2 = a_1 + d$ $a_3 = a_2 + d = (a_1 + d) + d = a_1 + 2d$ $a_4 = a_3 + d = (a_1 + 2d) + d = a_1 + 3d$ $a_5 = a_4 + d = (a_1 + 3d) + d = a_1 + 4d$ . . . $a_n = a_{n-1} + d = a_1 + (n-1)d$	a_1 given $a_2 = da_1$ $a_3 = da_2 = d(d\,a_1) = d^2a_1$ $a_4 = da_3 = d(d^2a_1) = d^3a_1$ $a_5 = da_4 = d(d^3a_1) = d^4a_1$ . . . $a_n = da_{n-1} = d^{n-1}a_1$

Now we have iterative formulas that will be very easy to use for computing any term in the sequence.

EXAMPLE 1

Classify each sequence as (i) arithmetic, (ii) geometric or (iii) neither. Then write the next three terms of each sequence given here.

(a) –1, 6, 13, 20, 27, . . .

(b) 22, 16, 10, 4, –2, . . .

(c) 2, 8, 32, 128, 508, . . .

(d) 1, 1, 2, 3, 5, 8, . . .

(e) $14, -7, \frac{7}{2}, \frac{-7}{4}, \ldots$

(a) –1, 6, 13, 20, 27, . . .

The pattern of this sequence is that we keep adding 7 to each previous term to get to the next term. So it is an arithmetic sequence.

The next three terms will be 27 + 7 = 34, then 34 + 7 = 41 and 41 + 7 = 48.

$\Rightarrow$ –1, 6, 13, 20, 27, 34, 41, 48, . . .

(b) 22, 16, 10, 4, –2, . . .

We keep subtracting 6 for this sequence, which is the same as repeatedly adding –6. Thus, it's an arithmetic sequence with a difference of –6.

The next three terms are: –2 – 6 = –8, then –8 – 6 = –14 and –14 – 6 = –20.

$\Rightarrow$ 22, 16, 10, 4, –2, –8, –14, –20, . . .

(c) 2, 8, 32, 128, 508, . . .

The repeated pattern here is that we multiply each successive term by 4. Therefore, this sequence is a geometric sequence with $d = 4$.

The next three terms will be: 508(4) = 2032, then 2032(4) = 8128, and 8128(4) = 32512.

$\Rightarrow$ 2, 8, 32, 128, 508, 2032, 8128, 32512, . . .

(d) 1, 1, 2, 3, 5, 8, . . .

The pattern for this sequence is quite different. When you see it, it's quite predictable. However, it does not fit the repeatable pattern for an arithmetic nor a geometric sequence. This is actually called the Fibonacci sequence, and it is quite famous. After the first two terms, we get each successive term by adding the two previous terms together.

The next three terms will be: 5 + 8 = 13, then 8 + 13 = 21 and 13 + 21 = 34.

$\Rightarrow$ 1, 1, 2, 3, 5, 8, 13, 21, 34, . . .

(e) $14, -7, \frac{7}{2}, \frac{-7}{4}, \ldots$

The pattern here seems to be multiplying each previous term by $-\frac{1}{2}$, which makes this a geometric sequence. The next terms will be: $\frac{-7}{4}\left(\frac{-1}{2}\right) = \frac{7}{8}$, then $\frac{7}{8}\left(\frac{-1}{2}\right) = \frac{-7}{16}$ and $\frac{-7}{16}\left(\frac{-1}{2}\right) = \frac{7}{32}$.

$\Rightarrow \quad 14, -7, \frac{7}{2}, \frac{-7}{4}, \frac{7}{8}, \frac{-7}{16}, \frac{7}{32}, \ldots$

EXAMPLE 2

Find the 30th term of the (a) arithmetic sequence and the (b) geometric sequence with a first term of 5 and $d = 2$.

(a) We know that $d = 2$ and $a_1 = 5$, and the formula for the arithmetic sequence terms is $a_n = a_1 + (n - 1)d$.

$\Rightarrow \quad a_n = 5 + (n - 1)2 = 5 + 2n - 2 = 3 + 2n$

We want to find the 30th term, which means we need to substitute 30 in for n in the above formula.

$\Rightarrow \quad a_{30} = 3 + 2(30) = 3 + 60 = 63$

(b) We'll do the same process as in part (a), except we need to use the formula for a geometric sequence term instead, which is $a_n = d^{n-1}a_1$.

$\Rightarrow \quad a_n = 2^{n-1}(5)$

To find the 30th term of the geometric sequence, we need to substitute 30 in for n.

$\Rightarrow \quad a_{30} = 2^{30-1}(5) = 5(2^{29}) = 5(536{,}870{,}912) = 2{,}684{,}354{,}560$

EXAMPLE 3

(a) If the first term of an arithmetic sequence is 5 and the 8th term is –23, find the 100th term of the sequence.

(b) If the first term of a geometric sequence is –3 and the 6th term is 96, find the 14th term of the sequence.

(a) We know we have an arithmetic sequence with $a_1 = 5$ and $a_8 = -23$. We also have the iterative formula for the nth term of an arithmetic sequence $a_n = a_1 + (n - 1)d$. Let's plug the information we have into the formula to solve for d.

$\Rightarrow \quad -23 = 5 + (8 - 1)d$ Simplify the right side of the equation.

$-23 = 5 + 7d$ Subtract 5 from both sides of the equation.

$-28 = 7d$ Divide both sides of the equation by 7 to get d by itself.

$-4 = d$

Now that we know what d is, we know the formula for any term of this sequence is

$$a_n = a_1 + (n-1)d = 5 + (n-1)(-4) = 5 - 4n + 4 \quad \Leftrightarrow \quad a_n = 9 - 4n.$$

To find the 100th term, then, we just need to substitute 100 in for n in the above formula for the nth term.

$\Rightarrow \quad a_{100} = 9 - 4(100) = 9 - 400 = -391$

(b) For a geometric sequence, we know the formula for the nth term in the sequence is $a_n = d^{n-1}a_1$. We don't know yet what d is. However, we can use the information given to compute d. We know $a_1 = -3$ and $a_6 = 96$.

So $a_n = d^{n-1}a_1 \Rightarrow 96 = d^{6-1}(-3)$ Simplify.

$96 = d^5(-3)$ Divide both sides of the equation by -3.

$-32 = d^5$ Take the fifth root of both sides of the equation.

$-2 = d$

Now, we know the formula for the nth term of this sequence is $a_n = (-2)^{n-1}(-3)$. We want to know the 14th term, so plug $n = 14$ into our formula.

$\Rightarrow \quad a_{14} = (-2)^{14-1}(-3) = (-2)^{13}(-3) = (-8192)(-3) = 24{,}576.$

The next idea we'd like to explore is to find the sum of the first n terms of an arithmetic or geometric sequence.

ARITHMETIC SEQUENCE SUM

$$S_n = a_1 + a_2 + a_3 + a_4 + \ldots + a_{n-1} + a_n$$

Since addition is commutative and associative, we can rearrange the numbers into any order and any grouping we'd prefer to add them. Let's group them in groups of two, adding the first and last term and then working our way into the middle.

$$a_1 + a_n = a_1 + (a_1 + (n-1)d) = 2a_1 + (n-1)d$$
$$a_2 + a_{n-1} = (a_1 + d) + (a_1 + (n-2)d) = 2a_1 + (n-1)d = a_1 + a_n$$
$$a_3 + a_{n-2} = (a_1 + 2d) + (a_1 + (n-3)d) = 2a_1 + (n-1)d = a_1 + a_n$$

Notice at this point, that all the two-termed sums are adding to the same amount, namely the sum of the first and last term in the sequence. How many of these groupings do we get? Since we're adding up n terms, in groups of two terms, we'll have $\frac{n}{2}$ groups of those same sums.

$\Rightarrow \quad S_n = \left(\frac{n}{2}\right)(a_1 + a_n)$ (Sum of the first n terms of an arithmetic sequence $d \neq 0$.)

We can see this technique with numbers in the following example.

EXAMPLE 4

Add the first 100 counting numbers together, i.e., 1 + 2 + 3 + 4 + . . . + 99 + 100.

This is the sum of an arithmetic sequence with the first term equal to 1; a common difference is 1, $n = 100$, and the final term is 100.

Following the idea given above, let's add the first and last term together and work our way toward the middle.

$1 + 100 = 101$
$2 + 99 = 101$
$3 + 98 = 101$
$4 + 97 = 101$
. . .
$48 + 53 = 101$
$49 + 52 = 101$
$50 + 51 = 101$

We have 101 added together 50 times total, so our final sum is 50(101) = 5050. If we had used the formula, we'd get

$$S_{100} = \left(\frac{100}{2}\right)(1 + 100) = 50(101) = 5050$$

GEOMETRIC SEQUENCE SUM

$$S_n = a_1 + a_2 + a_3 + a_4 + \ldots + a_{n-1} + a_n$$

Remember the explicit, or iterative, formula for each term of a geometric sequence is $a_n = d^{n-1}a_1$.

$\Rightarrow$ $S_n = a_1 + da_1 + d^2a_1 + d^3a_1 + \ldots + d^{n-1}a_1$ and we can multiply both sides of this equation by d to get $dS_n = da_1 + d^2a_1 + d^3a_1 + \ldots + d^{n-1}a_1 + d^na_1$. Now, let's subtract the two equations (vertically).

$\Rightarrow$ $S_n - dS_n = (a_1 + da_1 + d^2a_1 + d^3a_1 + \ldots + d^{n-1}a_1) - (da_1 + d^2a_1 + d^3a_1 + \ldots + d^{n-1}a_1 + d^na_1)$ On the right side of the equation, most of the terms add up to zero, except a couple of terms.

$\Leftrightarrow$ $S_n - dS_n = a_1 - d^na_1$ Factor both sides of the equation.

$\Leftrightarrow$ $S_n(1 - d) = a_1(1 - d^n)$ Solve for S_n by dividing both sides of the equation by $(1 - d)$.

$\Rightarrow$ $S_n = \dfrac{a_1(1 - d^n)}{1 - d}$ (Sum of the first n terms of a geometric sequence $d \neq 0, 1$.)

EXAMPLE 5

Find the sum of the first 7 terms of the geometric sequence given by

(a) $\frac{1}{5}, \frac{1}{15}, \frac{1}{45}, \ldots$

(b) $-2, -10, -50, -250, \ldots$

(a) $\frac{1}{5}, \frac{1}{15}, \frac{1}{45}, \ldots$

This is a geometric sequence with $a_1 = \frac{1}{5}$ and $d = \frac{1}{3}$. To find the sum of the first 7 terms, we can use the formula we derived, namely $S_n = \frac{a_1(1-d^n)}{1-d}$ with $n = 7$.

$$\Rightarrow S_7 = \frac{\frac{1}{5}\left(1-\left(\frac{1}{3}\right)^7\right)}{1-\frac{1}{3}}$$ Now we need to simplify this mess.

$$S_7 = \frac{\frac{1}{5}\left(1-\frac{1}{2187}\right)}{\frac{2}{3}} = \frac{\frac{1}{5}\left(\frac{2186}{2187}\right)}{\frac{2}{3}} = \frac{2186}{5(2187)} \cdot \frac{3}{2} = \frac{1093}{5(729)} = \frac{1093}{18225}$$

(b) $-2, -10, -50, -250, \ldots$

For this geometric sequence, we have $a_1 = -2$ and $d = 5$. Again, to find the sum of the first 7 terms, we use the formula $S_n = \frac{a_1(1-d^n)}{1-d}$ with $n = 7$.

$$\Rightarrow S_7 = \frac{-2(1-5^7)}{1-5}$$ Simplify this expression.

$$S_7 = \frac{-2(1-78125)}{-4} = \frac{-2(-78124)}{-4} = \frac{-78124}{2} = -39{,}062$$

EXAMPLE 6

Find the sum of the first 20 terms of the arithmetic sequence given by 2, 11, 20, 29, . . .

For the sum of an arithmetic sequence, we need to use the formula $S_n = \left(\frac{n}{2}\right)(a_1 + a_n)$. For this example, we have $a_1 = 2$, $d = 9$, and $n = 20$.

$$\Rightarrow S_{20} = \left(\frac{20}{2}\right)(2 + a_{20})$$

Notice that we don't yet know the 20th term of the sequence, and we need that to plug into the above formula. We can use the formula for the nth term of an arithmetic sequence, namely $a_n = a_1 + (n - 1)d$, to find the 20th term.

$$\Rightarrow a_{20} = 2 + (20 - 1)9 = 2 + 19(9) = 2 + 171 = 173$$

$$\Rightarrow S_{20} = \left(\frac{20}{2}\right)(2 + 173) = 10(175) = 1750$$

EXAMPLE 7

Emily attended a fun party last night with the intention of networking and letting everyone know that she's looking for a job now that she's graduated from college. This morning she awoke wondering if she could compute how many handshakes occurred at the party. She made a point of shaking every person's hand once while at the party, as did everyone else. There was a total of 42 people at the party. How many handshakes occurred at the party?

If there were 42 people total, and each person shook everyone else's hand once, then we'll count person 1 as shaking 41 hands (because person 1 cannot shake his or her own hand). Then, we'll count person 2 as shaking 40 hands (because we already counted person 1 and person 2 shaking hands). We'll count person 3 as shaking 39 hands (because now we've already counted persons 1 and 3 shaking hands, and persons 2 and 3 shaking hands). So, we'll count person 4 as shaking 38 hands, person 5 as shaking 37 hands, etc. all the way down to counting person 41 as shaking 1 hand and person 42 as shaking no hands.

This means we want to add the numbers: 41 + 40 + 39 + 38 + 37 + . . . + 2 + 1. Notice that this is simply the sum of the first 41 counting numbers. In other words, this is the sum of an arithmetic sequence with $d = 1$ and $a_1 = 1$.

$$\Rightarrow S_{41} = \left(\frac{41}{2}\right)(1 + 41) = \frac{41(42)}{2} = 41(21) = 861$$

So, there were 861 handshakes at the party of 42 people.

5.1 Homework

Classify each sequence as arithmetic, geometric, or neither. Then, write the next three terms of the sequence.

1. 2, 6, 10, 14, . . .
2. −1, 1, 0, 1, 1, 2, 3, . . .
3. $1, \frac{1}{2}, \frac{1}{4}, \frac{1}{8}, \ldots$
4. −3, 6, −12, 24, . . .
5. −3, −1, 1, 3, . . .
6. 1, 3, 9, 27, . . .
7. 17, 22, 27, 32, . . .
8. 1, 2, 2, 4, 8, 32, . . .
9. 125, 25, 5, 1, . . .
10. 10, 7, 4, 1, . . .
11. 90, 10, 80, −70, 150, −220, . . .
12. 2, −6, 18, −54, . . .

Find the formula for the nth term of the (a) arithmetic sequence and (b) geometric sequence with the given a_1 and d values.

13. $a_1 = 3, d = 2$
14. $a_1 = 1, d = -3$
15. $a_1 = -5, d = 4$
16. $a_1 = 0, d = \frac{1}{2}$
17. $a_1 = 2, d = -5$
18. $a_1 = -1, d = 2$
19. $a_1 = 100, d = -1$
20. $a_1 = 9, d = -4$
21. $a_1 = -2, d = \frac{3}{2}$
22. $a_1 = 8, d = \frac{1}{4}$
23. $a_1 = 0, d = 6$
24. $a_1 = -10, d = 3$
25. $a_1 = -6, d = 5$

Given the first term, a_1, of an arithmetic sequence and another term of the sequence, find the 50th term of the sequence.

26. $a_1 = 2, a_8 = 23$
27. $a_1 = 1, a_{15} = 57$
28. $a_1 = 14, a_2 = 19$
29. $a_1 = 8, a_5 = 0$
30. $a_1 = 5, a_{99} = 54$
31. $a_1 = -14, a_{20} = 43$
32. $a_1 = 0, a_{12} = -55$
33. $a_1 = 9\frac{1}{3}, a_{90} = 39$

Given the first term, a_1, of a geometric sequence and another term of the sequence, find the 20th term of that geometric sequence.

34. $a_1 = \frac{3}{2}, a_5 = \frac{243}{2}$
35. $a_1 = -10, a_4 = -80$
36. $a_1 = \frac{3}{2}, a_6 = \frac{3}{64}$
37. $a_1 = -7, a_{50} = 7$
38. $a_1 = 10, a_5 = \frac{1}{1000}$
39. $a_1 = 125, a_8 = \frac{1}{625}$
40. $a_1 = 3, a_2 = 9$
41. $a_1 = \frac{1}{16}, a_7 = 256$

Find the sum of the first 100 terms of the arithmetic sequence given.

42. 1, 4, 7 ,10, . . .
43. –3, –7, –11, –15, . . .
44. 5, 10, 15, 20, . . .
45. 90, 80, 70, 60, . . .
46. 2, –4, –10, –16, . . .
47. 8, 11, 14, 17, . . .
48. 1, 10, 19, 28, . . .
49. 2, 4, 6, 8, . . .

Find the sum of the first 10 terms of the geometric sequence given.

50. $2, 1, \frac{1}{2}, \frac{1}{4}, \ldots$
51. 1, 4, 16, 64, . . .
52. $\frac{2}{125}, \frac{2}{25}, \frac{2}{5}, 2, \ldots$
53. $9, 6, 4, \frac{8}{3}, \ldots$
54. 3, 6, 12, 24, . . .
55. 1, –5, 25, –125, . . .
56. –3, 3, –3, 3, . . .
57. $\frac{1}{27}, \frac{1}{9}, \frac{1}{3}, 1, \ldots$

5.2 Simple and Compound Interest

SIMPLE INTEREST

Let's say Joe invests $1,000 in an account that pays simple interest of 5%. Here is a table representing his earnings year by year.

Let P = the principal amount of money that's originally invested (which is $1,000 in this example),
t = the number of years we leave the money in the simple interest account,
S = the account value at any given time t,
r = the annual interest rate (5% for this example), and
I = the interest earned after t years.

NUMBER OF YEARS (t)	INTEREST ($I = Pr$)	ACCOUNT VALUE (S)
0	0	$\$1000 = P$
1	$0.05(1000) = \$50$	$1000 + 50 = \$1050 = 1000 + I = P + I$
2	$0.05(1000) = \$50$	$1050 + 50 = \$1100 = 1000 + 2I = P + 2I$
3	$0.05(1000) = \$50$	$1100 + 50 = \$1150 = 1000 + 3I = P + 3I$
4	$0.05(1000) = \$50$	$1150 + 50 = \$1200 = 1000 + 4I = P + 4I$
...	...	...
t	$0.05(1000) = \$50$	$1000 + tI = P + tI = P + t(Pr) = P(1 + rt)$

Notice that the account value increases exactly like an arithmetic sequence because we keep adding the same interest amount to the account value every year. In fact, earning simple interest is a perfect application example of an arithmetic sequence. Simple interest grows linearly. The growth of the account, then, is not ideal since we typically want our money to grow faster than that when we invest it.

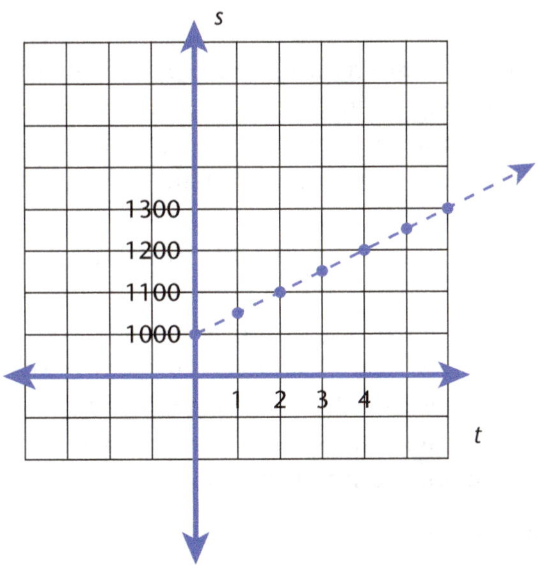

SIMPLE INTEREST FORMULA

For an account earning simple annual interest, at rate r, the value of the account after t years with P principal dollars originally invested is

$$S = P + t(Pr) \text{ or } S = P(1 + rt).$$

COMPOUND INTEREST

Let's look at the same $1,000 investment in an account earning compound interest, at the same annual 5% rate, but compounded yearly. All the variables will remain the same as above, and let

n = the number of compoundings per year, which will be 1 for this example. We'll see what happens to the money in this table.

# OF YEARS (t)	INTEREST ($I = Sr$)	ACCOUNT VALUE (S)	FORMULA
0	0	$1000	$S_0 = P$
1	0.05(1000) = $50	1000 + 50 = $1050	$S_1 = S_0 + I = S_0 + S_0r = S_0(1 + r) = P(1 + r)$
2	0.05(1050) = $52.50	1050 + 52.50 = $1102.50	$S_2 = S_1 + I = S_1 + S_1r = S_1(1 + r) = P(1 + r)(1 + r) = P(1 + r)^2$
3	0.05(1102.50) = $55.13	1102.50 + 55.13 = $1157.63	$S_3 = S_2 + I = S_2 + S_2r = S_2(1 + r) = P(1 + r)^2(1 + r) = P(1 + r)^3$
4	0.05(1157.63) = $57.88	1157.63 + 57.88 = $1215.51	$S_4 = S_3 + I = S_3 + S_3r = S_3(1 + r) = P(1 + r)^3(1 + r) = P(1 + r)^4$
...	...	...	...
t	0.05 S_{t-1}	...	$S_t = S_{t-1} + I = S_{t-1} + S_{t-1}r = P(1 + r)^{t-1}(1 + r) = P(1 + r)^t$

Notice that for this compounded interest account, we keep multiplying the previous value of the account by $(1 + r)$ to get the new value. This is exactly the behavior of a geometric sequence. Multiplying by $(1 + r)$ makes sense because it's saying that we multiply the account value by 1, thereby keeping the whole account value, plus the product of the value with the interest rate. Thus, we have the entire previous account value, plus the new interest earned.

You can see that a compounded interest account grows much more rapidly than a simple interest bearing account. It is the difference between exponential growth for compounded interest and linear growth for simple interest. The difference in the account values is not enormous for the first few years, but as time goes on the difference in the account values gets bigger and bigger, thereby making compounded interest accounts much more valuable for investments. On the contrary, if we are paying compounded interest on a credit card account, rather than earning the interest, we can see that it's much worse for us than paying simple interest.

The example we considered above only compounded once per year. If we compound four times per year instead (or quarterly), our interest rate every time we compound the interest would only be $\frac{0.05}{4} = 0.0125$.

And, we would compound four times per year instead of once. That would mean that in 10 years, we'd compound $4(10) = 40$ times. In general, if we compound n times per year, then the interest rate will be $\frac{r}{n}$ each time we compound, and the total number of compoundings in t years would be nt.

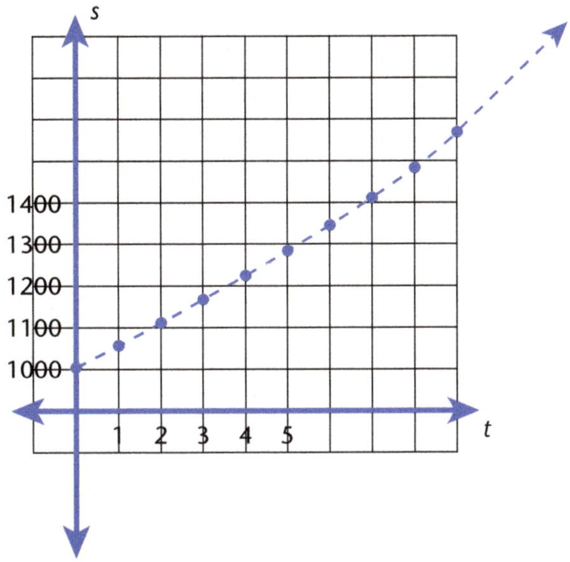

COMPOUND INTEREST FORMULA

For an account earning compounded annual interest, at rate r, with n compoundings per year, the value of the account after t years with P principal dollars originally invested is

$$S = P\left(1 + \frac{r}{n}\right)^{nt}.$$

What happens if we compound more and more frequently? Does the account value increase to become infinitely large? Let's look at a table of values to answer this question. We'll assume we invest \$100 for one year ($t = 1$) in an account earning compound interest with rate $r = 100\% = 1.0$, which is unreasonable in daily living but a reasonable assumption to make for this argument.

NUMBER OF COMPOUNDINGS PER YEAR (n)	FORMULA FOR ACCOUNT VALUE AFTER 1 YEAR	ACCOUNT VALUE AFTER 1 YEAR
1 (annually)	$S = 100\left(1+\frac{1}{1}\right)^{1}$	\$200.00
4 (quarterly)	$S = 100\left(1+\frac{1}{4}\right)^{4}$	\$244.14
12 (monthly)	$S = 100\left(1+\frac{1}{12}\right)^{12}$	\$261.30
52 (weekly)	$S = 100\left(1+\frac{1}{52}\right)^{52}$	\$269.26

NUMBER OF COMPOUNDINGS PER YEAR (n)	FORMULA FOR ACCOUNT VALUE AFTER 1 YEAR	ACCOUNT VALUE AFTER 1 YEAR
365 (daily)	$S = 100\left(1 + \frac{1}{365}\right)^{365}$	\$271.46
8,760 (hourly)	$S = 100\left(1 + \frac{1}{8760}\right)^{8760}$	\$271.81
525,600 (every minute)	$S = 100\left(1 + \frac{1}{525600}\right)^{525600}$	\$271.83
31,536,000 (every second)	$S = 100\left(1 + \frac{1}{31536000}\right)^{31536000}$	\$271.83

Notice that the account value doesn't actually grow unboundedly! It grows and reaches a limiting value. In fact, if you remember the value of e, you'll see that the limiting value is simply 100 times e, because we started with an initial principal of \$100.

CONTINUOUS COMPOUND INTEREST FORMULA

For an account earning interest compounded continuously, at rate r, the value of the account after t years with P principal dollars originally invested is

$$S = Pe^{rt}.$$

EXAMPLE 1

Given that we invest \$10,000 for four years at an annual interest rate of 7.5%.

(a) If it's simple interest, how much interest will we receive at the end of the four-year period? What will the account value be?

(b) If the interest is compounded twice yearly, what will the account be worth? How much interest will we receive at the end of the four-year period?

(a) We have $P = \$10{,}000$, $t = 4$ years, and $r = 0.075$. The simple interest earned will be

$$I = S - P = P(1 + rt) - P = P + Prt - P = Prt$$

$$\Rightarrow \quad I = 10{,}000\,(0.075)(4) = \$3{,}000 \text{ and } S = I + P = 3{,}000 + 10{,}000 = \$13{,}000$$

(b) For twice yearly compounding, $n = 2$ and the account will be worth

$$S = P\left(1 + \frac{r}{n}\right)^{nt} = 10000\left(1 + \frac{0.075}{2}\right)^{2(4)} \simeq \$13{,}424.71$$

The compounded interest earned will be the difference from the future account value and the principal invested.

$\Rightarrow \quad I = S - P \simeq 13424.71 - 10000 = \$3{,}424.71 \quad \Rightarrow \quad S = 10{,}000 + 3{,}000 = \$13{,}000$

From this example, you can start to see that compounded interest starts to have a dramatic positive effect on value fairly quickly, compared to simple interest.

EXAMPLE 2

Mary borrowed \$3,000 at an interest rate of 18%. How much interest is due in 65 weeks (a) if it's simple interest and (b) if the interest is compounded quarterly and (c) if the interest is compounded continuously?

We are given $P = \$3{,}000$ and $r = 0.18$. We want to know how much interest is due in 65 weeks, which is one whole year plus a portion of a year.

$$65 \text{ weeks} \left(\frac{1 \text{ year}}{52 \text{ weeks}} \right) = 1.25 \text{ years} \Rightarrow t = 1.25 \text{ years}$$

(a) For simple interest, the interest will be

$$I = Prt = 3000\,(0.18)\,(1.25) = \$675.$$

(b) For interest compounded quarterly, we have $n = 4$. And the interest will be

$$I = S - P = P\left(1 + \frac{r}{n}\right)^{nt} - P = 3000\left(1 + \frac{0.18}{4}\right)^{1.25(4)} - 3000 \simeq 3738.55 - 3000 = \$738.55$$

(c) For interest compounded continuously, we need to use the continuous compounding formula. The interest will still be the difference between the future value, S, and the original principal, P.

$$I = S - P = Pe^{rt} - P = 3000e^{0.18(1.25)} - 3000 \simeq 3756.97 - 3000 = \$756.97$$

We can see then that the interest owed on a simple interest bearing loan is much lower than the interest owed for a compounded interest loan. Thus we would want to earn compounded interest on our investments but pay simple interest on our loans, if possible.

EXAMPLE 3

Assume that we invest \$6,000 at 5% annual interest.

(a) If it's simple interest, how long will it take to double the value of the investment?

(b) If the interest is compounded every month, how long will it take to double in value?

(c) If the interest is compounded continuously, how long will it take to double in value?

We have $P = \$6{,}000$ and $r = 0.05$ for this problem. We want to know the doubling time given this information for a variety of types of interest. So we will need to solve equations for t, given $S = 2P = \$12{,}000$.

(a) For simple interest, we need to use the equation to solve for t.

$\Rightarrow\ 12000 = 6000(1 + 0.05t)$	Divide both sides of the equation by 6,000.
$2 = 1 + 0.05t$	Subtract 1 from both sides of the equation.
$1 = 0.05t$	Finally, divide both sides of the equation by 0.05 to isolate t.
$20 = t$	

This means that it will take 20 years for this simple interest account to double.

(b) For monthly compounding, $n = 12$. We need to use the compounded interest formula to solve for t.

$$S = P\left(1 + \frac{r}{n}\right)^{nt}$$

$\Rightarrow\ 12000 = 6000\left(1 + \frac{0.05}{12}\right)^{12t}$	Divide both sides of the equation by 6,000 and simplify inside the parentheses.
$2 = (1.0041\overline{6})^{12t}$	Take the natural logarithm of both sides of the exponential equation.
$\ln 2 = \ln(1.0041\overline{6})^{12t}$	Use logarithm properties to bring down the exponent.
$\ln 2 = (12t)\ln(1.0041\overline{6})$	Divide both sides of the equation by the coefficient of t.
$\frac{\ln 2}{12\ln(1.0041\overline{6})} = t$	Compute an approximate value with a calculator.
$t \simeq 13.9$ years	

Thus, it takes just less than 14 years to double the account value with this interest rate, compounding every month. It's a much better result than that for simple interest doubling time.

(c) For continuous compounding, we'd expect an even better result than in part (b). However, we don't expect a huge decrease in doubling time because of our previous finding that continuous compounding was, in fact, not enormously better than frequent compounding. We'll use the formula for continuous compounded interest, namely $S = Pe^{rt}$.

$\Rightarrow\ 12000 = 6000e^{0.05t}$	Divide both sides of the equation by 6,000.
$2 = e^{0.05t}$	Rewrite this exponential statement as its equivalent logarithmic statement, using the definition of the logarithm.
$\ln 2 = 0.05t$	Divide both sides of the equation by 0.05 to solve for t.
$\frac{\ln 2}{0.05} = t$	Compute an approximate value for t by plugging this into a calculator.
$t \simeq 13.86$ years	

So, you see, it's basically the same result as our result in part (b) for monthly compounding. Basically, it's going to take just under 14 years to double our account, with 5% interest, with frequent compounding. The result will be a bit better if the interest rate increases.

EXAMPLE 4

How much money do you need to invest now in order to retire with \$2,000,000 in 40 years if money compounds monthly at 8%? What if we have 45 years?

We want a future value of S = \$2,000,000, in t = 40 years, with an interest rate of r = 0.08, compounded monthly, n =12. We need to plug all those values into the compounded interest formula and solve for P.

$$\Rightarrow \quad 2000000 = P\left(1+\frac{0.08}{12}\right)^{12(40)}$$ Simplify the right side of the equation.

$$2000000 = P(1.00\bar{6})^{480} \Leftrightarrow 2000000 \simeq P(24.27338558)$$ Divide both sides of the equation by the coefficient of P to solve for P.

$$P \simeq \$82{,}394.77$$

If we had a little more than \$82,000 right now, we could invest it in such an account and just leave it there, knowing it would be worth \$2 million in 40 years when we want to retire.

Let's say we have 45 years before we want to retire. Then perform these same calculations with t = 45.

$$\Rightarrow \quad 2000000 = P\left(1+\frac{0.08}{12}\right)^{12(45)} \Leftrightarrow 2000000 \simeq P(36.16359935) \Leftrightarrow P \simeq \$55{,}304.23$$

Thus, by having an extra five years to earn compounded interest, we can decrease our current investment by almost \$30,000 to have the same \$2 million upon retirement!

EXAMPLE 5

(a) Find the value of an account after 3 years if it's compounded continuously and we originally invested \$5,000 at an interest rate of 10%.

(b) What interest rate must we find if we want to invest the same \$5,000 in a continuously compounded account for 3 years and have a future value of at least \$7,000?

(a) We know t = 3 years, P = \$5,000, and r = 0.10, and we need to use the continuous compounding formula to find S.

$$\Rightarrow \quad S = 5000e^{0.1(3)} \simeq \$6{,}749.29$$

(b) We still have t = 3 years and P = \$5,000, and now we want S = \$7,000. We need to solve for the unknown r.

$\Rightarrow \; 7000 = 5000e^{3r}$ Divide both sides of the equation by 5,000.

$1.4 = e^{3r}$ Rewrite this exponential statement as its equivalent logarithmic statement using the logarithm definition.

$\ln 1.4 = 3r$ Divide both sides of the equation by 3 to get r by itself.

$\frac{\ln(1.4)}{3} = r$ Calculate an approximate value of r by plugging this into a calculator.

$r \approx 0.112 = 11.2\%$

Therefore, we need an interest rate of at least 11.2% to turn the original $5,000 into $7,000 compounded continuously.

ANNUAL PERCENTAGE YIELD (APY)

The annual percentage yield is related to the annual interest rate, along with the compounding. Let's consider that a principal of $1,000 is invested at 6% interest compounded (a) quarterly or (b) monthly.

Then the account value after 1 year would be:

(a) $S = 1000\left(1+\frac{0.06}{4}\right)^{4(1)} = 1000(1.015)^4 \approx \$1061.36$ for quarterly compounding and

(b) $S = 1000\left(1+\frac{0.06}{12}\right)^{12(1)} = 1000(1.005)^{12} \approx \$1061.67$ for monthly compounding.

Thus, investing at 6% compounded quarterly is equivalent to investing the same money at 6.136% simple interest and investing at 6% compounded monthly is equivalent to investing the same money at 6.167% simple interest. The ***annual percentage yield*** is exactly that number that represents the percentage increase in the account after 1 year. For simple interest, the annual interest rate is exactly the same as the APY. For compounded interest accounts, the APY will be greater than the annual interest rate.

ANNUAL PERCENTAGE YIELD (APY) FORMULA

$\text{APY} = \left(1+\frac{r}{n}\right)^n - 1$ for periodic compounding

$\text{APY} = e^r - 1$ for continuous compounding

EXAMPLE 6

Which is the better investment deal?

(a) An account earning 10% simple interest.

(b) An account earning 9% interest compounded annually.

(c) An account earning 8.75% interest compounded quarterly.

(d) An account earning 8.6% interest compounded continuously.

We can answer these questions in a few different ways. Perhaps the most efficient way is to simply calculate the APY for each investment. That will give us a quick idea of the value of the investment after one year.

(a) For simple interest, the APY is exactly the same as the annual percentage rate. So, for simple interest of 10%, the APY = 0.10.

(b) We have $r = 0.09$ and $n = 1$.

$$\Rightarrow \text{APY} = \left(1+\frac{r}{n}\right)^n - 1 = \left(1+\frac{0.09}{1}\right)^1 - 1 = 1.09 - 1 = 0.09$$

Notice that compounding once per year gives us the original interest rate back for the APY.

(c) We have $r = 0.0875$ and $n = 4$.

$$\Rightarrow \text{APY} = \left(1+\frac{r}{n}\right)^n - 1 = \left(1+\frac{0.0875}{4}\right)^4 - 1 = 1.090413 - 1 = 0.090413$$

This means that this account will earn the same amount of interest in one year as a simple interest bearing account with 9.0413% interest.

(d) We have $r = 0.086$ and we need to use the APY formula for continuously compounded interest.

$$\Rightarrow \text{APY} = e^r - 1 = e^{0.086} - 1 \simeq 1.089809 - 1 = 0.089806$$

Thus, this account in one year earns the same interest as a simple interest account earning 8.9806% interest.

Based on these calculations, the best deal is the account earning 8.75% interest compounded quarterly. (We're assuming that the compounded interest will grow much more quickly than the simple interest account, even though the APY for the simple interest account is higher.)

5.2 Homework

SIMPLE INTEREST

1. $2,000 is deposited into a savings account for one year at a simple interest rate of 5%.

   a. What is the value of the savings account at the end of the year?
   b. What is the value of the savings account at the end of five years?
   c. What is the interest paid on this account at the end of five years?

2. $2,000 is deposited into a savings account for one year at a simple interest rate of 6.5%.

   a. What is the value of the savings account at the end of the year?
   b. What is the value of the savings account at the end of five years?
   c. What is the interest paid on this account at the end of five years?

3. Based on your answers to question 1 and 2 (above), what is the difference, in interest earned, over a five-year period of an investment of $2,000? Would you say that $1\frac{1}{2}$ percent difference in that interest rate on an investment like this makes much of a difference?

4. $10,000 is deposited into a savings account for five years at a simple interest rate of 12%.

   a. What is the value of the savings account at the end of the year?
   b. What is the value of the savings account at the end of the five year period?

5. $10,000 is deposited into a savings account for five years at a simple interest rate of 13.5%.

   a. What is the value of the savings account at the end of the year?
   b. What is the value of the savings account at the end of the five-year period?

6. Based on your answers to Problems 4 and 5 (above), what is the difference in the savings account value over a five-year period of an investment of $10,000? Would you say that 1.5 percent difference in that interest rate on an investment like this makes much of a difference?

7. Use the simple interest formula to calculate how long it would take to double a person's investment with a 5% interest rate.

8. Use the simple interest formula to determine how long it would take to double a person's investment with a 10% interest rate.

9. Use the simple interest formula to determine how long it would take to double a person's investment with a 25% interest rate.

10. "The Rule of 72" says that if one uses the interest rate (as a whole number, without the percent sign) being paid and divides it into 72 a person can calculate the approximate number of years it will take to double one's money. Referring to Problems 7, 8 and 9 (above), how close is "The Rule of 72" to the actual time it takes to double a person's money?

COMPOUND INTEREST

11. Which savings account pays more on $2,000 after one year?

    a. Savings account X: 4% annually with simple interest.
    b. Savings account Y: 3.5% annually, compounded quarterly.

Which of the given options would pay the greatest amount of interest? How much more would it pay?

12. $3,000 saved for 1 year at an annual interest rate of:

    a. 4% compounded quarterly.
    b. 4.2% compounded semi-annually.

13. $3,300 saved for 1 year at an annual interest rate of:

    a. 5% compounded quarterly.
    b. 5.1% compounded semi-annually.

14. $2,200 saved for 1 year at an annual interest rate of:

    a. 4% compounded monthly.
    b. 4.2% compounded quarterly.

15. $4,500 saved for 1 year at an annual interest rate of:

    a. 5.0% compounded monthly.
    b. 5.2% compounded quarterly.

16. $25,000 saved for 3 years at an annual interest rate of:

    a. 6% compounded quarterly.
    b. 6.1% compounded semi-annually.

17. $35,000 saved for 5 years at an annual interest rate of:

    a. 5% compounded monthly.
    b. 5.2% compounded semi annually.

18. $50,000 saved for 5 years at an annual interest rate of:

    a. 10.0% compounded monthly.
    b. 10.2% compounded quarterly.

19. How much interest would be earned from $1,000 invested at 5% for five years, compounded annually?

20. What is the value of an account into which $1,500 is invested for 4 years at 5% interest, compounded quarterly?

21. Compare the value, after 5 years, of $10,000 invested at 4.5% compounded monthly and the same investment at the same interest rate, compounded daily.

22. Create a table showing the value of an account into which $5,000 is invested at 4% for five years.
    - The first row of the table will show the value of this account with simple interest.
    - On the second row show the value of this account with interest compounded annually.
    - On the third row show the value of this account with interest compounded quarterly.
    - On the forth show the value of this account with interest compounded monthly.
    - On the final row show the value of this account with interest compounded daily.

23. How long would it take to double one's money with a 5% return compounded monthly?

24. How long would it take to double one's money with a 10% return compounded monthly?

25. How long would it take to double one's money with a 25% return compounded monthly?

26. Let's return to "The Rule of 72". You will remember that, according to this rule, if one uses the interest rate being paid on an account and divides it into 72, a person can calculate the approximate number of years it will take to double one's money. Referring to Problems 23, 24 and 25 (above), how close is "The Rule of 72" to the actual time it takes to double a person's money? Does the "The Rule of 72" more accurately approximate simple interest or compound interest?

27. Presume that your great grandfather purchased a common stock certificate from Ford Motor Company for $5 when stock was first issued in 1903. Suppose that every time a dividend was paid, your ancestor reinvested the return into more Ford stock (simulating compound interest). If returns on Ford common stock have averaged 5%, and dividends paid annually, how much is the value of the stock you inherited in 2010?

28. Susie's travel agency put up a bond of $5,000 in order to be granted a business license. The city holds Susie's money and pays 3% interest compounded quarterly. Susie only needs to keep the original $5,000 with the city in order to maintain her business license, so she withdraws the interest she has earned after 5 years. How much interest did she earn?

29. The Bryan Landscaping company earned $10,000 in its first year of business which it deposited in a cd account earning 6% compounded daily for four years. What is the value of this account after the four-year period?

30. Elaine's father was executed by lethal injection in Texas in 1969. Before his death, he had deposited $2,000 in a bank account earning 8% interest, compounded quarterly. Elaine just discovered its existence. What did she inherit from her late father in 2010?

31. The Calico Band was a warm up band for the Rolling Stones in the 1970s. The band never made it big. Only one of the band members has any money left. He deposited \$50,000 in a money market account earning 12% interest, compounded monthly in 1980. In 2010, what is the value of this account?

32. The XYZ department store issues a credit card for purchases of merchandise at its stores. It allows customers to make no payments on credit card balances for two years—although interest accrues on the unpaid balance. The company compounds interest monthly at an annual rate of 18%. If a customer purchases \$500 in goods at the XYZ store, what would the credit card balance be at the end of the two-year "payment free" period?

ANNUAL PERCENTAGE YIELD (APY)

33. Calculate the APY for the MENSA member credit card. It charges 12.5% interest on the unpaid balance—simple interest.

34. Calculate the APY for the Mistress Card that charges 12.1% interest on the unpaid balance compounded quarterly.

35. Calculate the APY for the MYOPIC card that charges 11.5% interest on the unpaid balance compounded continuously.

36. Based on the results from your calculations of APY from Problems 33, 34 and 35, do you think that there is trade-off between lower interest rates and more frequent compounding?

37. Calculate the APY for the Church of the Holy Shine credit card that charges 12% simple interest.

38. Calculate the APY for the Pawn Broker card that charges 21.5% interest on the unpaid balance compounded daily.

39. Calculate the APY for the We Gocha Mortgage lenders that charges 8.5% interest on the unpaid balance compounded monthly.

40. Mortgage lending companies usually charge a "loan origination fee" to the mortgage borrower. This fee varies, but is often 1% of the loan paid at the time the loan is made. Suppose a new homeowner purchases a house for \$200,000 at 5% interest over a 30-year period compounded monthly. Just for this problem, suppose that there is no down payment made and no payments are made for the 30-year term of the loan. What is the average APY for this mortgage loan? What is the total amount owed at the end of 30 years?

5.3 Future Value of Annuities

VOCABULARY

Annuity—A sequence (or plan) of payments made at fixed and regular payment intervals.

Ordinary Annuity—An annuity with payments made at the end of each payment period.

Annuity Due—An annuity with payments made at the beginning of each payment period.

Sinking Fund—A fund where a borrower makes periodic payments in order to produce a specified amount of money on a particular date in the future, used to pay a financial obligation.

The study of annuities is very important and a logical topic of study after compound interest. With the formula for compounded interest, we can figure out what amount of money we need to invest as a lump sum right now to achieve a specified financial goal in the future. However, typically, many people don't have a somewhat large sum of money to invest at once. More frequently, people invest for their future by investing a particular (and smaller) amount of money at regular intervals, like every year, quarter or month. We can use the compound interest formula to compute the future value of those periodic investments and thereby derive a formula to easily calculate the total future value of those investments.

FUTURE VALUE OF AN ORDINARY ANNUITY

Let's derive the formula for the future value of an ordinary annuity by looking at an example to set the foundation.

EXAMPLE 1

Suppose Jessica wants to create a college fund for her daughter, Olivia, who was just born. Jessica will deposit each of her annual \$2,000 bonus checks in this college account on Olivia's birthday each year (i.e., at the end of each year of her life) for the next 18 years. The account earns 8% interest, compounded yearly. How much money will Olivia have toward college when she turns 18 years old?

For this computation, we only need the compound interest formula. Basically, the \$2,000 Jessica invests at the end of the first year will be earning compound interest for 17 years. The \$2,000 she invests at the end of the second year will earn compounded interest for 16 years, and so on.

YEAR (OLIVIA'S AGE)	CONTRIBUTION ($)	NUMBER OF YEARS FOR COMPOUNDING (t)	FINAL VALUE OF CONTRIBUTION
1	$2,000	17	$S_1 = 2000\left(1+\frac{0.08}{1}\right)^{17} \simeq \$7400.04$
2	$2,000	16	$S_2 = 2000\left(1+\frac{0.08}{1}\right)^{16} \simeq \$6851.89$
3	$2,000	15	$S_3 = 2000\left(1+\frac{0.08}{1}\right)^{15} \simeq \$6344.34$
4	$2,000	14	$S_4 = 2000\left(1+\frac{0.08}{1}\right)^{14} \simeq \$5874.39$
5	$2,000	13	$S_5 = 2000\left(1+\frac{0.08}{1}\right)^{13} \simeq \$5439.25$
6	$2,000	12	$S_6 = 2000\left(1+\frac{0.08}{1}\right)^{12} \simeq \$5036.34$
7	$2,000	11	$S_7 = 2000\left(1+\frac{0.08}{1}\right)^{11} \simeq \$4663.28$
8	$2,000	10	$S_8 = 2000\left(1+\frac{0.08}{1}\right)^{10} \simeq \$4317.85$
9	$2,000	9	$S_9 = 2000\left(1+\frac{0.08}{1}\right)^{9} \simeq \$3998.01$
10	$2,000	8	$S_{10} = 2000\left(1+\frac{0.08}{1}\right)^{8} \simeq \$3701.86$
11	$2,000	7	$S_{11} = 2000\left(1+\frac{0.08}{1}\right)^{7} \simeq \$3427.65$
12	$2,000	6	$S_{12} = 2000\left(1+\frac{0.08}{1}\right)^{6} \simeq \$3173.75$
13	$2,000	5	$S_{13} = 2000\left(1+\frac{0.08}{1}\right)^{5} \simeq \$2938.66$
14	$2,000	4	$S_{14} = 2000\left(1+\frac{0.08}{1}\right)^{4} \simeq \$2720.98$
15	$2,000	3	$S_{15} = 2000\left(1+\frac{0.08}{1}\right)^{3} \simeq \$2519.42$
16	$2,000	2	$S_{16} = 2000\left(1+\frac{0.08}{1}\right)^{2} \simeq \$2332.80$
17	$2,000	1	$S_{17} = 2000\left(1+\frac{0.08}{1}\right)^{1} \simeq \$2160.00$
18	$2,000	0	$S_{18} = 2000\left(1+\frac{0.08}{1}\right)^{0} \simeq \$2000.00$

To find the total value of the college investment account on Olivia's 18th birthday, we need to add up the values from each year's investment contribution.

$$\Rightarrow \quad S = S_1 + S_2 + S_3 + \ldots + S_{18} \simeq \$74{,}900.51$$

For a total investment of ($2,000)18 = $36,000 over 18 years, the account grew to almost $75,000 with all the compounded interest. Not too shabby.

Remember that the compound interest formula basically just creates a geometric sequence of numbers. So, what we did in the last example was to add up many terms from a geometric sequence, which means we want the sum of a geometric series, and we already derived that formula in a previous section of this book. We'll use that formula for the sum of a geometric series, as well as some calculus, to derive this next formula.

FUTURE VALUE OF AN ORDINARY ANNUITY

Let N be the total number of payments made into the annuity and R be the regular payment amount. Then the future value of an ordinary annuity is

$$S = R\left(\frac{\left(1+\frac{r}{n}\right)^N - 1}{\frac{r}{n}}\right) = \frac{Rn}{r}\left(\left(1+\frac{r}{n}\right)^N - 1\right).$$

Another way to write this same formula is to define a new variable $r_c = \frac{r}{n}$, which is the interest rate for each compounding period. That transforms our formula to

$$S = R\left(\frac{(1+r_c)^N - 1}{r_c}\right).$$

(*Note:* The number of payments made per year must match with the number of compoundings per year in order to use this formula!)

EXAMPLE 2

Suppose that you can afford to put $500 per month into an investment account to save toward your retirement. You regularly invest this amount for 20 years in an account that you predict will earn 7.5% interest compounded monthly, without taking too much risk. What will the account be worth upon your retirement?

For this example, we have $R = \$500$, $N = 12(20) = 240$ payments made and $r_c = \frac{r}{n} = \frac{0.075}{12} = 0.00625$. This is an ordinary annuity, and we want to find the future value S.

$$\Rightarrow \quad S = R\left(\frac{(1+r_c)^N - 1}{r_c}\right) = 500\left(\frac{(1+0.00625)^{240} - 1}{0.00625}\right) \simeq \$276{,}865.36$$

Thus, for a total investment of $500(240) = $120,000 over those 20 years, the investment grows to be worth close to $300 thousand.

SINKING FUND

Next, let's look at an example of a sinking fund. This is when we set aside money in an investment account in order to be worth a pre-defined amount at some future time. Typically, we would do this in order to pay off some debt that will come due in the future. To find the necessary regular payment we need to make to the investment account in order to be worth S dollars some day, we can use the future value of an ordinary annuity formula and simply solve for R algebraically.

> **SINKING FUND PAYMENT FORMULA**
>
> The formula gives the regular payment, R, necessary to invest in an account to reach S dollars at the end of N compounding periods.
>
> $$R = S\left(\frac{r_c}{(1+r_c)^N - 1}\right)$$

EXAMPLE 3

Mark has a debt of \$28,000 to pay in 5 years. How much must he invest at the end of each quarter in an account earning 12% interest compounded quarterly to be able to pay off the debt?

We need to use the sinking fund formula. Or we can use the ordinary annuity formula and plug in everything we know, solving for the remaining variable. Either process turns out to be the same.

We have $S = \$28{,}000$, $r_c = \frac{r}{n} = \frac{0.12}{4} = 0.03$, and $N = 5(4) = 20$. Now, we can solve for R.

$$\Rightarrow \quad R = S\left(\frac{r_c}{(1+r_c)^N - 1}\right) = 28000\left(\frac{0.03}{(1+0.03)^{20} - 1}\right) \approx \$1{,}042.04$$

This means that the company must make a payment of \$1,042.04 into this account at the end of every quarter for five years in order to have enough to pay the loan off.

FUTURE VALUE OF AN ANNUITY DUE

An annuity due is basically the same as an ordinary annuity, with the only difference being when we make the payments into the account. We make the payments for an ordinary annuity at the end of each period; for an annuity due, we make the payments at the beginning of each period.

Let's say we invest at the end of each month of a calendar year. Then we would invest at the end of January, February, etc. until the end of December. If, on the other hand, we invest at the beginning of each month of the same calendar year, we would invest at the beginning of January, February, etc. until the beginning of December. Hopefully you can imagine that (1) some of the money invested in the annuity due sits in the account a bit longer and will therefore be earning interest for longer periods of time, which means it will be worth a bit more than the corresponding ordinary annuity; and (2) that investing at the end of January is basically equivalent to investing at the beginning of February, investing at the end of February is equivalent to investing at the beginning of March, etc. Thus, most of the payments for either annuity will have equivalent yields for those "inner" months or periods. The main difference between the two types of annuities then has to do solely with the first and last payments.

For an annuity due, then, we have

$$S = R\left(\frac{(1+r_c)^{N+1}-1}{r_c}\right) - R$$

which says mathematically that it's the same as investing $N + 1$ payments of R for an ordinary annuity, minus that last payment that we did make for an ordinary annuity but that we don't make for an annuity due (the very last payment that doesn't earn any interest since it was added to the account at the very end of the investment period).

Let's simplify this:

$$S = R\left(\frac{(1+r_c)^{N+1}-1}{r_c}\right) - R \Leftrightarrow S = R\left(\frac{(1+r_c)^{N+1}-1}{r_c} - 1\right) \quad \text{Get a common denominator.}$$

$$\Leftrightarrow S = R\left(\frac{(1+r_c)^{N+1}-1}{r_c} - \frac{r_c}{r_c}\right) \quad \text{Combine into one fraction.}$$

$$\Leftrightarrow S = R\left(\frac{(1+r_c)^{N+1}-(1+r_c)}{r_c}\right) \quad \text{Factor the numerator.}$$

$$\Leftrightarrow S = R(1+r_c)\left(\frac{(1+r_c)^{N}-1}{r_c}\right)$$

FUTURE VALUE OF AN ANNUITY DUE

Let N be the total number of payments made into the annuity, R is the payment amount and $r_c = \frac{r}{n}$ is the interest rate per compounding period. Then S_{due} is the future value of the annuity given by

$$S_{due} = R(1+r_c)\left(\frac{(1+r_c)^{N}-1}{r_c}\right) = S(1+r_c)$$

(where S = the future value of the corresponding ordinary annuity).

EXAMPLE 4

Find the future value of an account, earning 6% compounded monthly, with \$150 deposited at the beginning of every month for 8 years.

Since the deposits are made at the beginning of every month, this is an annuity due. We have $R = \$150$, $N = 8(12) = 96$, and $r_c = \frac{0.06}{12} = 0.005$.

$$\Rightarrow S_{due} = R(1+r_c)\left(\frac{(1+r_c)^{N}-1}{r_c}\right) = 150(1+0.005)\left(\frac{(1+0.005)^{96}-1}{0.005}\right) \approx \$18{,}516.40$$

EXAMPLE 5

George and Mary have a friendly contest between them to see who can earn the most money in the next 5 years. They each can only invest a total of \$10,000 in an account earning 10% interest compounded monthly, but neither of them has that much money up front. George decides to invest \$5,000 at the beginning of the 5 years, and then add \$85 at the end of each month to his account. Mary invests \$170.00 at the end of every month for the entire 5 years. Who has more money at the end of the 5 years?

For each of them, we have this data: $r_c = \dfrac{0.10}{12} = 0.008\overline{3}$, $t = 5$ years.

GEORGE

(1) The \$5,000 invested at the beginning of the 5-year period will just be sitting there earning compound interest. So, that will grow to be worth

$$S_1 = P\left(1+\frac{r}{n}\right)^{nt} = 5000(1+0.008\overline{3})^{(12)} = 5000(1.008\overline{3})^{60} \simeq \$8{,}226.54$$

(2) He also invests in an ordinary annuity every month with $N = 5(12) = 60$ and $R = \$85$. That will grow to be worth:

$$S_2 = R\left(\frac{(1+r_c)^N - 1}{r_c}\right) = 85\left(\frac{(1+0.008\overline{3})^{60} - 1}{0.008\overline{3}}\right) \simeq \$6{,}582.15$$

Altogether, George earned \$8226.54 + \$6582.15 = \$14,808.69.

MARY

She invests all her money in an ordinary annuity with $N = 5(12) = 60$ and $R = \$170.00$.

$$S = R\left(\frac{(1+r_c)^N - 1}{r_c}\right) = 170\left(\frac{(1+0.008\overline{3})^{60} - 1}{0.008\overline{3}}\right) \simeq \$13{,}164.30$$

Thus, George wins the contest! (George invested a total of \$10,100 and Mary invested \$10,200.) This shows the power of simply leaving money in a compounded interest account for as long as possible to gain the best results.

EXAMPLE 6

You graduate from college at the age of 25 and because you're tired of living with little money, you decide not to invest toward your retirement for the first five years of your career. This allows you to enjoy more of your money for that time interval. After that five-year period, you invest \$400 at the end of every month in an account earning 9% interest compounded monthly for the next 35 years.

(a) How much will your retirement account be worth at age 65?

(b) How much more would your account have been worth if you had only invested the $400 per month for the first eight years and then let the money sit there earning compound interest for those next 32 years?

(c) How much would the account be worth if you had invested the $400 monthly for the entire 40 years?

(a) You're investing in an ordinary annuity with $N = 35(12) = 420$, $R = \$400$, and $r_c = \frac{0.09}{12} = 0.0075$.

$$\Rightarrow \; S = R\left(\frac{(1+r_c)^N - 1}{r_c}\right) = 400\left(\frac{(1+0.0075)^{420} - 1}{0.0075}\right) \simeq \$1{,}176{,}713.79$$

(b) We need to break this into two parts for this calculation: (1) the ordinary annuity investment for the first 8 years and (2) a compounded interest account, with that final annuity amount as the principal amount, for 32 years.

(1) The ordinary annuity investment has input values of $R = \$400$, $N = 8(12) = 96$ and $r_c = \frac{0.09}{12} = 0.0075$.

$$\Rightarrow \; P = R\left(\frac{(1+r_c)^N - 1}{r_c}\right) = 400\left(\frac{(1+0.0075)^{96} - 1}{0.0075}\right) \simeq \$55{,}942.47$$

(2) Next, you would take that amount and use it as the principal for a compounded interest account, with $n = 12$ and $t = 32$ years.

$$\Rightarrow \; S = P(1+r_c)^{nt} = 55{,}942.47(1+0.0075)^{12(32)} \simeq \$985{,}922.29$$

(c) This is another ordinary annuity calculation with $N = 40(12) = 480$, $R = \$400$ and $r_c = \frac{0.09}{12} = 0.0075$.

$$\Rightarrow \; S = R\left(\frac{(1+r_c)^N - 1}{r_c}\right) = 400\left(\frac{(1+0.0075)^{480} - 1}{0.0075}\right) \simeq \$1{,}872{,}528.11$$

Let's put all this information together in a table:

STRATEGY	TOTAL INVESTMENT	FINAL INVESTMENT VALUE
(a)	$400(420) = $168,000	$1,176,713.79
(b)	$400(96) = $38,400	$985,922.29
(c)	$400(480) = $192,000	$1,872,528.11

Notice the difference between strategies (a) and (b). For less than a $40,000 investment, the account grew to almost $1 million in strategy (b). In strategy (a), the investment was almost 4.5 times more than for (b), and the payoff at the end is not substantially larger. Of course, the best route to take is strategy (c) for the maximum payout at retirement.

EXAMPLE 7

A $50,000 machine will have a market value of $6,000 in 15 years, which is the usable lifespan of this machine. A new machine is expected to sell for $68,000 at that time. The company sets up a sinking fund to pay for the new machine in the future. Payments are made to the fund twice each year. If the fund earns 7% interest compounded twice yearly, how much should each payment be?

Basically, the company needs to invest enough money to be worth $68,000 – $6,000 = $62,000. That is, they will need $68,000 to purchase the new machine, but they will be able to sell their current (used) machine at that point for $6,000 to put toward the new machine. So, they only need to save $62,000. Thus, we need to calculate the payments needed for S = $62,000, with $r_c = \frac{0.07}{2} = 0.035$ and $N = 15(2) = 30$.

$$\Rightarrow \quad R = S\left(\frac{r_c}{(1+r_c)^N - 1}\right) = 62000\left(\frac{0.035}{(1+0.035)^{30} - 1}\right) \simeq \$1{,}201.02$$

This means we need to deposit $1,201.02 into the sinking fund at the end of each half-year to save enough money to pay for the new machine in 15 years.

5.3 Homework

Note: If no other specification is made, assume the annuity is an ordinary annuity (with payments made at the end of each compounding period).

1. Calculate the future value of an ordinary annuity that pays 7% interest compounded quarterly if quarterly payments of $250 are made for one year.

2. Calculate the future value of an ordinary annuity that pays 7% interest compounded quarterly if two years of $250 quarterly payments are made.

3. Suppose a 5-year ordinary annuity is offered by a bank that pays 5% compounded annually. A customer decides to make regular payments of $100 per year into this annuity. What is the future value of this annuity?

4. For question number 3 (above), how many years will the first year's payment earn interest? Also answer this same question for the 5th payment.

5. What is the future value of an annuity that pays 11%, compounded quarterly, into which quarterly payments of $2,000 are made for 20 years?

6. Rubin-the-Dog's owner wants to bury Rubin in style—he was such a good dog. So, he sets up an annuity, in Rubin's name, in which he deposits $100 a month for 4 years at 5% interest, compounded monthly. What will be the future value of Rubin's funeral annuity?

7. Julie can sell her dirty socks on Ebay and earn $50 a month. If she deposits this money into an annuity for 3 years at 7% interest, compounded monthly, what is the future value of her Sock annuity?

8. Assume that you plan on buying a condominium in Hawaii five years from now. You estimate that you can save $2,500 per year. You plan on depositing the money at the end of the year for each of the 5 years in a bank account that earns 5% interest compounded annually. What is the most that you will be able to offer for your condo at the end of the 5 years?

9. Considering problem 8, how would your answer change if the interest rate paid by the bank were increased to 6%?

10. How would your answer to problem 6 change if you found that you could deposit your money into a money market account that paid 10% interest?

11. Juan wants to buy his dream car: a 1970 Chevy Corvette, electronic fuel injected aluminum block ZL1 engine with a removable hard top and painted cherry red. Scouting through E-bay, it believes he can locate and buy this car for about $24,000. Juan only pays cash for what he owns, so he decides to set up his own ordinary annuity. He can afford to put away $5,000 at the end of each year and can earn 6%, compounded annually with a long-term CD at his local bank. Will Juan be able to buy his dream car in 5 years?

12. How would the answer to problem 11 change if Juan put away the same $5,000 per year, but divided it in 12 equal parts and deposited this amount at the end of every month for 5 years (presume Juan can earn 6% interest compounded monthly)?

13. Mo sells beer at his bar, but he'd secretly like to be a professional dancer. It will cost $3,000 for lessons from the best teacher in Bountiful Heights. If Mo deposits $150 a month from beer sales into an annuity paying 4.5% interest compounded monthly for two years, will Mo have enough to become a dancer?

14. What is the future value of an ordinary annuity into which $2,000 a month is deposited at 10% interest, compounded monthly, for 10 years?

15. Saja is saving for college. She is 10 years old and plans to go to college when she's 18. She saves her allowance, $300 a month (she has rich parents). She has the family butler deposit this into an annuity at 8% interest compounded monthly. How much will Saja have in her college account when she's 18?

16. The Washington Senators, a baseball team, are doing so poorly that they are taking some of the games' ticket sales and putting them into an annuity to try to buy some better players. If they deposit $25,000 a month for 2 years at 7% interest, compounded monthly, what will the team have to bargain with at the end of the 2-year period?

17. Suppose the moment a barrel of Swine Vineyard is corked it's worth $3,500. Suppose that the value of wine increases by 8% annually as it ages, but increases proportionally with each month it ages. Another barrel of the same quality of wine is corked every month. What is the value of Swine Vineyard's wine at the end of a 4-year period?

18. Suppose you've joined the military. You can retire in 20 years and receive a full veteran's pension. However, you've determined that the pension will not be sufficient to live on. So, you decide to save for your retirement. You can afford to save $300 per month. The military will match your retirement savings dollar for dollar. What will be the future value of your retirement savings if it earns 6.5%, compounded monthly?

SINKING FUND

19. Jeri is engaged and will get married in three years. She wants a large, expensive wedding that will cost about $25,000. In order to save for the wedding, she has found a relatively safe investment that pays 9% compounded monthly. What monthly payment must she make to this account in order to pay for her wedding?

20. Creative Ideas International Company predicts it will have a tax bill of $15,000 in one year. What monthly payments must be deposited into an account that pays 6% compounded monthly in order to pay the taxes at the end of the year?

21. Jaclyn wants to participate in a university study abroad program in two years. The estimated total cost is $9,400, including travel expenses, lodging, food, books and tuition. She can make quarterly payments into an account that pays 7.5% interest, compounded quarterly. What is her quarterly payment amount that will help her save enough money for the trip?

22. Refer to question 21. If Jaclyn instead finds an account that pays 7.5% interest compounded monthly, what will her monthly payments be to save for her trip? How much money will she save with this account compared to the account from question 21?

23. Coen, at the age of 5, decides that he wants to save his money to go to Disneyland when he's 10 years old. He needs to save a total of $1000. His parents (Coen's "bank") will pay 6.5% interest compounded weekly. How much weekly allowance must Coen put into his savings account in order to afford his Disneyland trip in 5 years?

24. Eight-year-old Cayce hears about his younger brother's Disneyland savings (from problem 23). He decides that he will invest his weekly allowance in the same type of account, through his parents, to save for a car. He wants to have a total of $4,000 by the time he's 16. What will his weekly payment need to be to reach his goal?

FUTURE VALUE OF AN ANNUITY DUE

25. Review problem 18. What is the future value of your retirement savings if the $300 payments to your retirement fund are made at the beginning of the period?

26. What is the future value of a 20-year savings account that pays 6% interest, compounded monthly where payments of $100 are made to the account at the beginning of each month?

27. If Craig puts $250 at the beginning of every month into a fund that pays 7.5% interest compounded monthly, how much will it be worth in 10 years?

28. Beth wants to save for her son's college education. Andrew, Beth's son, wants to attend an expensive private college. She deposits her entire $15,000 annual bonus into an account that earns 6.2% interest compounded yearly on January 1st every year for 12 years. How much money will Andrew have for college?

29. As a single mother, Molly doesn't start saving for retirement until she's done raising her two children, at age 45. For the next 25 years, she invests $600 per month, at the beginning of each month, in an account that pays 8.4% interest, compounded monthly. At age 70, how much money will she have for retirement?

30. Refer to problem 29. If Molly had instead only saved $75 per month, at the beginning of each month, starting at age 20 (in the same interest bearing account), how much money would she have upon retirement at age 70?

5.4 Present Value of Annuities

VOCABULARY

Present Value of an Annuity—The lump sum of money that needs to be in an account now earning compounded interest in order to make regular withdrawals later. It is the sum of the present values of all payments over a specified time.

NOTE: We can still ask for the **present value of an ordinary annuity** (withdrawals made at the end of the period) **or an annuity due** (withdrawals made at the beginning of the period), just as we did with the future value of annuities.

Deferred Annuity—The present value of an ordinary annuity whose first withdrawal/payment is deferred for a specified time, at which point regular withdrawals are made.

Here's the basic idea. When we're saving for our retirement, we want to know how much money to invest monthly, for example, into an investment account in order to yield some future value, which is the future value of an annuity calculation. Then, upon retirement, we want to make specified monthly withdrawals from that investment account and need to know what the present value of the account needs to be in order to make those withdrawals. Thus, to successfully save for retirement, we would need to know how to calculate the future and present value of an annuity to know how much to save and what we can afford to withdraw.

Let's derive the formula by looking at an example first.

EXAMPLE 1

Frank is calculating when he can retire. He estimates that he'll need to withdraw \$10,000 at the end of every quarter from his retirement account for living expenses for 25 years. The account earns 6.5% interest compounded quarterly. What amount of money needs to be in the investment account upon his retirement to achieve his financial goal?

Notice that this is "backwards" of the types of questions in the last section. When calculating the future value of an annuity, you need to know the amount of money that regular payments into the account will produce in the future. In this example case, we want to know how much money is necessary to start with that will produce regular payments or withdrawals from the account.

Let's start with the formula for compounded interest. For this example, we have the values $r = 0.065$ and $n = 4$, and the overall time for quarterly withdrawals is 25 years. This means Frank will make $4(25) = 100$ withdrawals throughout his retirement.

$$\Rightarrow \quad S = P\left(1 + \frac{0.065}{4}\right)^{4t} \quad \Leftrightarrow \quad S = P(1.01625)^{4t}$$

We'll look at this process incrementally.

For the first withdrawal that's made only one quarter after the beginning of retirement, the principal amount that's in the account will have only that one quarter to earn interest. We'll call this principal amount P_1. To get the specified payment from the account, we need the account to be worth \$10,000 at that point, i.e., the withdrawal amount $S = \$10,000$. For this first payment (or withdrawal), $t = \frac{1}{4}$ year.

$\Rightarrow\ 10000 = P_1(1.01625)^{4\left(\frac{1}{4}\right)}$ Divide both sides of the equation by 1.01625^1 to solve for P_1.

$$\frac{10000}{(1.01625)^1} = P_1 \Leftrightarrow P_1 \simeq \$9840.10$$

For the second quarterly payment, we need enough money in the account (in addition to what we pulled out of the account for the first payment) to be worth \$10,000 after 2 quarters. Now, $t = \frac{1}{2}$ year.

$\Rightarrow\ 10000 = P_2(1.01625)^{4\left(\frac{1}{2}\right)} \Leftrightarrow 10000 = P_2(1.01625)^2$ Divide both sides of the equation by 1.01625^2 to solve for P_2.

$$\frac{10000}{(1.01625)^2} = P_2 \Leftrightarrow P_2 \simeq \$9682.75$$

For the third quarterly payment, we additionally need enough principal in the account to be worth \$10,000 after 3 quarters. For this payment, $t = \frac{3}{4}$ year.

$\Rightarrow\ 10000 = P_3(1.01625)^{4\left(\frac{3}{4}\right)} \Leftrightarrow 10000 = P_3(1.01625)^3$ Divide both sides of the equation by 1.01625^3 to solve for P_3.

$$\frac{10000}{(1.01625)^3} = P_3 \Leftrightarrow P_3 \simeq \$9527.92$$

As you can probably see from here, we need to continue in this pattern for the full 100 quarterly payments (withdrawals) and then add up all the necessary principal amounts to get the final principal, aka present value, of the investment account needed to retire within the given parameters.

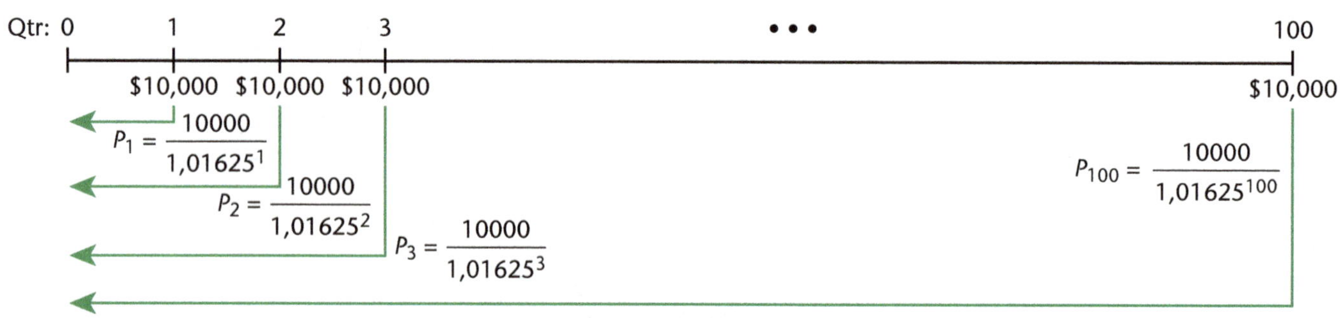

$\Rightarrow\ P = P_1 + P_2 + P_3 + P_4 + \ldots + P_{99} + P_{100}$ This will give us the total amount we need today to produce all of the required quarterly payments for 25 years.

$$P = \frac{10000}{(1.01625)^1} + \frac{10000}{(1.01625)^2} + \frac{10000}{(1.01625)^3} + \ldots + \frac{10000}{(1.01625)^{99}} + \frac{10000}{(1.01625)^{100}}$$

Notice that this is the sum of a geometric sequence, which we studied in a previous section of this chapter, because getting from any term to the next term in the sum simply requires multiplying by $\frac{1}{1.01625}$. We have a formula for this sum: Given $S_n = a_1 + a_2 + a_3 + a_4 + \ldots + a_{n-1} + a_n$ the sum is $S_n = \frac{a_1(1-d^n)}{1-d}$. For our example, $P = S_n$ and the first term is $a_1 = \frac{10000}{1.01625}$, $n = 100$; the common quotient that's repeatedly multiplied from one term to the next is $d = \frac{1}{1.01625} = 1.01625^{-1}$. Substitute all these values into the sum formula.

$$\Rightarrow P = \left(\frac{10000}{1.01625}\right)\frac{(1-(1.01625^{-1})^{100})}{1-1.01625^{-1}}$$ Simplify the denominator by distributing the 1.01625.

$$P = 10000\left(\frac{(1-(1.01625^{-1})^{100})}{1.01625-1}\right) = \frac{10000(1-(1.01625^{-1})^{100})}{0.01625}$$

This formula will allow us to directly and quickly find the present value necessary to produce all the 100 withdrawals from the investment/retirement account for Frank. Plugging this all into a calculator, we get

$$P \simeq \$492{,}615.03$$

Thus, if Frank's retirement account has a little less than \$500,000 in it, he'll be able to meet his retirement financial goal for living expenses.

This leads to the formula for the present value of an ordinary annuity.

PRESENT VALUE OF AN ORDINARY ANNUITY

Let R = the withdrawal/payment amount, n = the number of compoundings per year, r = the annual interest rate, t = total number of years for withdrawals/payments, N = the total number of withdrawals/payments = nt, $r_c = \frac{r}{n}$ = the interest rate per compounding period. Then the present value P of the account is

$$P = \frac{R(1-(1+r_c)^{-N})}{r_c}.$$

(*Note:* The number of compoundings per year must match the number of payments/withdrawals made per year.)

Similarly, the present value of an annuity due will need to be slightly higher than the corresponding ordinary annuity, with only $N-1$ withdrawals, because withdrawal of the first payment will occur at the very beginning of the overall time. But, since it's withdrawn on the first day of the account, it will have no time to earn any interest. Thus,

$$P_{due} = P_{N-1} + R$$

where P_{N-1} is the present value of the ordinary annuity with the same parameters, except with one fewer payment, and R is the one other payment withdrawn from the account at the first day of the account.

$$\Rightarrow P_{due} = P_{N-1} + R = \frac{R(1-(1+r_c)^{-(N-1)})}{r_c} + R$$ Get a common denominator for both terms.

$$P_{due} = \frac{R(1-(1+r_c)^{-(N-1)})}{r_c} + \frac{R(r_c)}{r_c}$$ Combine into one fraction and factor out R in the numerator.

$$P_{due} = \frac{R(1-(1+r_c)^{-(N-1)} + r_c)}{r_c}$$ Rearrange the terms in the numerator.

$$P_{due} = \frac{R((1+r_c)-(1+r_c)^{-(N-1)})}{r_c}$$ Factor out the $(1 + r_c)$ from both terms in the numerator.

$$P_{due} = \frac{R(1+r_c)(1-(1+r_c)^{-N})}{r_c}$$

$$\Rightarrow P_{due} = (1+r_c)\left(\frac{R(1-(1+r_c)^{-N})}{r_c}\right) = (1+r_c)P$$

In other words, the present value of an annuity due is just $(1 + r_c)$ times the present value of the ordinary annuity.

PRESENT VALUE OF AN ANNUITY DUE

Let R = the withdrawal/payment amount, n = the number of compoundings per year, r = the annual interest rate, t = total number of years for withdrawals/payments, N = the total number of withdrawals/payments = nt, $r_c = \frac{r}{n}$ = the interest rate per compounding period. Then the present value P_{due} of the account is

$$P_{due} = \frac{R(1+r_c)(1-(1+r_c)^{-N})}{r_c} = (1+r_c)P,$$

where P is the present value of the corresponding ordinary annuity.

EXAMPLE 2

Find the present value of a college fund that pays \$2,000 at the end of each month from an account that earns 9% interest compounded monthly for 6 years.

This is an ordinary annuity because the payments occur at the end of each month. We just need to plug in the correct values to our formula for the present value of an ordinary annuity. We have:

$$r = 0.09,\ n = 12,\ r_c = \frac{r}{n} = \frac{0.09}{12} = 0.0075,\ R = \$2000,\ t = 6 \text{ years},\ N = 12(6) = 72$$

$$\Rightarrow P = \frac{R(1-(1+r_c)^{-N})}{r_c} = \frac{2000(1-1.0075^{-72})}{0.0075} \approx \$110{,}953.70$$

This means that just over \$110,000 invested at the start of college in this account will yield \$2,000(72) = \$144,000 worth of monthly payments to the college student.

EXAMPLE 3

Tracy received an inheritance of $200,000 from his grandmother. How much can he withdraw from this account at the end of each quarter for the next 30 years, if his money compounds quarterly with 6% interest?

For this particular example, we know the present value (or the beginning value) of the account is $200,000. We want to know what R (the payment) is for this ordinary annuity, with

$$r = 0.06,\ n = 4,\ r_c = \frac{r}{n} = \frac{0.06}{4} = 0.015,\ t = 30,\ N = 4(30) = 120, \text{ and } P = \$200{,}000.$$

$$\Rightarrow \quad P = \frac{R(1-(1+r_c)^{-N})}{r_c} \quad \Leftrightarrow \quad 200{,}000 = \frac{R(1-1.015^{-120})}{0.015}$$

Multiply both sides of the equation by the reciprocal of the coefficient of R to solve for R.

$$200{,}000\left(\frac{0.015}{(1-1.015^{-120})}\right) = R \quad \Leftrightarrow \quad R \simeq \$3{,}603.70$$

Thus, Tracy can withdraw a total of 120 ($3,603.70) = $432,444.00 in payments through the 30 years, from a starting balance of $200,000 in this account.

EXAMPLE 4

Karen receives a lottery prize of $1,200,000 with payments of $5,000 at the beginning of every month for 20 years. If the money is in an account earning 5% interest compounded monthly, what is the real value of the lottery prize today?

This refers to an annuity due because the payments occur at the beginning of every month. We know the payments and all other necessary information for compounding, but we need to calculate the present value. We have:

$$r = 0.05,\ n = 12,\ r_c = \frac{r}{n} = \frac{0.05}{12} = 0.0041\overline{6},\ t = 20 \text{ years},\ N = 12(20) = 240, \text{ and } R = \$5000.$$

Plug all the necessary values in to the formula for the present value of an annuity due.

$$\Rightarrow \quad P_{due} \frac{R(1+r_c)(1-(1+r_c)^{-N})}{r_c} = \frac{5000(1.0041\overline{6})(1-(1.0041\overline{6})^{-240})}{0.0041\overline{6}} \simeq \$760{,}784.48$$

This means that Karen receives a total of $5,000 (240) = $1,200,000 over a period of 20 years of payments. However, the lottery organization only had to set aside $760,784.48 in the investment account to produce the prize money, which is a reasonable deal for them.

EXAMPLE 5

Suppose Brenda receives an inheritance of \$20,000, on her 20th birthday. However, according to the will of the donor, she is not to receive the money until she is 30 years old. She plans to leave the money in the interest bearing account, earning 8% interest compounded quarterly, and only make quarterly withdrawals, starting at the end of the first quarter after her birthday, for 15 years total. How much will each withdrawal be?

We need to break this computation into two pieces.

1. Calculate the future worth of the \$20,000 after 10 years of earning compounded interest with no withdrawals.
2. Using that value for the then-present value of the account, we can use the formula for the present value of an ordinary annuity to compute the quarterly payment amounts.

(1) For this computation, we only need the formula for compounded interest, with

$$P = \$20{,}000,\ r = 0.08,\ n = 4 \text{ and } t = 10 \text{ years.}$$

$$\Rightarrow\ P = P\left(1+\frac{r}{n}\right)^{nt} = 20000\left(1+\frac{0.08}{4}\right)^{4(10)} = 20000(1.02)^{40} \simeq \$44{,}160{,}79$$

(2) Now, we need to use the formula for the present value of an ordinary annuity with

$$P = \$44{,}160.79,\ r_c = \frac{r}{n} = \frac{0.08}{4} = 0.02,\ N = 4(15) = 60 \text{ total payments.}$$

$$\Rightarrow\ P = \frac{R(1-(1+r_c)^{-N})}{r_c} \Leftrightarrow 44160.79 = \frac{R(1-1.02^{-60})}{0.02}$$

Multiply both sides of this equation by the reciprocal of the coefficient of R to solve for R.

$$44160.79\left(\frac{0.02}{1-1.02^{-60}}\right) = R \Leftrightarrow R \simeq \$1{,}270.42$$

This means that the quarterly payment will be just over \$1,270 for the 15 years of collecting payments from the investment. In total, Brenda will receive \$76,225.20 from those payments, and her original investment was only \$20,000.

This example leads us into the idea of a deferred annuity. Notice that we had to perform two calculations to get today's value of the account—(1) the future value of the account at the end of earning compounded interest with no withdrawals made, and (2) the present value at that time of the remaining period when withdrawals took place.

Basically, if we substitute in the values in the present value of an ordinary annuity formula, we'll have:

$$P_{annuity} = \frac{R(1-(1+r_c)^{-N})}{r_c} \Leftrightarrow P(1+r_c)^m = \frac{R(1-(1+r_c)^{-N})}{r_c}$$

This is assuming that P is the principal of the account when started, i.e., the present value of the deferred annuity, and m = the number of compounding periods the money is left in the account before withdrawals take place. Solving for P, we get:

PRESENT VALUE OF A DEFERRED ANNUITY

Let P = present value of a deferred annuity, where it's deferred (with no withdrawals made) for m compounding periods, at which point N regular withdrawals of $\$R$ take place.

$$P = \frac{R(1-(1+r_c)^{-N})}{r_c(1+r_c)^m}$$

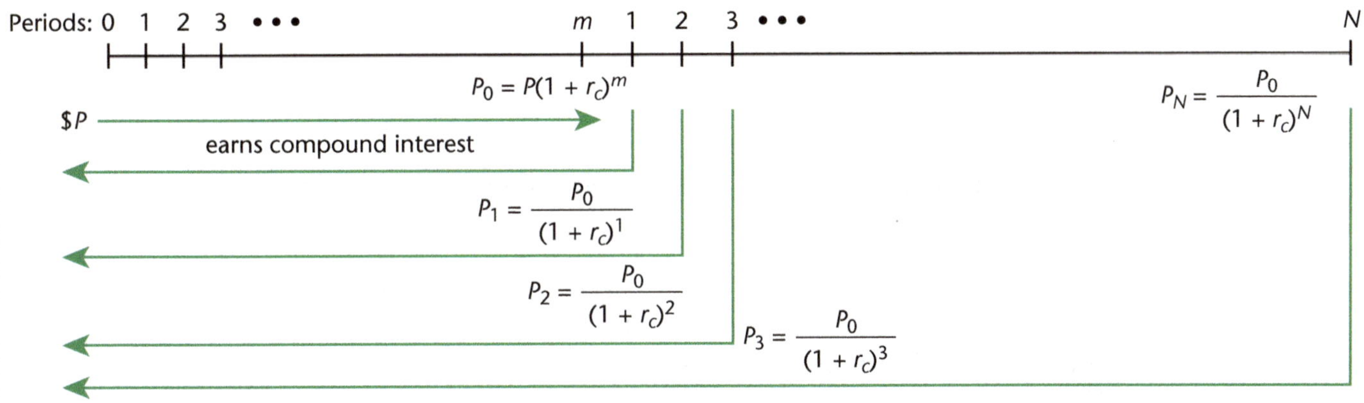

EXAMPLE 6

Upon graduation from college, Luce's parents decide to give her a large cash gift, since their business is doing well. They're forward-thinking parents and want to invest in their daughter's retirement. They give her exactly enough to create monthly retirement payments of $4,000, paid at the end of each month, starting in 40 years. The payments will last for 20 years total. After shopping around, they found an extremely safe investment account that pays 5.4% interest, compounded monthly. How much money will they put in this account for Luce upon graduation?

This is a deferred annuity because the money will simply be earning compounded interest for the first 40 years, and then the payments to Luce will occur monthly for the next 20 years. We have

$$r = 0.054,\ n = 12,\ r_c = \frac{r}{n} = \frac{0.054}{12} = 0.0045,\ R = \$4{,}000,\ N = 20(12) = 240,\ m = 40(12) = 480.$$

Substitute the necessary values into the formula for the present value of a deferred annuity.

$$\Rightarrow\ P = \frac{R(1-(1+r_c)^{-N})}{r_c(1+r_c)^m} = \frac{4000(1-1.0045^{-240})}{0.0045(1.0045^{480})} \simeq \$67{,}942.84$$

So they put just under $68,000 into the investment account for their daughter, where it will earn compounded interest until she is ready to retire. Then, she will receive a total of 4000(240) = $960,000 in payments over 20 years!

5.4 Homework

Note: If no other specification is made, assume the annuity is an ordinary annuity (with deposits/payments made at the end of each compounding period).

1. What is the present value of an annuity of $100 per year for 3 years earning a 5% interest rate compounded annually?

2. What is the present value of an annuity of $100 per month, compounded monthly, for 3 years earning a 5% interest rate?

3. Susie wants to set up a college fund for her daughter Mary. She calculates that Mary will need to withdraw $2,500 four times a year for the next four years. How much will Susie need to have in Mary's college fund if the fund pays 6%, compounded quarterly?

4. Della Mae's husband has left her $350,000 upon his untimely death. She has deposited this amount in a savings account that pays 4.5% interest compounded monthly. How much can Della Mae withdraw monthly if she intends to access this account for the next 25 years?

5. In reference to problem 4, suppose Della Mae is diagnosed with cancer and now has a life expectancy of 10 years. How much can she withdraw monthly and just exhaust the account?

6. The U.S. Bureau of Reclamation is considering building a dam and reservoir to hold water. The construction of the dam and reservoir will take four years. The winning bid for the construction has been awarded and contracted and will cost two billion dollars spread equally over the four years of construction, with payments made every quarter. What is the present value (PV) of the dam and reservoir if the interest rate is 4%, compounded quarterly?

7. The dam and reservoir in problem 6 will deliver water to farmers as soon as construction is completed (year 5). 40,000 acre feet (a standard measure of water) will be sold every quarter for $200 an acre foot. Sales of water have been contracted for 20 years. What is the present value of water sales if the interest rate is 4% compounded quarterly. For simplicity, suppose that part of the water deal requires that the purchasers of water make payments from the beginning of the construction of the dam even though water has not yet been delivered.

8. Businesses, and even the federal government, decide on whether to construct a project based on whether the benefits of a project are greater than the costs. If your answer to problem 6 is the cost of the dam and reservoir and your answer to problem 7 is a calculation of the benefits of the project, should the government build this project?

9. Lucy would like to retire from her job with a retirement fund from which she can draw one half of her current pay. She estimates that she'll live for 20 years after she retires. She currently earns $50,000 a year. How much money needs to be in a retirement account that earns 7% interest compounded monthly? She will want to withdraw money for her living expenses monthly.

10. Upon accepting his new job, ACME Hair Products Inc. put $225,000 in a retirement account for Palmer. It pays 5.5%, compounded monthly. Palmer cannot withdraw any of this money for 20 years while it continues to earn interest. Palmer intends to make monthly withdrawals for 18 years once he retires. How much will he be able to withdraw each month?

11. Katheryne won the state lottery grand prize of $1 million. She can choose to take a lump sum payment or she can accept annual payments over 20 years. The lump sum payment will be calculated by the state lottery board as if it were a 20-year annuity at 8% interest compounded annually. What is the lump sum (present value) value of Katheryne's lottery winnings? What will her winnings from the lottery be if she accepts the lump sum payment and the IRS takes 30% of her payment as an income tax in that year?

12. In reference to problem 11, suppose Katheryne won the "powerball" lottery with two of her friends. The winnings total $70 million to be shared equally among the three friends. What will her winnings from the lottery be if she accepts the lump sum payment (with the same assumptions as in problem 11)?

13. What is the present value of an annuity of $10,000 per year for 2 years earning a 4.5% interest rate compounded annually?

14. Ulysses wants to set up a college fund for himself. He calculates that he will need to withdraw $1,250 four times a year. He will need to do this for seven years (he wants to go to law school). How much will Ulysses need to have in his college fund if the fund pays 4.5%, compounded quarterly?

15. Amelia just inherited $250,000 as a death benefit from her mother's life insurance. She has deposited this amount in a savings account that pays 10% interest compounded monthly. How much can Amelia withdraw monthly if she intends to access this account for the next 20 years?

16. In reference to problem 15, suppose Amelia has taken a look at some actuarial tables and finds out that since her mother lived so long, her own life expectancy is only 5 more years. How much can she withdraw monthly and just exhaust the account?

17. General Aristotle is considering building a nuclear powerplant to supply electricity to his community. Construction of the powerplant will take three years. It will cost $4.5 billion spread equally over the three years of construction, with payments made each quarter. What is the present value (PV) of the project if the interest rate is 6%, compounded quarterly?

18. The powerplant in problem 17 will deliver electricity as soon as construction is completed (year 4). 5,800 gigawatt hours will be sold every year for $30,000 per gigawatt hour. The plant's useful life is 30 years. What is the present value of the sales of electricity if the interest rate is 4% compounded quarterly (for simplicity, assume that, during construction, the general taxes the people the same as the earnings from electricity sales)?

19. Scott would like to have a retirement fund from which he can withdraw $30,000 a year for 20 years. How much money needs to be in Scott's retirement account if it earns 6.5% interest compounded monthly? He will withdraw money from this account monthly.

20. Immediately upon being elected, a new U.S. Senator is guaranteed a retirement fund of $500,000. A good money manager can earn 8% interest, compounded quarterly. The average senator will live for 15 years after retirement. How much will a senator be able to withdraw each quarter after retiring from his or her job on Capital Hill?

21. Monte won the big $3 million prize on the TV show "Go Big or Go Home." But, the small print states the she doesn't get this amount of money as a lump sum payment. Instead, she can choose to take a lump sum payment or she can accept annual payments over 10 years. The lump sum payment will be calculated by the TV producers as a present value, as if it were a 10-year annuity at 3.5% interest compounded annually. What is the lump sum payment of Monte's game winnings? What will her winnings from the TV game show be if she accepts annual payments for 10 years?

22. In reference to problem 22, suppose that, accounting for the fact that the IRS will require income tax payments on the winnings Monte receives, the effective interest rate she will expect to be used to calculate her lump sum winnings is raised to 6.5%. What will her effective winnings be if she accepts the lump sum payment?

23. Jasmine wants to be a contestant on the TV game show "The Greatest Race." In order to train, get in shape and participate in the game show, she will need to set up a fund. She calculates that she can just survive on withdrawing $1,000 every month for two years. How much money will Jasime need if the fund pays 4.75%, compounded monthly?

24. What is the present value of an ordinary annuity of $300 per month for 5 years earning a 4.5% interest rate compounded monthly?

25. Aunt Brenda decides she wants to pay for her niece, Kristin, to get a college education. She is going to put one lump sum payment into an account earning 4.8% interest, compounded monthly, when Kristin completes Kindergarten. Brenda wants the money to provide payments of $1,000 every month for five years, starting when Kristin graduates high school. How much money does Aunt Brenda need to deposit to achieve her goal?

26. You receive a large inheritance from your grandfather who just passed away. You decide that you're going to set aside enough money to provide for your retirement and then spend the rest traveling across the world. According to your calculations, you will need $5000 per month to retire in 30 years, and you predict that you will receive that monthly payment for another 30 years. How much do you need to deposit into an account earning 6.3% interest, compounded monthly?

27. Nasser has $10,000 leftover from his college fund. He is going to invest it in an account earning 5.1% compounded quarterly and leave the money there for the next five years, at which point, he is going to make quarterly withdrawals for three years to supplement his income while he spends all his free time writing a novel. How much money will he get every quarter, during the three years of pulling money out of that account?

28. Jaclyn has shopped around for a safe investment account earning 3.9% interest, which compounds monthly, because she just received a re-enlistment bonus from the Navy for $50,000 (she has a tough job in Nuclear Power). Her plan is to invest the entire $50,000 in the account and leave it there for the next 35 years, at which time she will begin making monthly withdrawals for 30 years. What will her monthly retirement withdrawals be?

29. Andrew is getting divorced and he will owe his ex-wife alimony of $500 per month for two years. However, the divorce will not be final for 8 months. He decides to just set aside enough money now to produce those 24 alimony payments. If he puts his money in an account earning 6% interest, compounded monthly, what should his lump sum deposit be to produce the alimony payments in the future? How much money will he save by paying his alimony this way?

5.5 Amortization and Loans

VOCABULARY

Installment Payments—Payments made toward a loan that are all the same amount.

Amortization—Paying back a loan in installment payments.

Amortization Schedule—Basically a spreadsheet giving all information for the amortization of a loan, including interest paid per payment and principal paid per payment.

Loan Payoff Amount (or unpaid balance of loan)—The present value needed to produce all the remaining payments. (Most likely, this is a different amount than the remaining principal owed.)

When the bank gives an amortized loan, it is basically giving a lump sum of money to an investment (the person receiving the loan) where it makes regular, periodic withdrawals (the payments made by the receiver of the loan). In essence, this is simply a situation where we know the present value of an annuity, i.e., the loan amount, and want to calculate the regular payment amount, made from the "account," which is the person receiving the loan to the bank. Because this is ultimately describing a present value of an annuity problem, we already know the formula we use to compute this.

AMORTIZATION FORMULAS

Let r = annual interest rate, n = the number of compoundings per year, $r_c = \frac{r}{n}$ = the interest rate per payment, t = number of years for payments, N = total number of payments = nt, k = number of payments made so far, S = loan amount and R = payment amount.

(a) Periodic Payment of Amortized Loan: $R = S\left(\frac{r_c}{1-(1+r_c)^{-N}}\right)$

(b) Total Interest Paid: $NR - S$ (the total money paid minus the loan amount)

(c) Loan Payoff Amount: Payoff $= S_{N-k} = R\left(\frac{1-(1+r_c)^{-(N-k)}}{r_c}\right)$

If k payments have occurred so far and a total of N payments are due, then there are $(N-k)$ payments missing from the loan. (This is simply the formula for the present value of an ordinary annuity needed to produce those $(N-k)$ payments that will be paid as a lump sum.)

EXAMPLE 1

Katie works hard and saves her work and gift money during junior high and high school. She wants to buy a car for herself when she graduates from high school. Suppose she can save up \$4,000 by the time she graduates from high school and she wants to purchase a car valued at \$15,000. If she can get a loan with a special interest rate of 3.6%, compounded monthly, for 6 years, how much will her monthly payments be?

We need to use the formula for the loan payment of an amortized loan, since she wants the payments to be the same every month. We have:

$$r = 0.036,\ n = 12,\ r_c = \frac{r}{n} = \frac{0.036}{12} = 0.003,\ t = 6 \text{ years},\ S = \$15{,}000 - \$4{,}000 = \$11{,}000,\ N = 6(12) = 72$$

$$\Rightarrow R = S\left(\frac{r_c}{1-(1+r_c)^{-N}}\right) = 11000\left(\frac{0.003}{1-1.003^{-72}}\right) \simeq \$170.10$$

EXAMPLE 2

Scott found a lot of land to build his home on. It will cost \$40,000, and he will amortize his loan over 8 years with yearly payments. Create an amortization schedule for his loan, and answer the following questions if the interest rate on his loan is 7.8%, compounded annually.

(a) What is his yearly payment?

(b) What is the interest for the first payment?

(c) How much principal is paid in the first payment?

(a) To calculate the yearly payment, we simply need to plug in the necessary values from the following list into the payment formula for an amortized loan.

$$r = 0.078,\ n = 1,\ r_c = \frac{r}{n} = \frac{0.078}{1} = 0.078,\ t = 8 \text{ years},\ N = 8 \text{ payments},\ S = \$40{,}000$$

$$\Rightarrow R = S\left(\frac{r_c}{1-(1+r_c)^{-N}}\right) = 40000\left(\frac{0.078}{1-1.078^{-8}}\right) \simeq \$6{,}907.85$$

Now, let's create that amortization schedule.

PERIOD	PRINCIPAL AT THE BEGINNING OF THE PERIOD	INTEREST FOR THE PERIOD	PAYMENT (AT THE END OF THE PERIOD)	PRINCIPAL PAID (WITH THE PAYMENT)	UNPAID PRINCIPAL (AFTER PAYMENT)
1	$40,000	40,000(0.078) = $3,120.00	$6,907.85	6,907.85 – 3,120.00 = $3,787.85	40,000 – 3,787.85 = $36,212.15
2	$36,212.15	36,212.15(0.078) ≃ $2,824.55	$6,907.85	6,907.85 – 2,824.55 = $4,083.30	36,212.15 – 4,083.30 = $32,128.85
3	$32,128.85	32,128.85(0.078) ≃ $2,506.05	$6,907.85	6,907.85 – 2,506.05 = $4,401.80	32,128.85 – 4,401.80 = $27,727.05
4	$27,727.05	27,727.05(0.078) ≃ $2,162.71	$6,907.85	6,907.85 – 2,162.71 = $4,745.14	27,727.05 – 4,745.14 = $22,981.91
5	$22,981.91	22,981.91(0.078) ≃ $1,792.59	$6,907.85	6,907.85 – 1,792.59 = $5,115.26	22,981.91 – 5,115.26 = $17,866.65
6	$17,866.65	17,866.65(0.078) ≃ $1,393.60	$6,907.85	6,907.85 – 1,393.60 = $5,514.25	17,866.65 – 5,514.25 = $12,352.40
7	$12,352.40	12,352.40(0.078) ≃ $963.49	$6,907.85	6,907.85 – 963.49 = $5,944.36	12,352.40 – 5,944.36 = $6,408.04
8	$6,408.04	6,408.04(0.078) ≃ $499.83	$6,907.85	6,907.85 – 499.83 = $6,408.02	6,408.04 – 6,408.02 = $0.02*

*Notice that the final balance may be plus or minus a few pennies from zero. This is why it's common practice to make the final payment a few pennies less or more than the regular payment to account for those round-off errors accrued over some number of payments.

(b) Notice the interest for the first payment is the full loan amount, $40,000, multiplied by the periodic interest rate of 7.8% which gives us $3,120.

(c) The principal paid in any payment is the payment amount minus the interest paid. For the first payment, this is $6,907.85 – $3,120.00 = $3,787.85 as the principal applied to the loan balance.

EXAMPLE 3

Mike is buying a home for his family and has the choice of a 15-year, a 20-year or a 30-year amortized loan with monthly payments. The amount of his loan originally is $210,000, with an interest rate of 5.7% compounded monthly.

(a) Compute the finance charge for each loan type.

(b) How much savings is there if he goes with the 15-year loan?

(a) The finance charge is different wording for the "total interest paid," and we know that the formula is $NR - S$. First, we need to calculate the payments for each type of loan, so we have the values of R. We know these values:

$$S = \$210{,}000,\ r = 0.057,\ n = 12,\ r_c = \frac{0.057}{12} = 0.00475$$

that are fixed for all the different loan types.

LOAN TYPE	N	$R = 210{,}000\left(\frac{0.00475}{1-(1.00475)^{-N}}\right)$	FINANCE CHARGE ($NR - S$)
15-year	15(12) = 180	$R = 210{,}000\left(\frac{0.00475}{1-(1.00475)^{-180}}\right) \simeq \$1{,}738.24$	180(1,738.24) – 210000 = \$102,883.20
20-year	20(12) = 240	$R = 210{,}000\left(\frac{0.00475}{1-(1.00475)^{-240}}\right) \simeq \$1{,}468.39$	240(1,468.39) – 210000 = \$142,413.60
30-year	30(12) = 360	$R = 210{,}000\left(\frac{0.00475}{1-(1.00475)^{-360}}\right) \simeq \$1{,}218.84$	360(1,218.84) – 210000 = \$228,782.40

(b) With the 15-year loan, it is quite a savings.

Compared with the 20-year loan, the 15-year loan would save 142,413.60 – 102,883.20 = \$39,530.40. Compared with the 30-year loan, the 15-year loan would save 228,782.40 – 102,883.20 = \$125,899.20.

EXAMPLE 4

A ski resort purchases a large piece of equipment for a chairlift for \$300,000. Suppose the ski resort takes out a loan for this equipment, making quarterly payments with an interest rate of 4.8% compounded quarterly for 10 years.

(a) What are the quarterly payments?

(b) How much interest does the resort pay during the first year?

(c) Find the unpaid balance (loan payoff amount) after three years.

(d) If the ski resort pays off the loan at the end of three years, how much money do they save by not making the rest of the payments?

(a) We need to plug these values into the formula for the loan payment of an amortized loan.

$$r_c = \frac{0.048}{4} = 0.0012,\ N = 4(10) = 40,\ S = \$300{,}000$$

$$\Rightarrow\ R = S\left(\frac{r_c}{1-(1+r_c)^{-N}}\right) = 300000\left(\frac{0.012}{1-1.012^{-40}}\right) \simeq \$9{,}487.51$$

(b) To calculate the interest for the first year, which is from 4 payments, we need a small amortization schedule for those first four payments.

PERIOD	PRINCIPAL AT THE BEGINNING OF THE PERIOD	INTEREST FOR THE PERIOD	PAYMENT (AT THE END OF THE PERIOD)	PRINCIPAL PAID (WITH THE PAYMENT)	UNPAID PRINCIPAL (AFTER PAYMENT)
1	\$300,000	300,000(0.012) = \$3,600.00	\$9,487.51	9,487.51 – 3,600.00 = \$5,887.51	300,000 – 5,887.51 = \$294,112.49
2	\$294,112.49	294,112.49(0.012) ≃ \$3,529.35	\$9,487.51	9,487.51 – 3,529.35 = \$5,958.16	294,112.49 – 5,958.16 = \$288,154.33
3	\$288,154.33	288,154.33(0.012) ≃ \$3,457.85	\$9,487.51	9,487.51 – 3,457.85 =\$6,029.66	288,154.33 – 6,029.66 = \$282,124.67
4	\$282,124.67	282,124.67(0.012) ≃ \$3,385.50	\$9,487.51	9,487.51 – 3,385.50 = \$6,102.01	282,124.67 – 6,102.01 = \$276,022.66

To compute the total interest paid for the first year, we need to add the four interest payments from the first four quarterly payments made in that first year.

$$\Rightarrow \text{ Total interest for first year} \simeq 3600 + 3529.35 + 3457.85 + 3385.00 = \$13{,}972.20$$

(c) We need to use the formula for the loan payoff amount to calculate the unpaid balance after the first three years. We have the variable values that were notated in part (a), and we also have $k = 4(3) = 12$ payments that are actually made in the first three years.

$$\Rightarrow \text{ Payoff} = S_{N-k} = R\left(\frac{1-(1+r_c)^{-(N-k)}}{r_c}\right) = 9487.51\left(\frac{1-(1+0.012)^{-(40-12)}}{0.012}\right) \simeq \$224{,}495.21$$

(d) If the ski resort pays the loan in quarterly payments for the first three years, and then pays off the loan with a lump sum, it will pay a total of

$$9{,}487.51(12) + 224{,}495.21 = \$338{,}345.33.$$

If, instead, the ski resort continues to pay the loan in installment payments for the full 10 years, it will pay a total of

$$9{,}487.51(40) = \$379{,}500.40.$$

The difference between the two proposed plans is

$$379{,}500.40 - 338{,}345.33 = \$41{,}155.07,$$

which is how much the ski resort would save by paying off the loan after only 3 years.

5.5 Homework

1. Palmer's Tires is considering purchasing a new flat bed tow truck for $100,000. It will borrow from Roosevelt Savings and Loan which will charge 6% interest compounded yearly and will finance the loan over a 5-year period. The loan will be repaid in 5 equal payments at the end of each year. What will the payment amount be?

2. Referring to problem 1, how much interest will Palmer's Tires avoid paying to Roosevelt Savings and Loan if it pays off the loan at the end of the fourth year instead taking the loan to the full 5-year term?

3. For problem 1, develop a loan amortization schedule for the 5-year loan.

4. The Sanchez Family is considering refinancing the family home. Currently, the mortgage terms and conditions include an interest rate of 5.87%, compounded monthly, and there are 22 years left on a 30-year loan. The remaining principal is equal $190,000. With these terms, what is the monthly mortgage payment?

5. In reference to problem 4, the Sanchez family has an opportunity to refinance the home. The mortgage lender has offered a 5% interest rate, compounded monthly, on a 15-year loan. The principal amount need to finance would be $190,000. With these terms, what would be the monthly mortgage payment?

6. Also in reference to problem 4, how much interest would the Sanchez family save if they went with a 15-year, 5% interest, mortgage loan? Do you think the Sanchez family should go with the 15-year loan?

7. Sky has taken out a student loan of $30,000 for her first Freshman year of university study at Whatsamatta U. She's majoring in forensic psychology. The student loan is to be repaid at 8% interest, compounded monthly, over a 10-year period in equal monthly payments. What will Sky's payments be?

8. In reference to Sky's student loan (problem 7), how much of the first payment will be interest? How much will be principal? How much interest will Sky save if she pays off her student loan at the end of 5 years?

9. Vic's Landscaping Company is borrowing $500,000 on a 5-year, 6.5% amortized term loan. Payments will be made every year. At the third year, what percent of the payment made will be an interest payment?

10. Alpine Church has purchased land to build its new church in Syracuse, Utah. It had to finance $150,000 for the purchase of the land. The interest rate on the loan is 6.5%, compounded monthly for a 10-year period. What is the monthly payment that must be made by Alpine Church? What is the total interest paid?

11. In reference to problem 10, how much interest would Alpine Church avoid if it paid off the loan for the land in five years of equal monthly payments?

12. In reference to problem 10, how much interest would Alpine Church avoid if it paid off the loan for the land in three years of equal monthly payments? Based on your answers to problems 10–12, is there a trade-off between interest charges on a loan and liquidity?

13. Baliwood Pictures Inc., with movie studios in New Delhi, India, has taken out a loan in order to finance its latest heart-tug picture. The loan is for $7.2 million. The terms are 6% interest over a 5-year period. Compounding and payments will be made quarterly. What will the payment amount be?

14. Referring to problem 13, how much interest will Baliwood Pictures avoid paying if the movie is a big success and the movie company can afford to pay off the loan in three years of equal payments instead taking the loan to the full 5-year term?

15. For problem 13, develop a loan amortization schedule for the first 2 years of the 5-year loan (remember, payments are made quarterly).

16. The Tennessee Valley Authority (TVA) has arranged to borrow $650 million from the U.S. Treasury to finance a new dam to provide water and power to the farmers of Tennessee Valley. The interest rate is 4.2% for a 30-year term. Payments are made to the treasury at the end of the year and compounding is computed annually. Calculate the annual payment amount.

17. In reference to problem 16, the TVA can borrow from private sources for a 15-year term at 5.5%. Payments and compounding would be made quarterly. Compute the quarterly payment.

18. Also in reference to problem 16, what is the difference in the amount of interest the TVA would pay if they finance through the U.S. Treasury as compared to borrowing from private sources?

19. Mike wants to renovate his flat in Newark, NJ. He will take out a loan for $20,000. Since he's mostly unemployed, he will have to take out the loan from the local loan shark at 18% per month financed over a 2-year period. Interest is compounded monthly and payments are made at the end of each month. What will Mike's monthly payment be?

20. In reference to problem 19, what will the total interest be on this loan at the end of the 2-year loan period?

21. In reference to Mike's renovations (problem 19), how much interest will Mike save if he pays off the loan shark in one year?

22. Florecita's Tupperware business needs to borrow $5,000 on a 3-year, 6.5% amortized term loan, compounded annually, in order have an inventory of products on hand for immediate sale to food storage aficionados. Payments will be made every year. At the end of the first year, what percent of the payment made will be an interest payment?

23. Piano by Wendy is a piano retailer who has been working out of a home in Layton, Utah. Business has been so brisk lately that she has decided to open a store in Roy. She wishes to have a store built and will finance the construction cost of $750,000. The interest rate on the loan is 7.5%, compounded monthly for a 5-year period. What is the monthly payment that must be made by this piano company?

24. In reference to problem 23, how much interest would Piano by Wendy avoid if she paid off the loan in equal payments in 2 years?

25. Antonio is considering buying a house for \$400,000. He can get a mortgage with a 5.5%, 30-year fixed interest rate, compounded monthly. What will his monthly payments be at the end of each month in order to pay off his house at the end of the 30-year loan period?

26. Antonio decides to save money by paying his mortgage payment at the beginning of each month instead of the end of the month. What will Antonio's monthly mortgage payment be? How does this compare with your answer for problem 25?

27. Antonio decides to save even more money on his mortgage by making an extra payment every year (so, 13 payments every year). What will Antonio's monthly payment need to be to pay off his mortgage over the 30-year repayment period?

28. What will monthly payments be on an automobile purchased for \$15,000 at 9.5% interest, billed and compounded monthly, financed for 5 years?

29. Paris has purchased a Malibu Beach house for a cool \$4.1 million. She has secured a jumbo mortgage with a 7.5%, 15-year fixed interest rate, compounded monthly. What will her monthly payments be at the end of each month?

30. Paris' TV career hasn't been doing as well lately. She decides to save money by paying her mortgage payment at the beginning of each month instead of the end of the month. What will Paris' monthly mortgage payment be? How does this compare with your answer for problem 29?

31. Paris buys a new Bentley automobile for \$95,000. What will her monthly be to pay off the car loan over 5 years at 9.5% interest, compounded monthly?

32. What will monthly payments be on a used Volvo automobile purchased for \$14,000 at 11.5% interest, billed and compounded monthly, financed for 60 months?

33. Newark Car Parts owes a noted New Jersey mobster \$100,000, and has to pay the total amount in full by the end of the year. Payments must be made monthly. What will the payments be at 28% annual interest—compounded monthly?

Chapter 5 Formulas

5.1	Arithmetic Sequence	$a_n = a_1 + (n - 1)d$
	Geometric Sequence	$a_n = d^{n-1}\, a_1$
	Sum of Arithmetic Sequence	$S_n = \frac{n}{2}(a_1 + a_n)$
	Sum of Geometric Sequence	$S_n = \frac{a_1(1 - d^n)}{1 - d}$

r = annual interest rate, n = number of compoundings per year, t = time (in years), $N = nt$, $r_c = \frac{r}{n}$
S = account balance (at either future or present time, depending on formula),
P = principal, R = regular payment/deposit, m = number of compounding periods money is deferred,
k = number of payments that have been made already

5.2	Simple Interest	$S = P(1 + rt)$
	Compound Interest	$S = P(1 + r_c)^N$
	Continuous Compound Interest	$S = Pe^{rt}$
	APY (periodic compounding)	$APY = \left(1 + \frac{r}{n}\right)^n - 1$
	APY (continuous compounding)	$APY = e^r - 1$
5.3	Future Value of Ordinary Annuity	$S = R\left(\frac{(1 + r_c)^N - 1}{r_c}\right)$
	Future Value of Annuity Due	$S_{due} = R(1 + r_c)\left(\frac{(1 + r_c)^N - 1}{r_c}\right)$
	Sinking Fund Payment	$R = S\left(\frac{r_c}{(1 + r_c)^N - 1}\right)$
5.4	Present Value of Ordinary Annuity	$P = R\left(\frac{1 - (1 + r_c)^{-N}}{r_c}\right)$
	Present Value of Annuity Due	$P_{due} = R(1 + r_c)\left(\frac{1 - (1 + r_c)^{-N}}{r_c}\right)$
	Present Value of Deferred Annuity	$P = R\left(\frac{1 - (1 + r_c)^{-N}}{r_c(1 + r_c)^m}\right)$
5.5	Periodic Payment of Amortized Loan	$R = S\left(\frac{r_c}{1 - (1 + r_c)^{-N}}\right)$
	Total Interest Paid	$NR - S$
	Loan Payoff	$S_{N-k} = R\left(\frac{1 - (1 + r_c)^{-(N-k)}}{r_c}\right)$

Chapter 5 Review

5.1

Find the formula for the nth term and the first three terms of the (a) arithmetic sequence and (b) geometric sequence with the given a_1 and d values.

1. $a_1 = -4, d = 3$
2. $a_1 = \frac{1}{2}, d = \frac{5}{2}$
3. $a_1 = 25, d = -2$
4. $a_1 = 1, d = 6$

Given the first term, a_1, of an (a) arithmetic and (b) geometric sequence, and another term of the sequence, find the 12th term of that sequence.

5. a. $a_1 = 3, a_7 = 12$
   b. $a_1 = 3, a_7 = 0.000003$
6. a. $a_1 = \frac{-1}{5}, a_4 = \frac{-19}{5}$
   b. $a_1 = \frac{-1}{5}, a_4 = -25$
7. a. $a_1 = -6, a_{10} = 30$
   b. $a_1 = -6, a_{10} = \frac{-3}{256}$

Find the sum of the first 80 terms of the arithmetic sequence given.

8. 1, 5, 9, 13, . . .
9. 21, 15, 9, 3, . . .
10. −10, −7, −4, −1, . . .

Find the sum of the first 12 terms of the geometric sequence given.

11. 2, 8, 32, 128, . . .
12. $-4, -2, -1, -\frac{1}{2}, \ldots$
13. 625, 375, 225, 135, . . .

5.2

14. Hector has saved $8,000 from his tips over the last year. If he invests it in an account paying 11% simple annual interest, how long will it take to be worth $13,000? If, instead, he invests it in an account paying 11% interest compounded annually, how long will it take to be worth $13,000?
15. If Betty only saved $6,000 of her tips from last year, and she puts that money in an account paying 9% interest compounded monthly, how much will it be worth in 4 years?
16. What lump sum of money should be deposited in an account that will earn 7.2% interest compounded monthly to grow to $250,000 for retirement in 30 years?
17. How much more interest will be earned if $10,000 is invested for 8 years at 6% compounded continuously, instead of at 6% compounded quarterly?

18. If my money is invested at 8% compounded twice per year, what is the annual percentage yield (APY)? What is the APY if the money is compounded monthly instead?

5.3

19. Find the future value of an ordinary annuity of \$500 paid quarterly for 6 years, if the interest is 10% compounded quarterly.

20. Find the future value of an annuity due paying \$1200 each month for 4 years if the interest rate is 8.4% compounded monthly.

21. How much will need to be invested at the end of each year at 9% interest, compounded annually, to pay off a debt of \$30,000 in 10 years?

22. Lori's Baked Goods company establishes a sinking fund to discharge a debt of \$56,000 due in 8 years by making equal deposits twice yearly (semiannually), at the end of each period. If the investment pays 6.8% compounded semiannually, what is the amount of each deposit?

23. Margie retired from the post office and bought a small diner in town. She realizes that she needs an additional \$120,000 at the end of 15 years in order to truly retire from working altogether. She finds a safe investment account that pays 7.8% compounded monthly. What do her monthly investment payments, at the end of each month, need to be to reach her goal?

24. (Refer to problem 23.) If Margie can only afford to deposit \$280 per month, how many years will it take for her to reach her goal of saving \$120,000?

5.4

25. Jesse, in retirement, inherits \$93,000 upon his mother's death. He invests it at 6.6% compounded monthly in an ordinary annuity that pays out for the next 12 years. What amount does Jesse receive from this annuity every month?

26. A trust provides \$8,000 to a county library at the beginning of each 3-month period for the next 4 years. If money is worth 6.9% compounded quarterly, find the amount in the trust when it began.

27. As a result of a court settlement, an accident victim is awarded \$1.5 million. The attorney takes a third of this amount, another third is used for medical expenses already incurred, and the remaining third is invested in an account that earns 7.8% interest compounded monthly. What amount will the accident victim receive at the end of each month, if it pays for the next 10 years?

28. Find the present value of an annuity of \$3,200 paid at the end of each quarter for 5 years after being deferred for 3 years, if money is worth 8% compounded quarterly.

29. A couple received an inheritance of \$125,000 the year they turned 46 and invested it in a fund that earns 6.3% compounded monthly. If this amount is deferred for 16 years, until they retire, how much will it provide at the end of each month for the next 20 years after they retire?

5.5

30. Sean's parents loan him \$10,000 for a trip to Australia. He plans to amortize the loan and pay it back to them in 10 equal quarterly payments. If their interest rate is 6%, compounded quarterly, what is his periodic payment?

31. Frank purchases \$25,000 worth of equipment for his restaurant by making a \$4,000 down payment and amortizing the rest with monthly payments over the next 5 years. The interest rate on the debt is 8.4%, compounded monthly.
    a. What is his monthly payment amount?
    b. Find the total amount paid over the life of the loan.
    c. What is the total interest paid?

32. An investor interested in purchasing an apartment building determines that she can make payments of \$3000 per month. If a loan is available at 12%, compounded monthly, for 25 years, how much can she afford to pay for the building?

33. A young couple is ready to buy their first home. They have \$16,000 to use as a down payment, and their budget can support a monthly mortgage payment of \$1200. What is the most expensive home they can buy if they can borrow money for 30 years at a fixed 5.7% interest rate, compounded monthly?

34. My friend, Kathy, just bought a home, borrowing \$160,000 at a fixed 6.6% interest rate, compounded monthly, for 20 years. What monthly payment does she make? If she decides to sell the house after making exactly 24 monthly payments, what is the unpaid balance?

Final Review

1. Find the inverse of the following matrix, if possible. If it's not possible, then explain why.

$$A = \begin{bmatrix} 1 & -2 \\ 0 & 5 \end{bmatrix}$$

2. For $A = \begin{bmatrix} 1 & 0 & 2 \\ 3 & 2 & 1 \\ 4 & 0 & 3 \end{bmatrix}$, $B = \begin{bmatrix} 5 & 3 \\ 1 & 2 \end{bmatrix}$ and $C = \begin{bmatrix} 4 & 2 \\ 3 & 5 \end{bmatrix}$, perform the indicated matrix operations, if possible. If not possible, explain why.

   a. $A + A^T$

   b. BC

3. Use Gauss-Jordan Elimination to solve the following system.

$$\begin{aligned} 2x - 4y + 2z &= -4 \\ 4x - 9y + 7z &= 2 \\ -2x + 4y - 3z &= 10 \end{aligned}$$

4. Given the arithmetic sequence

   $-2, 1, 4, 7, 10, \ldots$

   a. Find the 100th term.

   b. Find the sum of the first 100 terms.

5. How much would have to be invested at the end of each year at 6% interest compounded annually to pay off a debt of $80,000 in 10 years?

6. A lottery prize worth $1,000,000 is awarded in payments of $10,000 five times a year for 20 years. Suppose the money is worth 20% compounded 5 times per year. What is its present value?

7. For $f(x) = \sqrt{1 - x}$ and $g(x) = x^2 + 1$

   a. State the domain for both functions.

   b. Find $g \circ f$ and state the domain of this new function.

   c. Find $\frac{f}{g}$.

8. Solve the equation.

$$\frac{2x}{x-1} + \frac{1}{3} = \frac{5}{6} + \frac{2}{x-1}$$

9. Find the equation of the line passing through the points $(-1, 1)$ and $(2, 3)$.

10. Graph the linear inequality.

    $-4x < 6y$

11. Graph the system of inequalities and shade the solution region. Label all vertices for the solution (shaded) region.

$$\begin{aligned} 3x + 4y &\geq 12 \\ x - y &\geq 2 \\ x &\leq 6 \\ y &\geq 0 \end{aligned}$$

12. Find the maximum of the objective function $P = 2x + y$ subjected to the following constraints.

$$x \geq 0$$
$$y \geq 0$$
$$x + y \leq 10$$
$$2x + 3y \leq 24$$

13. If the cost of production for a product is given by $C(x) = x^2 + 11x + 84$ and the revenue is given by $R(x) = 30x$,

a. Find the profit function $P(x)$.

b. Find the break-even point(s).

14. If 100 feet of fence is used to fence in a rectangular yard, then the resulting area is given by $A(x) = x(50 - x)$ where x feet is the width of the rectangle and $(50 - x)$ feet is the length. Determine the width and length that give the maximum area.

15. Let $f(x) = -(x - 1)^2 + 4$.

a. Solve $f(x) = 0$ to find the x-intercepts.

b. Find the vertex of the parabola.

c. Sketch the graph, showing the vertex and x-intercepts.

16. Suppose that the population of Smalltown, USA grows according to the formula $P(t) = 3200e^{0.025t}$ where time t is measured in years.

a. What is the initial population of the town (at $t = 0$)?

b. How long will it take the population to double?

c. What is the population after one year?

17. Rewrite $\log_2 32 = x$ in exponential form and solve for x.

18. Solve for x.

$$|3 - 4x| = 13$$

19. The Utah Company manufactures a certain product that has a selling price of \$40 per unit. Fixed costs are \$1,600 and variable costs are \$20 per unit. Determine the least number of units that must be sold for the company to have a profit of no less than \$5,000.

20. A rectangular plot of land has an area of 18,000 square feet. If its length is five times its width, how much fencing would be required to surround the property?

21. For the following functions, find the following.

$$f(x) = \frac{x + 1}{3x^2 + 20x + 25} \qquad g(x) = -x$$

a. The domain of $f(x)$

b. The domain of $g(x)$

c. $f(-1)$

d. $f(0)$

e. $(f \circ g)(2)$

22. Graph the function $y = |x - 2| - 1$ and determine the x- and y-intercepts.

23. The students at a university buy 3,000 graphing calculators per year when they cost \$50 each, and they buy 2,000 calculators per year when they cost \$100 each. Let P be the price per calculator and Q be the quantity of calculators sold. Assuming the relationship between P and Q is linear, give an equation expressing P in terms of Q.

24. Find the value for x which maximizes the quadratic function

$$f(x) = -x^2 + 11x - 24.$$

25. Solve the following equations.

a. $\ln(2x + 7) = 0$

b. $e^{2x} = 9$

c. $\log x + \log 3 = 2$

26. Jeremy wants to make one savings deposit today so that in 7 years, he will have \$16,000. Given an interest rate of 4% compounded semiannually (twice a year), how much money should Jeremy deposit?

27. Brittany is 25 years old and she plans to retire when she turns 60. When she retires, she would like to have \$1,000,000 of savings. She is going to achieve the savings by contributing to a sinking fund between now and her retirement, with equal monthly payments paid at the end of each month. Assume the interest rate is 6% per year, compounded monthly. How much should her monthly payments be?

28. Given the matrices A and B, perform the indicated operations or state that it's not possible. If it's not possible, explain why.

$$A = \begin{bmatrix} 1 & 5 & -1 \\ 2 & 3 & 0 \end{bmatrix} \qquad B = \begin{bmatrix} 0 & 1 & 1 \\ 1 & 0 & -1 \\ 2 & 1 & 0 \end{bmatrix}$$

a. AB

b. BA

c. B^{-1}

29. Determine if the system of equations below has any solutions. If a solution exists, find it.

$$x + y + z = 6$$
$$x + 2y + 3z = 14$$
$$2x + y + 2z = 10$$

30. Maximize the objective function $z = 4x - 3y$ subject to the constraints:

$$x \geq 0$$
$$y \geq 0$$
$$x + y \leq 4$$
$$3x + 2y \geq 6$$

31. Find all solutions to the following equation and inequality.

a. $|3x + 2| > 20$

b. $|-2x + 6| = |3x + 5|$

32. The Cordova Company sells bicycles. They pay \$164 for each bicycle and have monthly fixed costs of \$5500. If they sell every bicycle for \$185, how many bicycles do they need to sell each month to have a profit of at least \$5000?

33. Solve for x. $x^2 + 3 = 4x$

34. For the following functions, find the following.

$f(x) = x^2 - 1 \qquad g(x) = 2x - 1$

a. Domain of $\left(\frac{g}{f}\right)(x)$

b. $(f \circ g)(x)$

c. $g^{-1}(x)$

d. $(f - g)(x)$

35. Find the x- and y-intercepts and the vertex of $f(x) = 2x^2 - 4x - 6$ algebraically. Use this information to sketch a graph of $f(x)$.

36. A gardener has two fertilizers that contain different concentrations of nitrogen. One is 7% nitrogen and the other is 13% nitrogen. How many pounds of each should she mix together to obtain 27 pounds of 9% concentration?

37. Solve for x.

a. $\log_2 32 = x$

b. $27^{x-2} = 9^{3x+9}$

c. $\log_{10} 2x + \log_{10} (x - 5) = 2$

38. Suppose that \$2000 is put into an account earning 6% interest compounded quarterly. How many years will it take for the balance in the account to reach \$3000?

39. You are buying a \$220,000 house with a down payment of \$25,000. If the interest rate is 6%, compounded monthly, determine the size of the monthly payments (at the end of the month) you must make over the next 30 years to pay off the house.

40. Mr. Johnson wants to invest in his new grandson's education. He will deposit a lump sum of money now into an account earning 4.8% interest, compounded quarterly. He wants that money to produce quarterly payments of \$12,000 each for his grandson for a total of five years, starting 18 years from now. How much money must Mr. Johnson put into the account now to achieve his goals for his grandson's college education?

41. Given the matrices A and B, perform the indicated operations or state that it's not possible. If it's not possible, explain why.

$$A = \begin{bmatrix} 1 & 3 \\ -2 & 2 \\ 4 & 1 \end{bmatrix} \quad B = \begin{bmatrix} 0 & 2 & 1 & -1 \\ 3 & 0 & 4 & 2 \end{bmatrix} \quad C = \begin{bmatrix} 4 & -5 & 6 & 0 \\ 7 & 3 & 1 & 4 \end{bmatrix}$$

a. AB

b. BA

c. $2B - C$

42. Follow the steps below to solve the system of linear equations using an inverse matrix.

$$2x - 5y = 2$$
$$-x + 3y = 1$$

a. Write the system above as a matrix equation, i.e. in the form $A\begin{bmatrix} x \\ y \end{bmatrix} = b$.

b. Find A^{-1}.

c. Use A^{-1} to solve the system of equations.

43. Maximize the objective function $P = x + 2y$ subject to the constraints:

$$x \geq 0$$
$$y \geq 0$$
$$x + y \geq 5$$
$$x + y \leq 12$$
$$x \leq 10$$

44. Find all solutions to the following equation.

$$\frac{2x + 5}{x + 7} = 1 + \frac{x - 11}{2x + 14}$$

45. Write the equation of the line through the point (1, 2) that is perpendicular to the line through the points (3, 2) and (–1, –6). Write the answer in slope-intercept form.

46. If $f(x) = -2x^2 - x + 4$ and $g(x) = x - 5$, find each of the following, simplifying as far as possible.

a. $g(f(-3))$

b. $f(g(x))$

47. Find the inverse, $f^{-1}(x)$, of the function $f(x) = \sqrt[5]{x^3 - 5} + 1$.

48. The daily profit from the sale of x units of a product is $P(x) = 10x - 0.5x^2 + 150$ dollars.

a. Find the domain of $P(x)$.

b. Find the x-intercept(s) and y-intercept.

c. Find the maximum value of $P(x)$. (Hint: Use the formula for the vertex of a parabola.)

49. If the demand function for a commodity is given by the equation $p^2 + 5q = 200$ and the supply function is given by $40 - p^2 + 3q = 0$, find the equilibrium quantity and price.

50. Suppose $\ln x = 3$, $\ln y = 5$ and $\ln z = 2$. Use logarithmic properties to find $\ln \sqrt{\frac{xy}{z^2}}$.

51. Solve for x.

a. $500 = 600 - 600e^{-0.4x}$

b. $8 + \log(5x) = 2\log x + 7$

52. If \$1,000 is invested at 2% per month, the future value S at any time t (in months) is given by $S = 1{,}000(1.02)^t$. When will the investment double?

53. A small business loan is for \$50,000 for 10 years at 9% interest, compounded twice per year. The loan was taken out 6½ years ago. Find the payoff amount after the company makes the 14th semi-annual payment of \$3843.81.

54. Grandparents decide they want to pay for their first grandchild's college education. To do this, they anticipate needing \$200,000 in 18 years. If they have an investment that earns 11%, compounded monthly, how much should the grandparents deposit at the beginning of each month to reach their goal?

55. Given the matrices *A*, *B* and *C*, perform the indicated operations or state that it's not possible. If it's not possible, explain why.

$$A = \begin{bmatrix} 1 & 2 & 3 \\ 1 & 1 & 2 \\ 3 & 1 & 4 \end{bmatrix}, B = \begin{bmatrix} 2 & 1 & 3 \\ 1 & 1 & 1 \end{bmatrix}, C = \begin{bmatrix} 0 & -3 & 4 \\ 2 & -1 & 5 \\ 3 & 0 & 1 \end{bmatrix}$$

a. $A + 3C$

b. AB

c. BA

56. Solve the following system of equations using any method.

$$x - 2y + z = 7$$
$$3x + y - z = 2$$
$$2x + 3y + 2z = 7$$

57. A company manufactures one type of table and one type of chair. It takes 2 hours to make a table and 6 hours to make a chair. The company has a maximum of 1800 total work hours to use in manufacturing each day, and it can deliver at most 400 total pieces (tables and chairs combined) per day.

a. Find the equations for the constraints. Then, graph the feasible region. (Make sure to define what your variables represent.)

b. If the company sells the table for \$60 and the chair for \$30, how many of each piece should it produce per day to maximize its sales?

58. Given the line $2x + y = 4$:

a. Find the equation of the line through the point (–2, 1) that is parallel to the given line. (Give answer in slope-intercept form.)

b. Find the equation of the line through the point (0, 4) that is perpendicular to the given line. (Give answer in slope-intercept form.)

59. Solve each equation. (Give *exact* answers.)

a. $3x^2 + 1 = -5x$

b. $x^2 - 13x + 36 = 0$

60. The supply and demand for a product are given by $2p - q = 27$ and $(2p - 5)q = 363$, respectively. Find the equilibrium quantity and price.

61. Joe has a total of \$2500 to invest into two accounts, earning simple interest. The accounts have annual interest rates of 5% and 10%. The total interest made from his investments in one year is \$200.

a. Write the system of equations that describe this scenario. Use *x* to represent the amount invested at 5% and *y* to represent the amount invested at 10%.

b. Write the matrix equation for this system of equations.

c. Solve the system of equations.

62. A chocolate truffle maker plans to make truffles for a street fair. She makes two kinds of truffles: a dark one that she sells for \$1 a piece, and a lighter chocolate truffle that she sells for 80 cents a piece. She needs 2 teaspoons of heavy whipping cream and 10 grams of chocolate to make a dark truffle and 4 teaspoons of heavy whipping cream and 5 grams of chocolate to make a light truffle. She has a total of 1600 teaspoons of heavy whipping cream and 5000 grams of chocolate that she can use for these truffles.

    a. Find the inequalities for the constraints. Then, graph the feasible region and label the corner points. (Make sure to define what your variables represent.)

    b. How many truffles of each kind should she make to maximize revenue and what is that maximum revenue?

63. Solve for x.

    a. $\log_3 (x^2) - \log_3 (x + 3) = \log_3 (x - 2)$

    b. $3^{x^2 + x - 3} = \frac{1}{27}$

64. Sketch the graph of the given functions in the xy-plane. Clearly label (1) two points on each graph and (2) any asymptotes.

    a. $y = 3^{-2x}$

    b. $y = \log(x + 1)$

65. Paul borrows \$40,000 from the bank at an APR of 12% compounded monthly. If the payments on this loan are made at the end of each month, what are his payments if Paul wants to pay off his debt in 4 years? What is the total amount that Paul pays to the bank?

66. a. Dave has a paper route that earns him \$1000 per year, which he deposits, at the end of each year, in an account yielding 5% interest compounded annually. He does this for five years, from the age of 14 to 18. If after that he makes no deposits or withdrawals, but leaves the money in the account, how much money does he have when he is 65 years old?

    b. Dave's brother, Leo, wins a prize of \$3,000 in a contest at age 14. He puts this in an account yielding 6% interest compounded annually. If he also makes no withdrawals or deposits, but leaves the money in the account, how much money does he have when he is 65 years old?

67. The startup costs for a street vendor are \$2000, and his cost per unit to produce food is \$1. Demand for his product dictates that the price at which q thousand units would be sold is $4 - q$. Therefore, if q is the number of thousands of units sold, and cost and revenue are measured in thousands of dollars, the cost function is $C(q) = 2 + q$ and the revenue function is $R(q) = (4 - q)q$.

    a. At which price(s) will the vendor break even? What quantities (in thousands) will he sell at these prices?

    b. What is the maximum profit? How many thousand units are sold to attain this profit? What is the price?

68. Given the matrices A, B, and C, compute the following, if possible. If it's not possible, state the reason why.

$$A = \begin{bmatrix} 2 & 1 & 3 \\ 0 & 5 & -1 \end{bmatrix}, B = \begin{bmatrix} 4 & 0 \\ -3 & -2 \end{bmatrix}, C = \begin{bmatrix} 7 & 0 \\ 2 & 6 \\ -1 & 4 \end{bmatrix}.$$

    a. AB

    b. B^{-1}

    c. BA

    d. $3A - C^T$

69. Solve for x. (Show all work without a calculator.)

a. $\ln x + \ln (x + 2) = \ln 35$

b. $5(2^{3x}) - 15 = 305$

70. Marty and Paul buy a house for $175,000. They have a $25,000 down payment and expect to amortize the rest of the debt with monthly payments over the next 30 years. The interest on the debt is 5.1% compounded monthly.

a. What are the monthly payments?

b. Find the total amount of house loan payments (excluding down payment).

c. Find the total amount of interest paid on this loan.

71. Kilam is 20 years old. Investing in mutual funds, he can earn 8% interest, compounded quarterly.

a. In order to have $50,000 in the account to buy a sports car when he is 30 years old, how much money must he deposit at the end of each quarter?

b. When Kilam turns 30, he decides not to buy the car, but leaves his money in the account (still earning compounded interest), now *withdrawing* $2,000 at the end of each quarter. After how many quarters does he run out of money?

Midterm 1

Covering sections 1.1–1.8, 2.1

1. Solve the equation $\frac{x+4}{x+2} = \frac{1}{5} + \frac{x-1}{x+2}$.

2. Solve this inequality and graph the solutions on the real number line $2x + 1 \geq -3(x - 3)$

3. For the system of inequalities.

   $2x + 2y \geq 0$
   $y \leq -3x + 6$
   $y - x \leq 6$

   a. Graph the solution set and **label the corners** (vertices).

   b. Maximize the objective function $f = 5x - 12y$ with the constraints in (a). List the point where the maximum occurs and the maximum value of f.

4. Write an equation of the line that goes through the point (3, –7) and is perpendicular to $x + 4y = 28$. (Put answer in slope-intercept form.)

5. Solve the system of equations.

   $4x + y + z = -7$
   $5x - 3y = -18$
   $5z = 20$

6. The selling price for an hour of expert tutoring is $75 and the total cost for that hour of private tutoring is given by $C(x) = 40x + 1620$ where x is the number of private tutoring hours purchased at this tutoring company.

   a. Find the revenue function, $R(x)$.

   b. Find the value of $C(30)$. Describe the meaning of this number.

   c. How many hours of tutoring need to be sold in order to break even?

   d. What is the marginal cost?

7. Given $A = \begin{bmatrix} 4 & 0 & -5 \\ 1 & 3 & -8 \end{bmatrix}$, $B = \begin{bmatrix} 1 & 6 \\ 2 & -3 \\ 0 & 9 \end{bmatrix}$, *and*

   $C = \begin{bmatrix} -2 & 6 \\ 5 & -1 \\ 10 & 0 \end{bmatrix}$ answer the following questions.

   If the computation is impossible, state that and the reason why it's not possible.

   a. Find B^T.

   b. Find c_{32}.

   c. Find $2C - B$.

   d. Find $A + B$.

8. If the supply function for a product is $100p = 3q - 2450$ and the demand function is $p + 2q = 2513$, what is the equilibrium **point**?

9. Answer the following questions.

   a. Does this graph represent a relation or a function? Explain your reasoning.

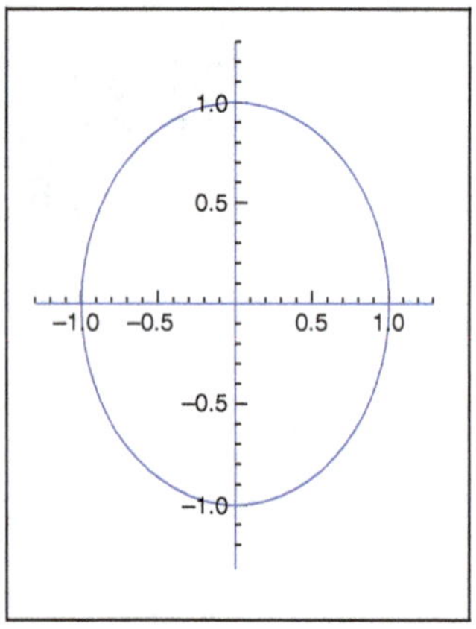

   b. Given the function $f(x) = \dfrac{4 - x}{\sqrt{x + 2}}$

      (i) What is the domain?

      (ii) find $f(5)$. (Give exact value, not approximate value.)

      (iii) Find $f(y + w)$. (Don't simplify.)

   c. Set up the inequality for this problem. (Don't solve anything; just write the inequality that represents this situation.)

      You are offered a new sales job. They are offering you a choice of the following two pay scales.

      Option 1: You earn a monthly salary of \$1200 AND sales commission of 6% of all your sales that exceed the first \$1500 of sales.

      Option 2: You earn a monthly salary of \$2500 AND sales commission of 3.5% of the total sales you make.

   Set up the inequality that will help you make the decision of which option to take.

Midterm 2

Covering sections 2.1–2.5, 3.1–3.3

1. Solve the system of linear equations following the steps below.

   $x - y + z = 2$
   $x - y + 2z = 1$
   $-x + 3z = -11$

   a. Write the corresponding matrix equation, $AX = B$, for this system of equations, i.e. write what A, X, and B are.

   b. Write the augmented matrix for this system of equations.

   c. Solve the system of equations either by

      (i) finding A^{-1} and using that to solve the system of equations, OR

      (ii) using Gauss-Jordan elimination to solve the system of equations.

2. Given the matrices

$$A = \begin{bmatrix} 3 & 0 \\ -4 & 8 \end{bmatrix} \quad B = \begin{bmatrix} 1 & 9 \\ -2 & 5 \end{bmatrix} \quad C = \begin{bmatrix} 1 & 2 \\ 7 & -5 \\ 3 & 10 \end{bmatrix}$$

$$D = \begin{bmatrix} 2 & 1 & 5 \\ -3 & 8 & 19 \end{bmatrix},$$

   calculate each of the following. If a computation is not possible, state that and give the reason why.

   a. A^{-1}

   b. D^TC

   c. $17A - 52D$

   d. $3A + B$

   e. DC

3. Juan and his partner need to borrow $40,000 to start their cupcake delivery company. Three supportive family members have enthusiastically offered to lend them money because the family believes in this business model and have tasted their yummy cupcakes. However, no one person has all the money to loan them. Juan's older sister is willing to loan money at 5% interest, while Juan's younger sister is willing to loan money at 7.5% interest. His parents are offering their loan to him at 4% interest. Juan can pay exactly 4.85% interest overall on the borrowed amount. He also thinks it's best that the amount of money he borrows from his parents matches the total of what he gets from his two sisters. How much money should Juan and his partner borrow from each family member?

   a. Set up the system of linear equations to represent this situation.

      Let x = amount Juan's older sister loans them, y = amount Juan's parents loan them, and z = the amount Juan's younger sister loans them.

   b. Solve the problem (in whatever way you'd like to use) to answer the final question.

4. A local brewery sells great beer and wants to maximize their daily profit. The demand function is given by: $p = 15 - 0.25x$ where p is the price of one pint of beer in dollars and x is the number of pints sold. The cost function is given by $C(x) = 0.5x + 80$, again in dollars.

a. What is the revenue function? (**Hint:** Revenue is the price times the number of pints sold at that price.)

b. What is the profit function?

c. How many pints need to be produced and sold to break even?

d. How many prints must the company sell to maximize profit?

e. What should the price of one pint be to maximize profit?

f. What is the maximum profit?

5. Find the inverse matrix for $A = \begin{bmatrix} 1 & 0 & 1 \\ 1 & 0 & 0 \\ 0 & 1 & 1 \end{bmatrix}$.

6. Given the function $y = f(x) = \frac{1}{2}(x-3)^2 - 4$,

answer the following questions.

a. Find the vertex.

b. Find the axis of symmetry.

c. Is this parabola concave up or down?

d. Find the y-intercept.

e. Find the x-intercept(s).

f. Sketch the graph. Make sure to label at least three points that your parabola goes through.

7. If you're doing Gauss-Jordan elimination and you get this augmented matrix

$$\left[\begin{array}{ccc|c} 1 & -5 & 8 & 2 \\ 0 & 1 & 3 & 9 \\ 0 & 0 & 0 & 0 \end{array}\right]$$ what is the solution to the original system of equations?

Midterm 3

Covering sections 3.4, 3.6, 3.7, 4.1–4.5, 5.1

1. Given $f(x) = \dfrac{3x - 1}{x + 4}$

   a. Find the inverse function, $f^{-1}(x)$

   b. Fill in the domain/range table. Use interval notation.

	Domain	Range
$f(x)$		
$f^{-1}(x)$		

2. a. Use logarithmic properties to expand this expression completely.

   $$\log_6\left(\frac{\sqrt[3]{x+4}}{x^3}\right)$$

   b. Use logarithmic properties to condense this expression into one logarithmic expression.
   $\log(x + 8) - 2\log x + 3\log(x - 1)$

   c. Rewrite $\log_5 w = 3$ as its equivalent exponential statement.

   d. Rewrite $4^K = L$ as its equivalent logarithmic statement.

   e. Is this statement true or false?
   $\log(x + y) = \log x + \log y$

   f. Which of these two statements is true (circle the one that is true)?
   $3^{m+p} = 3^m + 3^p$ or $3^{m+p} = 3^m \cdot 3^p$

3. Given this graph of $y = f(x)$, tell which function it is. (**Hint:** It has a horizontal asymptote at $y = 0$.)

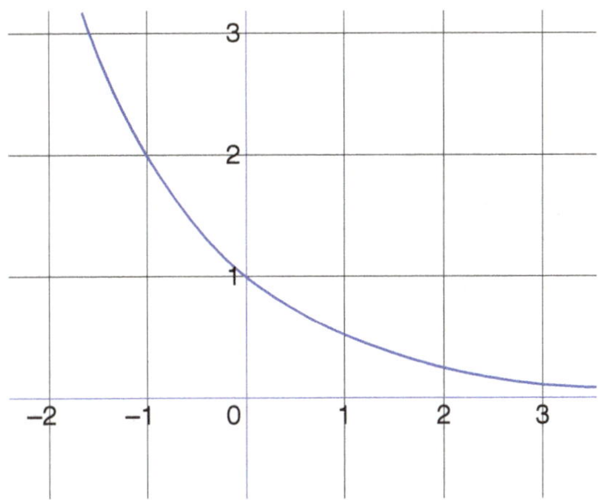

4. Solve the following equations. Give the EXACT answers, not the approximated, calculator values.

   a. $\log_2(x + 5) - \log_2(x + 1) = 5 - \log_2 4$

   b. $6(3^{4-x}) + 10 = 46$

5. a. Given the arithmetic sequence 1, 5, 9, 13, …

   (1) find the common difference, d.
   (2) find the formula for the nth term of the sequence.
   (3) find the sum of the first 50 terms of this sequence.

b. Given the geometric sequence 3, 12, 48, 192, …

(1) find the common ratio.
(2) find the formula for the nth term of the sequence.
(3) find the sum of the first 7 terms of this sequence.

6. Given the function $f(x) = (x-1)^3 + 5, g$

$g(x) = \dfrac{2}{x+5}, h(x) = \sqrt{x}$. answer the following

questions. Give all numerical answers in EXACT form, no calculator approximations will receive credit.

a. $(f+g)(-4)$

b. $f(h(x))$

c. domain of $g(x)$

d. $(gh)(4)$

e. $\left(\dfrac{f}{g}\right)(2)$

f. $(h \circ g)(1)$

7. Sketch the graph of the following functions and answer the questions asked.

a. $y = 3^{x+2}$

Fill in this table of points.

x	y
–2	
–1	
–3	

Where is the Horizontal Asymptote (HA)? (give line equation, not just a number)

How has this function been transformed from the parent graph of $y = 3^x$?

Sketch the graph.

b. $y = \log_2(-x) + 3$

Notice the $y = \log_2 x$ is the parent/base graph for this function.

Fill in this table of points to capture all two of the transformations in this function.

(x, y) on parent graph of $y = \log_2 x$	Horizonal Relfection	Vertical Shift
(1,)		
(2,)		
(0.5,)		

Where is the Vertical Asymptote (VA)? (give line equation, not just a number)

What is the domain of this function?
Sketch the graph.

Final Exam 1

1. If three less than two times a number is equal to six more than the number, then what is the number?

2. Find the equation of the line passing through the points (0, 1) and (1, 3). *(put your answer in the form $y = mx + b$)*

3. Consider the following system of inequalities:

   $y \le -2x + 7$

   $y \ge x - 5$

   $x \ge 0$

   a. Graph the solution set and label the corner points with coordinates.

   b. Calculate the maximum and minimum of the objective function $P = 2x - 3y$ given the constraints above.

4. Solve the following system of equations using any method.

   $x + 2y + 3z = 3$

   $y + 4z = 11$

   $5x + 6y = -25$

5. Consider the matrices:

$$A = \begin{bmatrix} 7 & 10 \\ 5 & 6 \end{bmatrix}, B = \begin{bmatrix} 4 & 9 \\ 7 & -2 \\ -5 & 6 \end{bmatrix}, C = \begin{bmatrix} 1 & 2 \\ 0 & -1 \end{bmatrix}$$

   Calculate all of the following matrices. If a computation is impossible, state that and the reason why.

   a. Calculate B^T.

   b. Calculate $3C - A$.

   c. Calculate A^{-1}.

   d. Calculate AC.

6. The supply equation for a product is given by $p = 2q^2 - 24q + 4$ and the demand equation is given by $p = -8q + 100$.

   a. If the price increases, will consumers demand more or less product? Circle your answer.

   More Less

   b. Find the market equilibrium.

7. An auto repair shop sells car repair services. Suppose that the shop has monthly fixed costs of \$100 dollars and variable costs of $x + 20$ dollars per service, where x is the total number of services sold. The revenue function is $R(x) = (60 - x)x$ in dollars.

   a. Find the cost function.

   b. What kind of function is the revenue function? Circle your answer.

   Linear Quadratic

   c. How many services need to be sold to maximize revenue?

   d. Find the profit function.

   e. Is it most valuable for a business owner to maximize revenue, or is there something better to maximize (or minimize)? Explain your thinking.

8. Consider the following functions:

$$f(x) = \frac{5 - 3x}{2}, \quad g(x) = x^3 + 1, \quad h(x) = \sqrt{x - 4}$$

a. What is the domain of $h(x)$?

b. Find $h(g(x))$.

c. Find $(f + h)(5)$.

d. Find $(gf)(1)$.

e. Find $f^{-1}(x)$.

9. This question is about the graphs of various functions we studied in class.

a. Graph $y = \log_2(x - 4) + 3$. Label at least two points on the graph with coordinates for full credit.

b. Graph $y = 2^{x-3} + 4$. Label at least two points on the graph with coordinates for full credit.

c. Graph $y = -x^2 + 1$. Label the vertex and x-intercepts with coordinates for full credit.

10. Solve the following equations.

a. $x = \log(100^{1011})$. Your answer should be an integer. (**Hint:** A calculator will not help you.)

b. $2(e^{2x-5}) - 4 = 10$. Give the **exact** answer, with **no calculator** approximation.

c. $\log_2(x) + \log_2(x - 1) = 1$. Check the domain for full credit.

11. An investor has $10,000 to invest. Suppose the money is invested at a rate of 7% compounded monthly. How many years will it take the investor to earn $15,000? *(round your answer to two decimal places)*

12. Magnolia owes $5,000 in student debt, to be paid off in regular payments over 4 years. If the interest rate is 3%, compounded monthly, what should her monthly payments be? *(round your answer to two decimal places)*

13. Tony and Maria have a son. They decide to enroll in a savings plan on his second birthday to save for his college education. They want to have $250,000 saved by the time their son turns 18. Assume their investment earns 6% every year, compounded quarterly, and that they will make deposits at the end of each quarter. How large must their quarterly deposits be to reach their investment goal? *(round your answer to two decimal places)*

14. Imagine Tony and Maria from the previous problem have reached their goal. Their son is starting college and their savings plan contains $250,000. At the same annual interest rate of 6% compounded quarterly, how much money should their son withdraw from the account at the beginning of each quarter if he plans to spend down the entire account in 4 years? *(round your answer to two decimal places)*

Final Exam 2

1. a. Graph the region of the plane (the feasible region) that satisfies the following inequalities.

   $x \geq 0$
   $y \geq 0$
   $x + y \leq 5$
   $5y - 3x \leq 15$

   Make sure to label the vertices of the region and shade in the desired region.

   b. Find the maximum and minimum value of the objective function $P = 5 - x + y$ subject to the constraints in part (a) (above).

2. The supply equation for IKEA tables is $3p - 2q = 200$ and the demand equation is $p = 200 - 2q$. (Remember that p = price and q = quantity.)

   a. What is the equilibrium quantity and price?

   b. If the market price of IKEA tables is \$200 per table, then how many will the company make (i.e. how many tables will IKEA be willing to supply)?

   c. If the market price of IKEA tables is \$50 per table, then how many tables will customers buy (i.e. how many tables will be demanded)?

3. Given the matrix $A = \begin{bmatrix} 0 & 1 & 1 \\ 1 & 1 & 1 \\ 1 & 2 & 3 \end{bmatrix}$

   a. Find the inverse matrix A^{-1}. (*Note: To get credit for this problem, you must show all your work here, using Gauss–Jordan elimination or elementary row operations, not just use a formula to give your final answer.*)

   b. Solve this system of equations. (*Note that you can choose to use the inverse matrix from part a for this problem, or you can solve this using Gauss-Jordan elimination, substitution, or some other method, but not by guessing.*)

   $y + z = 0$
   $x + y + z = 1$
   $x + 2y + 3z = 2$

4. Use these matrices to compute the following or state why that computation is not possible.

$$A = \begin{bmatrix} 5 & 2 \\ -2 & 3 \end{bmatrix},\ B = \begin{bmatrix} 7 & 4 \\ 0 & 0 \\ 1 & -9 \end{bmatrix},$$

$$C = \begin{bmatrix} 3 & 8 & -1 \\ 0 & 3 & 5 \end{bmatrix},\ D = \begin{bmatrix} -2 & 7 & 0 \\ 4 & 1 & -1 \end{bmatrix}$$

   a. $D - B^{T}$

   b. AD

   c. Using the matrices listed above, answer the following true/false questions.

   We can compute the product of CA.

   True or False (circle one)

   We can compute the product of AC

   True or False (circle one)

5. For this quadratic function $y = -x^2 + 2x + 3$, answer the following questions.

   a. Find the vertex and axis of symmetry. (*Be sure to report your vertex as a point with two coordinates, and list the axis of symmetry as a line equation.*) (**Hint:** To answer this question most efficiently, you'll need to complete the square.)

   b. Find the x- and y-intercepts. If there are none, then write NONE as your answer. (*Be sure to list them as points, with two coordinates.*)

   c. Is the parabola concave up or down?

   Concave up concave down (circle one)

   d. Graph the parabola, using all the information you've listed above. Label at least three points on the graph.

6. A tailor manufactures customized jumpers. Suppose the tailor sells each jumper for \$100. Call x the number of jumpers sold by the tailor. The business has fixed costs of \$1,500 and each jumper manufactured costs $(x + 20)$ dollars.

   a. Find the revenue function.

   b. Find the cost function.

   c. Find the profit function.

   d. How many jumpers need to be sold to maximize the profit?

   e. What is the maximum possible profit?

7. Solve these equations. If there is no solution, write NS or NONE. (*Note: Be Sure to check your answers and state only the valid solutions.*)

   a. $\log_2 x - \log_2(x + 3) = 2$

   b. $2e^{3x+1} - 5 = 0$ (*Give EXACT answer—no calculator approximations.*)

8. If I invest \$5,000 in an account earning 6% interest, how long will it take to double my investment, with each of these situauations? (*Use your calculator for this question.*)

   a. The account earns interest compounded quarterly.

   b. The account earns interest compounded continuously.

9. Given this graph for a function $y = f(x)$, graph $y = f(x + 1) - 2$.

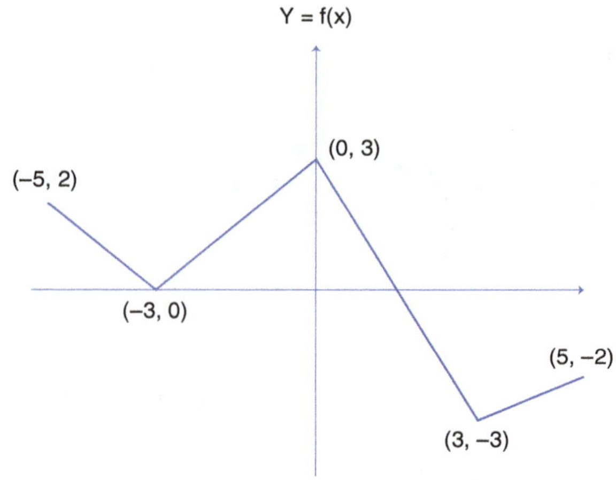

10. Alex purchases a home for \$350,000. He takes out a loan for this, making quarterly payments with an interest rate of 4.5% compounded quarterly for 20 years.

    a. What are his quarterly payments?

    b. Find the unpaid balance after 10 years.

11. A company is expected to replace one of their machines in 15 years. Market studies expect the cost of such a machine to be \$68,000 in 15 years. The company sets up a sinking fund to pay for the new machine in the future. Payments are made to the fund twice each year. If the fund earns 7% interest compounded twice yearly, how much should each payment be?

12. Given functions $f(x) = \dfrac{1}{x} + 5$ and $g(x) = x^2 + 2x - 3$, compute the following.

    a. Domain of $f(x)$ and $g(x)$.

    b. $(f \circ g)(x)$

    c. $\left(\dfrac{g}{f}\right)(-1)$

    d. $f^{-1}(x)$

Final Exam 3

1. Given the following system of inequalities:

$y \le -3x + 26$

$y \le \frac{4}{7}x + 1$

$y \ge 0$

$x \ge 0$

$x \le 8$

   a. Graph the solution set and label the vertices (corner points) with coordinates.

   b. Calculate the maximum and minimum of the objective function $P = 3x - 2y$ given the constraints above.

2. Given the supply equation is $45p - 30q = 600$ and the demand equation is $p = 200 - q$:

   a. What quantity is the producer willing to supply at $25?

   b. What quantity will consumers demand at $25?

   c. What do the results of (a) and (b) communicate about the market?

3. Sam has $100,000 to invest in three different mutual funds. They wish to invest this money and make an 8% profit over the next year. Over the last year, the first fund had a 5% return, the second fund had a 9% return, and the third fund had a 12% return. Sam likes the second fund the most so they plan to invest $10,000 more in the second fund than the third fund.

   Let:

   - x represent the money invested in the first fund.
   - y represent the money invested in the second fund.
   - z represent the money invested in the third fund.

   Write, but **do not solve**, the system of equations Sam would use to determine how much to invest in each fund to achieve their goal.

4. Consider this system of linear equations we are looking to solve.

$2x + y - z = 2$

$x + 2y + 3z = 3$

$-3x - 4y + 2z = 0$

   We have already done the work to get this upper triangular augmented matrix for the system above:

$$\left[\begin{array}{ccc|c} 1 & 2 & 3 & 3 \\ 0 & 1 & -4 & -5 \\ 0 & 0 & 1 & 1 \end{array}\right]$$

   Use this matrix to finish solving the system. If a unique solution exists, write it as an ordered triple.

5. Find the inverse of the matrix

$$A = \begin{bmatrix} 1 & 1 & 0 & 0 \\ 0 & 1 & 0 & 0 \\ 0 & 0 & 1 & 0 \\ 0 & 1 & 0 & 1 \end{bmatrix}.$$

6. For parts *a-d*, refer to the following matrices.

$$A = \begin{bmatrix} 2 & 0 \\ 1 & 3 \end{bmatrix}, B = \begin{bmatrix} 1 & 2 \\ 1 & 4 \\ 3 & 1 \end{bmatrix}, \text{ and } C = \begin{bmatrix} 1 & 2 & 3 \\ 3 & 1 & 2 \end{bmatrix}.$$

Compute or **explain why the computation is impossible**.

a. Find AC.

b. Find B^T.

c. Find AB.

D. Find $C^T + B$.

7. Given $f(x) = 4x^3 + 2$, $g(x) = \frac{2}{x}$, $h(x) = \sqrt{2x + 1}$, find the following.

a. $g(h(x))$

b. $(f \circ g)(x)$ **Do not simplify**.

C. domain of $h(x)$

d. $\left(\frac{f}{g}\right)(1)$

e. $(gh)\ (4)$

f. $f^{-1}(x)$

8. Graph each function. Fill in the information about the function. If it asks for an asymptote, be sure to draw that asymptote.

a. $f(x) = 2^x + 1$

vertex:
axis of symmetry:
concave up or down?
y-intercept:

b. $f(x) = \log_3(x)$

vertex:
axis of symmetry:
concave up or down?
y-intercept:

c. $f(x) = (x - 1)^2 - 2 = x^2 - 2x - 1$

vertex:
axis of symmetry:
concave up or down?
y-intercept:

Note: No need to find the x-intercept.

9. A company's revenue function is given by $R(x) = 4000x - 30x^2$ and their cost function is given by $C(x) = 85750 + 700x$.

a. Find the profit function for this problem.

b. Find the maximum profit.

c. **Explain how** you would find the break-even point(s), but **do not do the calculation.**

10. Indicate whether the following statements are true or false.

a. We use the vertical line test to see if a relation is a function.

b. $f^{-1}(f(x)) = x = f(f^{-1}(x))$ (assuming f and f^{-1} both exist)

c. $\frac{\ln 5x}{\ln 10x} = \ln\frac{1}{2}$

d. $\log_b(m + n) = \log_b m + \log_b n$

e. The sequence 2, 6, 24, 120, . . . is an arithmetic sequence.

11. Solve the following equations.

a. $\log(x + 2) + \log(x - 1) = 1$ (**Hint:** Check that your solutions are in the domain for full credit.)

b. $3(5^{2x}) - 7 = 11$ (Give the exact value, not a calculator approximation.)

12. **Read instructions carefully!!**

Suppose that you can put \$600 per month into an investment account to save toward your retirement. Your plan is to do this for 20 years in an account that will earn 6.3% interest compounded monthly. What will the account be worth upon your retirement?

a. Write the formula you plan to use to solve this problem.

b. How much money will be in the account when you retire in 20 years?

Your coworker, whose salary is the same as yours, also has a retirement plan. She will have \$300,000 in her retirement fund, also in 20 years, but she gets 7% interest compounded monthly.

c. Write the formula you would use to find out how much money your coworker is putting in her account each month.

d. How much money is your coworker putting in her account each month?

e. Who is saving more of their salary per month in the retirement plan? (circle one)

You Your Colleague

f. Who has more money to retire? (circle one)

You Your Colleague

13. Suppose you take out a 15-year loan in the amount of $275,000 to purchase a home. The annual interest rate is 4.5% compounded monthly.

a. Write the formula used to compute your monthly loan payment.

b. Using the formula in (a), compute your monthly loan payment.

c. Seven years after taking out the home loan, you sold your successful small business. You are trying to decide whether or not to pay off your home loan with the business sale proceeds or perhaps invest that money elsewhere. You calculate that your loan payoff amount is $169,338.35. How much will you have paid in total if you make the monthly payments for 7 years and then pay off the home loan at the end of those 7 years?

d. How much would you pay in total if you paid monthly payments for the full 15 years?

e. So, how much money will you save if you pay off the home loan early?

14. You are offered three possible accounts for your savings.

(1) an account earning 4.6% simple interest
(2) an account earning 4.1% interest compounded continuously
(3) an account earning 4.2% interest compounded monthly

You want to invest $1000 for 5 years. (This is one-time lump sum investment.)

a. How much will each account be worth after the 5 years? *For full credit, show all your steps.*

b. Which account yields the most money?

Homework Answers

CHAPTER 1

Section 1.1

1. $x = 2$
2. $x = -1$
3. $x = 3$
4. $x = -1$
5. $x = 5$
6. $x = 2$
7. $x = \frac{16}{5}$
8. $x = 6$
9. $x = -3$
10. $x = 1$
11. $x = 2$
12. $x = \frac{26}{3}$
13. $x = \frac{-9}{4}$
14. $x = 0$
15. $x = \frac{4}{5}$
16. $x = \frac{17}{13}$
17. $x = 32$
18. $x = \frac{5}{3}$
19. $x = \frac{-29}{2}$
20. $x = \frac{13}{5}$
21. $x = 45$
22. $x = 48$
23. $x = \frac{-35}{12}$
24. $x = -47$
25. $x = -25$
26. $x = -13$
27. $x = 15$
28. $x = \frac{29}{2}$
29. $x = \frac{5}{2}$
30. $x = 13$
31. $x = 2$
32. $x = -7$
33. $x = \frac{-11}{12}$
34. $x = \frac{-1}{6}$
35. $x = \frac{-21}{20}$
36. $x = 18$
37. $x = 2$
38. $x = 210$
39. $x = 0$
40. $x = 2$
41. $x = 2$
42. $x = 8$
43. $x = 12$
44. $x = 7$
45. $x = 8$
46. $x = \frac{23}{2}$
47. $x = 8$
48. $x = 9$
49. $x = -1$
50. $x = \frac{24}{5}$
51. $x = 3$
52. $x = 3$
53. $x = 5$
54. yes
55. no
56. yes, a constant
57. yes
58. no
59. yes
60. no
61. 17
62. 7
63. 5
64. 31 weeks
65. $c = 0.20m + 5$ and \$7.40 for 12 miles
66. $c = 0.15x + 80$ and \$113 for 220 miles
67. $f = 8w + 25$ and 145 sections after 15 weeks
68. 70 feet by 280 feet
69. 10 inches by 2 inches
70. \$7,500 and \$2,500
71. \$5,200 and \$8,800
72. \$600,000 and \$400,000
73. 95°F
74. 10°C
75. –40°
76. 26% annual interest rate
77. 8 years
78. \$7,500
79. 100 years
80. $\frac{1}{2}$ liter
81. 52, 54, and 56
82. \$32,000
83. –45°C
84. 14 ft, 16 ft

85. $x = 1$
86. 18.5 ft, 33.5 ft
87. $l = 45\text{m}$, $w = 40\text{m}$
88. 111.11 miles
89. 11, 12, 13
90. $\frac{32}{7}$ pounds
91. 12.6
92. 11.9 to 19.18
93. 12.5 ft, 17.5 ft
94. 98
95. $800.00
96. $1420.25
97. $115,646.26
98. $1500

Section 1.2

1. $x < 2$

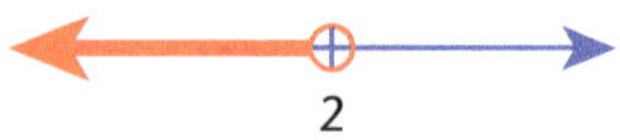

2. $x \geq 6$

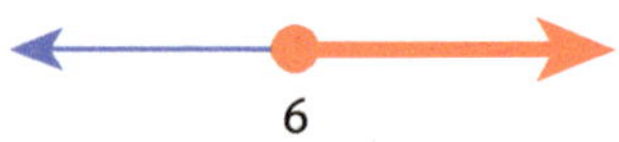

3. $x \geq 0$

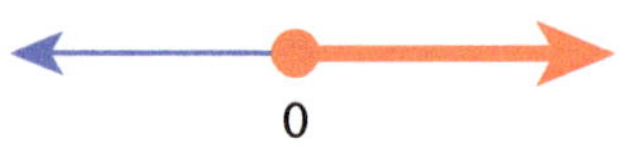

4. $x > 3$

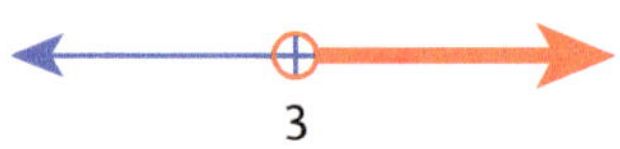

5. $x > -6$

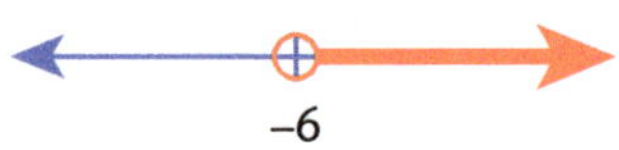

6. $x \geq 5$

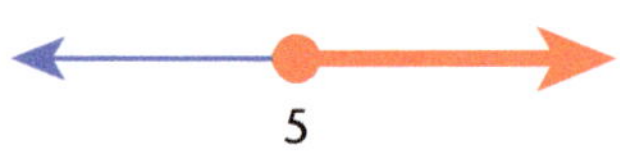

7. $x \leq 3$

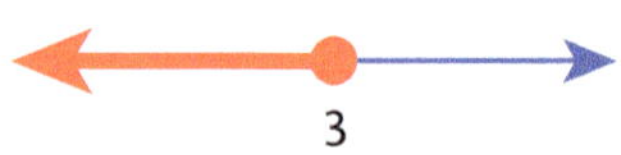

8. $x > \frac{1}{3}$

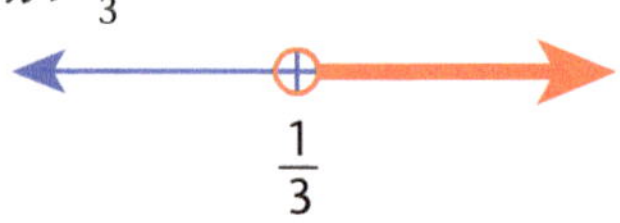

9. $x \leq \frac{-6}{5}$

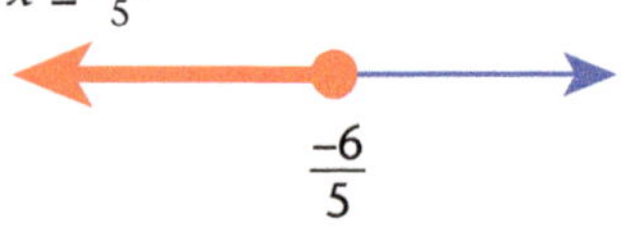

10. $x \geq -6$

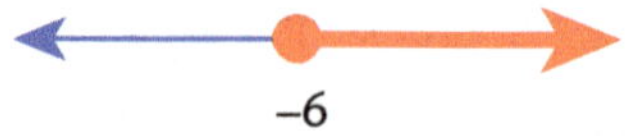

11. $x < \frac{15}{4}$

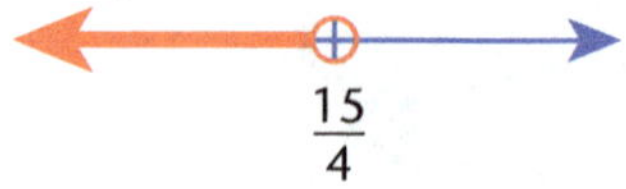

12. $x > \frac{-6}{25}$

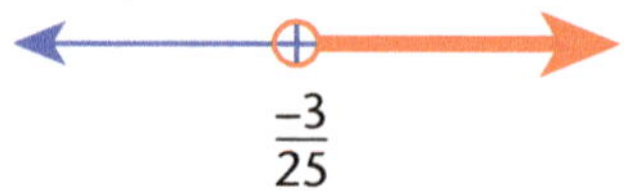

13. $x > -14$

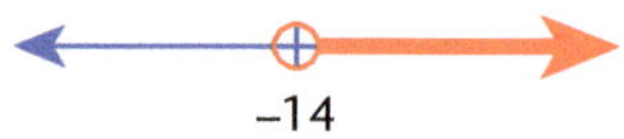

14. $x \geq 4$

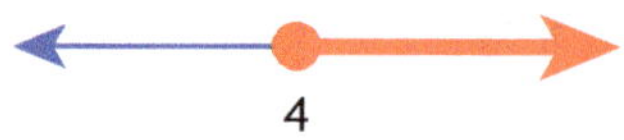

15. $x \geq 2$

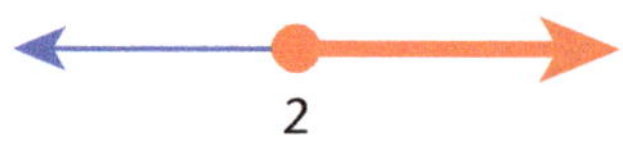

16. $x < \frac{15}{2}$

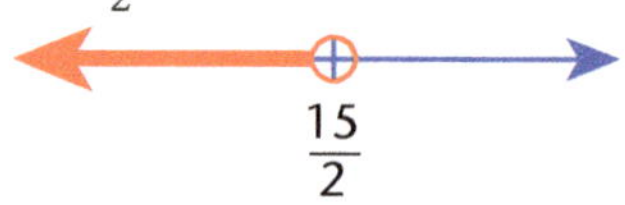

17. $x \leq \frac{-5}{3}$

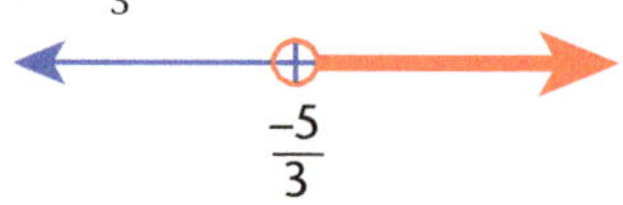

18. $x > \frac{15}{11}$

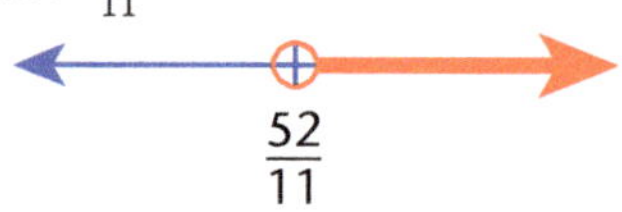

19. $x \leq -48$

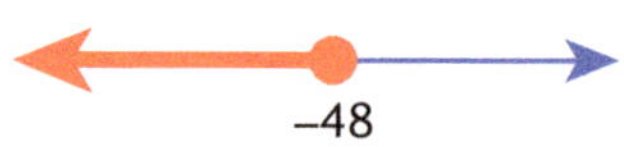

20. $x < 3$

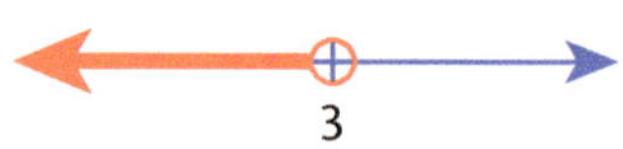

21. $x \leq 6$

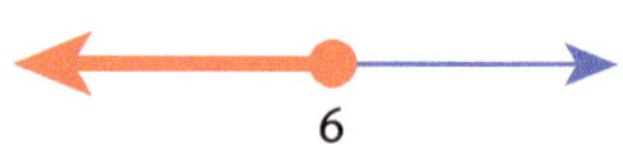

22. $x < 7$

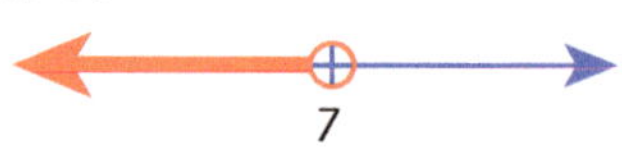

23. $x \leq -7$

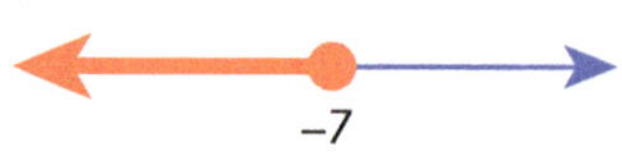

24. $x \geq 5$

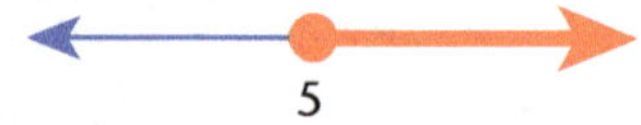

25. $x > 11$

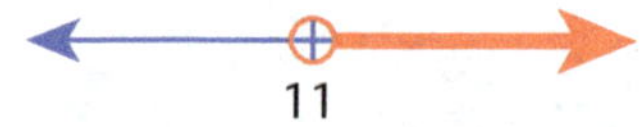

26. $-1 \leq x \leq 2$

27. $\frac{-1}{2} < x < 6$

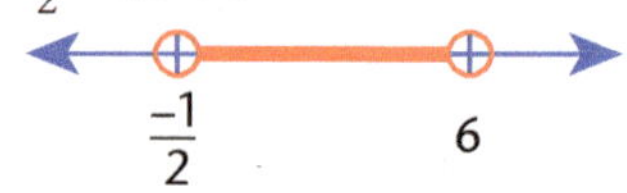

28. $x < 16$

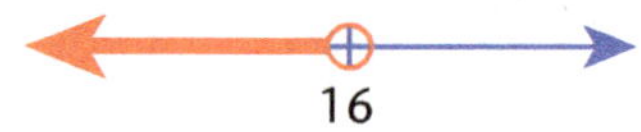

29. $2 \leq x$

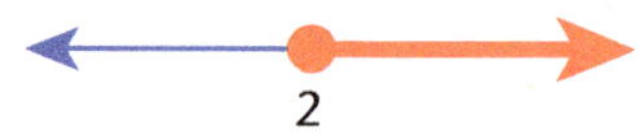

30. $-9 \leq x \leq \frac{15}{2}$

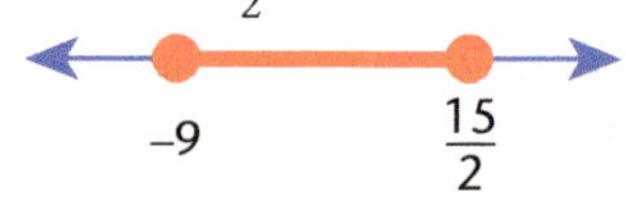

31. $17 < x \leq 37$

32. $\frac{-33}{8} \leq x < \frac{3}{2}$

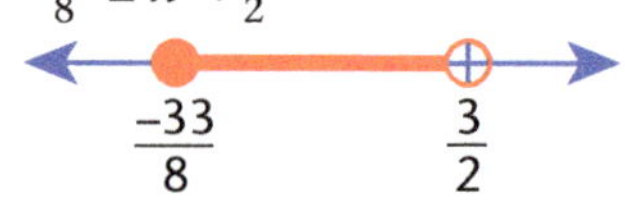

33. $x < -12$

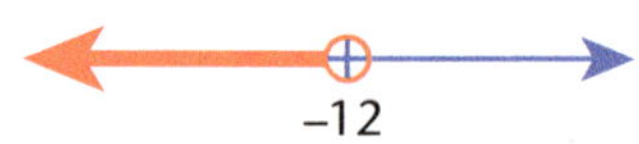

34. $x \geq 6$

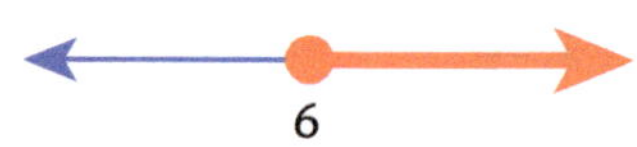

35. $y \leq 2$

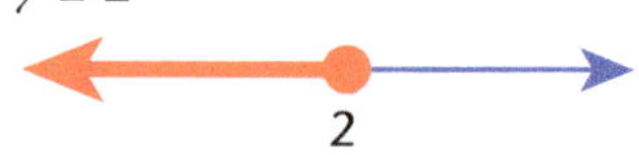

36. $x \leq \frac{19}{4}$

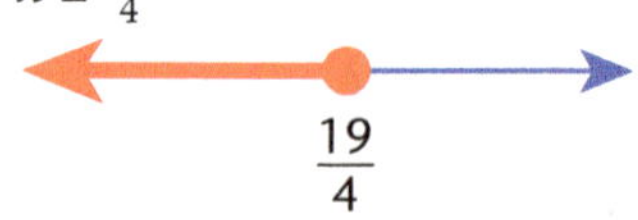

37. $-4 \leq x < 4$

38. $w > 8$

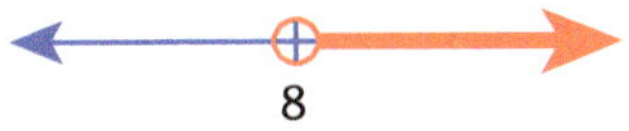

39. $x < \frac{113}{9}$

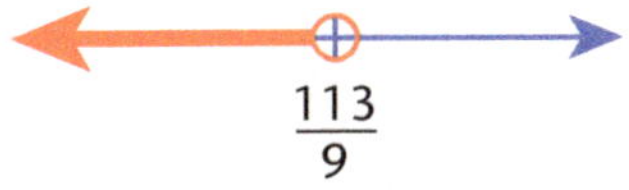

40. $x \geq -12$

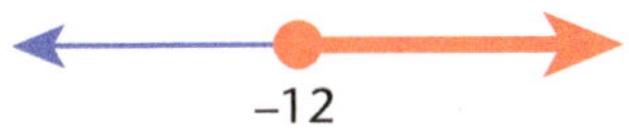

41. $y \geq \frac{-41}{5}$

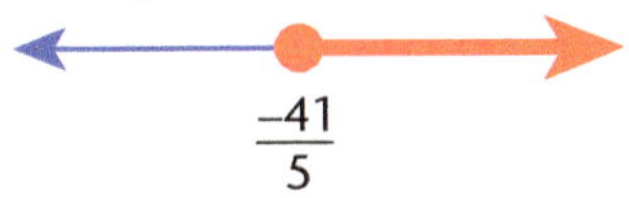

42. $q < -1$

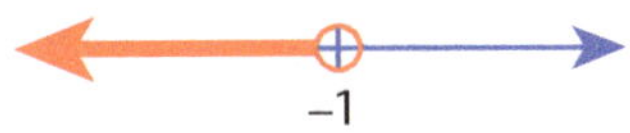

43. $x > \frac{37}{7}$

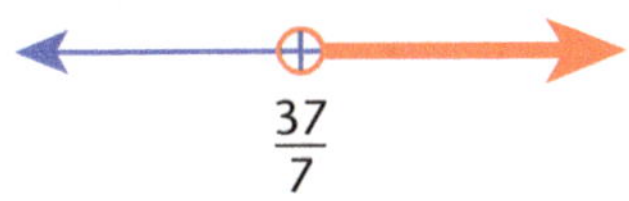

44. $y > -36$

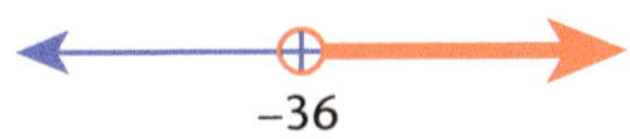

45. $v > \frac{-5}{2}$

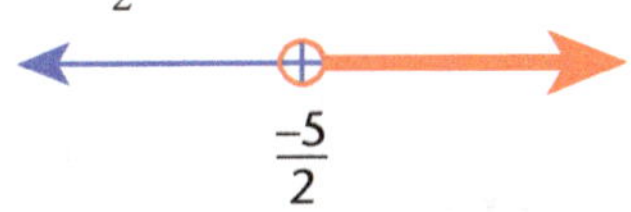

46. $x < 8$
47. $x > -4.3$
48. $7 < x < 12$
49. $-11 < x < -4$
50. $-3 < 3x < 13$
51. no more than 5 or $x \leq 5$
52. at least 4 or $x \geq 4$
53. $3 \leq x < 10$
54. no more than 3 or $x \leq 3$
55. A score of 94 out of 100 or better.
56. At least \$3,000 in the riskier investment and no more than \$7,000 in the safer investment.
57. At least \$57142.86 in the riskier investment and no more than \$42857.14 in the safer investment.
58. $-46.1 \leq C° \leq -11.67$
59. Total sales \$32,000, at least.
60. \$5,000 < sales < \$20,000
61. $x > 80$
62. $t \leq 28.85$ °F
63. $35 \geq x$
64. $x > 2954$
65. $x > 110{,}000$
66. $52.74 \leq H \leq 58.29$
67. $x < 3600$
68. $x > 40{,}000$
69. sales of \$250,000 per year or \$20,833 per month

Section 1.3

1. 1
2. –1
3. –3
4. –10
5. $-\frac{1}{2}$
6. $\frac{2}{3}$
7. undefined, vertical line
8. $\frac{1}{2}$
9. 0, horizontal line
10. 2
11. $-\frac{1}{3}$
12. $-\frac{1}{17}$
13. $-\frac{3}{4}$
14. $\frac{7}{10}$
15. $-\frac{4}{3}$
16. parallel
17. perpendicular
18. neither
19. perpendicular
20. neither
21. perpendicular
22. $m = -1$, y-intercept is (0, 2)
23. $m = -6$, y-intercept is (0, –1)
24. $m = \frac{3}{2}$, y-intercept is (0, –3)
25. $m = \frac{5}{4}$, y-intercept is (0, 3)
26. $m = -\frac{8}{3}$, y-intercept is (0, –3)
27. $m = \frac{7}{10}$, y-intercept is $(0, -\frac{2}{5})$
28. $m = 8$, y-intercept is (0, 12)
29. $m = 3$, y-intercept is (0, –12)
30. DNE (vertical line)
31. $m = -2$, y-intercept is (0, 17)
32. $m = \frac{7}{5}$, y-intercept is $(0, \frac{14}{5})$
33. $m = 0$, y-intercept is $(0, \frac{4}{17})$
34. $m = \frac{1}{6}$, y-intercept is $(0, -\frac{5}{6})$
35. $m = \frac{4}{9}$, y-intercept is (0, –4)
36. $m = -\frac{3}{2}$, y-intercept is $(0, -\frac{3}{4})$
37. $m = -\frac{5}{4}$, y-intercept is (0, 4)
38. $m = 3.8$, y-intercept is (0, 0)
39. $m = -\frac{8}{7}$, y-intercept is $(0, -\frac{24}{7})$
40. $y = x - 2$

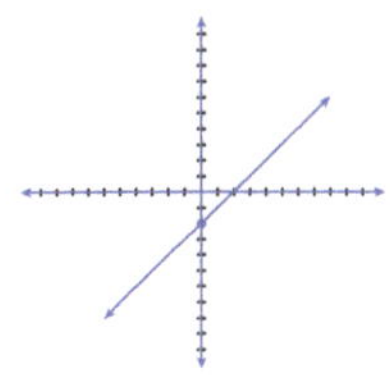

41. $y = -x + 9$

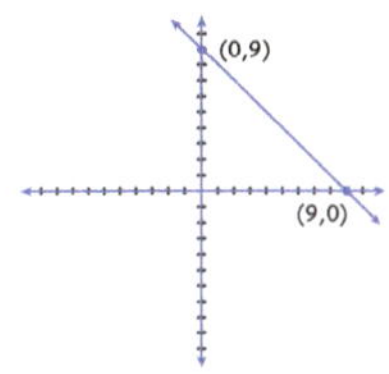

42. $y = 3x - 8$

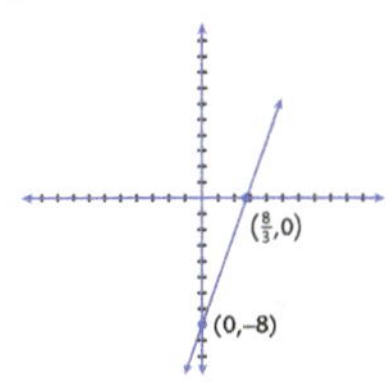

43. $y = 2$

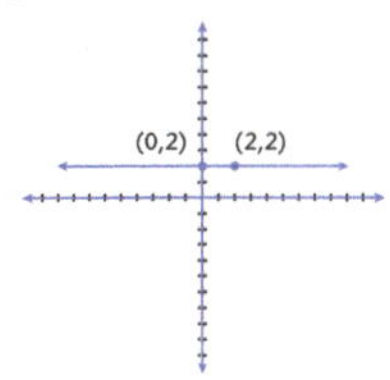

44. $y = \frac{2}{3}x + 1$

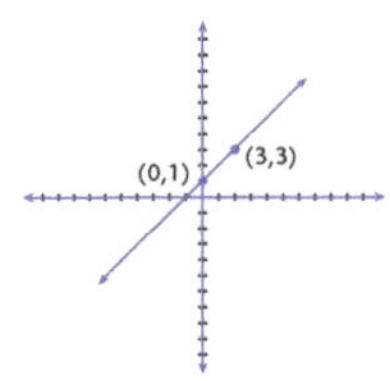

45. $y = -\frac{1}{2}x - 4$

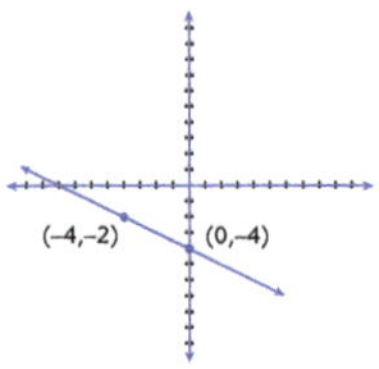

46. $y = -\frac{7}{3}x - 5$

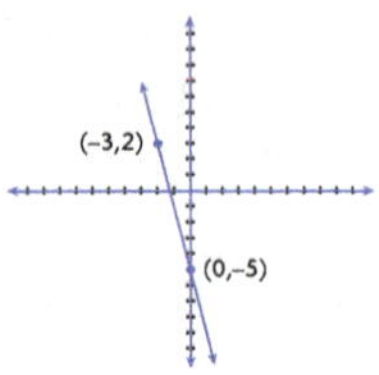

47. $y = 5x + 8$

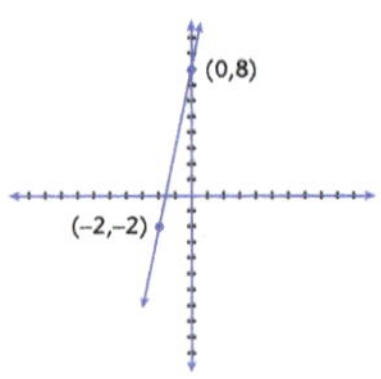

48. $y = \frac{4}{5}x + 28$

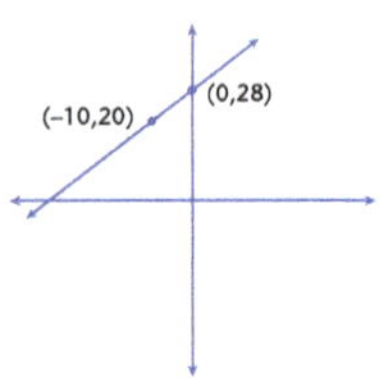

49. $y = 3x - 3$

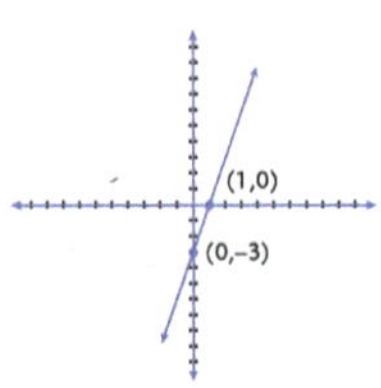

50. $y = 5$

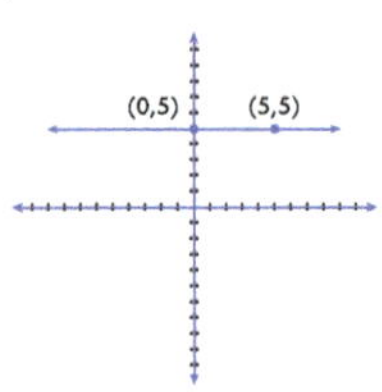

51. $x = 2$

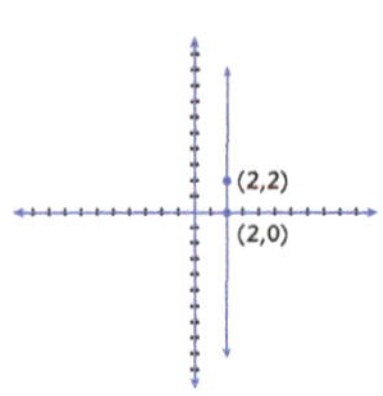

52. $y = 7x + 6$

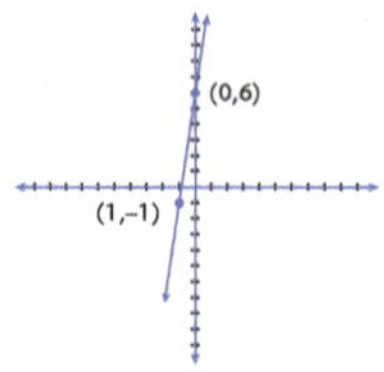

53. $y = 2x - 1$

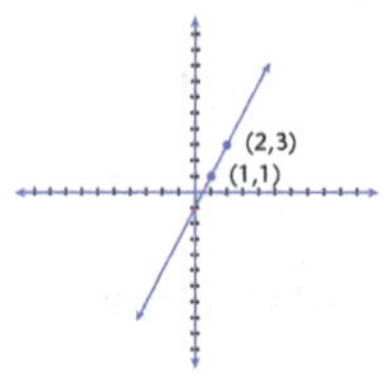

54. $y = -x + 10$

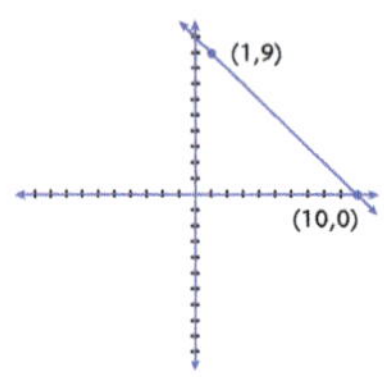

55. $y = -3x - 2$

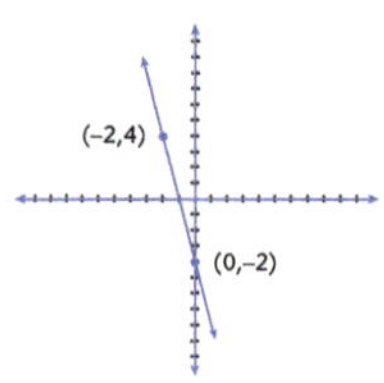

56. $y = x + 1$

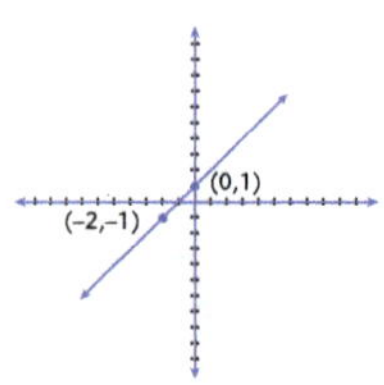

57. $y = -\frac{1}{3}x - \frac{2}{3}$

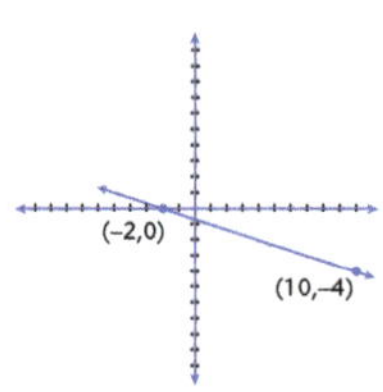

58. $y = \frac{5}{2}x - \frac{1}{2}$

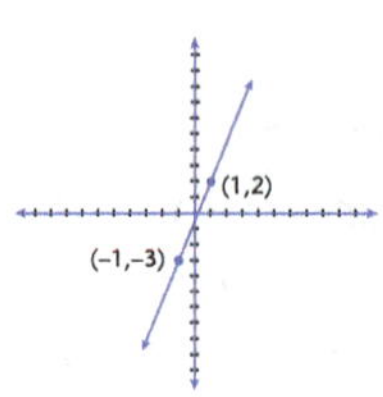

59. $y = 4x - 10$

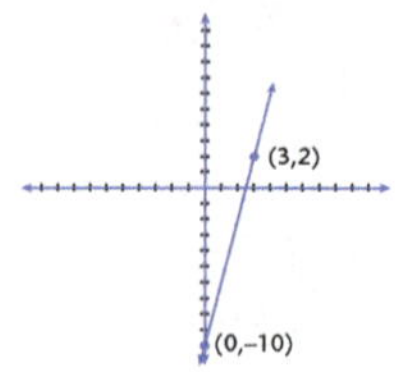

60. $y = 5x - 21$

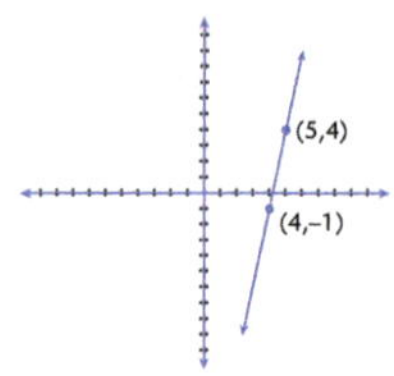

61. $y = \frac{3}{2}x - 2$

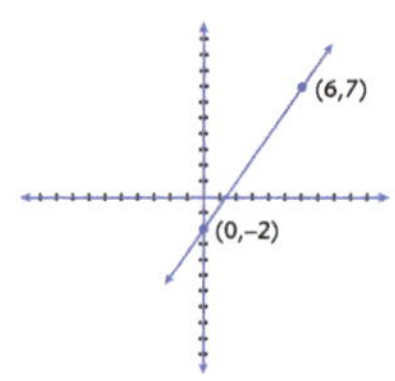

62. $y = -\frac{1}{3}x + \frac{19}{3}$

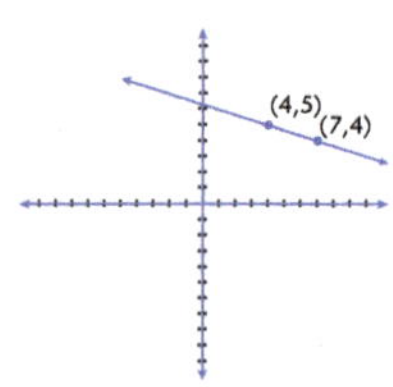

63. $y = -x + 4$

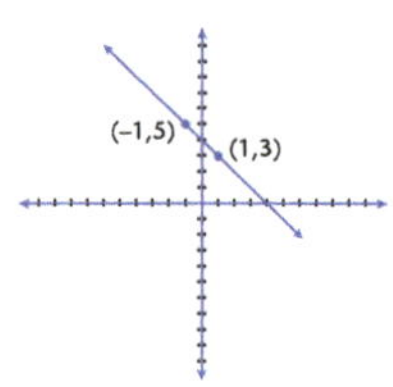

64. $y = -\frac{9}{2}x - \frac{3}{2}$

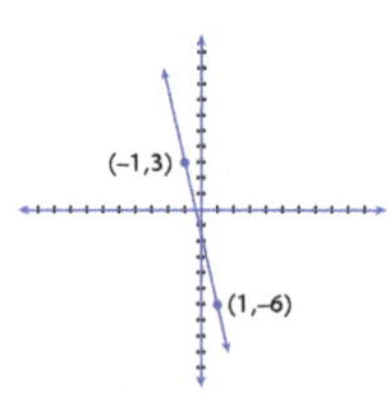

65. $y = -\frac{1}{3}x - 4$

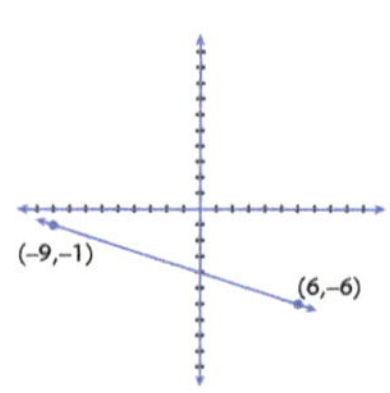

66. $y = \frac{3}{2}x + \frac{17}{2}$

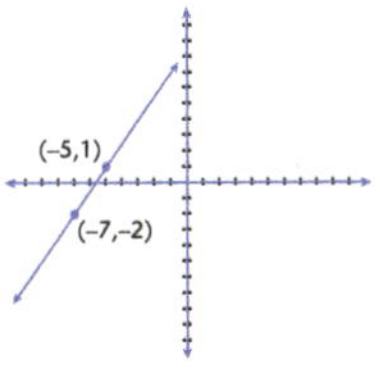

67. $y = \frac{1}{4}x - \frac{1}{4}$

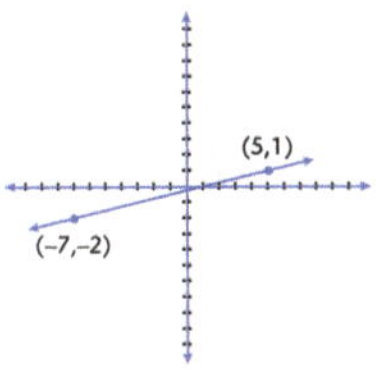

68. $y = \frac{1}{3}x + \frac{11}{3}$

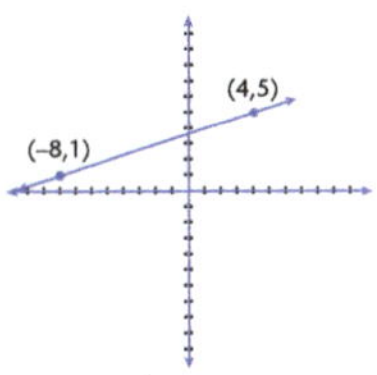

69. $y = \frac{7}{12}x + \frac{16}{3}$

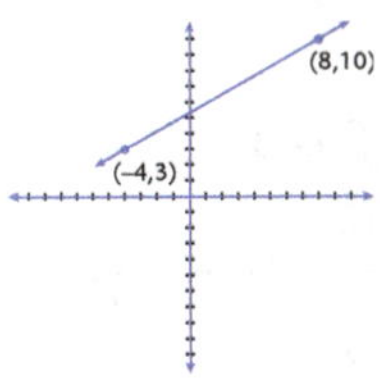

70. $y = \frac{1}{2}x + \frac{7}{2}$

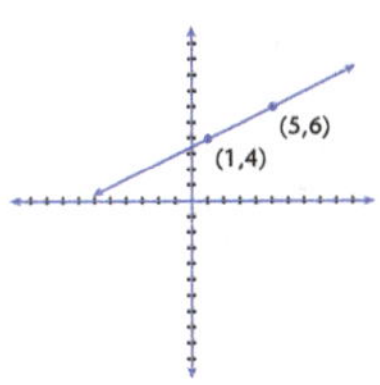

71. $y = x$

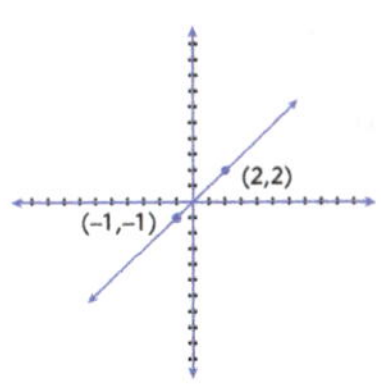

72. $y = \frac{3}{7}x - 3$

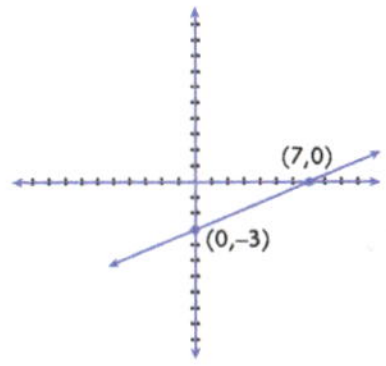

73. $y = -\frac{4}{3}x$

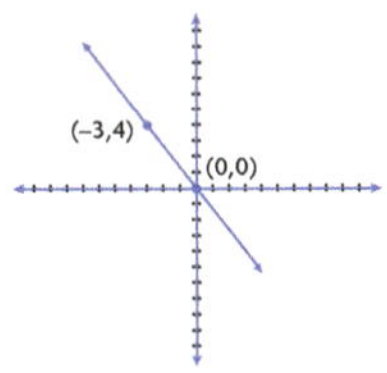

74. $y = -13x - \frac{15}{4}$

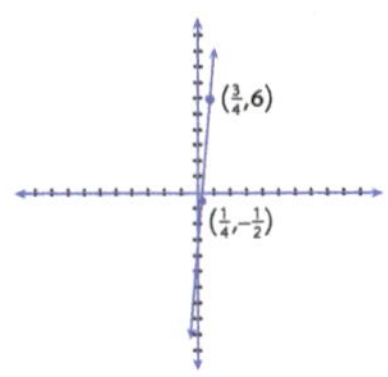

75. parallel
76. parallel
77. perpendicular
78. perpendicular
79. parallel
80. $y = -\frac{3}{4}x + \frac{1}{2}$
81. $y = \frac{-2}{7}x + \frac{50}{7}$
82. $y = 4x - 16$
83. $y = -\frac{1}{2}x + \frac{17}{3}$
84. $y = 2x + 3$
85. $y = -\frac{5}{2}x - \frac{35}{2}$
86. $y = \frac{4}{3}x - \frac{11}{3}$
87. $y = 2x - 6$
88. $y = \frac{-1}{8}x - \frac{11}{8}$
89. $y = -\frac{1}{2}x + \frac{5}{2}$
90. $y = \frac{-2}{3}x + \frac{16}{3}$
91. $y = 5x + 12$
92. $y = 0.25x + 7$, 0.25 is the slope and this is the cost for each additional mile the bicycle is ridden.
93. $y = 0.38x + 100$, 100 is the initial cost to rent the car, or the cost to rent if the car is driven zero miles.
94. $y = 1.5x + 2$, the slope is 1.5 or the \$1.50 per game bowled, the y-intercept is (0, 2) or \$2.00 for the shoe rental.
95. $y = 50x - 4500$, or \$500.00 profit on a day when the plant creates 100 widgets.
96. $y = 0.05x - 100$, or \$400.00 return on an investment of \$10,000.00.
97. $y = .15x + 15$
98. $°C = \frac{5}{9}(°F - 32)$; 70°F is 21.1°C
99. $y = -487.5x + 5200$
100. $P = 17x - 3400$
101. $R = 3.2x - 0.2$
102. $y = -18x + 360$
103. $y = \frac{2}{7}x - \frac{3464}{7}$ $\left(\text{or } y = \frac{2}{7}x + 68\right)$

Section 1.4

1. (2, 1)
2. $(1, 1 - \sqrt{2})$
3. (1, –2)
4. (1, –2)
5. (–1, 5)
6. (0, –2)
7. (–2, 2)
8. (4, 1)
9. no solution
10. (–2, 0)
11. (–9, 6)
12. same lines
13. (0, 1)

14. (2, –1)
15. (1, –2)
16. (4, 7)
17. (2, –1)
18. (1, 1)
19. (2, –1)
20. (1, –5)
21. (4, –3)
22. same lines
23. (3, 2)
24. no solution
25. (–17, 11)
26. (6, 7)
27. (2, –10)
28. (–6, –14)
29. (0, 6)
30. (–55, 60)
31. (3, –3)
32. (7, 2)
33. same lines
34. (–9, 20)
35. (5, –2)
36. no solution
37. (–4, –1)
38. $\left(\frac{67}{11}, -\frac{57}{11}\right)$
39. same lines
40. same lines
41. $\left(\frac{9}{4}, \frac{21}{2}\right)$
42. $\left(\frac{1}{2}, \frac{3}{4}\right)$
43. no solution
44. no solution
45. (4, 6)
46. (–3, 1)
47. $\left(\frac{13}{14}, \frac{31}{14}\right)$
48. $\left(\frac{4}{5}, -\frac{4}{5}\right)$
49. no solution
50. $\left(\frac{1}{7}, -\frac{8}{21}\right)$
51. no solution
52. same lines
53. $\left(\frac{2}{11}, -\frac{9}{11}\right)$
54. $\left(-\frac{15}{8}, -\frac{1}{2}\right)$
55. (1, 1, 3)
56. (4, –27, 13)
57. (1, –3, –7)
58. (–2, 11, 2)
59. (5, 1, –3)
60. (–3, 0, 7)
61. (0, 5, –4)
62. (6, –1, 3)
63. (6, 2, 0)
64. no solution
65. no solution
66. 1035 student tickets and 400 adult tickets
67. mom: 48 cupcakes, daughter: 32 cupcakes
68. 9 liters
69. 61.5 ft by 68.5 ft
70. 85 gallons
71. 18 years old
72. 27 years
73. 4 miles
74. 108 days
75. 296.22 ounces of brand 206 and 70.78 ounces of brand 822
76. Jack's team: 57 points, Dylan's team: 49 points
77. 3500 @ $15, 4,500 @ $25
78. 70 dimes, 30 quarters

Section 1.5

1. is a function
2. not a function
3. is a function
4. is a function
5. not a function
6. is a function
7. is a function
8. is a function
9. not a function
10. is a function
11. is a function
12. is a function
13. is a function
14. is a function, only the principle square root is used here
15. not a function
16. is a function
17. is a function
18. is a function
19. not a function
20. is a function
21. is a function
    domain: {3,4,5,6,7,10}
    range: {2,3,4,6,9,10}
22. is a function
    domain: {1,4,5,7,8,9}
    range: {2,5,9,10}
23. not a function
24. not a function
25. is a function
    domain: {1,6,7,8,9,10}
    range: {1,2,3,9,10}
26. is a function
    domain: {2,4,5,6,7,8}
    range: {1,3,4,5,7,9}
27. is a function
    domain: {3,4,5,6,7,8}
    range: {2,5,6,8,9,10}
28. is a function
    domain: {1,5,6,7,9,10}
    range: {1,2,6,10}
29. is a function
    domain: {1,2,3,4,6,9}
    range: {2,4,5,8,9}
30. is a function
    domain: {1,2,3,4,7,8}
    range: {3,4,6,8}
31. is a function
    domain: $x \in \mathbb{R}$
    range: $y \in \mathbb{R}$
32. is a function
    domain: $x \in \mathbb{R}$
    range: $y \in \mathbb{R}$
33. is a function
    domain: $x \in \mathbb{R}$, $x \geq -10$
    range: All real numbers greater than or equal to zero

34. not a function
35. is a function
domain: $x \in \mathbb{R}, x \neq 3$
range: $y \in \mathbb{R}$
36. is a function
domain: $x \in \mathbb{R}, x \neq 5$
range: $y \in \mathbb{R}$
37. is a function
domain: $x \in \mathbb{R}, x \neq 0$
range: $y \in \mathbb{R}$
38. is a function
domain: $x \in \mathbb{R}\ x \leq 2$
range: All real numbers greater than or equal to zero
39. is a function
domain: $x \in \mathbb{R}, -1 \leq x \leq 3$
range: All real numbers greater than or equal to zero
40. is a function
domain: $x \in \mathbb{R}, x \neq 1, x \neq 4$
range: $y \in \mathbb{R}$
41. $f(5) = 165$
42. $f(10) = \sqrt{35}$
43. $f(-8) = 6$
44. $f(-5) = 8$
45. $f(3) = \frac{\sqrt{3}}{6}$
46. $f(2u - 6) = 51 - 14u$
47. $f(8 - 2a) = 4a^2 - 30a + 63$
48. $f(b + 2) = b^2$
49. $f\left(\frac{3w-1}{w-1}\right) = \frac{2w-1}{3w-2}$
50. $f(p + 2) = p^3 + 6p^2 + 5p$
51. $g(4) = \sqrt{14}$
52. $h(2) = -5$
53. $h(-2) = 4$
54. $g(5) = 21$
55. $f(y + h) = y^2 + 2hy + h^2 - 4$
56. $g(w^2) = w^4 - 2w^2 + 3$
57. $h(h(x)) = 16x - 15$
58. $x \in \mathbb{R}$
59. $x \in \mathbb{R}, x \neq 4$
60. $x > 6$
61. $x \geq 5$
62. $x \in \mathbb{R}, x \neq \pm 4$
63. $x \in \mathbb{R}, x \neq 2$
64. $x \geq \frac{1}{2}$
65. $-5 \leq x < 5$
66. $x > 2$ or $-2 < x < 2$
67.

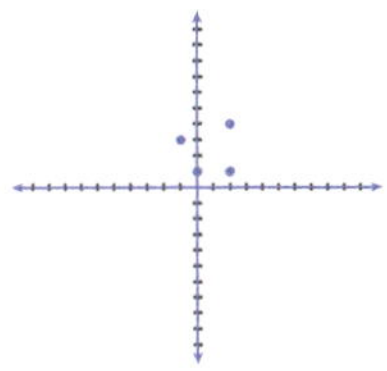

not a function

68.

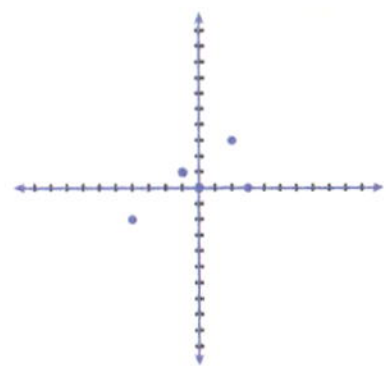

it is a function

69.

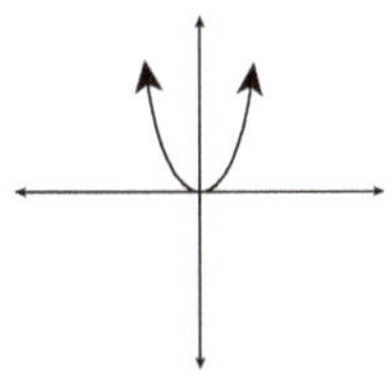

it is a function

70.

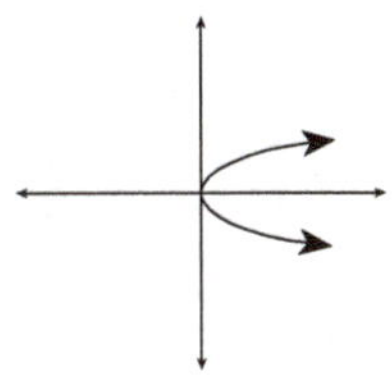

not a function

71. yes
72. no
73. yes
74. no

Section 1.6

1. a. \$8,900.00
   b. \$3.50
   c. \$7.95
   d. $4.45b - 8900$
   e. 2,000
2. a. \$7,500.00
   b. \$2.00
   c. \$8.00
   d. $6e - 7500$
   e. 1,250
3. a. 66,600
   b. 95,800
   c. 65,000
   d. 80,800
   e. price = \$420.00, quantity = 65,800
4. a. 1228
   b. 500
   c. 1172
   d. 1900
   e. ticket price = \$65.00, ticket quantity = 1200
5. a. $p = \frac{-7}{400000}q + 105$
   b. 3,142,857
   c. $75000p = q + 1000000$
   d. \$72
   e. price = \$53, quantity = 2,972,973
6. a. $q = 5150 - 40p$
   b. 3,150
   c. $p = \frac{1}{50}q + 30$
   d. \$54
   e. price = \$73.89, uniforms = 2194
7. a. $C(x) = 1.6x + 55{,}000$
   b. $R(x) = 11x$
   c. $P(x) = 9.4x - 55{,}000$
   d. 5851
   e. \$57,800
8. a. 7132
   b. 7050
   c. \$58,750
   d. \$12
   e. \$10.05
9. a. \$41
   b. \$41
   c. \$12,000
   d. \$60
   e. \$11,000
10. a. $C(x) = 0.15x + 175$
    b. 2166
11. \$6397
12. \$7250

13. \$29,853
14. 50
15. a. $C(x) = 28x + 12{,}000$,
 $R(x) = 45x$
 b. 706
 c. $MC = \$28$, $MR = \$45$
 d. $MP = \$17$
 e. $x = 800$, $R = \$36{,}000$
16. $C(x) = 29x + 4600$
17. $q = 21$, $p = \$178$
18. $q = 10$, $p = \$260$
19. $p = \frac{-1}{5000}q + \frac{8}{5}$
20. $p = 0.000375q - 1.75$
21. 72 units
22. (350, \$5)
23. a. $C(x) = 65000 + 10x$,
 $R(x) = 50x$,
 $P(x) = 40x - 65000$
 b. 1625 phones
24. a. $R(x) = 50x$,
 $C(x) = 10000 + 20x$
 b. $P(x) = 30x - 10000$
 c. about 333 bracelets
25. a. demand: 44 toys; supply: 3.5 toys (demand is much bigger than supply at this price)
 b. (35, \$30)
26. (35, \$30)
27. a. $C(b) = 35000 + 2700b$,
 $R(b) = 5200b$,
 $P(b) = 2500b - 35000$
 b. 14 battles
28. a. $C(x) = 1500 + 0.5x$,
 $R(x) = 3x$,
 $P(x) = 2.5x - 1500$
 b. 600 hot dogs

Section 1.7

1.
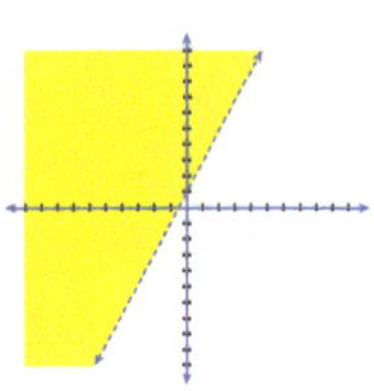

2.
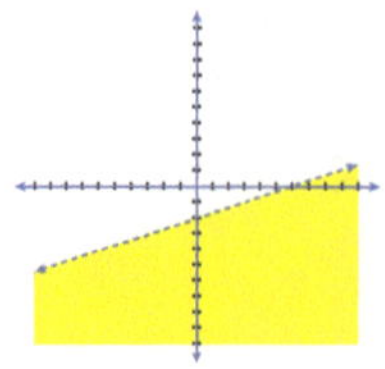

3.
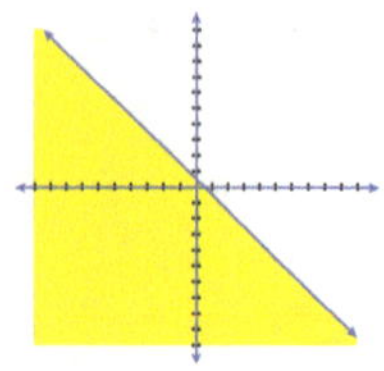

4.
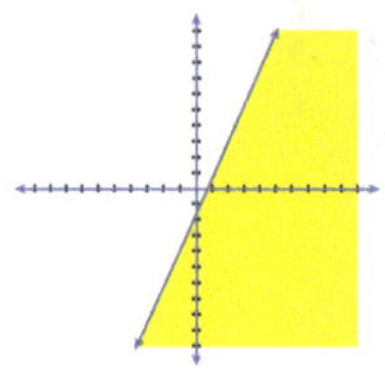

5.
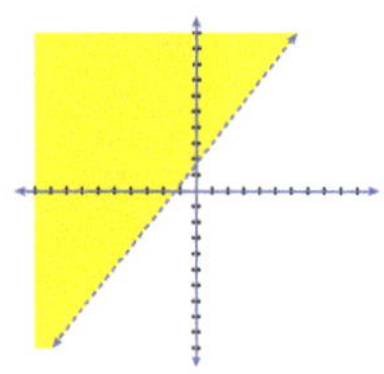

6.
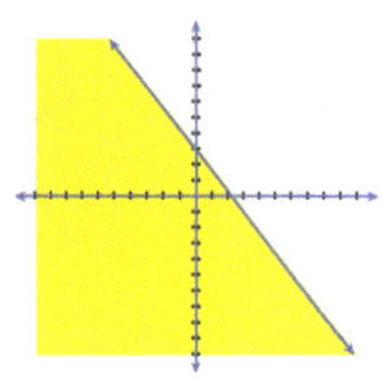

7.
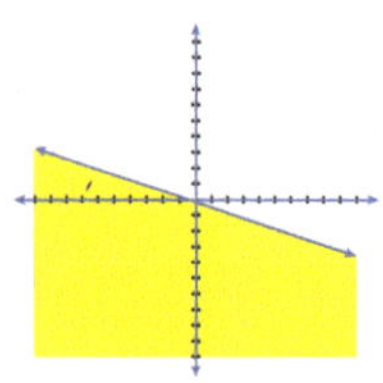

8.
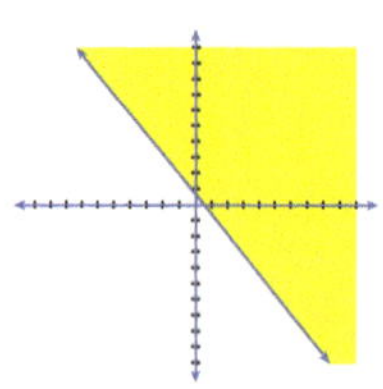

9.
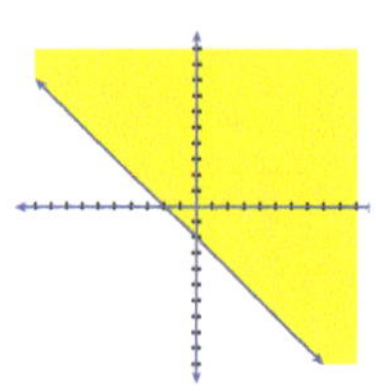

10.
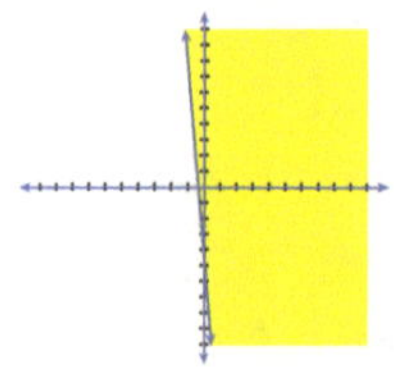

11.
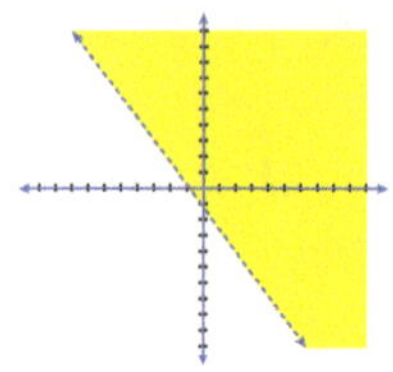

12.
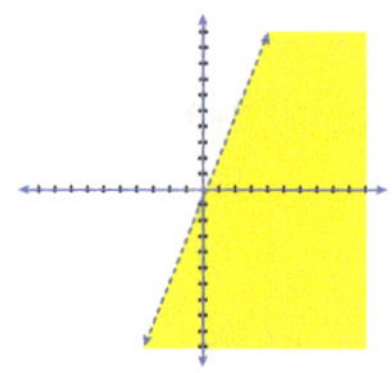

13.
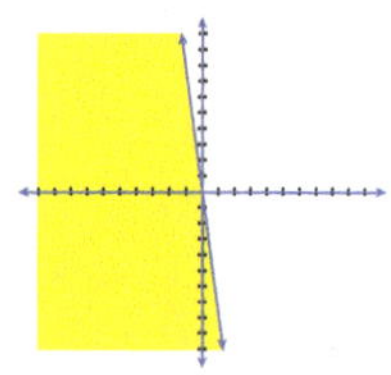

14.
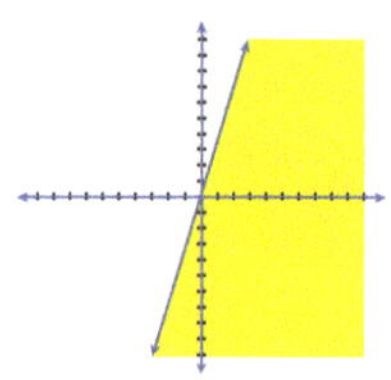

15.
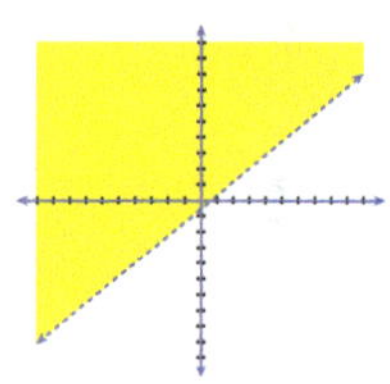

16.
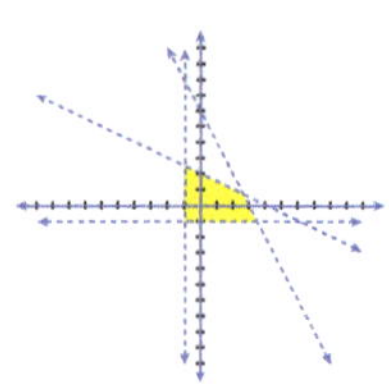

17.
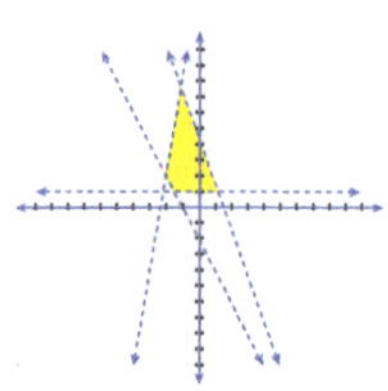

18. 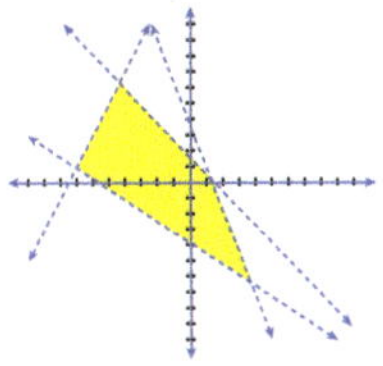

19. 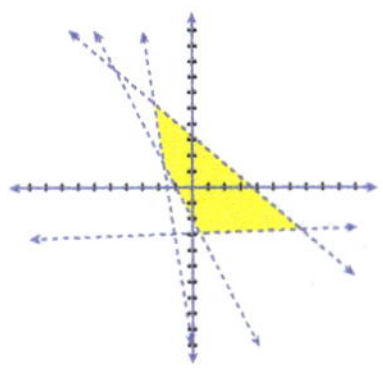

20. 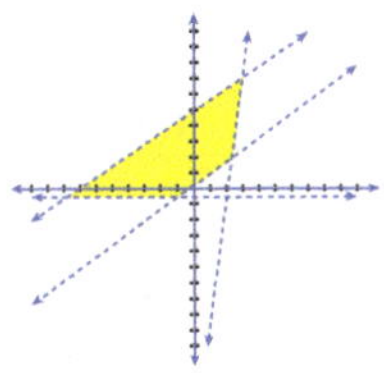

21. 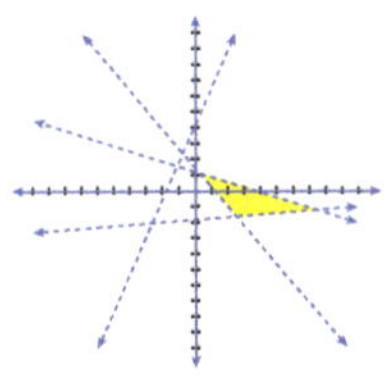

22. 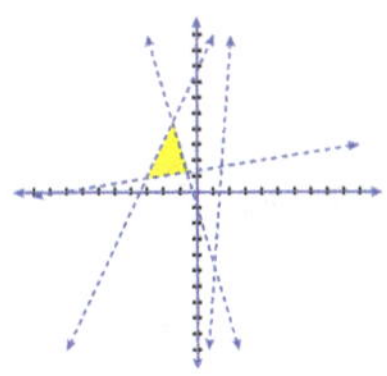

23. 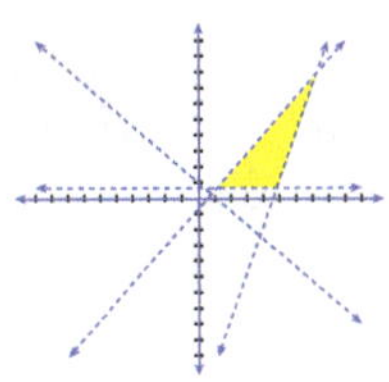

24. 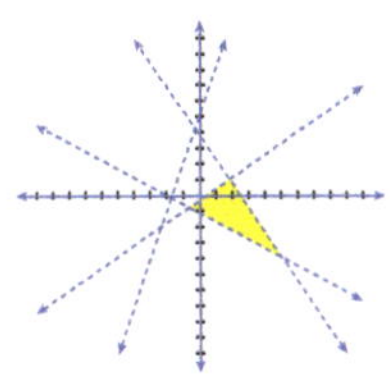

25. 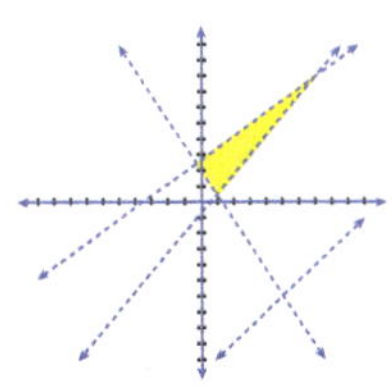

26. 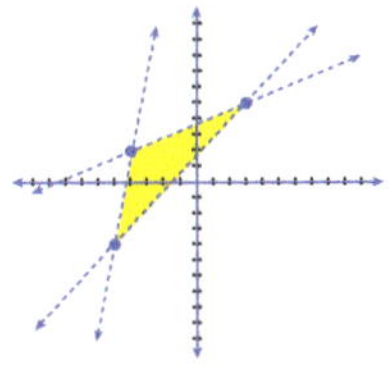

27. 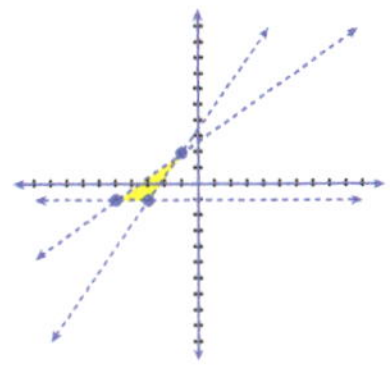

28. 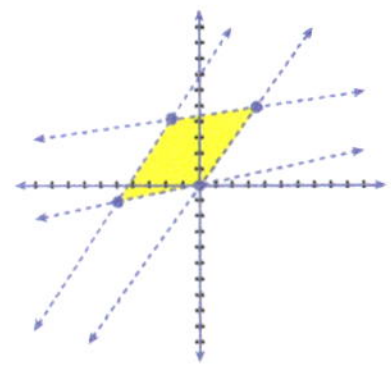

29. 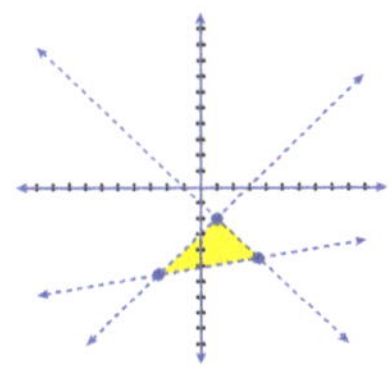

30. 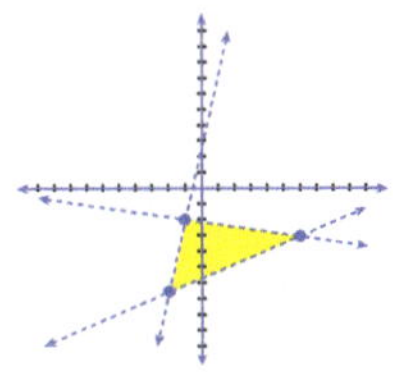

31. 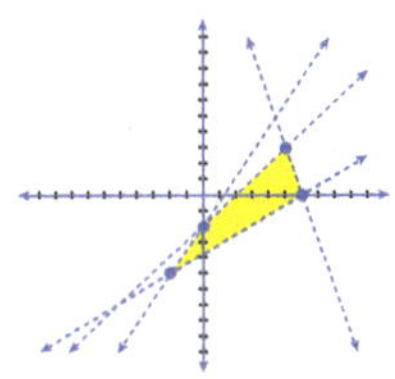

32. 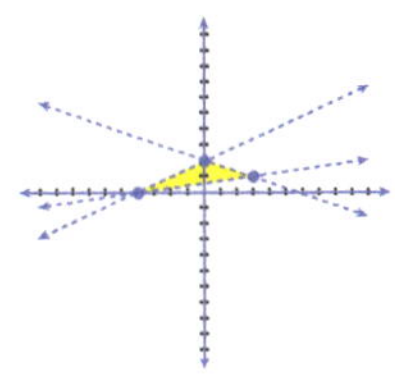

33. 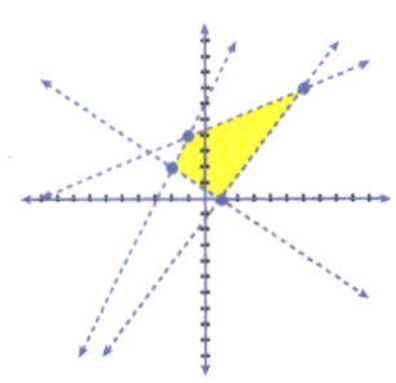

34. $c \geq 0$
$t \geq 0$
$c + t \leq 1000$
$c \geq t + 200$

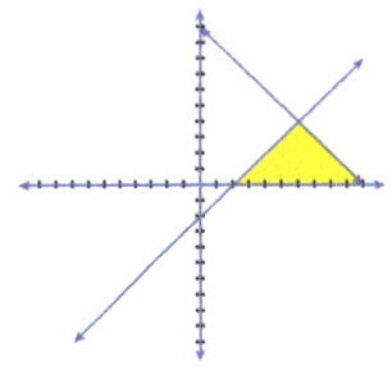

35. $b \geq 0$
$c \geq 10$
$b + c \leq 100$

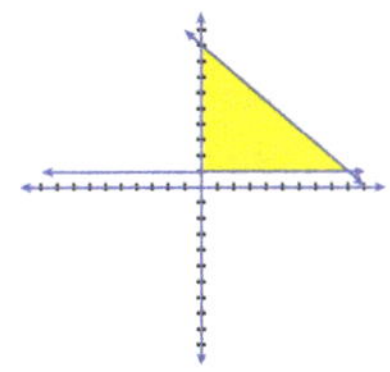

36. $x \geq 1000$
$y \geq 2000$
$x + y \leq 10000$
$y \geq 2x$

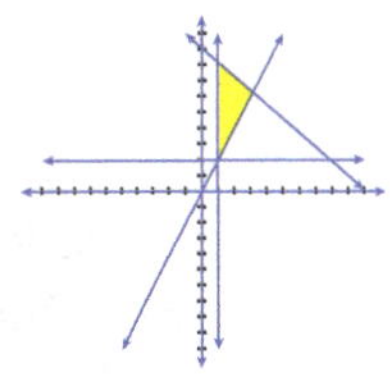

37. $c \geq 0$
$t \geq 0$
$c + t \leq 8000$
$\frac{c}{12000} + \frac{t}{6000} \leq 1$ or

$t \leq -\frac{1}{2}c + 6000$

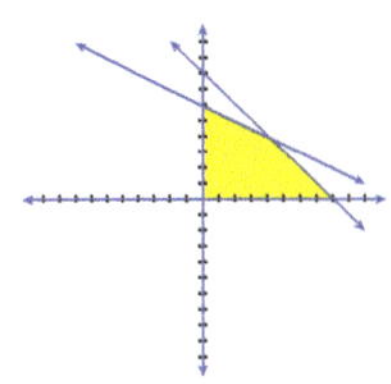

38. $c \geq 0$
$b \geq 0$
$c + b \leq 9000$
$400c + 200b \leq 2400000$ or
$c \leq \frac{-1}{2}b + 6000$

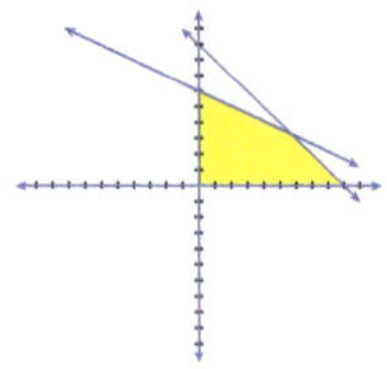

39. r = number of reserved seats,
g = number of general admission seats
$50r + 35g \geq 80000$
$r \geq 0$
$g \geq 0$
$r + g \leq 3000$
$r \leq 2000$

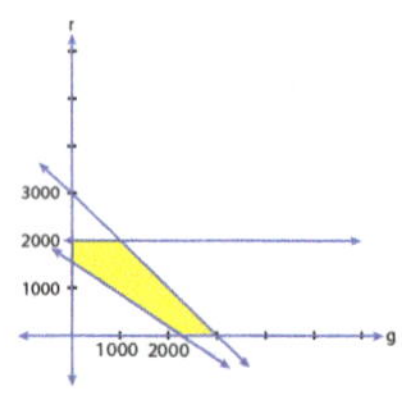

40. y = number of large tables,
x = number of small tables
$y \geq 0, x \geq 0$
$2y + 3x \leq 12$
$\frac{5}{4}x + 2y \leq 10$

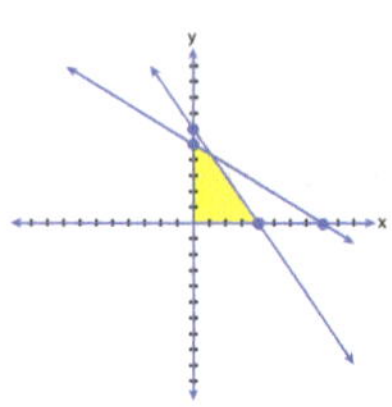

Section 1.8

1. Maximum at (3, 9) of 12
Minimum at (0, 0) of 0

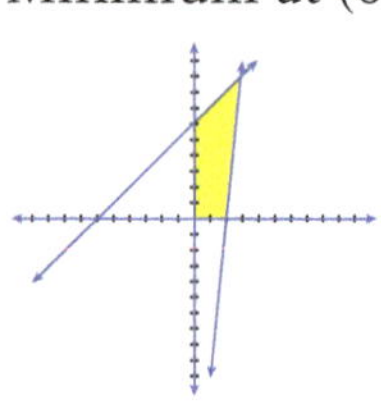

2. Maximum at (4, 0) of 4
Minimum at (0, 7) of –21

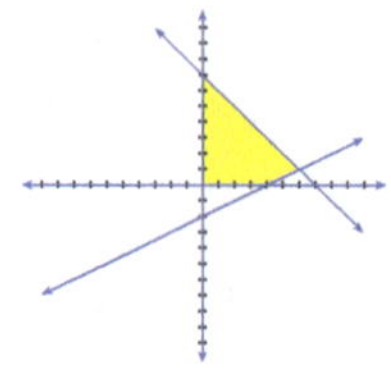

3. Maximum at (0, 1) of 1
Minimum at (3, 0) of –6

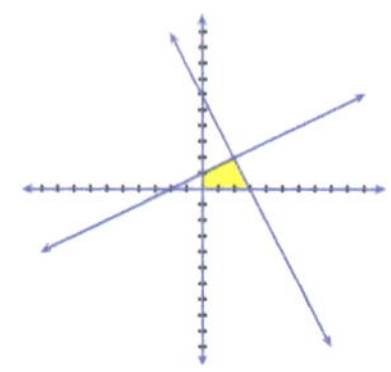

4. Maximum at (9, 0) of 90
Minimum at $(\frac{6}{5}, \frac{39}{5})$ of –42.6

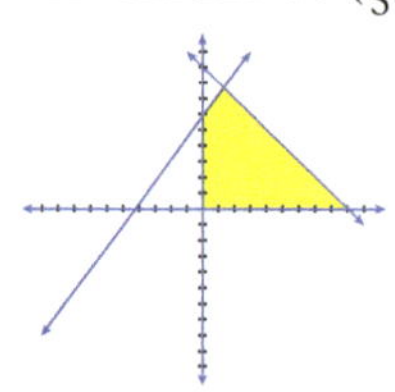

5. Maximum at (0, 3) of 18
Minimum at $(\frac{3}{8}, 0)$ of –3

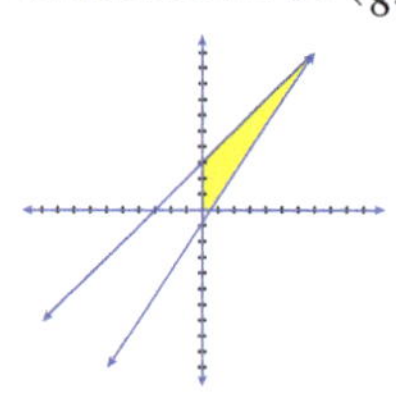

6. Maximum at (2, 8) of 16
Minimum at (–2, –1) of –9

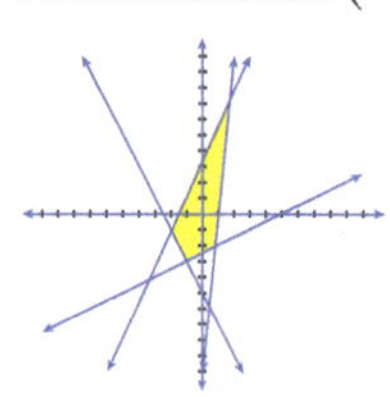

7. Maximum at (3, 8) of 37
Minimum at (1, –2) of –11

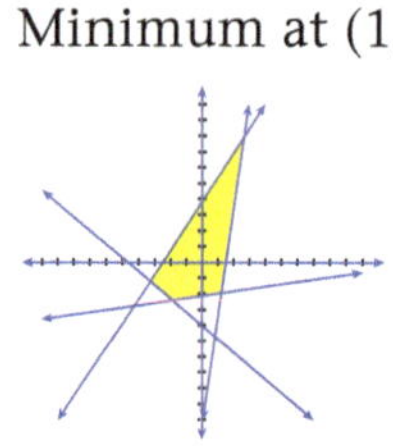

8. Maximum at (2, 7) of 17
Minimum at (1, –2) of –8

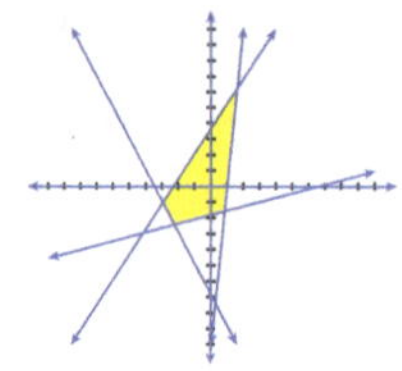

9. Maximum at (1, 1) of 11
Minimum at (–3, –3) of –33

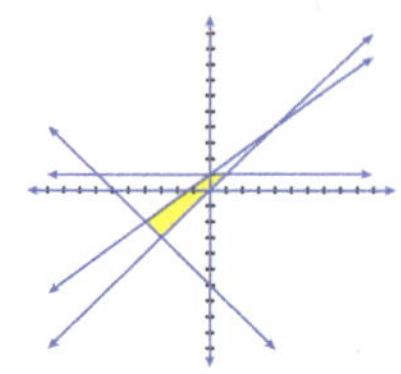

10. Maximum at (0, 1) of 8
Minimum at (4, –2) of –52

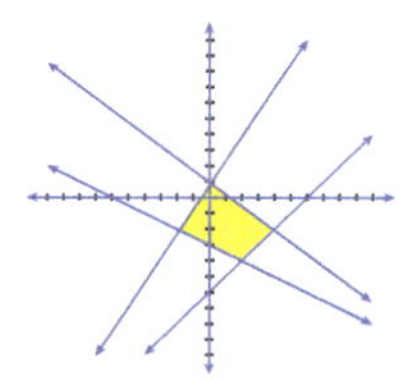

11. Maximum at (–5, –4) of 41
Minimum at (3, 1) of –26

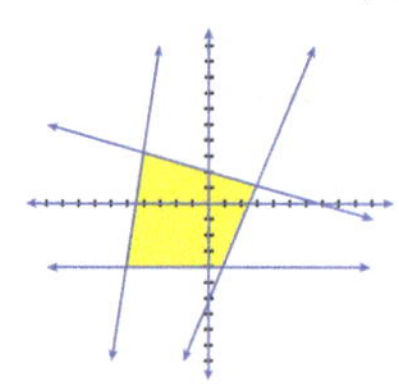

12. Maximum at (4, 4) of 8
Minimum at (–1, –1) of –2

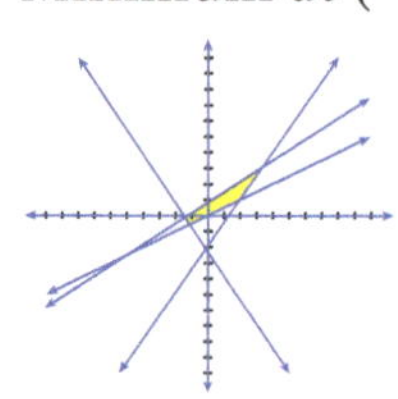

13. Maximum at (0, 3) of 21
Minimum at (4.5, –3) of –48

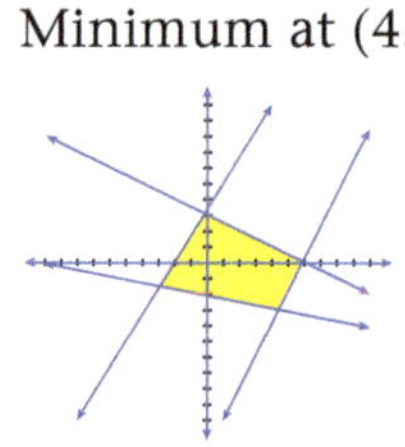

14. Maximum at (–1, 3) of 34
Minimum at (1, –6) of –61

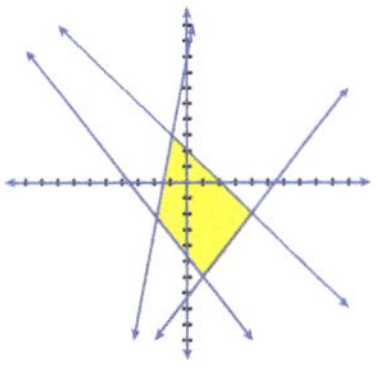

15. Maximum at (–2, 6) of 48
Minimum at (1, –4) of –33

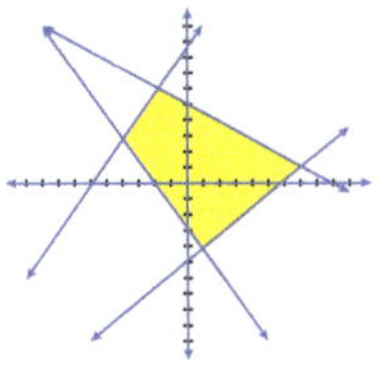

16. Maximum at (0, –1) of –1
Minimum at (–7, –4) of –67

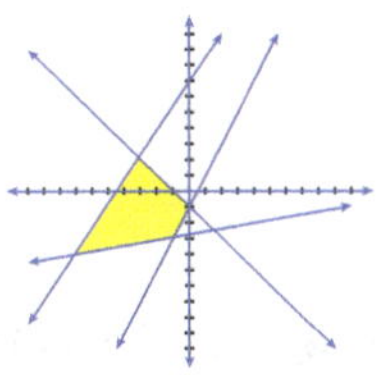

17. Maximum at (5, 4) of 53
Minimum at (–5, –4) of –53

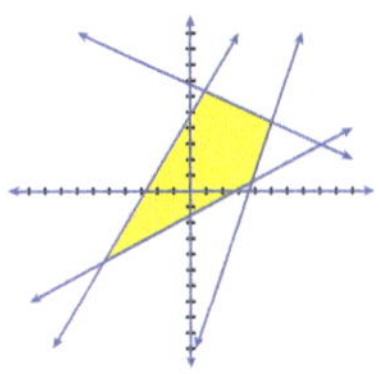

18. Maximum at (–1, 1) of 11
Minimum at (2, –3) of –28

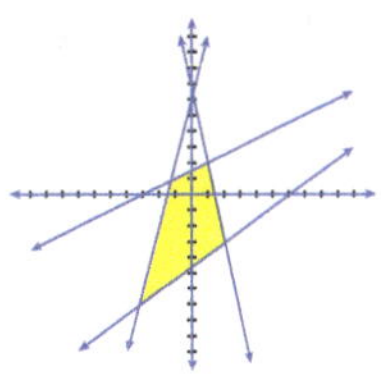

19. Maximum at (4, 3) of 7
Minimum at (–3, –2) of –5

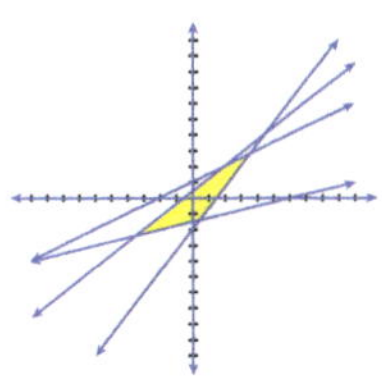

20. No maximum
Minimum at (0.5, –1.5) of –5.5

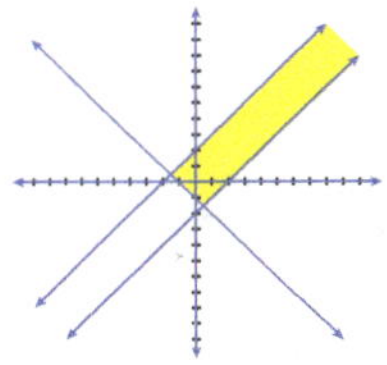

21. No maximum
Minimum at (1.25, 2.375) of –3.125

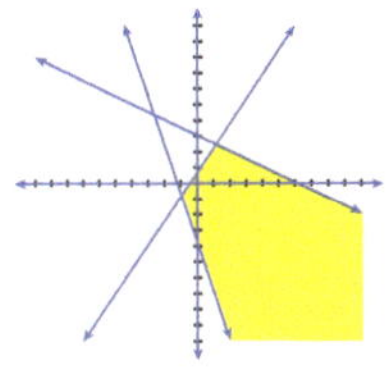

22. No maximum
No minimum

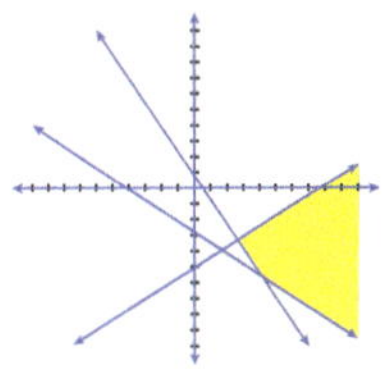

23. No maximum
Minimum at $(\frac{3}{8}, \frac{-13}{4})$ of –27.5

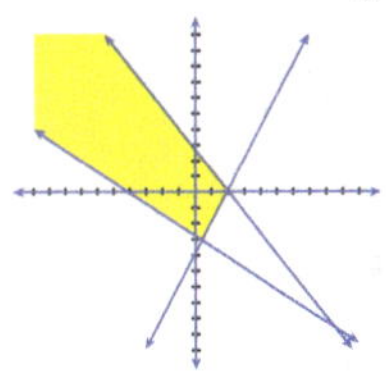

24. No maximum
Minimum at (–7, –1.5) of 14

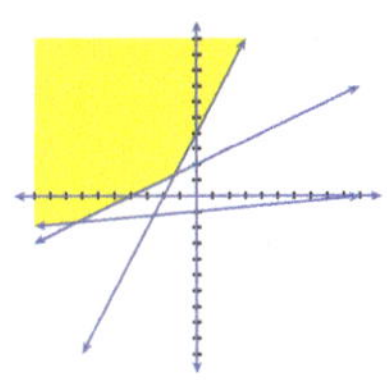

25. Maximum at $(\frac{20}{19}, \frac{45}{19})$ and (2, 0) of 5
Minimum at (0, 0) of 0

26. Maximum at (1, –1) of 0
Minimum at (–4, –6) of –90

27. Maximum at (–3, 5) of 49
Minimum at (9, 3) of –57

28. Maximum at (0, 4) of 16
Minimum at (1, 0) of 3

29. $c \geq 0$
$t \geq 0$
$c + t \leq 1000$
$c \geq t + 200$
The maximum profit comes when 600 cars and 400 trucks are manufactured for a total of $1,200,000 in profit.

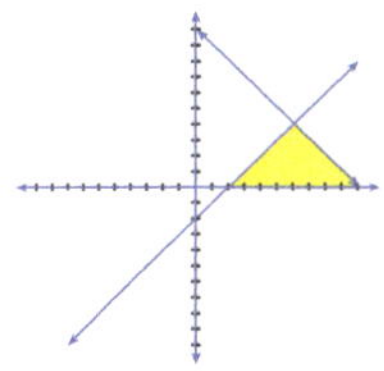

30. $b \geq 0$
$c \geq 10$
$b + c \leq 100$
The maximum profit comes when 90 acres of beans and 10 acres of corn are planted for a total profit of $245,000.

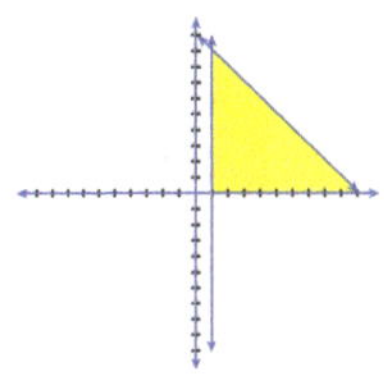

31. $x \geq 1000$
$y \geq 2000$
$x + y \leq 10000$
$x + y \geq 9000$
$y \geq 2x$
The minimum gasoline cost is when 3,000 gallons are purchased from Refinery X and 6,000 gallons are purchased from Refinery Y for a total cost of $10,200.

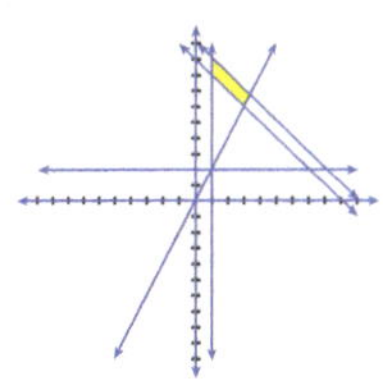

32. $t \geq 0$

$c + t \leq 8000$

$\frac{c}{12000} + \frac{t}{6000} \leq 1$ or

$t \leq -\frac{1}{2}c + 6000$

The maximum profit is when the plant manufactures 4,000 cars and 4,000 trucks for a total profit of \$20,000,000.

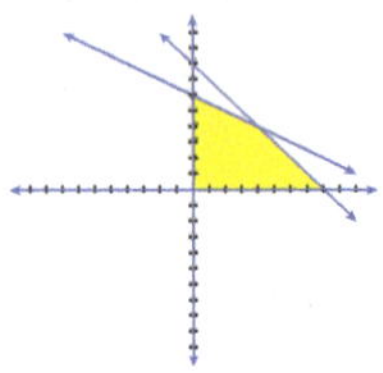

33. $c \geq 0$

$b \geq 0$

$c + b \leq 9000$

$400c + 200b \leq 2400000$ or

$c \leq \frac{-1}{2}b + 6000$

The maximum profit comes when 6,000 acres of beets and 3,000 acres of corn are planted for a total profit of \$34,950,000.

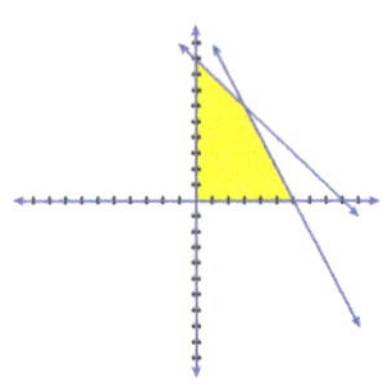

34. Maximum profit of \$4000 per month when 800 of toy 1 are produced/sold and 400 of toy 2 are produced/sold.

CHAPTER 2

Section 2.1

1. 2×2
2. 3×3
3. 3×4
4. 4×3
5. 4×4
6. 3×3
7. 2×6
8. 2×1
9. 1×2
10. 6×2
11. $\begin{bmatrix} 6 & 9 \\ 15 & 12 \end{bmatrix}$
12. $\begin{bmatrix} 1 & 5 \\ 8 & 0 \end{bmatrix}$
13. $\begin{bmatrix} -8 & 1 \\ -2 & -1 \end{bmatrix}$
14. $\begin{bmatrix} 13 & -2 \\ -3 & 4 \end{bmatrix}$
15. $\begin{bmatrix} 15 & 12 \\ 30 & 3 \end{bmatrix}$
16. $\begin{bmatrix} -9 & -10 \\ -29 & 24 \end{bmatrix}$
17. $\begin{bmatrix} -6 & 3 \\ 5 & 4 \end{bmatrix}$
18. $\begin{bmatrix} -1 & 22 \\ 35 & -4 \end{bmatrix}$
19. $\begin{bmatrix} -18 & -1 \\ -15 & 16 \end{bmatrix}$
20. $\begin{bmatrix} -13 & 17 \\ 20 & -2 \end{bmatrix}$
21. $\begin{bmatrix} 12 & 6 & 9 \\ 15 & -18 & 12 \\ 6 & -9 & 12 \end{bmatrix}$
22. $\begin{bmatrix} 9 & -1 & 5 \\ 1 & -4 & 1 \\ 1 & -6 & 2 \end{bmatrix}$
23. $\begin{bmatrix} 0 & -1 & 1 \\ 0 & 1 & 1 \\ 2 & 1 & -8 \end{bmatrix}$
24. $\begin{bmatrix} -16 & 11 & 3 \\ 23 & -15 & -8 \\ -11 & 10 & 32 \end{bmatrix}$
25. $\begin{bmatrix} 6 & -4 & 10 \\ 2 & -2 & -14 \\ -8 & -2 & 4 \end{bmatrix}$
26. $\begin{bmatrix} -11 & 10 & -15 \\ 2 & -3 & 14 \\ 3 & 2 & 22 \end{bmatrix}$
27. $\begin{bmatrix} -5 & 11 & -15 \\ 2 & -6 & 22 \\ 8 & -3 & 22 \end{bmatrix}$
28. $\begin{bmatrix} 0 & 9 & -1 \\ 13 & -13 & 18 \\ 9 & -2 & 8 \end{bmatrix}$
29. $\begin{bmatrix} 36 & -16 & 24 \\ -9 & 0 & -11 \\ 1 & -19 & -24 \end{bmatrix}$
30. $\begin{bmatrix} 13 & 2 & 20 \\ 21 & -19 & 4 \\ 5 & -5 & -6 \end{bmatrix}$
31. $\begin{bmatrix} 0 & 7 \\ 4 & 2 \\ -3 & 1 \end{bmatrix}$
32. not possible, matrices are different sizes
33. not possible, matrices are different sizes
34. $\begin{bmatrix} 0 & 1 & 1 \\ -4 & 4 & -5 \end{bmatrix}$
35. $\begin{bmatrix} 1 & 2 \\ -2 & 1 \\ 0 & 0 \end{bmatrix}$
36. not possible, matrices are different sizes
37. $\begin{bmatrix} 4 & 0 \\ -3 & 4 \\ 2 & -4 \end{bmatrix}$
38. $\begin{bmatrix} 0 & -4 \\ 1 & 4 \\ 1 & -5 \end{bmatrix}$
39. $\begin{bmatrix} 7 & 5 & 3 \\ -3 & 9 & -11 \end{bmatrix}$
40. $\begin{bmatrix} 0 & -4 \\ 1 & 4 \\ 1 & -5 \end{bmatrix}$

41. not possible, matrices are different sizes
42. not possible, matrices are different sizes
43. $\begin{bmatrix} 5 \\ 3 \\ 1 \end{bmatrix}$
44. $\begin{bmatrix} -1 & 5 & 5 \end{bmatrix}$
45. not possible, matrices are different sizes
46. $\begin{bmatrix} 5 & 5 & 5 \end{bmatrix}$
47. $\begin{bmatrix} 1 & 1 & 0 \end{bmatrix}$
48. $\begin{bmatrix} 4 \\ -2 \\ 2 \end{bmatrix}$
49. $\begin{bmatrix} 3 & 7 & 6 \end{bmatrix}$
50. $\begin{bmatrix} -1 \\ 5 \\ 5 \end{bmatrix}$

Section 2.2

1. $\begin{bmatrix} -6 \\ -3 \end{bmatrix}$
2. not possible to multiply
3. $\begin{bmatrix} 24 & 42 \\ 12 & 21 \end{bmatrix}$
4. $\begin{bmatrix} 45 \end{bmatrix}$
5. $\begin{bmatrix} 3 & 73 & 53 \end{bmatrix}$
6. not possible to multiply
7. $\begin{bmatrix} -46 & -30 & -40 \\ 31 & 9 & 19 \end{bmatrix}$
8. not possible to multiply
9. $\begin{bmatrix} 36 & 24 \\ 24 & 16 \end{bmatrix}$
10. $\begin{bmatrix} -58 \\ -64 \\ 20 \end{bmatrix}$
11. not possible to multiply
12. $\begin{bmatrix} -63 & 69 \\ -56 & 80 \\ 0 & 30 \end{bmatrix}$
13. $\begin{bmatrix} 36 & 24 \\ -12 & -8 \end{bmatrix}$
14. $\begin{bmatrix} -63 & -56 & 0 \\ 69 & 80 & -30 \end{bmatrix}$
15. not possible to multiply
16. $\begin{bmatrix} 42 & -66 \end{bmatrix}$
17. not possible to multiply
18. $\begin{bmatrix} 49 & -7 \\ 0 & 36 \end{bmatrix}$
19. $\begin{bmatrix} 36 & -12 \\ -12 & 4 \end{bmatrix}$
20. not possible to multiply
21. $\begin{bmatrix} 1 & 0 \\ 0 & 1 \end{bmatrix}$
22. $\begin{bmatrix} 1 & 0 \\ 0 & 1 \end{bmatrix}$
23. $\begin{bmatrix} -60 & 23 \\ -7 & 3 \\ -12 & 5 \end{bmatrix}$
24. not possible to multiply
25. $\begin{bmatrix} -5 & 5 & 10 \\ 2 & -16 & 2 \\ 46 & 19 & 23 \end{bmatrix}$
26. not possible to multiply
27. $\begin{bmatrix} 35 & 22 \\ 12 & 7 \\ 17 & 10 \end{bmatrix}$
28. $\begin{bmatrix} 77 & 45 & 8 \\ 45 & 42 & -13 \\ 8 & -13 & 29 \end{bmatrix}$
29. not possible to multiply
30. $\begin{bmatrix} -20 & 55 \\ 28 & -56 \\ -16 & 49 \end{bmatrix}$
31. $x = 11, y = 36$
32. $X = \begin{bmatrix} 4 & 2 & -1 \\ 9 & -9 & 8 \end{bmatrix}$
33. $X = \begin{bmatrix} 9 & -10 \\ -11 & 12 \end{bmatrix}$
34. $X = \begin{bmatrix} 9 & 4 \\ -7 & -2 \end{bmatrix}$
35. $X = \begin{bmatrix} -4 & -7 \\ -2 & 12 \end{bmatrix}$
36. $x = 5$
37. $x = -5$
38. $x = -4$
39. $x = -7$

Section 2.3

1. $\left[\begin{array}{ccc|c} -4 & 8 & -9 & 61 \\ -5 & 9 & 4 & 55 \\ -6 & -2 & -1 & -5 \end{array}\right]$
2. $\left[\begin{array}{cc|c} 5 & 1 & 37 \\ -1 & 3 & 49 \end{array}\right]$
3. $\left[\begin{array}{ccc|c} 1 & 1 & -2 & 3 \\ 0 & 1 & 5 & 1 \\ 1 & 0 & -1 & 5 \end{array}\right]$
4. $\left[\begin{array}{cc|c} 3 & -1 & 4 \\ 1 & 2 & 5 \end{array}\right]$
5. $\left[\begin{array}{cc|c} 1 & 0 & 4 \\ 0 & 1 & -3 \end{array}\right], (4, -3)$
6. $\left[\begin{array}{cc|c} 1 & 0 & -5 \\ 0 & 1 & 4 \end{array}\right], (-5, 4)$
7. $\left[\begin{array}{cc|c} 1 & 0 & 2 \\ 0 & 1 & 6 \end{array}\right], (2, 6)$
8. $\left[\begin{array}{cc|c} 1 & 0 & 3 \\ 0 & 1 & 2 \end{array}\right], (3, 2)$
9. $\left[\begin{array}{cc|c} 1 & 0 & 9 \\ 0 & 1 & 6 \end{array}\right], (9, 6)$
10. $\left[\begin{array}{cc|c} 1 & 0 & 4 \\ 0 & 1 & 1 \end{array}\right], (4, 1)$
11. $\left[\begin{array}{cc|c} 1 & 0 & -2 \\ 0 & 1 & 3 \end{array}\right], (-2, 3)$
12. $\left[\begin{array}{cc|c} 1 & 0 & 3 \\ 0 & 1 & 2 \end{array}\right], (3, 2)$
13. no solution
14. $\left[\begin{array}{cc|c} 1 & 0 & -2 \\ 0 & 1 & 5 \end{array}\right], (-2, 5)$

15. $\left[\begin{array}{cc|c}1&0&-2\\0&1&7\end{array}\right], (-2,7)$

16. $\left[\begin{array}{cc|c}1&0&\frac{46}{31}\\0&1&\frac{-857}{217}\end{array}\right], \left(\frac{46}{31}, \frac{-857}{217}\right)$

17. $\left[\begin{array}{cc|c}1&0&5\\0&1&-4\end{array}\right], (5,-4)$

18. no solution

19. $\left(\frac{-34}{5}, \frac{-7}{5}\right)$

20. $\left[\begin{array}{cc|c}1&0&-2\\0&1&-3\end{array}\right], (-2,-3)$

21. same line

22. same line

23. $\left[\begin{array}{cc|c}1&0&\frac{17}{14}\\0&1&-\frac{69}{14}\end{array}\right], \left(\frac{17}{14}, -\frac{69}{14}\right)$

28. $\left[\begin{array}{cc|c}1&0&\frac{9}{10}\\0&1&\frac{37}{20}\end{array}\right], \left(\frac{9}{10}, \frac{37}{20}\right)$

25. no solution

26. no solution

27. same line

28. same line

29. no solution

30. $\left[\begin{array}{ccc|c}1&0&0&1\\0&1&0&-2\\0&0&1&2\end{array}\right], (1,-2,2)$

31. $\left[\begin{array}{ccc|c}1&0&0&10\\0&1&0&7\\0&0&1&1\end{array}\right], (10,7,1)$

32. $\left[\begin{array}{ccc|c}1&0&0&1\\0&1&0&-5\\0&0&1&-6\end{array}\right], (1,-5,-6)$

33. $\left[\begin{array}{ccc|c}1&0&0&3\\0&1&0&1\\0&0&1&2\end{array}\right], (3,1,2)$

34. no solution

35. no solution

36. $\left[\begin{array}{ccc|c}1&0&0&1\\0&1&0&-4\\0&0&1&2\end{array}\right], (1,-4,2)$

37. $\left[\begin{array}{ccc|c}1&0&0&-5\\0&1&0&2\\0&0&1&-2\end{array}\right], (-5,2,-2)$

38. Many solutions, $(5a-2, 4a-3, a)$

39. $\left(\frac{1}{7}a+\frac{1}{7}, \frac{11}{7}a-\frac{3}{7}, a\right)$

40. $\left[\begin{array}{ccc|c}1&0&0&5\\0&1&0&-1\\0&0&1&3\end{array}\right], (5,-1,3)$

41. $\left[\begin{array}{ccc|c}1&0&0&-3\\0&1&0&4\\0&0&1&1\end{array}\right], (-3,4,1)$

42. $(7-a, a-4, a)$

43. $(-a-3, 2a+4, a)$

44. no solution

45. no solution

46. $\left(\frac{5}{2}a-4, a, -5\right)$

47. $\left(\frac{15}{7}, -\frac{12}{7}, -\frac{13}{7}\right)$

48. $(1, 1, -1)$

49. $\left(\frac{26}{5}, \frac{-18}{5}, -4\right)$

50. no solution

51. no solution

52. $\left(\frac{1}{2}, \frac{1}{3}, \frac{1}{6}\right)$

53. (13a + 3, 8a + 1, a)

54. $(3, 4, -1)$

55. $\left[\begin{array}{ccc|c}1&0&0&\frac{3}{7}\\0&1&0&\frac{5}{7}\\0&0&1&\frac{2}{7}\end{array}\right], \left(\frac{3}{7}, \frac{5}{7}, \frac{2}{7}\right)$

56. $\left[\begin{array}{ccc|c}1&0&0&\frac{6}{7}\\0&1&0&-\frac{27}{35}\\0&0&1&\frac{32}{35}\end{array}\right], \left(\frac{6}{7}, -\frac{27}{35}, \frac{32}{35}\right)$

Section 2.4

1. $\frac{1}{51}\begin{bmatrix}9&-6\\4&3\end{bmatrix} = \begin{bmatrix}\frac{3}{17}&\frac{-2}{17}\\\frac{4}{51}&\frac{1}{17}\end{bmatrix}$

2. $-\frac{1}{32}\begin{bmatrix}0&-8\\-4&9\end{bmatrix} = \begin{bmatrix}0&\frac{1}{4}\\\frac{1}{8}&-\frac{9}{32}\end{bmatrix}$

3. no inverse exists, using Gauss-Jordan elimination on an augmented matrix results in a row of all zeros for the left portion of the matrix, or, if using the formula, the determinant value, $ad-bc$, is zero so thedivision by this value fails

4. $\frac{1}{7}\begin{bmatrix}-3&-2\\-4&-5\end{bmatrix} = \begin{bmatrix}-\frac{3}{7}&-\frac{2}{7}\\-\frac{4}{7}&-\frac{5}{7}\end{bmatrix}$

5. not a square matrix, no inverse exists

6. $-\frac{1}{14}\begin{bmatrix}3&-5\\2&-8\end{bmatrix} = \begin{bmatrix}-\frac{3}{14}&\frac{5}{14}\\-\frac{1}{7}&\frac{4}{7}\end{bmatrix}$

7. $\begin{bmatrix}3&-4\\-5&7\end{bmatrix}$

8. $\frac{1}{3}\begin{bmatrix}1&-2\\-2&7\end{bmatrix} = \begin{bmatrix}\frac{1}{3}&-\frac{2}{3}\\-\frac{2}{3}&\frac{7}{3}\end{bmatrix}$

9. $\frac{1}{27}\begin{bmatrix}7&-5\\4&1\end{bmatrix} = \begin{bmatrix}\frac{7}{27}&-\frac{5}{27}\\\frac{4}{27}&\frac{1}{27}\end{bmatrix}$

10. $-\frac{1}{4}\begin{bmatrix}5&7\\7&9\end{bmatrix} = \begin{bmatrix}-\frac{5}{4}&-\frac{7}{4}\\-\frac{7}{4}&-\frac{9}{4}\end{bmatrix}$

11. $$\frac{1}{15}\begin{bmatrix} 2 & -\frac{1}{2} \\ -6 & 9 \end{bmatrix} = \begin{bmatrix} \frac{2}{15} & -\frac{1}{30} \\ -\frac{2}{5} & \frac{3}{5} \end{bmatrix}$$

12. $$\frac{1}{24}\begin{bmatrix} 4 & -6 \\ -\frac{2}{3} & 7 \end{bmatrix} = \begin{bmatrix} \frac{1}{6} & -\frac{1}{4} \\ -\frac{1}{36} & \frac{7}{24} \end{bmatrix}$$

13. $$\frac{1}{10}\begin{bmatrix} 2 & 0 \\ -4 & 5 \end{bmatrix} = \begin{bmatrix} \frac{1}{5} & 0 \\ -\frac{2}{5} & \frac{1}{2} \end{bmatrix}$$

14. no inverse exists, using Gauss-Jordan elimination on an augmented matrix results in a row of all zeros for the left portion of the matrix, or, if using the formula, the determinant value, $ad - bc$, is zero so the division by this value fails

15. $$-\frac{4}{5}\begin{bmatrix} -\frac{1}{4} & -\frac{2}{3} \\ -\frac{3}{4} & 3 \end{bmatrix} = \begin{bmatrix} \frac{1}{5} & \frac{8}{15} \\ \frac{3}{5} & -\frac{12}{5} \end{bmatrix}$$

16. $$\begin{bmatrix} -2 & -1 & 0 \\ 34 & 10 & -1 \\ -31 & -9 & 1 \end{bmatrix}$$

17. $$\frac{1}{2}\begin{bmatrix} -12 & 10 & 18 \\ -25 & 21 & 37 \\ -9 & 7 & 13 \end{bmatrix} = \begin{bmatrix} -6 & 5 & 9 \\ -\frac{25}{2} & \frac{21}{2} & \frac{37}{2} \\ -\frac{9}{2} & \frac{7}{2} & \frac{13}{2} \end{bmatrix}$$

18. no inverse exists, using Gauss-Jordan elimination on an augmented matrix results in a row of all zeros for the left portion of the matrix

19. $$\begin{bmatrix} 2 & -1 & 0 \\ -1 & 2 & -1 \\ 0 & -1 & 1 \end{bmatrix}$$

20. $$\begin{bmatrix} \frac{5}{4} & -\frac{7}{4} & 1 \\ -\frac{1}{2} & \frac{3}{2} & -1 \\ 0 & -1 & 1 \end{bmatrix}$$

21. $$\begin{bmatrix} 1 & 2 & -2 \\ -1 & 6 & -5 \\ 0 & -1 & 1 \end{bmatrix}$$

22. $$\begin{bmatrix} 1 & -3 & 1 \\ 1 & -2 & 5 \\ 1 & -2 & 6 \end{bmatrix}$$

23. no inverse exists, using Gauss-Jordan elimination on an augmented matrix results in a row of all zeros for the left portion of the matrix

24. $$\begin{bmatrix} 1 & 1 & 1 \\ 4 & 5 & 5 \\ 5 & 8 & 9 \end{bmatrix}$$

25. $x = \frac{23}{10}, y = \frac{1}{10}$

26. $x = \frac{5}{7}, y = \frac{25}{7}$

27. (10, 20)

28. (0, 1, 2)

29. (1, 2, –1)

30. (8, –6, –2)

Section 2.5

1. original message is

$$\begin{bmatrix} 20 & 19 & 19 & 0 & 3 & 20 \\ 8 & 0 & 0 & 19 & 18 & 0 \\ 9 & 9 & 1 & 5 & 5 & 0 \end{bmatrix},$$

encrypted message is

$$\begin{bmatrix} 31 & 46 & 22 & -23 & -18 & 20 \\ -94 & -130 & -82 & 65 & 48 & -80 \\ 62 & 75 & 59 & -28 & -17 & 60 \end{bmatrix}$$

2. the inverse of the encryption matrix is

$$\begin{bmatrix} \frac{2}{3} & \frac{2}{3} & 1 \\ \frac{10}{3} & \frac{7}{3} & 2 \\ \frac{7}{3} & \frac{4}{3} & 1 \end{bmatrix}$$

decrypted message is

$$\begin{bmatrix} 4 & 14 & 0 & 12 & 1 & 15 \\ 15 & 15 & 20 & 12 & 14 & 14 \\ 0 & 20 & 5 & 0 & 25 & 5 \end{bmatrix},$$

or "Do not tell anyone"

3. original message is

$$\begin{bmatrix} 9 & 15 & 0 & 9 & 13 & 19 & 5 \\ 7 & 18 & 20 & 19 & 5 & 1 & 0 \\ 14 & 5 & 8 & 0 & 19 & 7 & 0 \end{bmatrix}$$

encrypted message is

$$\begin{bmatrix} -30 & -7 & 4 & 10 & -46 & -32 & -5 \\ 20 & -4 & -52 & -30 & 43 & 61 & 15 \\ 23 & -11 & -24 & -29 & 41 & 31 & 5 \end{bmatrix}$$

4. the inverse of the encryption matrix is

$$\begin{bmatrix} -\frac{4}{5} & \frac{2}{5} & -1 \\ -1 & 0 & -1 \\ -\frac{3}{5} & -\frac{1}{5} & 0 \end{bmatrix}$$

decrypted message is

$$\begin{bmatrix} 13 & 19 & 5 & 1 & 2 & 14 & 5 \\ 5 & 1 & 0 & 19 & 5 & 0 & 14 \\ 19 & 7 & 8 & 0 & 5 & 19 & 20 \end{bmatrix},$$

or "Message has been sent"

5. original message is

$$\begin{bmatrix} 13 & 20 & 5 & 20 & 23 \\ 5 & 0 & 0 & 0 & 15 \\ 5 & 13 & 1 & 20 & 0 \end{bmatrix}$$

encrypted message is

$$\begin{bmatrix} -36 & -27 & -9 & -20 & -91 \\ 44 & 34 & 13 & 20 & 114 \\ 11 & 14 & 8 & 0 & 31 \end{bmatrix}$$

6. the inverse of the encryption matrix is

$$\begin{bmatrix} -8 & -7 & 3 \\ 2 & 2 & -1 \\ -9 & -8 & 3 \end{bmatrix}$$

decrypted message is

$$\begin{bmatrix} 14 & 13 & 5 & 20 & 20 & 8 & 5 \\ 15 & 1 & 0 & 0 & 0 & 18 & 0 \\ 0 & 11 & 9 & 1 & 20 & 5 & 0 \end{bmatrix},$$

or "No make it at three"

7. 19 student tickets and 37 adult tickets

8. 771 tickets for the first movie and 871 tickets for the second movie
9. 44 cakes, 65 pies, and 13 boxes of cookies
10. 100 acres of beans, 300 acres of rice, 600 acres of corn
11. 10 sets of resistors, 20 sets of transistors, and 40 sets of computer chips
12. 6 gallons of solution A and 4 gallons of solution B
13. 14 pounds of peanuts, 24 pounds of almonds, and 12 pounds of pistachios
14. 18 pounds of peanuts, 9 pounds of cashews, and 13 pounds of pistachios
15. 25 gallons spray X, 2 gallons spray Y, 16 gallons spray Z
16. 3 pounds commercial brand X, 5 pounds commercial brand Y, 4 pounds commercial brand Z
17. \$7,500 in company A, \$2,500 in company B
18. \$20,000 in bonds, \$5,000 in stocks
19. \$42,727.28 in the first fund, \$33,636.36 in the second fund, \$23,636.36 in the third fund
20. \$360,000 in municipal bonds, \$160,000 in blue chip stocks, \$80,000 in growth or speculative stocks
21. \$500,000 in the first fund, \$375,000 in the second fund, \$125,000 in the third fund
22. \$90,000 in the first lender, \$15,000 in the second lender, \$45,000 in the third lender
23. \$600,000 in the first lender, \$300,000 in the second lender, \$100,000 in the third lender
24. 10 years old
25. 15 nickels, 5 dimes, 30 quarters
26. 4 mocha lattes, 3 regular coffees, and 5 danishes
27. 9
28. a. $\begin{bmatrix} 0 & 1 & 1 & 1 \\ 1 & 0 & 1 & 0 \\ 1 & 1 & 0 & 0 \\ 1 & 0 & 0 & 0 \end{bmatrix}$
29. a. $\begin{bmatrix} 0 & 1 & 1 & 1 & 1 \\ 1 & 0 & 1 & 0 & 0 \\ 1 & 1 & 0 & 1 & 0 \\ 1 & 0 & 1 & 0 & 0 \\ 1 & 0 & 0 & 0 & 0 \end{bmatrix}$

    b. $\begin{bmatrix} 4 & 6 & 6 & 6 & 4 \\ 6 & 2 & 5 & 2 & 1 \\ 6 & 5 & 4 & 5 & 2 \\ 6 & 2 & 5 & 2 & 1 \\ 4 & 1 & 2 & 1 & 0 \end{bmatrix}$

    c. 3
30. a. $A = \begin{bmatrix} 0 & 1 & 1 & 0 & 1 & 0 \\ 1 & 0 & 0 & 0 & 0 & 0 \\ 1 & 0 & 0 & 1 & 0 & 0 \\ 0 & 0 & 1 & 0 & 0 & 1 \\ 1 & 0 & 0 & 0 & 0 & 1 \\ 0 & 0 & 0 & 1 & 1 & 0 \end{bmatrix}$

    b. $\begin{bmatrix} 0 & 3 & 4 & 1 & 4 & 1 \\ 3 & 0 & 0 & 1 & 0 & 1 \\ 4 & 0 & 0 & 3 & 1 & 1 \\ 1 & 1 & 3 & 0 & 1 & 3 \\ 4 & 0 & 1 & 1 & 0 & 3 \\ 1 & 1 & 1 & 3 & 3 & 0 \end{bmatrix}$

    c. 1
31. 2000 tons L
    4000 tons M
    5000 tons H
32. 10, 12, 14
33. 8, 12, 22
34. 100 front seats, 300 main seats, 100 balcony seats

CHAPTER 3

Section 3.1

1. $x = \pm 7$
2. $x = \pm 4$
3. $x = \pm 2$
4. $x = \pm 5$
5. $x = \pm\sqrt{15}$
6. $x = \pm 4$
7. $x = \pm 3$
8. $x = \pm\sqrt{10}$
9. $x = \pm\frac{4}{3}$
10. $x = 1 \pm\sqrt{6}$
11. $x = 7, -3$
12. $x = -4, 2$
13. $x = \pm 4$
14. $x = 5, -9$
15. $x = 19, 7$
16. $x = \pm\frac{2}{5}$
17. $x = -3, 4$
18. $x = -5, -2$
19. $x = 3, 8$
20. $x = -7, 9$
21. $x = -\frac{1}{2}, 2$
22. $x = -4, \frac{5}{3}$
23. $x = -\frac{3}{2}, \frac{2}{3}$
24. $x = -2, \frac{1}{2}$
25. $x = \frac{2}{5}, -2$
26. $x = 6, -\frac{1}{2}$
27. $x = \frac{2}{3}, -\frac{3}{4}$
28. $x = 3, -8$
29. $x = -\frac{4}{5}, \frac{3}{2}$
30. $x = -2, \frac{4}{5}$
31. $x = -2, 4$
32. $x = 2 \pm \sqrt{13}$

33. $x = -3 \pm \sqrt{5}$
34. $x = -\frac{1}{2} \pm \frac{\sqrt{17}}{2}$
35. $x = \frac{3}{2} \pm \frac{\sqrt{5}}{2}$
36. $x = \frac{1}{2} \pm \frac{\sqrt{69}}{6}$
37. $x = -\frac{5}{4} \pm \frac{\sqrt{17}}{4}$
38. $x = \frac{3}{10} \pm \frac{\sqrt{29}}{10}$
39. $x = 2, -1$
40. $x = \frac{-9}{2} \pm \sqrt{\frac{113}{2}}$
41. $x = \frac{-3}{4} \pm \frac{\sqrt{57}}{4}$
42. $x = \frac{11}{2} \pm \frac{\sqrt{143}}{2}$
43. $x = -\frac{1}{2} \pm \frac{\sqrt{5}}{2}$
44. $t = -5 \pm \sqrt{29}$
45. $y = 4 \pm \sqrt{31}$
46. $p = 2, -8$
47. $x = -\frac{1}{2} \pm \frac{\sqrt{41}}{2}$
48. $x = -\frac{1}{6} \pm \frac{\sqrt{28}}{12} = -\frac{1}{6} \pm \frac{\sqrt{7}}{6}$
49. $x = -\frac{3}{4} \pm \frac{\sqrt{17}}{4}$
50. $x = -4 \pm \frac{\sqrt{92}}{2} = -4 \pm \sqrt{23}$
51. $x = 1 \pm \frac{\sqrt{132}}{6} = 1 \pm \frac{\sqrt{33}}{3}$
52. $x = \frac{1}{6} \pm \frac{\sqrt{52}}{12} = \frac{1}{6} \pm \frac{\sqrt{13}}{6}$
53. $x = \frac{3}{2} \pm \frac{\sqrt{76}}{4} = \frac{3}{2} \pm \frac{\sqrt{19}}{2}$
54. $x = \frac{5}{4} \pm \frac{\sqrt{116}}{8} = \frac{5}{4} \pm \frac{\sqrt{29}}{4}$
55. $x = \frac{-2}{3} \pm \frac{\sqrt{13}}{3}$
56. $x = 1 \pm \sqrt{6}$
57. $y = \frac{2}{3}, \frac{3}{2}$
58. $x = 3 \pm \sqrt{5}$
59. $y = 0, -\frac{9}{14}$
60. $x = \frac{7}{2} \pm \frac{\sqrt{21}}{2}$
61. $x = \frac{1}{2} \pm \frac{\sqrt{65}}{2}$
62. $x = \frac{-1}{2} \pm \frac{\sqrt{5}}{2}$
63. $x = \frac{3}{2} \pm \frac{\sqrt{53}}{2}$
64. $x = -\frac{1}{2} \pm \frac{\sqrt{37}}{2}$
65. $x = 1 \pm \sqrt{3}$
66. $x = -3, -2$
67. $x = 3 \pm \sqrt{21}$
68. $x = \pm\sqrt{10}$
69. $x = -4 \pm \sqrt{6}$
70. $x = \frac{7}{2} \pm \frac{\sqrt{21}}{2}$
71. $x = -7, 8$
72. $x = \frac{-5}{3}, 2$
73. $x = -4 \pm \sqrt{6}$
74. $x = \frac{19}{2} \pm \frac{\sqrt{5}}{2}$
75. $x = 2 \pm \sqrt{8} = 2 \pm 2\sqrt{2}$
76. $x = -4, \frac{1}{2}$
77. $x = \frac{-17}{10}$
78. $y = 0, -\frac{8}{19}$
79. $x = 0, -1$
80. $x = 7, \frac{4}{3}$

Section 3.2

1. (0,–10), maximum
2. (–2,9), maximum
3. $\left(-\frac{1}{2}, -\frac{7}{4}\right)$, minimum
4. $\left(-\frac{3}{2}, -\frac{23}{4}\right)$, maximum
5. $\left(\frac{1}{4}, \frac{11}{2}\right)$, minimum
6. $\left(\frac{1}{2}, \frac{3}{4}\right)$, minimum
7. $\left(\frac{5}{14}, -\frac{59}{28}\right)$, maximum
8. $\left(\frac{1}{2}, -4\right)$, minimum
9. (–3,–9), maximum
10. (2,1), maximum
11. $\left(-\frac{1}{2}, \frac{55}{4}\right)$, minimum
12. (–1,7), minimum
13. $\left(\frac{17}{26}, -\frac{29}{52}\right)$, minimum
14. (–2,9), maximum
15. $\left(\frac{13}{12}, -\frac{145}{24}\right)$, minimum
16. $\left(-\frac{1}{3}, -\frac{4}{3}\right)$, minimum
17. (–3,–2), maximum
18. (–1,–1), maximum
19. $\left(\frac{1}{3}, \frac{2}{3}\right)$, minimum
20. $\left(\frac{3}{2}, \frac{57}{4}\right)$, maximum
21. (1,3), minimum
22. (–1,0), maximum
23. $\left(\frac{3}{2}, \frac{3}{4}\right)$, minimum
24. (–3,6), $x = -3$, concave up, no x-intercepts
25. (–9,9), $x = -9$, concave up, no x-intercepts

26. (1,8), $x = 1$, concave down, x-intercepts $\left(1 \pm \frac{2\sqrt{2}}{3},0\right)$
27. (1,0), $x = 1$, concave up, x-intercept (1,0)
28. (4,9), $x = 4$, concave down, x-intercepts (7,0) and (1,0)
29. (–1,–8), $x = -1$, concave up, x-intercepts $\left(-1 \pm \frac{2\sqrt{2}}{\sqrt{5}},0\right)$
30. (–8,–7), $x = -8$, concave up, x-intercepts $\left(-8 \pm \sqrt{\frac{7}{3}},0\right)$
31. (–7,–3), $x = -7$, concave down, no x-intercepts
32. (17,–8), $x = 17$, concave up, x-intercepts $\left(17 \pm \sqrt{2},0\right)$
33. (7,–16), $x = 7$, concave down, no x-intercepts
34. (16,–18), $x = 16$, concave up, x-intercepts $\left(16 \pm \frac{3\sqrt{26}}{13},0\right)$
35. (–5,15), $x = -5$, concave down, x-intercepts $\left(-5 \pm \frac{\sqrt{15}}{4},0\right)$
36. (11,–5), $x = 11$, concave up, x-intercepts $\left(11 \pm \sqrt{5},0\right)$
37. (4,7), $x = 4$, concave up, no x-intercepts
38. (12,7), $x = 12$, concave down, x-intercepts $\left(12 \pm \frac{\sqrt{14}}{6},0\right)$
39. (2,–3), $x = 2$, concave up, x-intercepts $\left(2 \pm \sqrt{\frac{3}{2}},0\right)$
40. (1,2), $x = 1$, concave up, no x-intercepts
41. (–10,1), $x = -10$, concave down, x-intercepts (–9,0) and (–11,0)
42. (3,2), $x = 3$, concave down, x-intercepts $\left(3 \pm \frac{\sqrt{6}}{3},0\right)$
43. (–1,–4), $x = -1$, concave down, no x-intercepts
44. time is $\frac{25}{8} = 3.125$ seconds high points is $\frac{625}{4} = 156.25$ feet
45. at time 0 the object is at (0,0), after one second the object is at (1,84), the average slope between these two points is $\frac{84-0}{1-0} = 84$ ft/sec
46. time is ≈ 9.42 seconds high point is ≈ 470.81 feet
47. at time 0 the object is at (0,0), after one second the object is at (1,94.69), the average slope between these two points is $\frac{94.69-0}{1-0} = 94.69$ ft/sec
48. number of items is 4825 profit is about $21,855.63
49. profit is $800 number of tickets is 100
50. People buying tickets is given by the function $B(x) = -0.01x + 9$, with x being the number of tickets sold. (When the price is $4.00 then 500 people will buy and the buyers decrease by 100 for every dollar increase.) The total sales is given by $xB(x) = x(-0.01x + 9)$. The maximum for this function is at $p = \$4.50$, with a total sales of $2,025.
51. The area is given by $A = x(160 - 2x)$, the maximum area is when the side parallel to the barn has a length of 80 feet and the other 2 sides are both 40 feet.
52. The area of both corrals is given by $A = x(270 - 3x)$, the maximum area is when the side parallel to the barn has a length of 135 feet and the other 3 sides are all 45 feet.
53. The area of the playing field is given by $A = w(300 - w)$, the maximum area is when sides are 150 feet in length making a square.
54. $y = 6x^2 + 2$

Section 3.3

1. $x \approx 2599$
2. 900
3. 2399 or 1
4. 50
5. 64 or 16
6. 80 or 20
7. 20 or 80
8. 17.25 or 2.4
9. $q \simeq 3.67$, $p \simeq \$23.49$
10. $q = 3$, $p = \$43$
11. $q = 6$, $p = \$16$
12. $q = 10$, $p = \$30$
13. $q = 9.5$, $p = \$186.25$
14. $514
15. $166,666.67
16. a. $P(x) = -60x^2 + 1500x - 4000$
    b. $5,375
17. 20
18. $420
19. a. $R(x) = 25x$
    b. $P(x) = -x^2 + 20x - 96$
    c. $(x, C) = (x, R) = (12, \$300)$ and $(8, \$200)$
    d. 10 units
20. about $82.02

21. a 44 units
    b. \$736
22. a. $P(x) = -x^2 + 19x - 84$
    b. $(x, C) = (x, R) = (7, \$175)$ and $(12, \$300)$
23. width = length = 40 ft.
24. a. 30 bicycles
    b. \$400
    c. 10 or 50 bicycles
25. 20 or 80 units
26. $q = 4$, $p = \$68$
27. $x = 105$ units with profit = \$8525

Section 3.4

1. a. $x^3 - 3x^2 + x + 1$
   b. 3
   c. 1
   d. up
2. a. $-2x^2 + 12x + 40$
   b. 2
   c. -2
   d. down
3. a. $-p^3 - 4p^2 + p + 11$
   b. 3
   c. -1
   d. down
4. a. $-3y^3 + 12y^2 - 15y + 12$
   b. 3
   c. -3
   d. down
5. a. $12x^3 - 4x^2 - 3x + 12$
   b. 3
   c. 12
   d. up
6. a. $-2x^3 - x^2 + 4x - 12$
   b. 3
   c. -2
   d. down
7. a. $-3x^9 + x^7 + x^4 - 3x^2 + 11$
   b. 9
   c. -3
   d. down
8. a. $-31y^5 + y^2 - 27y$
   b. 5
   c. -31
   d. down
9. a. $20x^6 + 14x^4 - 19x^3 - 31$
   b. 6
   c. 20
   d. up
10. a. $3x^7 - 4x^4 + 3x^2 - 14x - 5$
    b. 7
    c. 3
    d. up
11. a. 2
    b. $x = -\frac{1}{2}, 4$
    c. $(0,-4)$
    d. $\left(-\frac{1}{2},0\right)(4,0)$
    e.

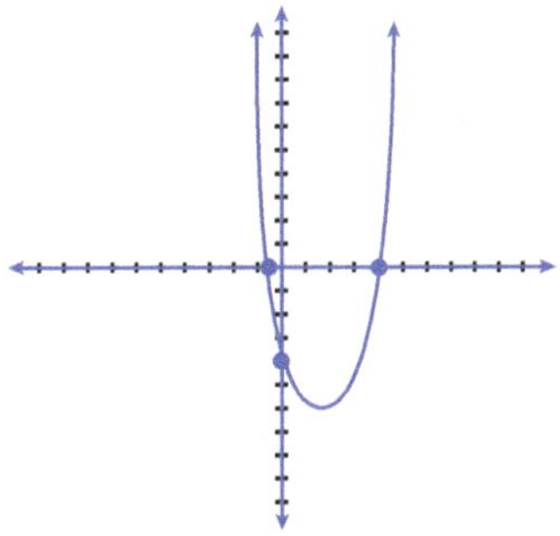

12. a. 2
    b. $x = \frac{1}{2}, -3$
    c. $(0,-3)$
    d. $\left(\frac{1}{2},0\right)(-3,0)$
    e.

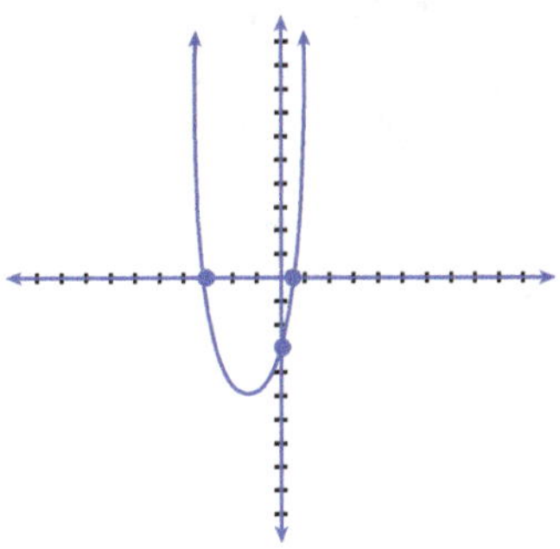

13. a. 3
    b. $x = 0, 5, -2$
    c. $(0,0)$
    d. $(0,0)$ $(5,0)$ $(-2,0)$
    e.

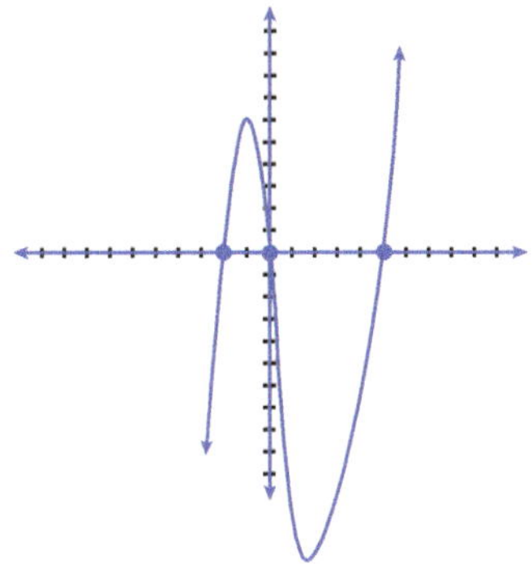

14. a. 2
    b. $x = 2, -1$
    c. $(0,-6)$
    d. $(2,0)$ $(-1,0)$
    e.

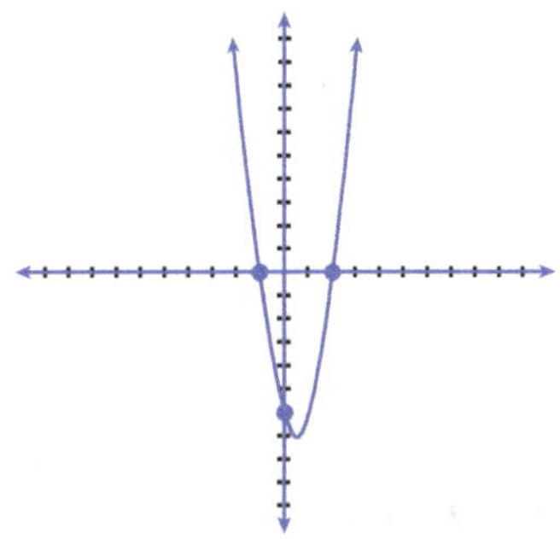

15. a. 3
    b. $x = 0, 1, -1$
    c. $(0,0)$
    d. $(0,0)$ $(1,0)$ $(-1,0)$
    e.

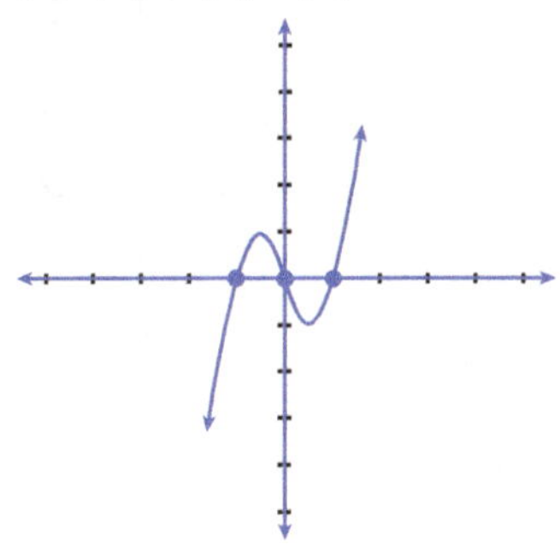

16. a. 4
 b. $x = 2, -2$
 c. (0,16)
 d. (2,0) (–2,0)
 e.

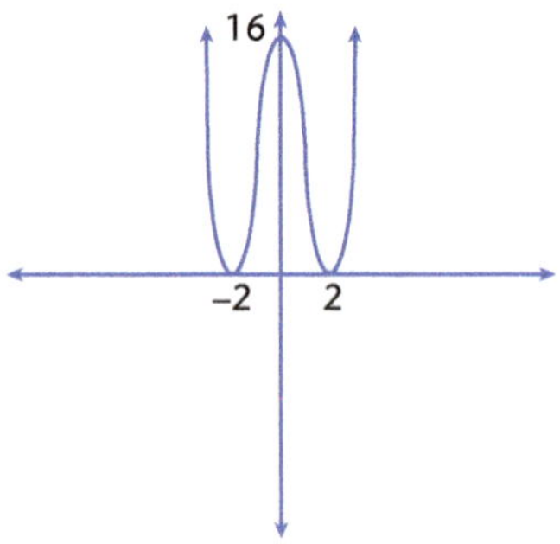

17. a. 3
 b. $x = 0, -1, 2$
 c. (0,0)
 d. (0,0) (–1,0) (2,0)
 e.

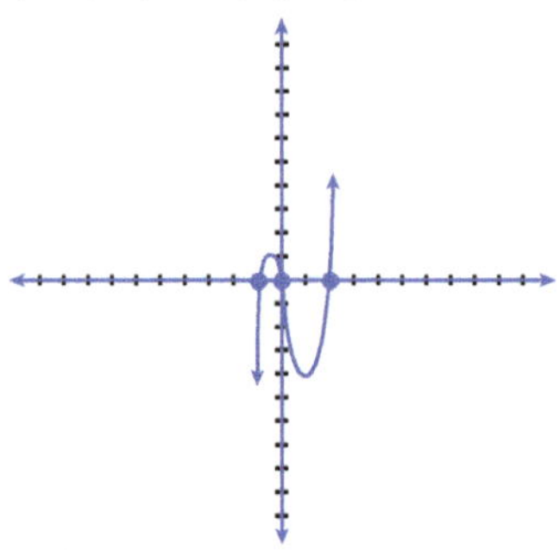

18. a. 4
 b. $x = 1$
 c. (0,2)
 d. (1,0)
 e.

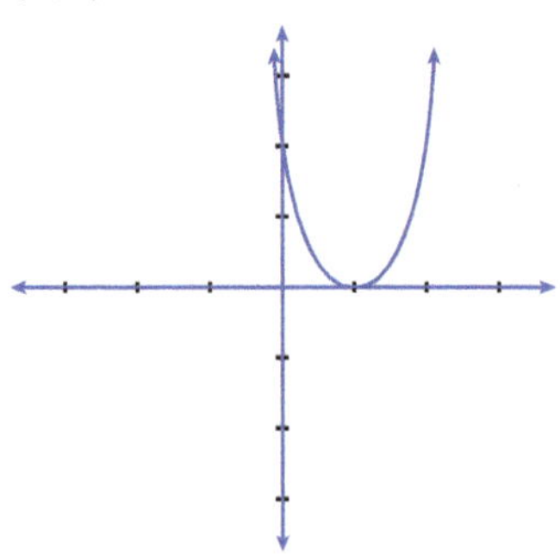

19. a. 3
 b. $x = 0, 2, -3$
 c. (0,0)
 d. (0,0) (2,0) (–3,0)
 e.

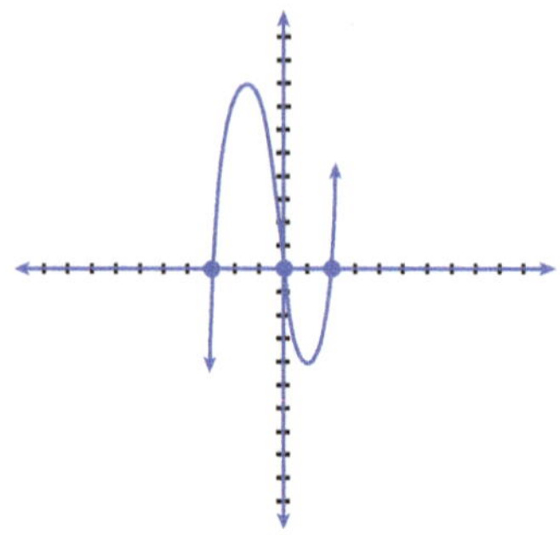

20. a. 2
 b. $x = 4, -4$
 c. (0,–80)
 d. (4,0) (–4,0)
 e.

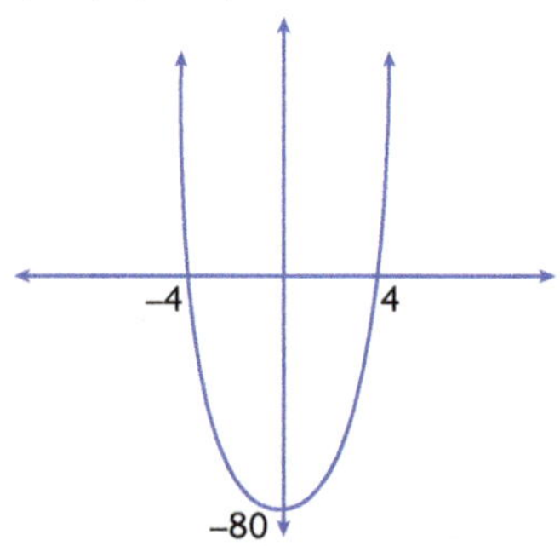

21.

x	0	1	–1	2	2
y	1	3	1	5	4

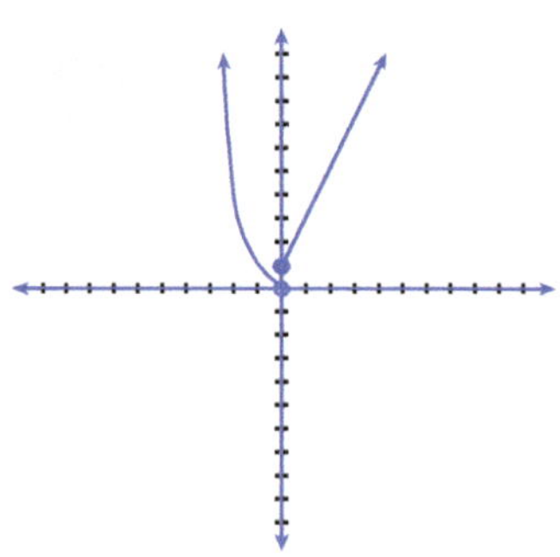

22.

x	1	0	–1	2	3
y	6	8	10	7	8

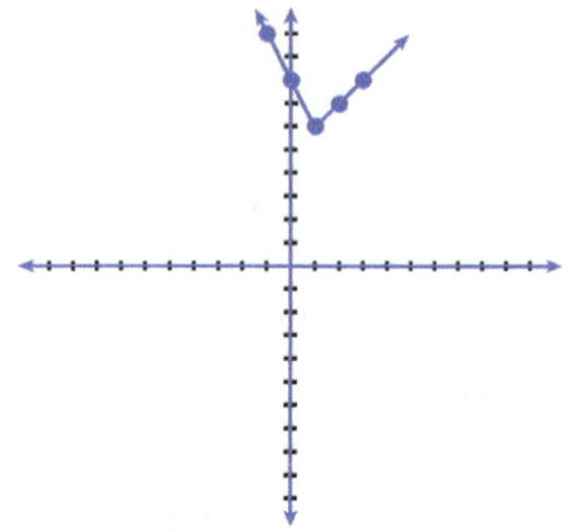

23.

x	–1	0	1	2	3	4
y	6	–1	0	3	4	5

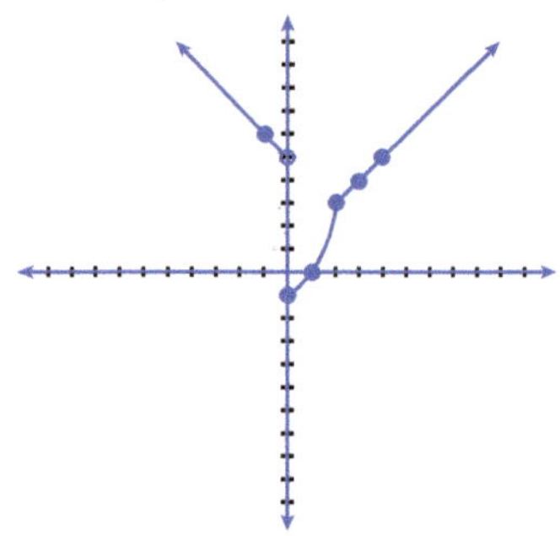

24.

x	–2	–1	0	1	2	3
y	1	4	6	6	4	9

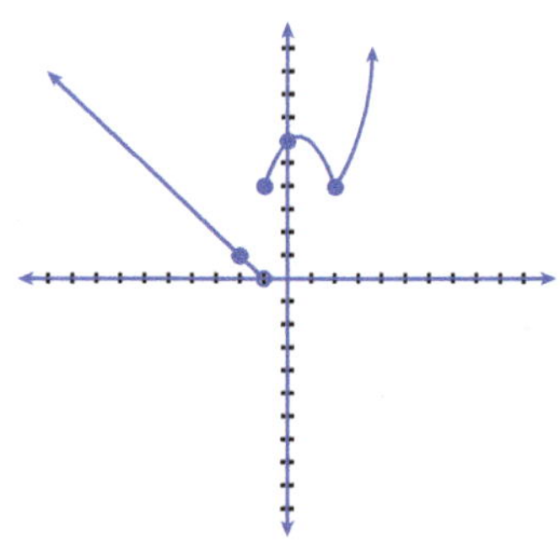

25.

x	–4	–3	–1	0	1	2
y	4	2	4	–1	1	3

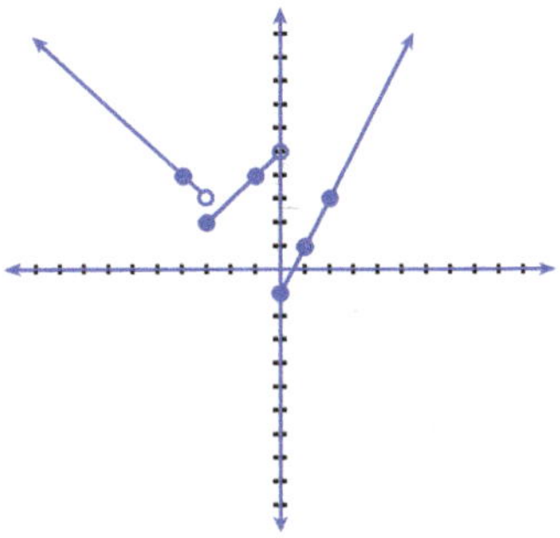

26.

x	–2	–1	0	1	2	3
y	1	1	0	–2	–4	–6

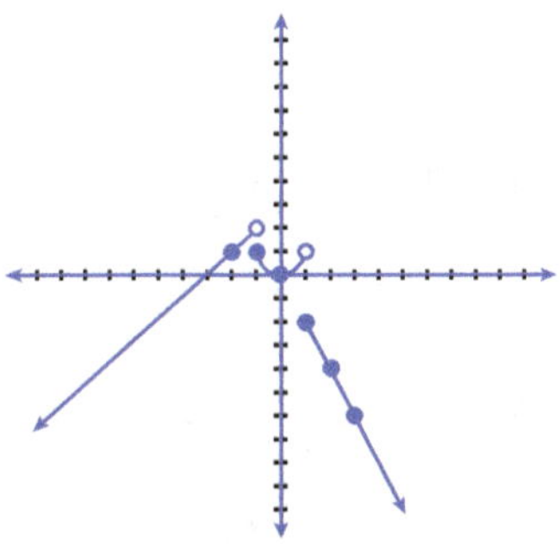

27.

x	–1	0	1	2	3	6
y	–1	0	–1	0	1	2

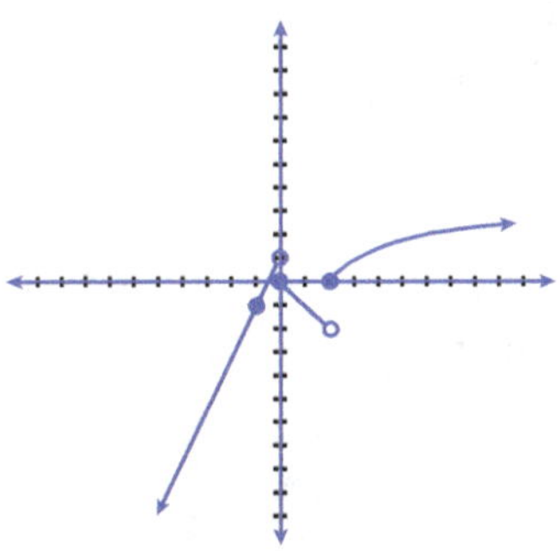

28.

x	–4	–3	0	1	3	4
y	–1	3	0	1	4	4

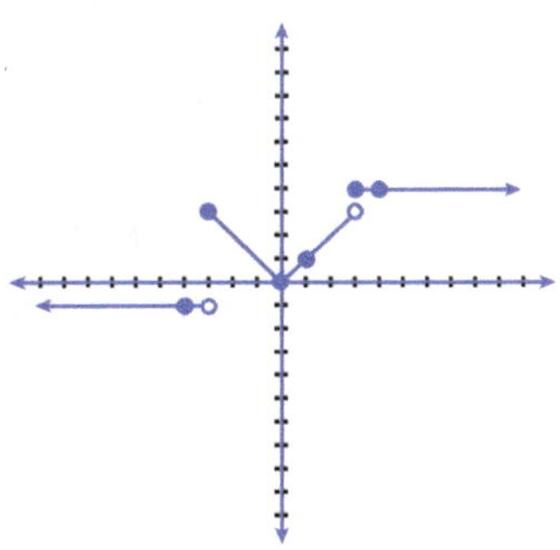

29.

x	–2	–1	0	1	2	3
y	0	1	0	–1	0	–3

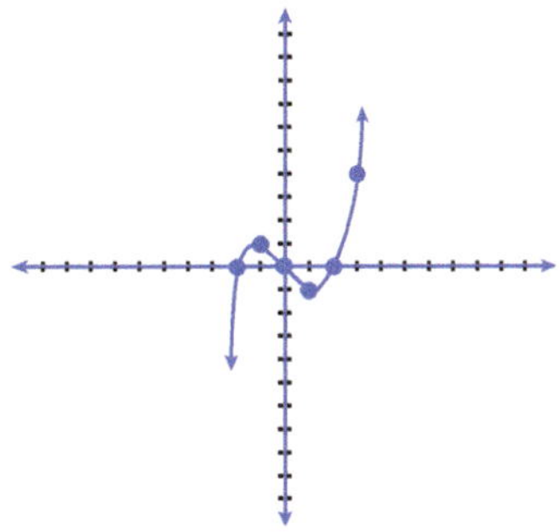

Section 3.5

1. VA: $x = 2$
   $x = -1$
   HA: x-axis, or $y = 0$
2. VA: $x = 0$
   HA: $y = 2$
3. VA: $x = 2$
   HA: x-axis, or $y = 0$
4. VA: none
   HA: $y = 1$
5. VA: $x = \pm 2$
   HA: $y = 2$
6. VA: $x = -\frac{1}{3}$
   HA: $y = -1$
7. VA: none
   HA: $y = 1$
8. VA: $x = -2$
   HA: x-axis, or $y = 0$
9. VA: $x = \pm 1$
   HA: x-axis, or $y = 0$
10. VA: $x = 1$
    HA: $y = -1$
11. D: $x \neq -2$
    VA: $x = -2$
    HA: x-axis, or $y = 0$
    y_I: $(0, \frac{1}{2})$
    x_I: none

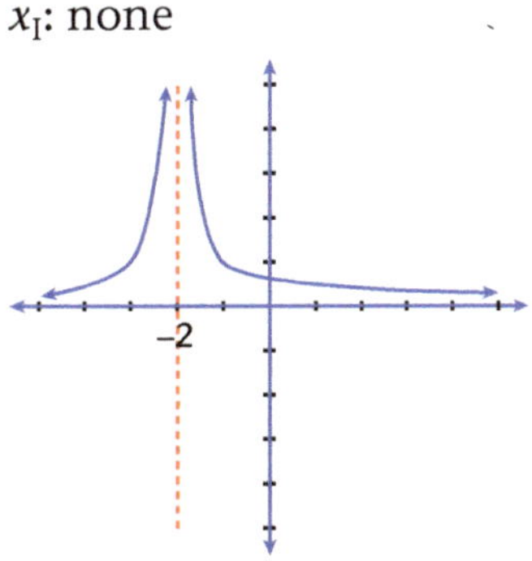

12. D: $x \neq 1$
    VA: $x = 1$
    HA: x-axis, or $y = 0$
    y_I: $(0,-1)$
    x_I: none

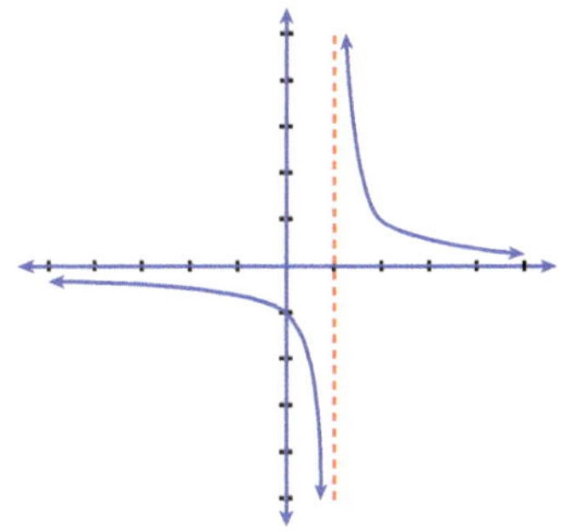

13. D: $x \neq \frac{1}{2}$
    VA: $x = \frac{1}{2}$
    HA: $y = \frac{3}{2}$
    y_I: $(0,-2)$
    x_I: $(-\frac{2}{3},0)$

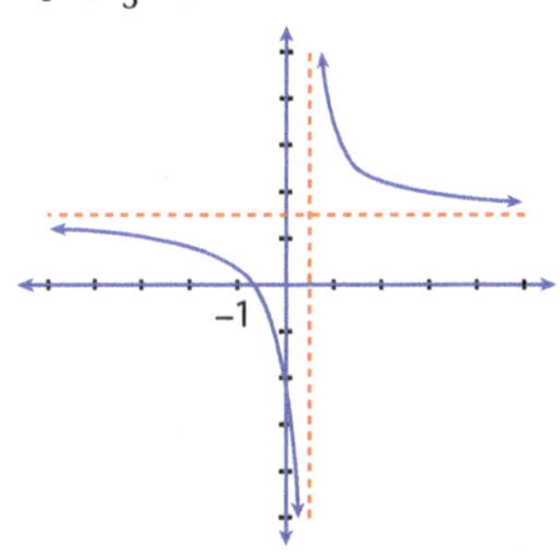

14. D: $(-\infty,\infty)$
    VA: none
    HA: x-axis
    y_I: $(0,5)$
    x_I: none

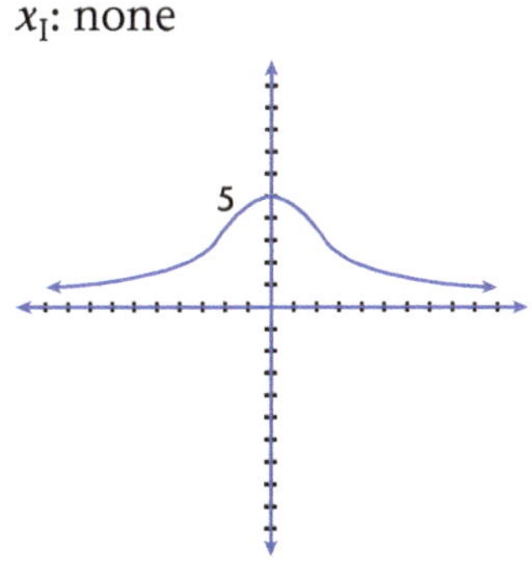

15. D: $x \neq 0$
$x \neq -4$
VA: $x = -4$, $x = 0$
HA: x-axis
y_I: none
x_I: $(\frac{1}{2},0)$

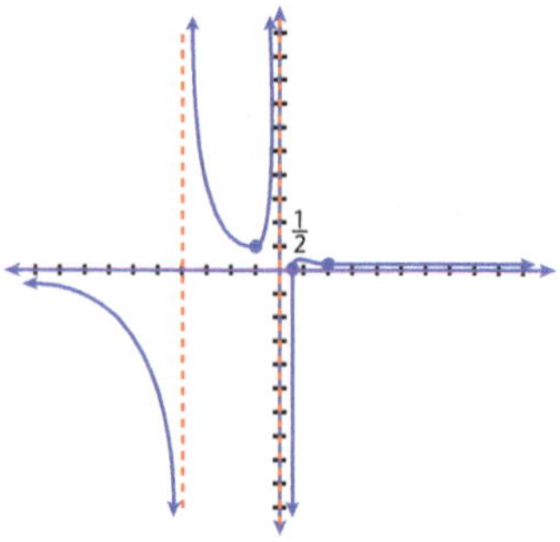

16. D: $x \neq 0$
VA: $x = 0$
HA: $y = 2$
y_I: none
x_I: $(-2,0)$

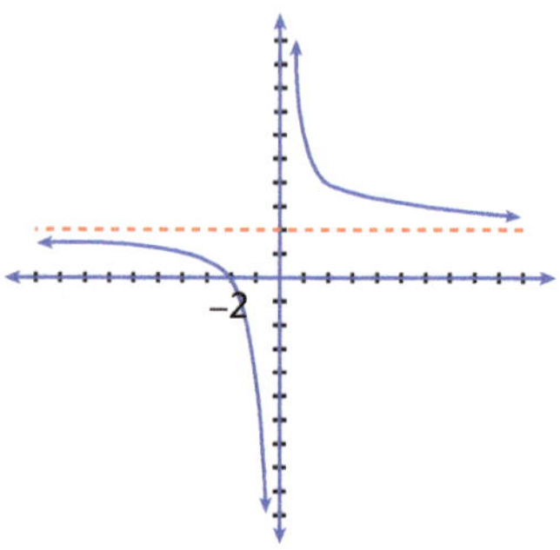

17. D: $x \neq -3$
VA: $x = -3$
HA: $y = 1$
y_I: $(0,-\frac{4}{9})$
x_I: $(2, 0)$ & $(-2, 0)$

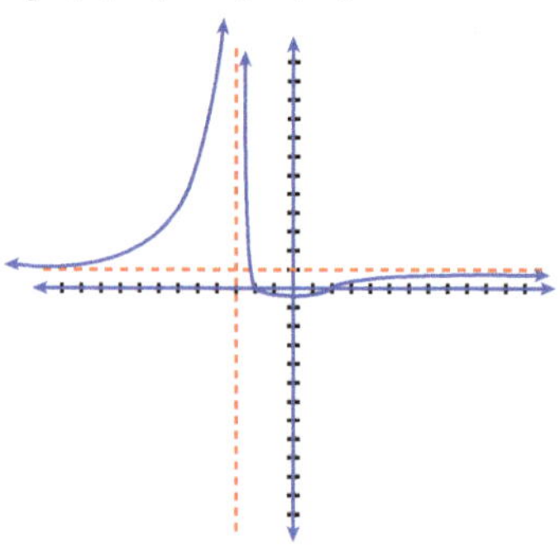

18. D: $x \neq -4$
VA: $x = -4$
HA: $y = 2$
y_I: $(0,-\frac{3}{2})$
x_I: $(3,0)$

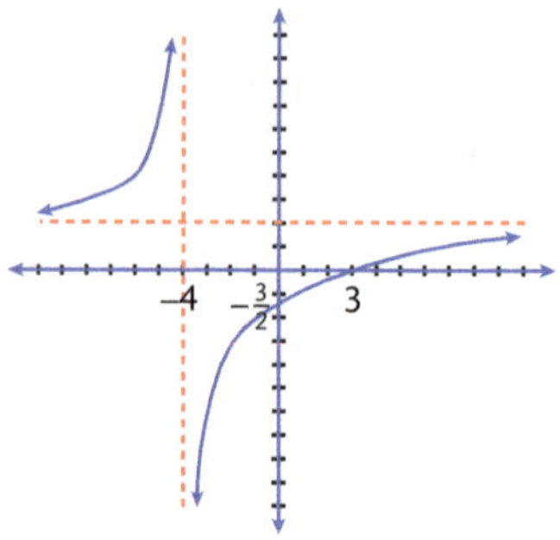

19. D: $x \neq \pm 2$
VA: $x = \pm 2$
HA: x-axis, or $y = 0$
y_I: $(0,0)$
x_I: $(0,0)$

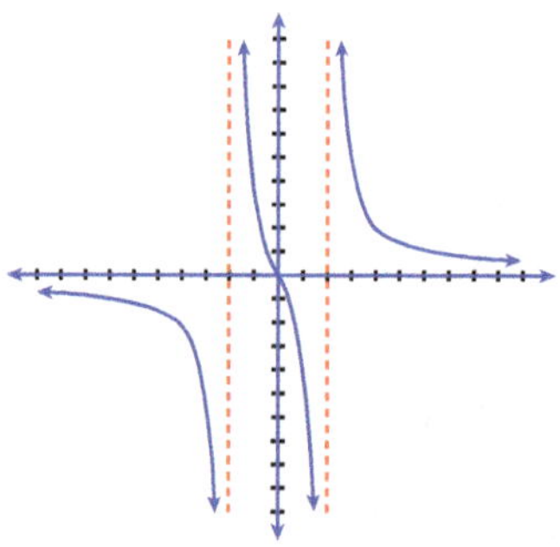

20. D: $x \neq 3, -\frac{1}{2}$
VA: $x = 3$, $x = -\frac{1}{2}$
HA: x-axis, or $y = 0$
y_I: $(0,-\frac{4}{3})$
x_I: $(-2, 0)$

21. D: $x \neq 0, -1, 2$
VA: $x = 0$, $x = -1$, $x = 2$
HA: $y = 0$
y_I: none
x_I: $(5, 0)$

22. D: $x \neq 0, 3$
VA: $x = 0$, $x = 3$
HA: $y = \frac{1}{2}$
y_I: none
x_I: $(-1, 0)$ & $(2, 0)$

23. D: $x \neq -5, -1$
VA: $x = -5$, $x = -1$
HA: $y = 0$
y_I: $(0, -\frac{56}{25})$
x_I: $(4, 0)$ & $(-7, 0)$

24. D: $x \neq 1, -1, -\frac{3}{2}$
VA: $x = 1$, $x = -1$, $x = -\frac{3}{2}$
HA: $y = \frac{1}{2}$
y_I: $(0, 0)$
x_I: $(0, 0)$ & $(6, 0)$

25. D: $x \neq -\frac{1}{3}, 5$
VA: $x = -\frac{1}{3}$, $x = 5$
HA: $y = 1$
y_I: $(0, 1)$
x_I: $(\frac{1}{3}, 0)$ & $(-5, 0)$

Section 3.6

1. a. none
b. none
c. yes; 2
d. no
e. no
f.

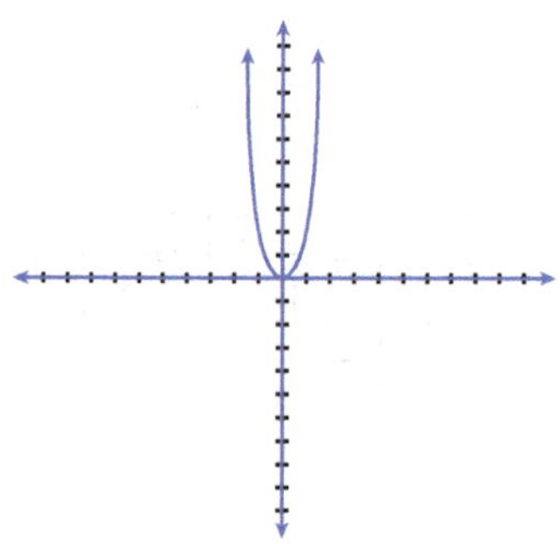

2. a. 3 down
   b. none
   c. yes; 2
   d. no
   e. no
   f.

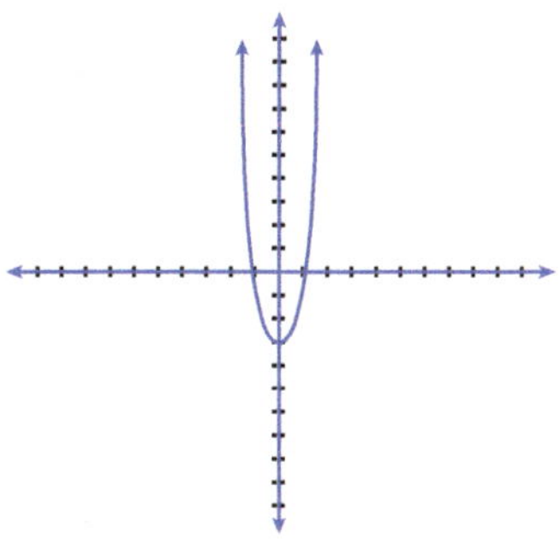

3. a. 2 down
   b. 1 left
   c. no
   d. no
   e. yes; vertical
   f.

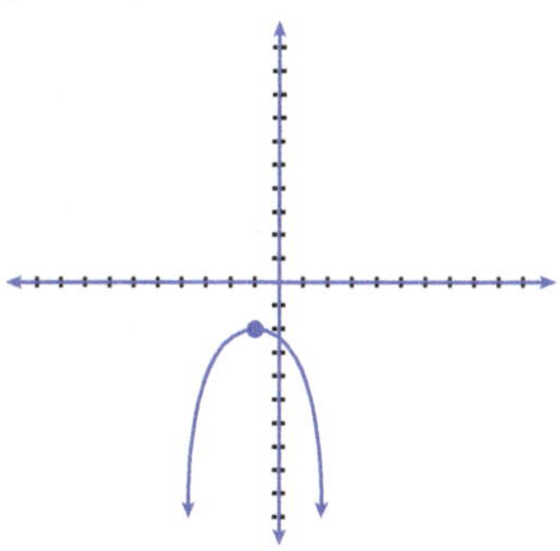

4. a. 5 up
   b. 2 right
   c. yes; 3
   d. no
   e. no
   f.

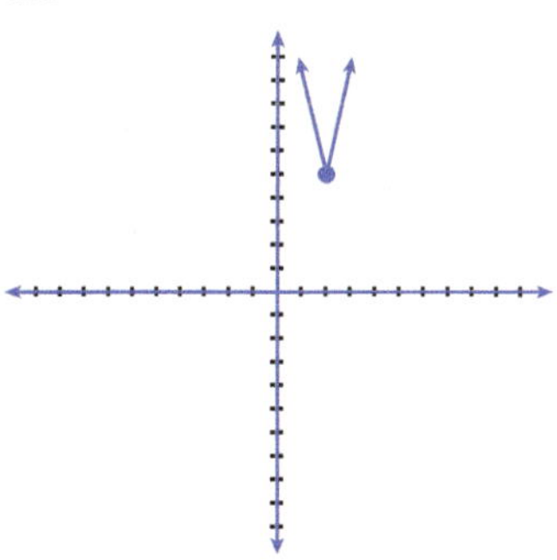

5. a. 3 down
   b. 1 right
   c. yes; 4
   d. no
   e. yes; vertical
   f.

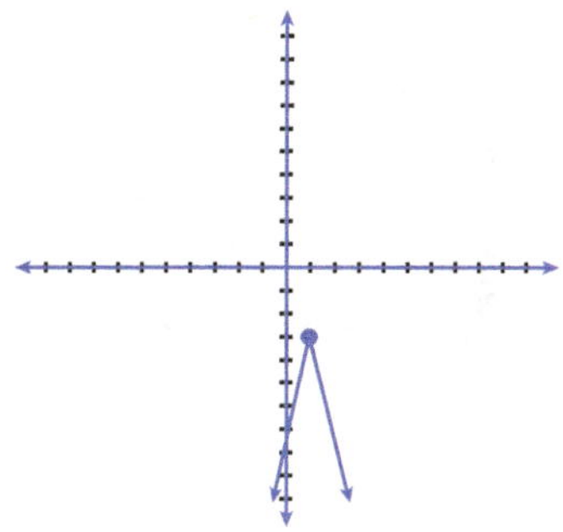

6. a. 4 down
   b. 3 left
   c. no
   d. no
   e. yes; horizontal
   f.

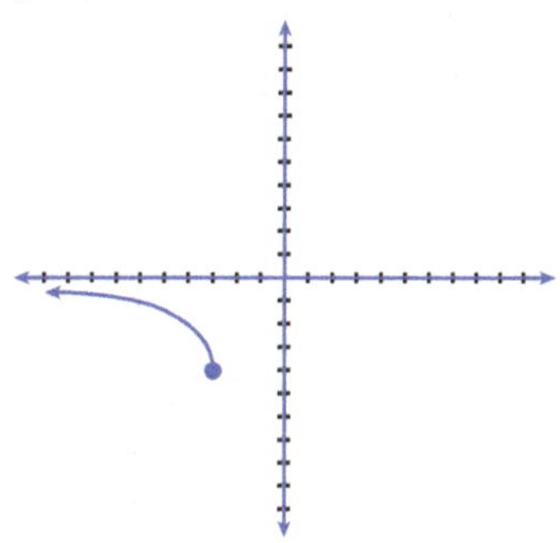

7. a. 1 up
   b. 2 right
   c. no
   d. no
   e. yes; horizontal
   f.

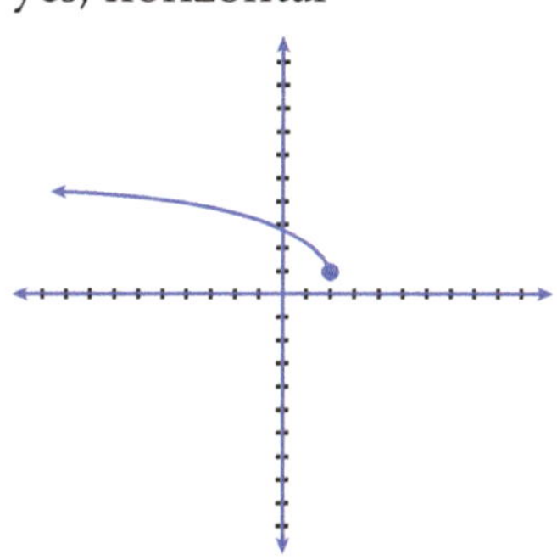

8. a. 3 down
   b. 3 left
   c. no
   d. yes; 2
   e. no
   f.

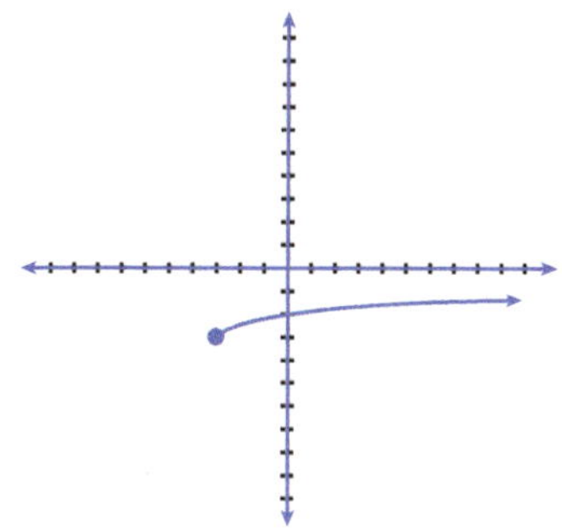

9. a. 4 up
   b. 2 right
   c. no
   d. yes; $\frac{1}{3}$
   e. yes, vertical
   f.

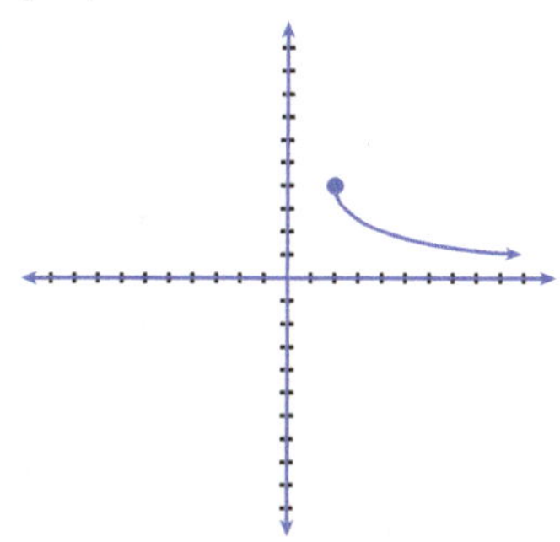

10. a. 1 up
    b. 2 right
    c. yes; 2
    d. no
    e. yes; vertical
    f.

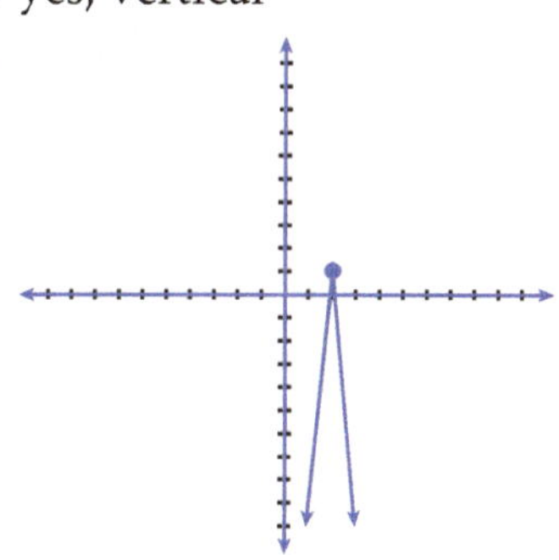

11. a. 3 up
    b. 2 right
    c. no
    d. no
    e. no
    f. 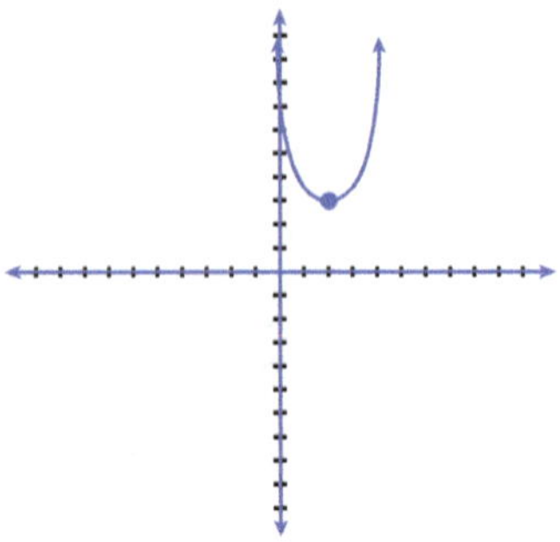

12. a. 4 up
    b. 1 left
    c. yes; 3
    d. no
    e. yes; vertical
    f. 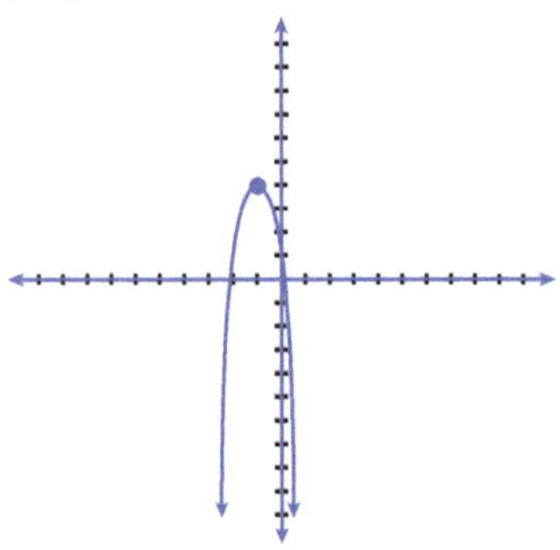

13. a. 9 up
    b. 2 left
    c. no
    d. no
    e. yes; vertical
    f. 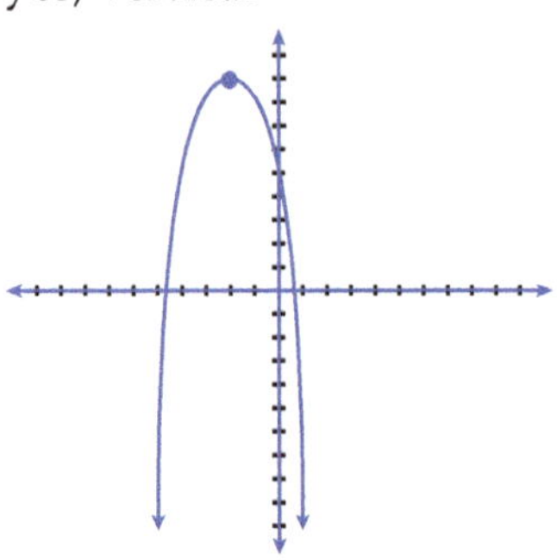

14. a. 5 up
    b. 3 right
    c. yes; $\frac{1}{2}$
    d. no
    e. no
    f. 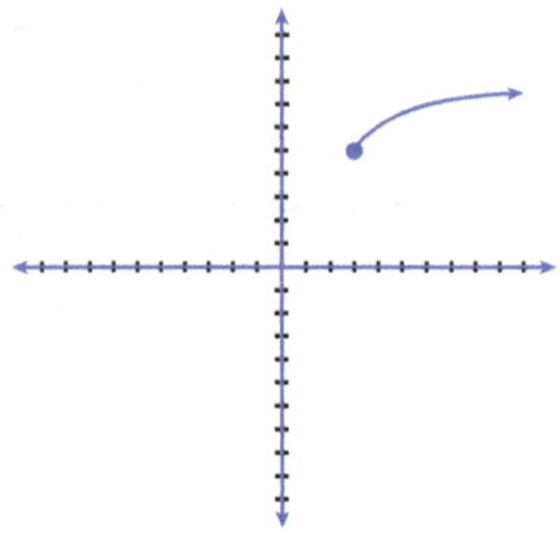

15. a. $\frac{1}{2}$ down
    b. 1 left
    c. yes; 4
    d. no
    e. yes; vertical
    f. 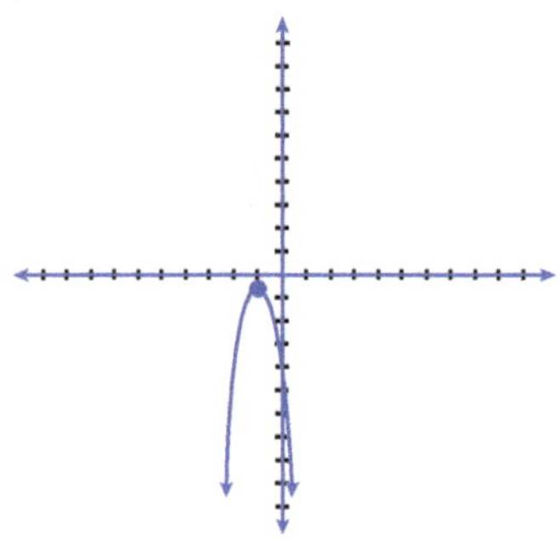

16. a. 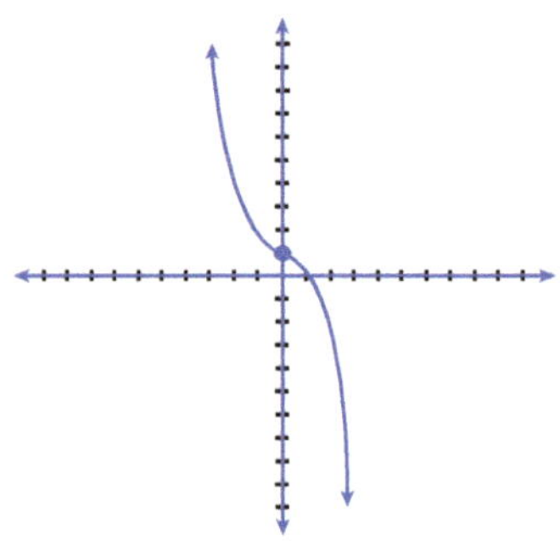
    b. 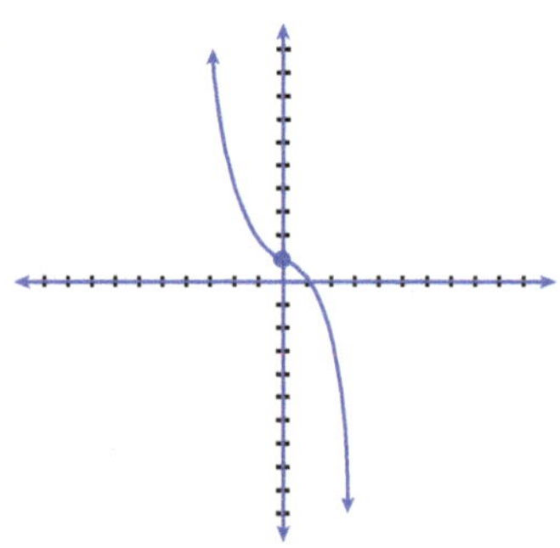
    c. 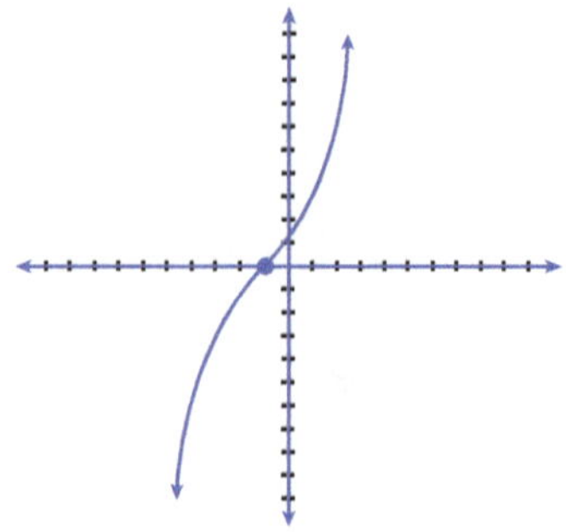
    d. 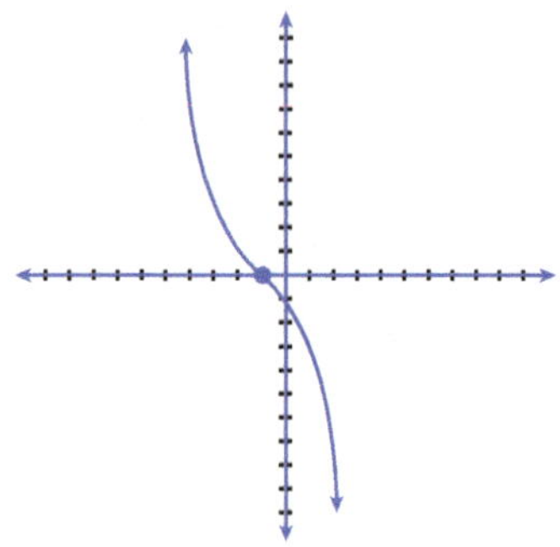

17. a. 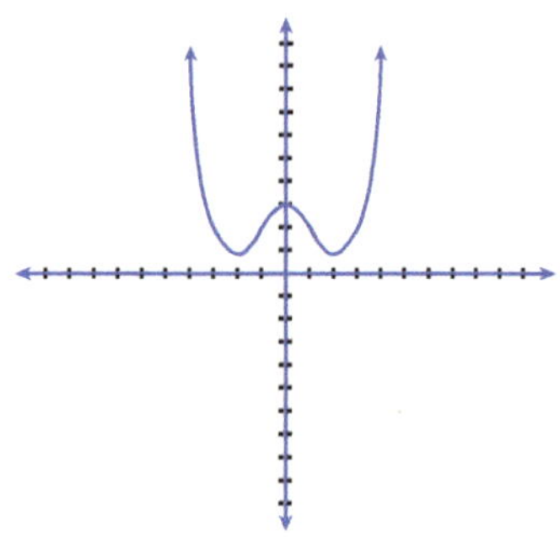
    b. 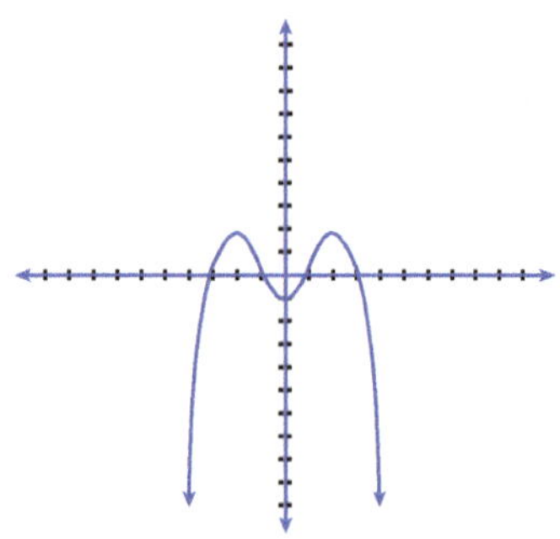
    c. 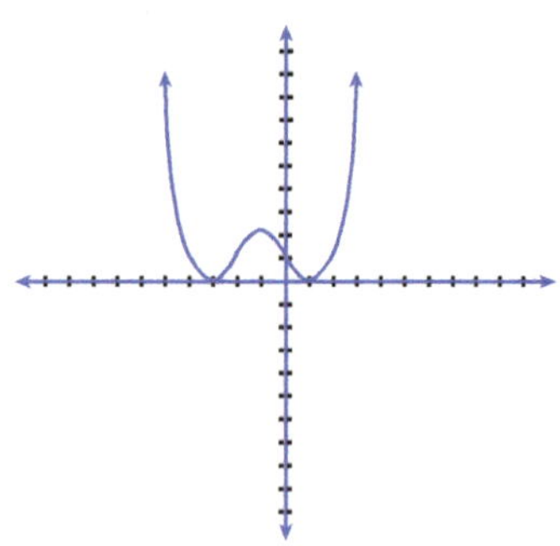

d. 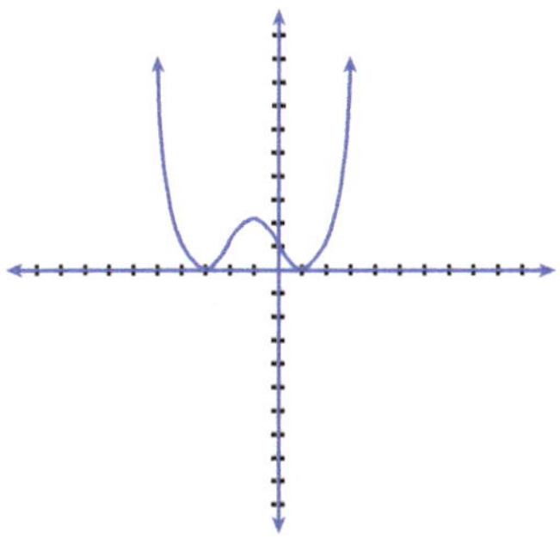

18. a. 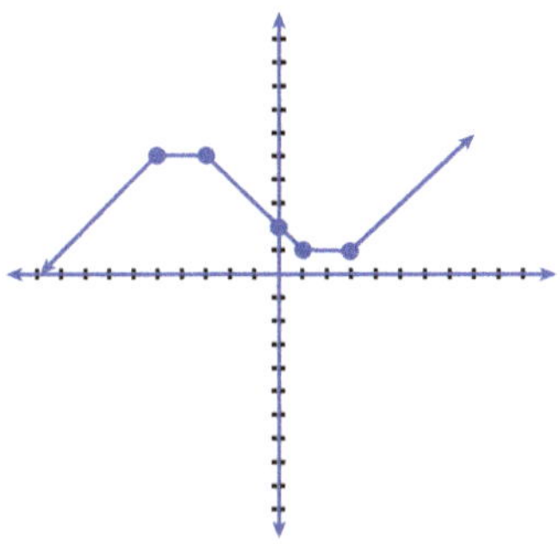

b. 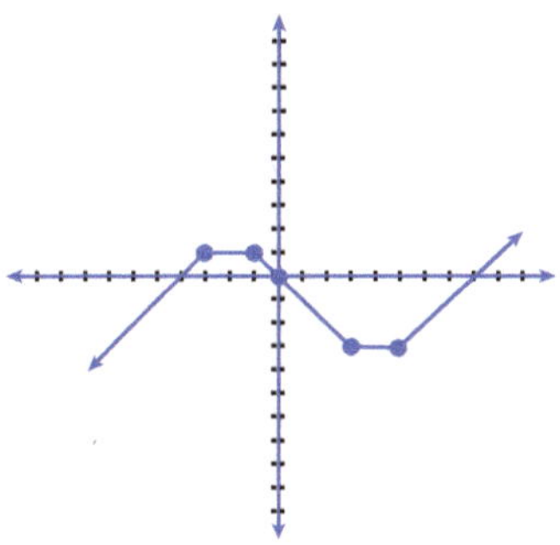

c. 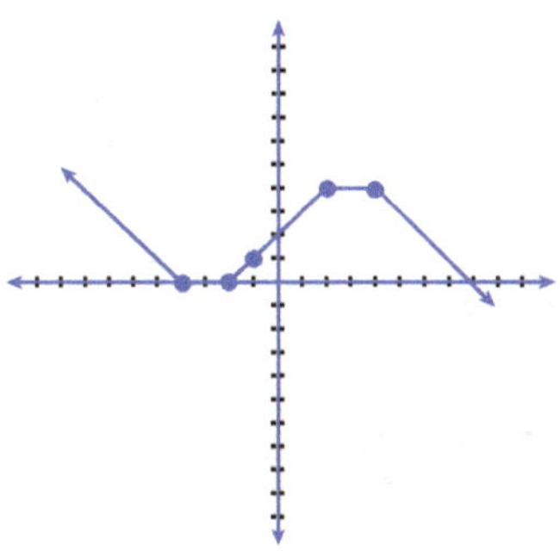

d. 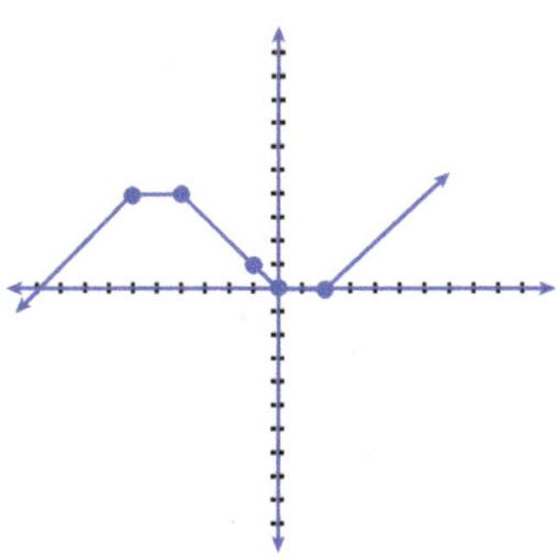

19. a. 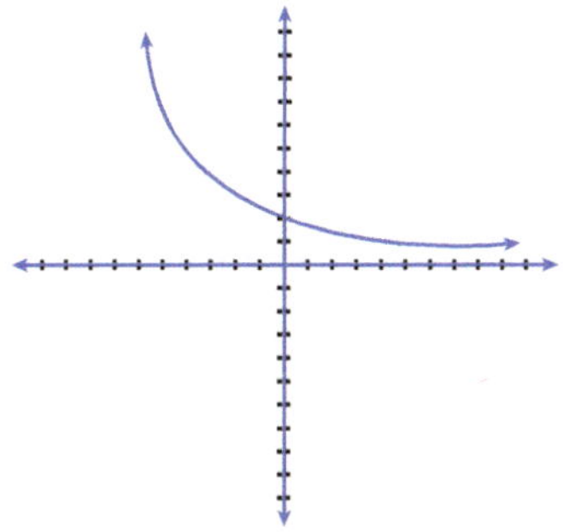

b. 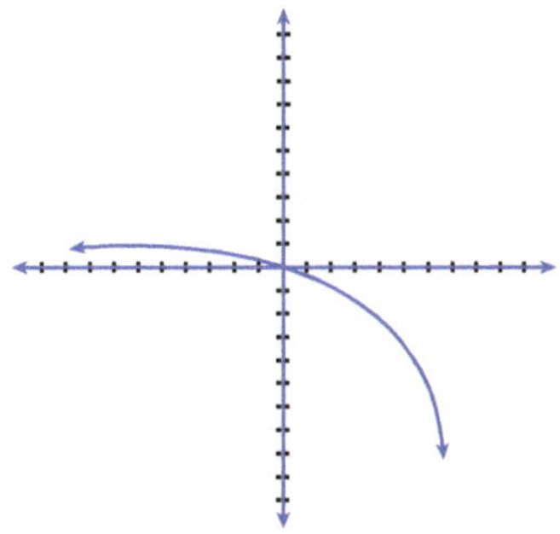

c. 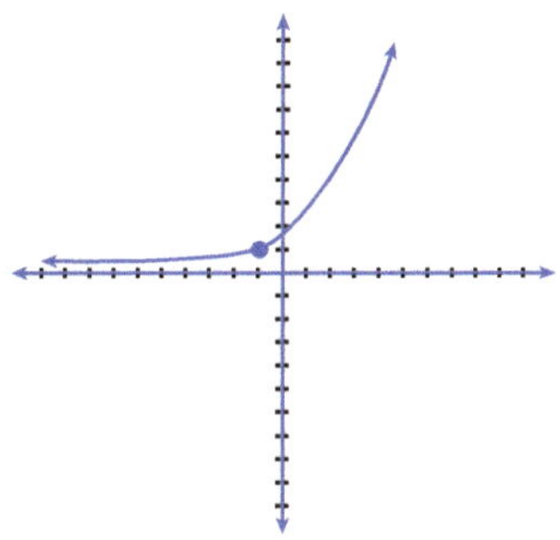

d. 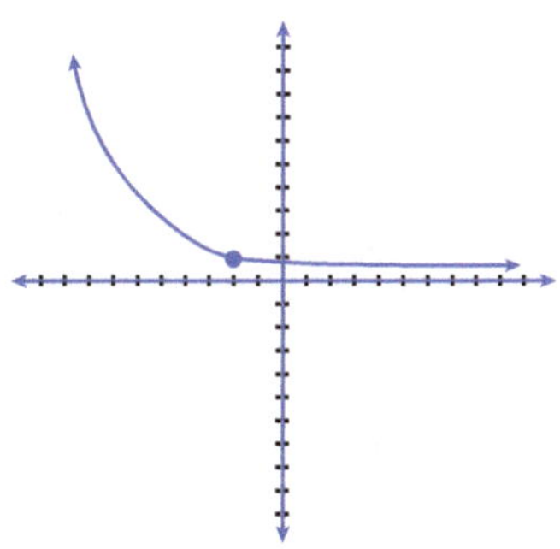

20. a. 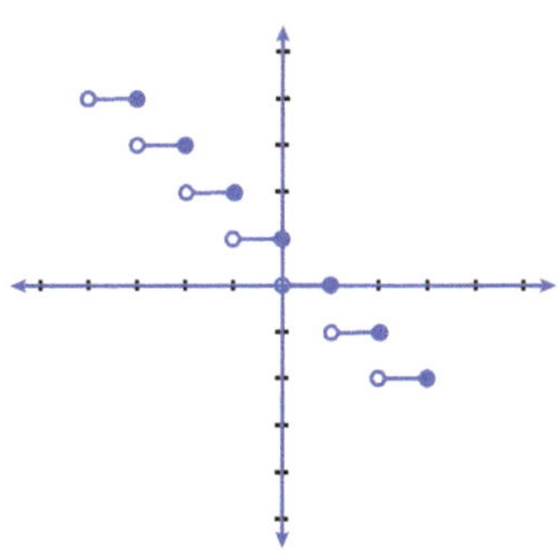

b. 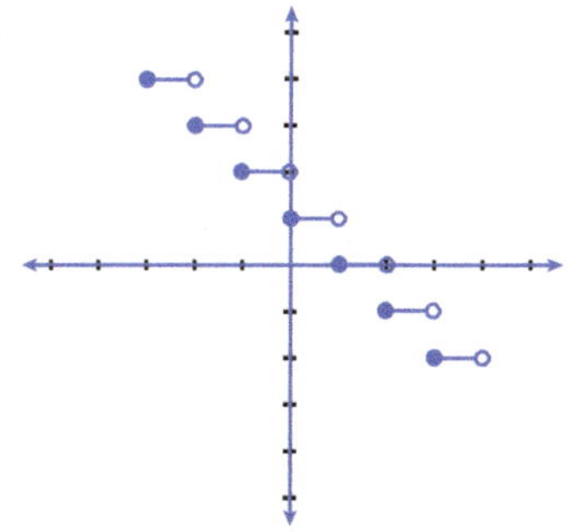

c. 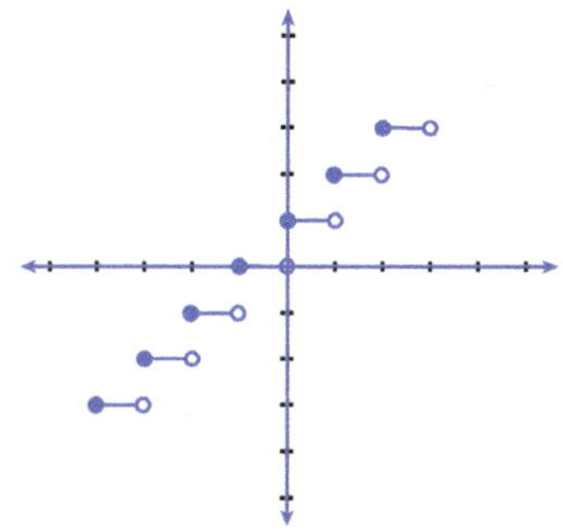

d. 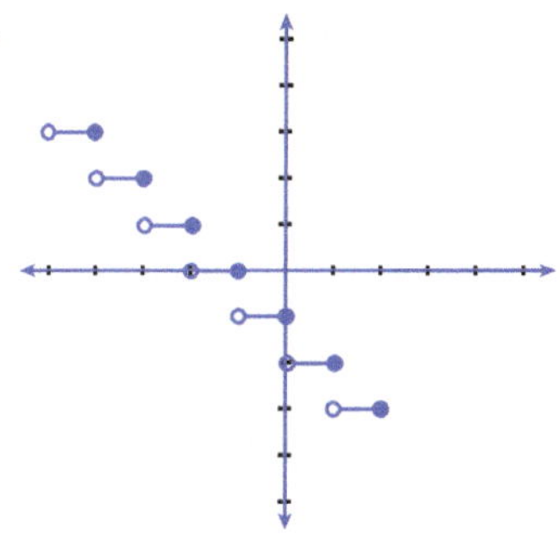

21. $f(x) = -(x-3)^2 + 1$
22. $g(x) = -2x + 3$
23. $h(x) = \sqrt{x+5} - 1$
24. $f(x) = -x^3 - 4$
25. $y(x) = \frac{3}{2}x - 2 + 3$
26. $f(x) = 2(x-3)^3 + 2$
27. $g(x) = -2(x+1)^2$
28. $f(x) = -x^2 + 5$

Section 3.7

1. a. $x^2 + x - 1$
   b. $x^3 + x^2 - 2x - 2$
   c. $-x^2 + x + 3$
   d. $\dfrac{x^2 - 2}{x + 1}$
   e. $x^2 + 2x - 1$
   f. $x^2 - 1$

2. a. $2x + 8$
   b. $x^2 + 8x + 15$
   c. 2
   d. $\frac{x + 3}{x + 5}$
   e. $x + 8$
   f. $x + 8$
3. a. $2x^2 - 3$
   b. $x^4 - 3x^2 - 4$
   c. -5
   d. $\frac{x^2 + 1}{x^2 - 4}$
   e. $x^4 - 8x^2 + 17$
   f. $x^4 + 2x^2 - 3$
4. a. $x^2 + 2x - 4$
   b. $2x^3 - x^2 - 6x + 3$
   c. $x^2 - 2x - 2$
   d. $\frac{2x - 1}{x^2 - 3}$
   e. $2x^2 - 7$
   f. $4x^2 - 4x - 2$
5. a. $\sqrt{x + 1} + \frac{2}{x}$
   b. $\frac{2}{x}\sqrt{x + 1}$
   c. $\frac{2}{x} - \sqrt{x + 1}$
   d. $\frac{x\sqrt{x + 1}}{2}$
   e. $\sqrt{\frac{2}{x} + 1}$
   f. $\frac{2}{\sqrt{x + 1}}$
6. a. $x^3 + \frac{2}{x^2 + 1}$
   b. $\frac{2x^3}{x^2 + 1}$
   c. $\frac{2}{x^2 + 1} - x^3$
   d. $\frac{x^3(x^2 + 1)}{2}$
   e. $\frac{8}{(x^2 + 1)^3}$
   f. $\frac{2}{x^6 + 1}$
7. a. $\sqrt{x + 1} + x + 5$
   b. $(5 + x)\sqrt{x + 1}$
   c. $x + 5 - \sqrt{x + 1}$
   d. $\frac{\sqrt{x + 1}}{x + 5}$
   e. $\sqrt{x + 6}$
   f. $5 + \sqrt{x + 1}$
8. a. $x^3 + x - 1$
   b. $x^4 - x^3$
   c. $-x^3 + x - 1$
   d. $\frac{x^3}{x - 1}$
   e. $(x - 1)^3$
   f. $x^3 - 1$
9. a. $-x + \frac{x - 4}{x - 2}$
   b. $\frac{-x^2 + 4x}{x - 2}$
   c. $\frac{x - 4}{x - 2} + x$
   d. $\frac{-x^2 + 2x}{x - 4}$
   e. $\frac{-x + 4}{x - 2}$
   f. $\frac{x + 4}{x + 2}$
10. a. $-x^2 + 1$
    b. $2x^2 - 6x^4$
    c. $1 - 5x^2$
    d. $\frac{2x^2}{1 - 3x^2}$
    e. $18x^4 - 12x^2 + 2$
    f. $1 - 12x^4$
11. a. $2x + 7$
    b. $x^2 + 5$
    c. $x^2 + 10x + 25$
    d. $2x^2 + 20x + 47$
    e. 8
    f. 27
    g. 5
12. a. $\frac{x^2}{4} - 1$
    b. $\frac{1}{2}\sqrt{x - 1}$
    c. $\sqrt{\frac{x - 2}{2}}$
    d. $\frac{1}{2}x - 2$
    e. 4
    f. $8\sqrt{2}$
    g. $\frac{1}{2}$
13. a. $(x^2 + 1)\sqrt{x^2 + 1}$
    b. $\frac{1}{x}\sqrt{1 + x^2}$
    c. $\frac{1}{\sqrt{x^2 + 1}}$ or $(x^2 + 1)^{-\frac{1}{2}}$
    d. $(x^2 + 1)^{-\frac{3}{2}}$
    e. $8 + \sqrt{5}$
    f. 9
    g. $\sqrt{2} - 1$
14. a. $6x - 3$
    b. $3x^2 - 2$
    c. $9x^2 - 12x + 4$
    d. $18x^2 - 24x + 9$
    e. 9
    f. 63
    g. 0
15. a. $\sqrt{1 + \frac{2}{x}}$
    b. $\frac{2}{x^2 + 3}$
    c. $3 + \frac{4}{x^2}$
    d. $2\sqrt{1 + \frac{1}{x^2}}$
    e. $\sqrt{3} + 1$
    f. 24
    g. -2

16. a. $\dfrac{2}{x+5} - 1$

b. $\dfrac{1}{3|x+2|+5}$

c. $3\left|\dfrac{2x+11}{x+5}\right|$

d. $6\left|\dfrac{2x+11}{x+5}\right| - 1$

e. $3\frac{1}{7}$

f. 75

g. $-\dfrac{53}{6}$

17. a. $x - 2$

b. $\sqrt{x-1}$

c. $\sqrt{x-1}$

d. $x - 2$

e. 4

f. 24

g. -1

18. a. $2x^{\frac{3}{2}}$

b. $27x^{-3}$

c. $\dfrac{3}{x^3}$

d. $\dfrac{2\sqrt{3}}{x^{\frac{3}{2}}}$

e. $8 + 2\sqrt{2}$

f. $2\sqrt{3}$

g. -2

19. $x^4 - 6x^3 + 14x^2 - 15x + 8$
20. $x^4 - 6x^3 + 17x^2 - 24x + 16$
21. $\dfrac{x^4 + 3x^2 - 4x + 5}{(x-2)(x^2 - 2x + 5)}$
22. $2x + h$
23. 3
24. $4x + 2h$
25. $f(x) = x + \dfrac{1}{x}$, $g(x) = x^2$
26. $h(x) = 2x - 4$, $g(x) = -x$
27. $f(x) = x + 4$, $g(x) = x^2$, $h(x) = \dfrac{1}{x} - 3$

CHAPTER 4

Section 4.1

1. a. yes

b. no

2. a. no

b. no

3. a. yes

b. no

4. a. yes

b. yes

5. a. yes

b. yes

6. a. yes

b. yes

7. no; restrict $x \geq 2$ or $x \leq 2$
8. yes
9. no; restrict $x \geq 0$ or $x \leq 0$
10. no; restrict $x \geq 0$ or $x \leq 0$
11. yes
12. yes
13. yes
14. yes
15. no; restrict $x \geq \dfrac{2}{3}$ or $x \leq \dfrac{2}{3}$
16. no; restrict $x \geq 0$ or $x \leq 0$
17. no; restrict $x \geq 2$ or $x \leq 2$
18. yes
24. $f^{-1}(x) = \dfrac{x+3}{2}$
25. $f^{-1}(x) = \dfrac{x^3}{64} - 2$
26. $f^{-1}(x) = \dfrac{x^5 - 1}{3}$
27. $g^{-1}(x) = \dfrac{5}{x+1}$
28. $p^{-1}(x) = \dfrac{125x^3 + 3}{2}$
29. $f^{-1}(x) = \dfrac{x-4}{2}$
30. $y^{-1}(x) = (x-2)^3 - 1$
31. $h^{-1}(x) = 3x + \dfrac{3}{2}$
32. $k^{-1}(x) = \dfrac{2x - 4}{x+3}$
33. $f^{-1}(x) = \dfrac{4}{x} - 2$
34. $h^{-1}(x) = \dfrac{1}{x-3}$
35. $m^{-1}(x) = \sqrt[3]{\dfrac{5+x}{1-x}}$
36. $g^{-1}(x) = \dfrac{x+2}{5x-4}$
37. $w^{-1}(x) = \dfrac{2 + 3x^5}{x^5 - 1}$
38. $f^{-1}(x) = \sqrt[3]{\dfrac{x+2}{5}}$
39. $y^{-1}(x) = \dfrac{243x^5 + 1}{2}$
40. $g^{-1}(x) = \dfrac{3x^3 - 1}{1 - 2x^3}$
41. $h^{-1}(x) = -\dfrac{1}{5}x + \dfrac{4}{5}$
42. $k^{-1}(x) = \dfrac{3x-1}{x}$
43. $n^{-1}(x) = \dfrac{7x}{4-2x}$
44. $f^{-1}(x) = (x-4)^2$ with the restricted domain of $x \geq 4$
45. No, they are only inverses if we restrict the domain of $g(x)$ to $x \geq -1$, which we didn't do.
46. $f^{-1}(x) = \sqrt{x-3} + 2$, **only** if $x \geq 2$ for $f(x)$
47. No, we can make them inverses if we restrict the domain of $f(x)$ to be $x \geq 0$.
48. $p^{-1}(x) = \sqrt{\dfrac{x-2}{3}} + 1$
49. $g^{-1}(x) = 4(x-5)^2 + 1$, $x \geq 5$
50. it does not pass the horizontal line test.
51. 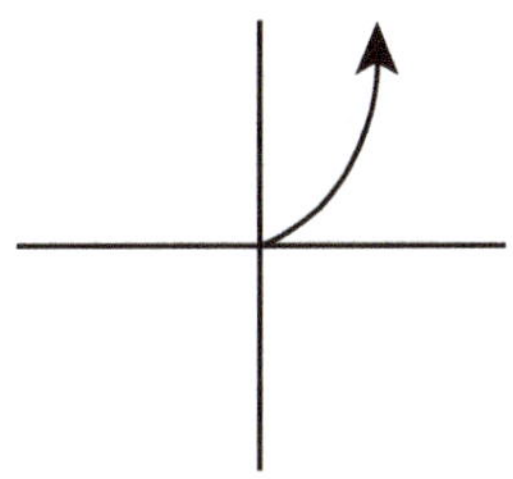

52. 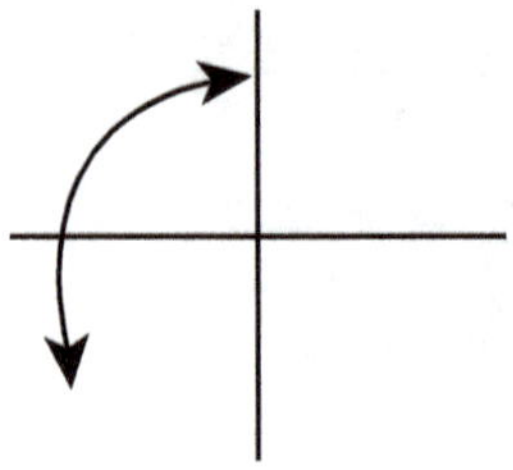

53. 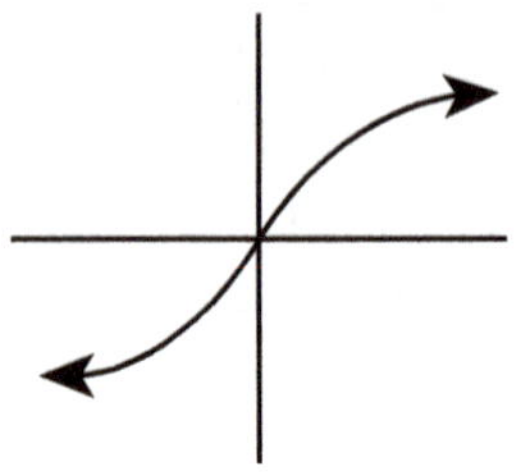

Section 4.2

1. 3527.145
2. 1024
3. 7.389
4. 0.007
5. 31.006
6. 36.462
7. 4.113
8. 23.141
9. 22.459
10. 522.735
11. g
12. a
13. b
14. h
15. e^{3x-1}
16. π^{-7}
17. $2^{(4x^2-4x)}$
18. 5^{-2x}
19. $\left(\frac{2}{3}\right)^x$
20. $2^x e^{-1.2x}$
21. 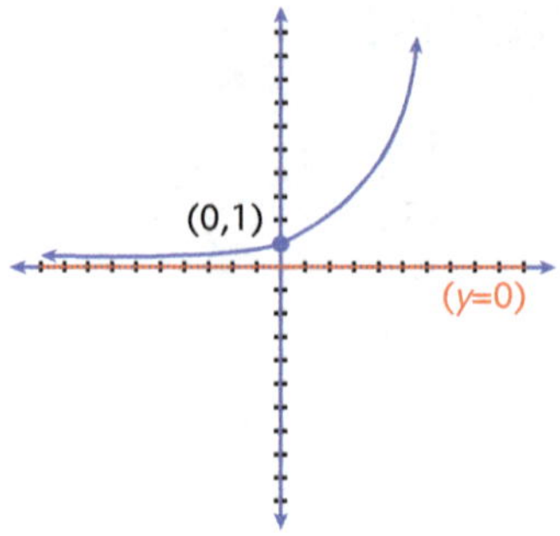

22. 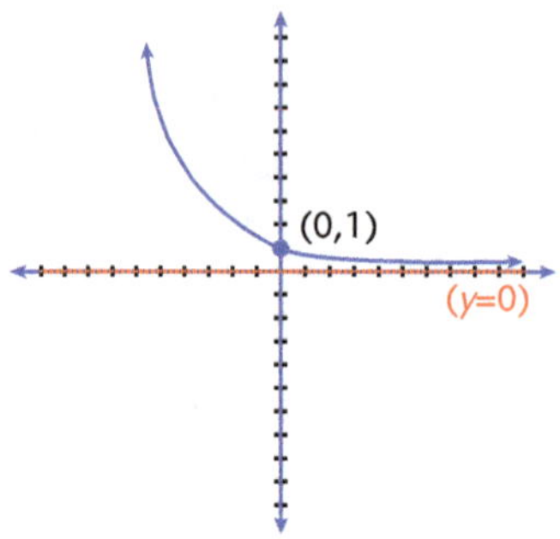

23. 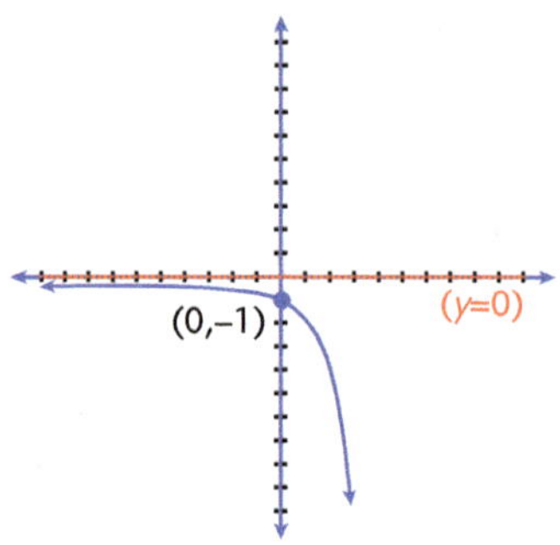

24. 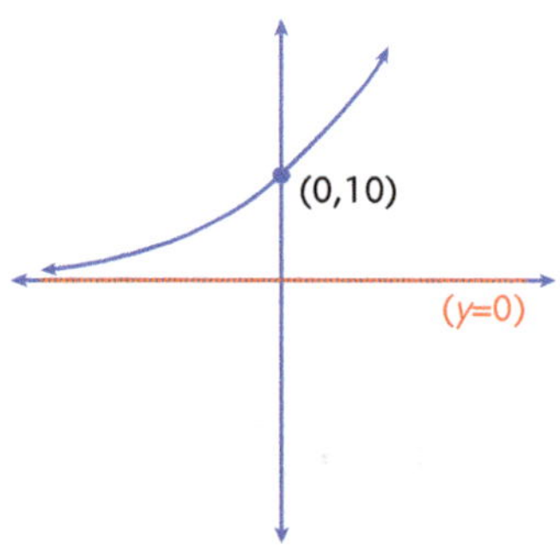

25. 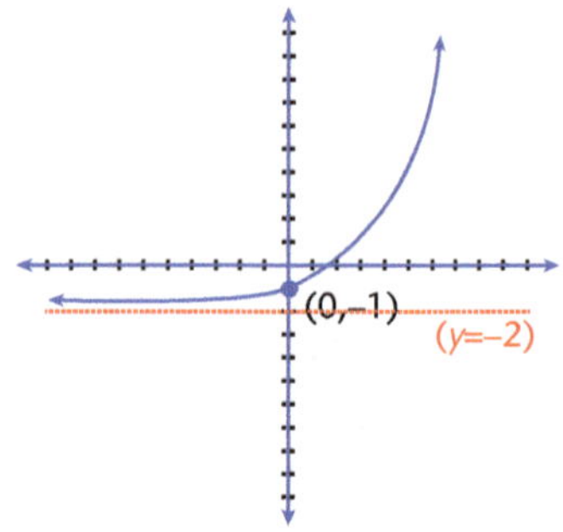

26. 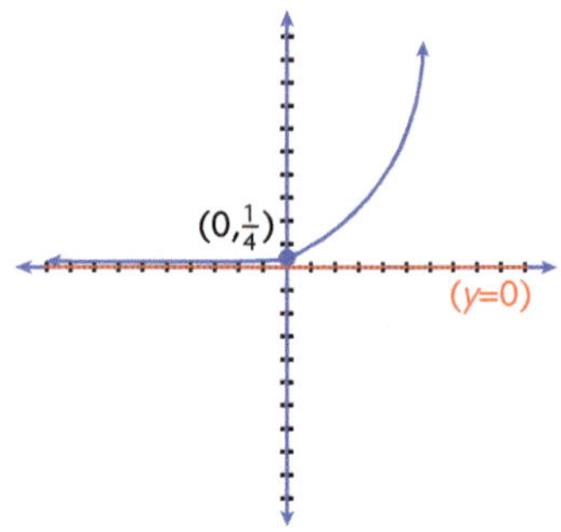

27. 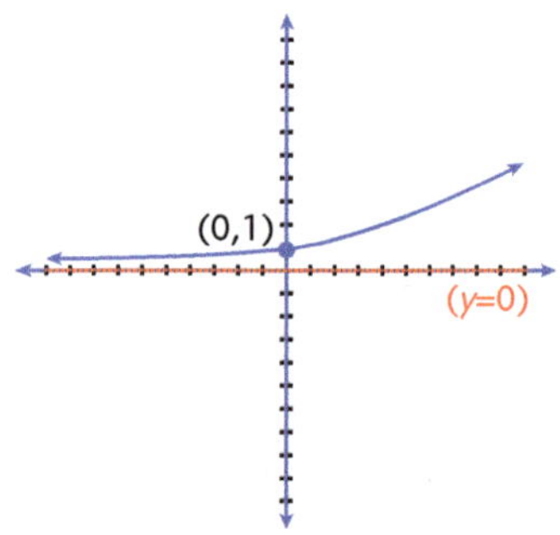

28. 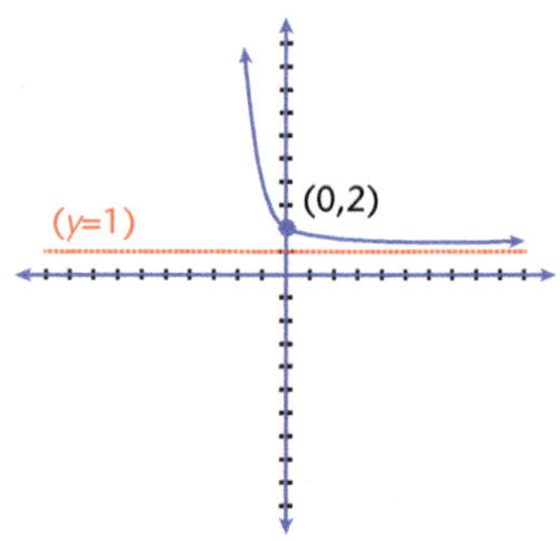

29.

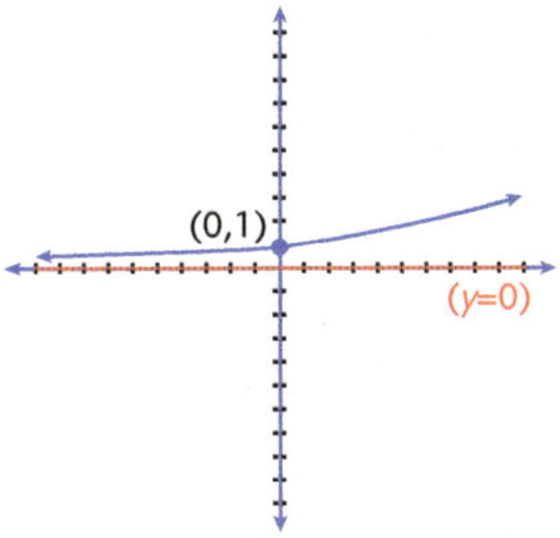

30.

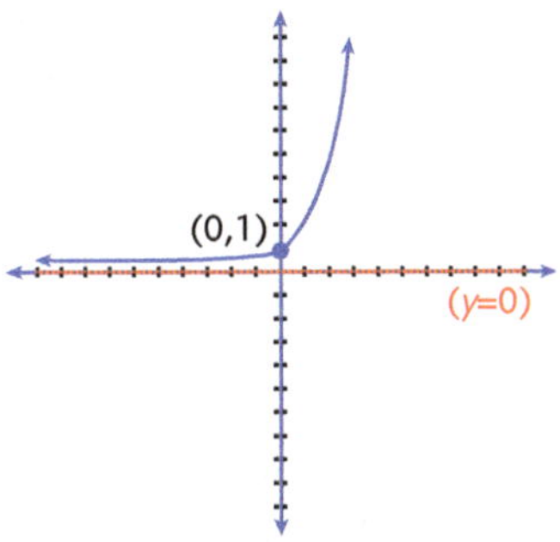

31. $p \approx \$499.76$
32. a. 24.21 lbs
    b. 25 lbs
33. 1.133 grams
34. \$2669.68
35. ~1121
36. 27297
37. 210,123 people
38. a. 6.96 billion
    b. 8.26 billion
39. In this context, it's the growth rate. Converted into a percentage, it means the population is growing 1.14% continuously.

Section 4.3

1. $3^3 = 27$
2. $2^5 = 32$
3. $e^0 = 1$
4. $4^{3/2} = 8$
5. $10^2 = 100$
6. $10^0 = 1$
7. $4^3 = 64$
8. $7^{-2} = \frac{1}{49}$
9. $16^{3/4} = 8$
10. $3^4 = 81$
11. $\log 0.001 = -3$
12. $\log_8 64 = 2$
13. $\log_9 27 = \frac{3}{2}$
14. $\log_{81} 3 = \frac{1}{4}$
15. $\ln 2 = x$
16. $\log_7 49 = 2$
17. $\ln a = x$
18. $\log_6 36 = 2$
19. $\ln 5 = -x$
20. $\log_2 128 = 7$
21. 2
22. –4
23. 0
24. 1
25. $\frac{1}{2}$
26. 3
27. 2
28. 4
29. $\frac{2}{3}$
30. $-\frac{1}{2}$
31. 0
32. 30 db
33. 80 db
34.

r	0.005	0.01	0.02	0.03	0.04
t	138	69	35	23	17

35. a. $x > 0$
    b. (1,0)
    c.

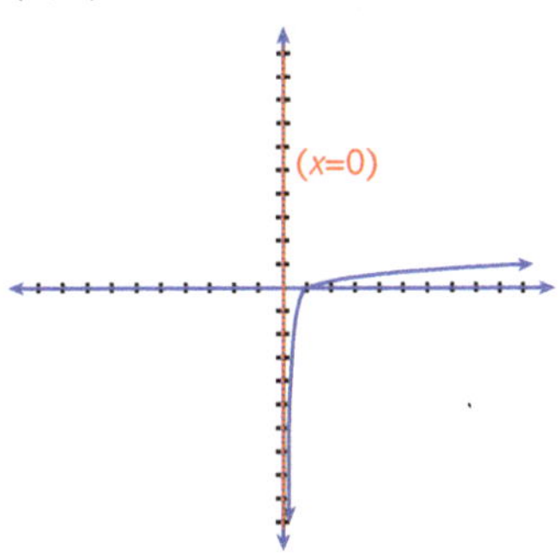

36. a. $x > -1$
    b. (0,0)
    c.

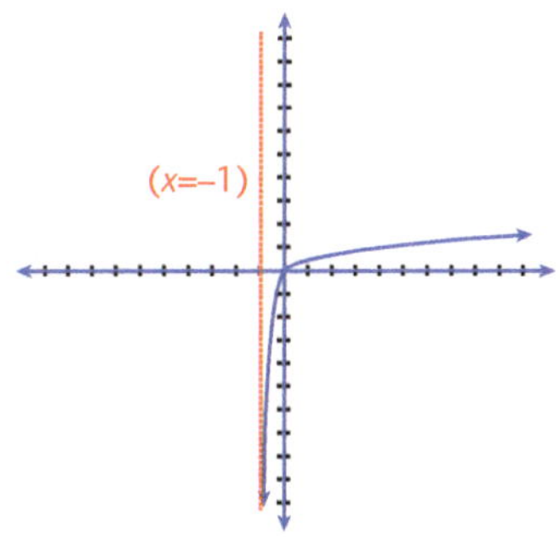

37. a. $x > 3$
    b. (4,0)
    c.

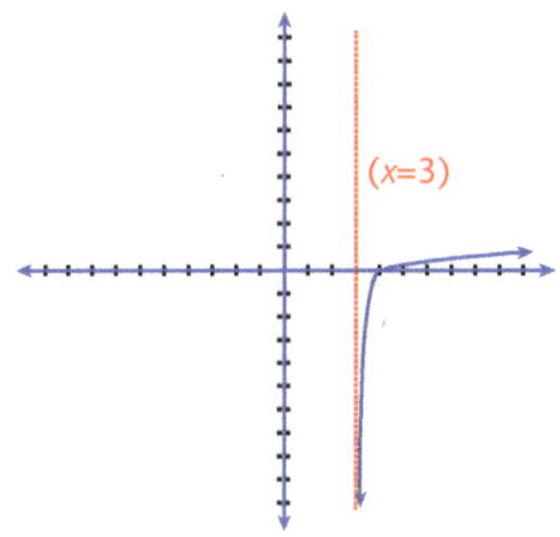

38. a. $x < 3$
    b. (2,0)
    c.

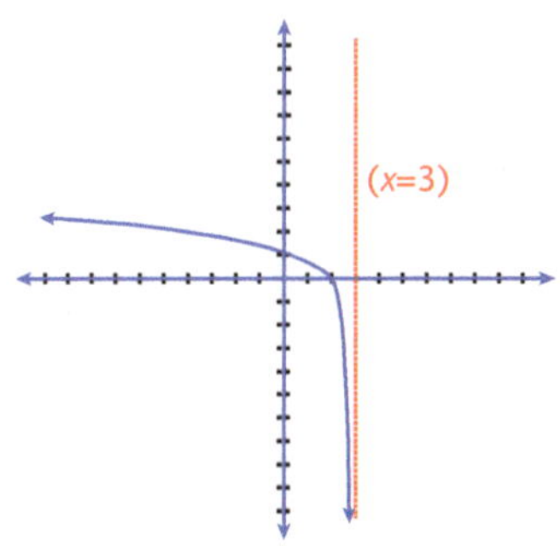

39. a. $x < 0$
    b. $\left(\frac{-1}{e^5}, 0\right)$
    c.

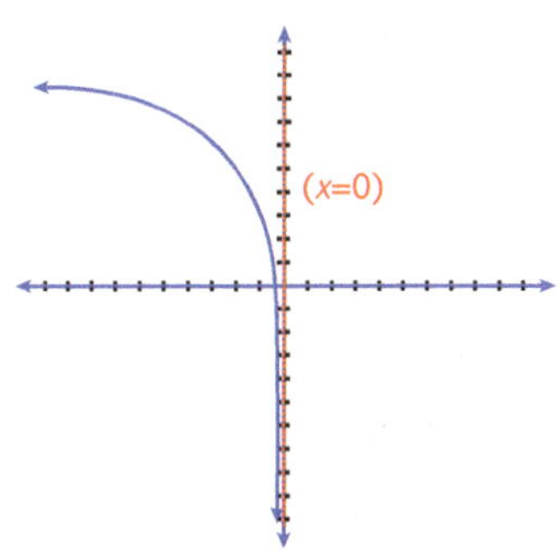

40. a. $x > 1$
 b. (2,0)
 c.

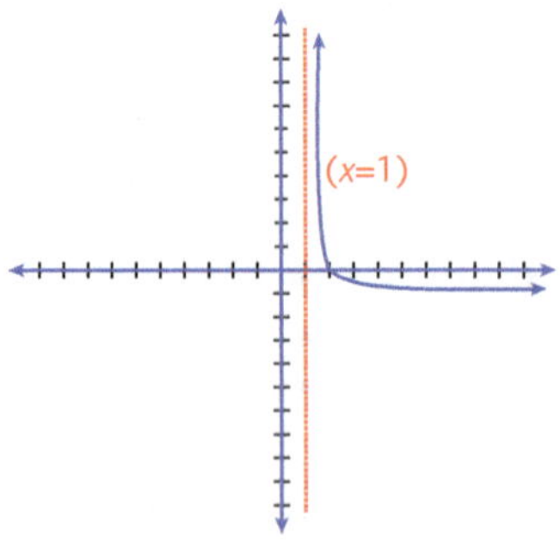

Section 4.4

1. $2 \log x + \frac{1}{2} \log (x^4 - 3x^3)$
2. $2 \log (x + 2) - \log (x - 1)$
3. $\ln x$
4. $x - e \ln x$
5. $\frac{3}{4} [\log_5 (x - 2) - \log_5 (x + 1)]$
6. $7 \log_3 x - \frac{1}{2} \log_3 (x + 1)$
 $- \frac{1}{2} \log_3 (x - 1)$
7. $\ln (e^x - e^{-x}) - 2x$
8. $2 \log x + \frac{1}{2} \log (x + 3)$
 $- \frac{1}{2} \log (x - 1)$
9. $\ln (x - 1) - \frac{1}{2} \ln (x + 1)$
10. $\frac{8}{3} \ln x + \frac{1}{3} \ln (2x - 3)$
 $- \frac{2}{9} \ln (4x + 2)$
11. $\log \left[\frac{(2x - 1)\sqrt[3]{x^4 - 3}}{\sqrt[4]{(2x - 1)^3}} \right]$
12. $\ln x^6$ or $6 \ln x$
13. $\log_5 (x^3 y)$
14. $\log_2 \left[\frac{2x^2 y^3 \sqrt{7}}{3} \right]$
15. $\ln (4x)$
16. $\log (x^4)$
17. $\ln \left(\frac{24x^2}{5} \right)$
18. $14 \ln x$
19. 0
20. $\log_2 6$
21. 3
22. 6
23. $\frac{8}{3}$
24. 4
25. $10 + e$
26. 1
27. 0
28. 2
29. –6
30. 9
31. 9.2
32. 7.7
33. –11.8
34. 5.6
35. 38.13
36. 41.292
37. –42.5
38. –3.8

Section 4.5

1. $x = \frac{1}{2} \ln 6 \simeq 0.896$
2. $x = \frac{\ln 4}{\ln \left(\frac{3}{7} \right)} \simeq 1.636$
3. $x = 1$
4. $x = \frac{1}{2} \left(1 + \frac{\ln 2}{\ln 3} \right) \simeq 0.815$
5. $x = \frac{1}{2} \ln 7 \simeq 0.973$
6. $x = \frac{\ln 3}{\pi \ln 2} \simeq 0.505$
7. $x = -3, 1$
8. $x = \frac{1}{2} \left(\frac{\ln 4}{\ln 31} \right) \simeq 0.202$
9. $x = 0, 1$
10. $x = \frac{\ln 4}{\ln \pi} \simeq 1.211$
11. $x = 100$
12. $x = 6$
13. $x = \frac{1}{2}, -\frac{1}{5}$
14. $x = \pi^2 \simeq 9.87$
15. $x = \pm e^2$
16. N.S.
17. $x = 1$
18. $x = 2, 1$
19. $x = 67$
20. $x = -2, 6$
21. $x = 2$
22. $x = 3 - \frac{\ln \left(\frac{5}{3} \right)}{\ln 4} \simeq 2.632$
23. $x = \pm \sqrt{4 + e^{\frac{\ln 3}{\ln 3 + 1}}} \simeq \pm 2.385$
24. $x = \pm \sqrt{2 + \ln \left(\frac{17}{12} \right)} \simeq \pm 1.532$
25. $x = \pm e^{\frac{15 \ln 10}{2 \ln 10 - 2}} \simeq \pm 572507.57$
26. $x = -2, 1$
27. $x = \frac{3}{2}$
28. $4 - \frac{\ln 15}{\ln 3} \simeq 1.535$
29. $x = 10^{\frac{5 \ln 5}{\ln 50}} \simeq 114.04$
30. N.S.
31. $x = 1$
32. $x = 0, 5$
33. $x = \log_5 3 \simeq 0.6826$
34. N.S.
35. $x = \ln 5 \simeq 1.609$
36. $x = 5$
37. $x = \log_2 3 \simeq 1.58496$
38. $x = 1, e^{\pm\sqrt{3}}$
39. 2.322
40. 2.262
41. 0.431
42. 0.434
43. 11
44. 1.585
45. 1.183

Section 4.6

1. 30.14 years, 100.1 years
2. 53.2 days
3. 1.334 grams
4. 99.99999%, 99.9984%
5. 11,480 years
6. 0.0976%
7. 1.178 days
8. 11.7 days, 19.6 days
9. 107 days

10.

ISOTOPE	HALF-LIFE	y_0	t_1	y_1	t_2 (10 HALF-LIVES)	y_2
Iodine-131	8 days	100 g	30 days	7.432 g	80 days	0.09765625 g
Strontium-90	28.9 days	10 g	90 days	1.155 g	289 days	0.009765625 g
Radium-226	1690 years	1000 g	100 years	959.82 g	16,900 years	0.97740686931 g
Potassium	1.3 billion years	1 g	1000 years	0.999999g	13 billion years	0.0009765625 g

11. 1.3%
12. 30.229 degrees
13. 20.9°C
14. 54.93 years
15. 63 years
16. 1240 years ago
17. 8.46 days
18. 49.44 hours
19. 10 watts per square meter
20. 111.76 decibels
21. 251 times larger
22. 10,000 times larger
23. 13.9 years
24. 3.48%
25. \$5; \$500

CHAPTER 5

Section 5.1

1. arithmetic; 18, 22, 26
2. neither; 5, 8, 13
3. geometric; $\frac{1}{16}, \frac{1}{32}, \frac{1}{64}$
4. geometric; –48, 96, –192
5. arithmetic; 5, 7, 9
6. geometric; 81, 243, 729
7. arithmetic; 37, 42, 47
8. neither; 256, 8192, 2097152
9. geometric; $\frac{1}{5}, \frac{1}{25}, \frac{1}{125}$
10. arithmetic; –2, –5, –8
11. neither; 370, –590, 960
12. geometric; 162, –486, 1458
13. a. $a_n = 2n + 1$
    b. $a_n = \frac{3}{2}(2^n)$
14. a. $a_n = -3n + 4$
    b. $a_n = (\frac{-1}{3})(-3)^n$
15. a. $a_n = 4n - 9$
    b. $a_n = (\frac{-5}{4})(4^n)$
16. a. $a_n = \frac{1}{2}n - \frac{1}{2}$
    b. $a_n = 0$
17. a. $a_n = -5n + 7$
    b. $a_n = \frac{-2}{5}(-5)^n$
18. a. $a_n = 2n - 3$
    b. $a_n = \frac{-1}{2}(2^n)$
19. a. $a_n = -n + 101$
    b. $a_n = -100(-1)^n$
20. a. $a_n = -4n + 13$
    b. $a_n = \frac{-9}{4}(-4)^n$
21. a. $a_n = \frac{3}{2}n - \frac{7}{2}$
    b. $a_n = \frac{-4}{3}\left(\frac{3}{2}\right)^n$
22. a. $a_n = \frac{1}{4}n + \frac{31}{4}$
    b. $a_n = 32\left(\frac{1}{4}\right)^n$
23. a. $a_n = 6n - 6$
    b. $a_n = 0$
24. a. $a_n = 3n - 13$
    b. $a_n = \frac{-10}{3}(3^n)$
25. a. $a_n = 5n - 11$
    b. $a_n = \frac{-6}{5}(5^n)$
26. 149
27. 197
28. 259
29. –90
30. $29\frac{1}{2}$
31. 133
32. –245
33. $25\frac{2}{3}$
34. $\frac{81^5}{2}$
35. $-5(16^5)$
36. $\frac{3}{16^5}$
37. 7
38. 10^{-18}
39. 5^{-16}
40. 3^{20}
41. 4^{17}
42. 14950
43. –20100
44. 25250
45. –40500
46. –29500
47. 15650

48. 44650
49. 10100
50. $\frac{1023}{256}$
51. 349,525
52. $\frac{4882812}{125}$ or 39062.496
53. $\frac{58025}{2187} \simeq 26.53$
54. 3069
55. −1627604
56. 0
57. $\frac{29524}{27} = 1093.\overline{481}$

Section 5.2

1. a. $2,100
   b. $2,500
   c. $500
2. a. $2,130
   b. $2,650
   c. $650
3. $150, not much difference
4. a. $11,200
   b. $16,000
5. a. $11,350
   b. $16,750
6. $750; Yes, a larger principal makes a difference
7. 20 years
8. 10 years
9. 4 years
10. The Rule of 72 seems to predict that a person's money will double more quickly than it actually does.
11. account a pays more
12. account b pays more $5.51
13. account b pays more $2.33
14. account b pays more $4.23
15. account b pays more $8.37
16. account b pays more $47.91
17. account b pays more $324.44
18. account b pays more $468.41
19. $1,276.28
20. $1,829.83
21. $12,517.96
    $12,523.05
22. Row 1 $6,000.00
    Row 2 $6,083.26
    Row 3 $6,100.95
    Row 4 $6,104.98
    Row 5 $6,106.95
23. about 14 yrs
24. about 7 yrs
25. Just under 3 years
26. The Rule of 72 estimates are better with compound interest than with simple interest.
27. $925.18
28. $805.92
29. $12,712.24
30. $51,458.62
31. $1,797,482.07
32. $714.75
33. 12.5%
34. 12.66%
35. 12.19%
36. yes
37. 12.00%
38. 23.98%
39. 8.84%
40. 5.116%; $893,548.86

Section 5.3

1. $1,026.56
2. $2,126.89
3. $552.56
4. 4 years; 0 year
5. $564,425.75
6. $5,301.50
7. $1,996.50
8. $13,814.08
9. $14,092.74
10. $5,872.27
11. $28,185.47; yes
12. $29,070.85
13. $3,759.60
14. $409,689.96
15. $40,160.57
16. $642,025.79
17. $197,224.70
18. $294,252.53
19. $607.49
20. $1,216.00
21. $1,100.04
22. $364.25; saves $58.41
23. $3.26
24. $7.34
25. $295,846.40
26. $46,435.11
27. $44,760.60
28. $271,897.20
29. $613,405.00
30. $698,257.72

Section 5.4

1. $272.32
2. $3,336.58
3. $35,328.16
4. $1,945.41
5. $3,627.34
6. $1,839,734,222
7. $439,105,648.90
8. No, the costs are greater than benefits
9. $268,713.56
10. $4,924.09
11. $490,907.37; $343,635.16
12. $11,454,505.31
13. $18,726.68
14. $29,880.72
15. $2412.55
16. $5311.76
17. $4,090,314,453
18. $3,031,972,708
19. $335,312.51
20. $14,383.98
21. $2,494,981.60; $3,000,000
22. $2,156,649.07
23. $22,852.17
24. $16,091.81
25. $29,967.86
26. $122,639.12

27. $1,164.70
28. $921.42
29. $10,840.16; $1,159.84

Section 5.5

1. $23,739.64
2. $1,343.75
3.

Period	Principle at Beginning of the Period	Interest for the Period	Payment	Principle Paid	Unpaid Principle
1	$100,000	$6,000	$23,740	$17,740	$82,260
2	$ 82,260	$4,936	$23,740	$18,804	$63,456
3	$ 63,456	$3,807	$23,740	$19,933	$43,523
4	$ 43,523	$2,611	$23,740	$21,129	$22,394
5	$ 22,394	$1,344	$23,740	$22,396	($2)

4. $1,283.29
5. $1,502.51
6. $68,336; yes
7. $363.98
8. $200; $163.98; $3,887.89 savings
9. 17.2%
10. $1,703; $54,386.36
11. $28,291
12. $38,882; yes
13. $419,369.30
14. $466,234.61
15.

Period	Principle at Beginning of the Period	Interest for the Period	Payment	Principle Paid	Unpaid Principle
1	$7,200,000	$108,000	$419,369	$311,369	$6,888,631
2	$6,888,631	$103,329	$419,369	$316,040	$6,572,591
3	$6,572,591	$ 98,589	$419,369	$320,780	$6,251,811
4	$6,251,811	$ 93,777	$419,369	$325,592	$5,926,220
5	$5,926,220	$ 88,893	$419,369	$330,476	$5,595,744
6	$5,595,744	$ 83,936	$419,369	$335,433	$5,260,311
7	$5,260,311	$ 78,905	$419,369	$340,464	$4,919,847
8	$4,919,847	$ 73,798	$419,369	$345,571	$4,574,275

16. $38,507,682.28
17. $15,979,938.71
18. $196,434,145.60
19. $3,669.09
20. $68,058.06
21. $26,442.21
22. 17%
23. $15,028.46
24. $91,714.93
25. $2,271.16
26. $2,260.80; save $10.36 per month for an overall savings of $3,729.60
27. $2,096.31
28. $315.03
29. $38,007.51
30. $37,771.44; monthly savings of $236.07 for an overall savings of $42,492.60.
31. $1,995.18
32. $307.90
33. $9,650.59

Review Answers

CHAPTER 1 REVIEW ANSWER KEY

1. $x = -5$
2. $x = \frac{62}{11}$
3. $x = -23$
4. $x = 0$
5. $x = \frac{7}{8}$
6. $x = \frac{-47}{3}$
7. $x = \frac{14}{13}$
8. N.S.
9. 10
10. $1722.60
11. 22, 24 and 26
12. $x \geq 15$

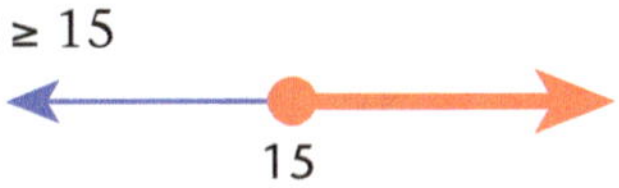

13. $x \geq 48$

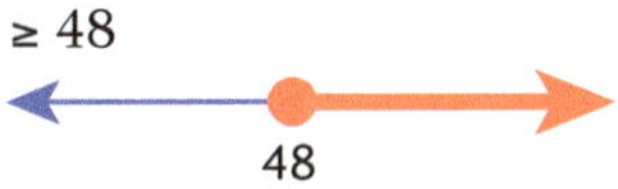

14. $x \leq \frac{24}{13}$

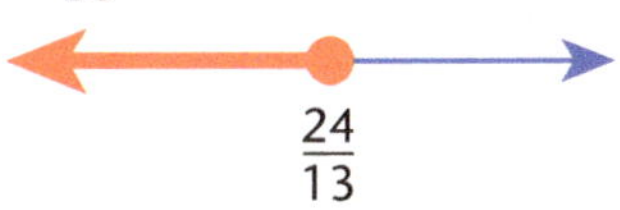

15. $-2 \leq x < 1$

16. $w > -11$

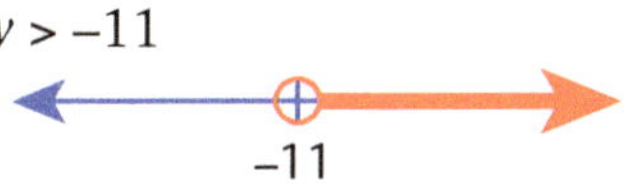

17. $x < \frac{3}{10}$

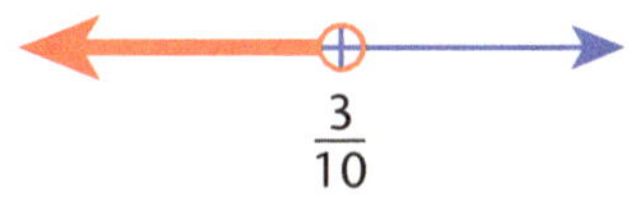

18. 71
19. $5,000
20. $\frac{-1}{3}$
21. 0
22. undefined slope
23. $y = 3x + 11$
24. $y = -2x + 5$
25. $y = -\frac{1}{2}x + 2$
26. $y = -5x + 9$
27. $x = 10$
28. $53.75
29. (1, 1)
30. (–3, 0)
31. (2, 9)
32. N.S.
33. same line; infinitely many solutions
34. N.S.
35. $\left(\frac{1}{2}, -\frac{1}{3}\right)$
36. same line; infinitely many solutions
37. (5, 3, 0)
38. (–1, 7, –3)
39. 35 hats, 65 party favors
40. not a function
41. is a function; domain: $x \in \mathbb{R}$, $x \neq 5, -\frac{1}{2}$
42. is a function; domain: $x \in \mathbb{R}$, $x \geq -5$
43. is a function; domain: $x \in \mathbb{R}$, $x > 4$

44. **a.** 7
b. $2h^2 - 4h + 1$

45. **a.** $\dfrac{9}{\sqrt{2}}$ or $\dfrac{9\sqrt{2}}{2}$
b. $\dfrac{9}{\sqrt{3w+5}}$

46. **a.** $-\dfrac{9}{4}$
b. $\dfrac{2x^2+1}{x^2-8}$

47. **a.** $\sqrt{3}$
b. $\sqrt{2-y-h} + \sqrt{y+h+1}$

48. **a.** $\sqrt{5}$
b. $\sqrt{x^4 + 2x^2 - 3}$

49. (150, \$21)

50. \$190; 360 chairs

51. (a) 201 shirts; (b) \$2;
(c) (210, \$18)

52. $p = -\dfrac{-1}{800}q + 2$; \$0.75 each

53. 400 books; \$18,000; \$18,000

54. 77 jars; \$237.50; \$725

55. 10 students; \$1,110

56. (a) $P = 3.5x - 310$;
(b) (88.6, \$575.70)

57.

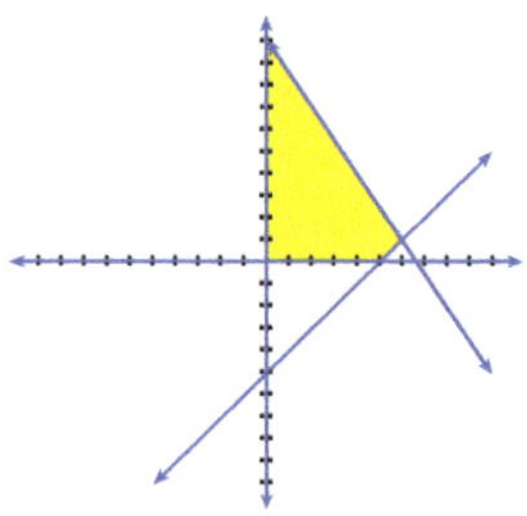

58.

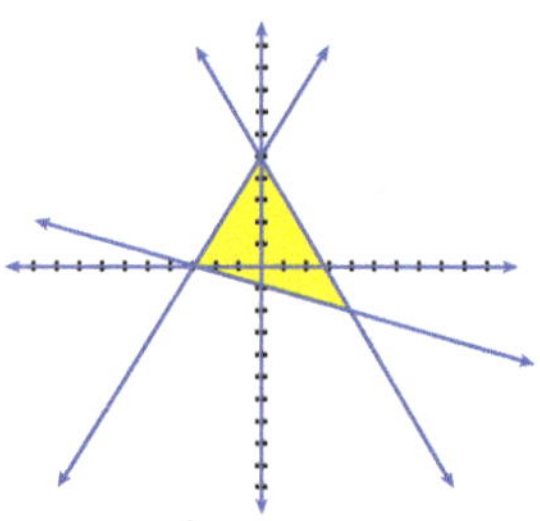

59.

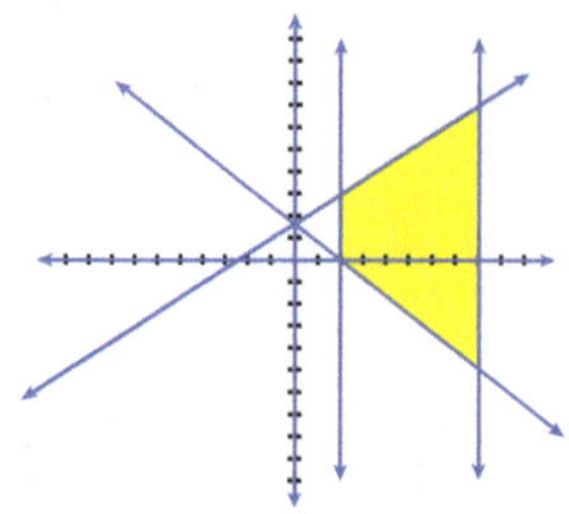

60.

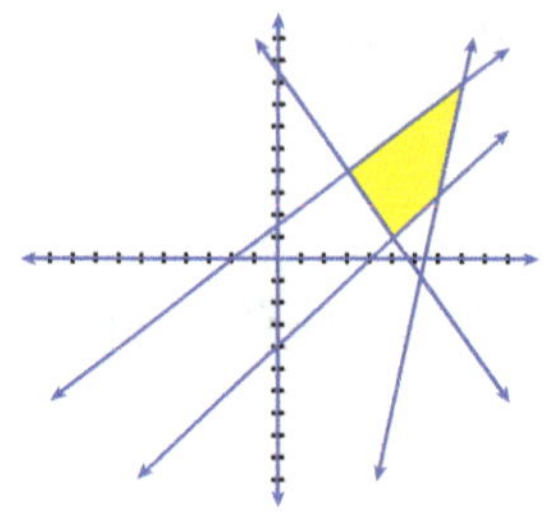

61.

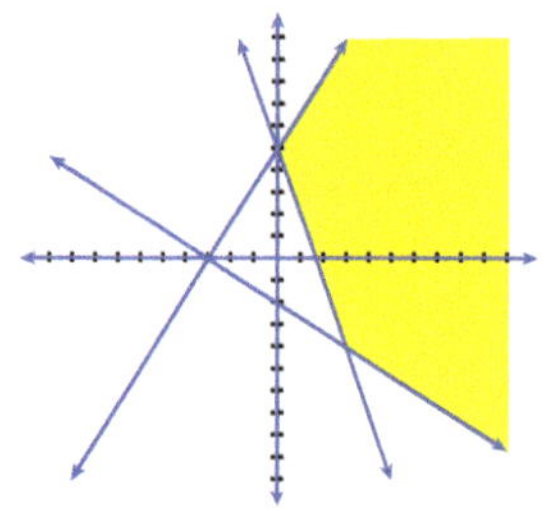

62. $x \geq 0,\ y \geq 0$
$x + y \leq 65$
$10x + 6y \geq 590$

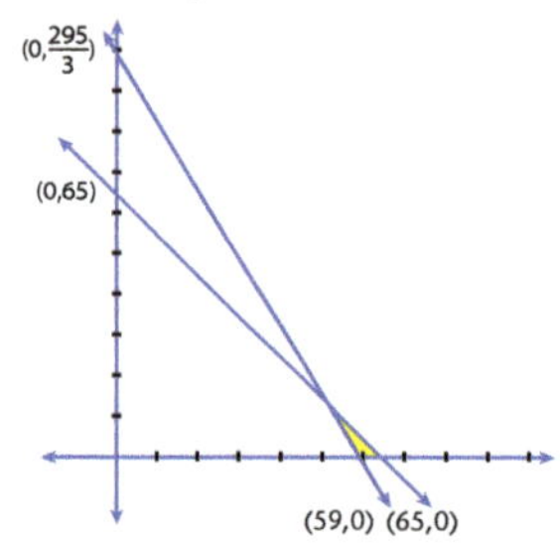

63. maximum of 40 at (0, 8)
minimum of 11 at (2, 3)

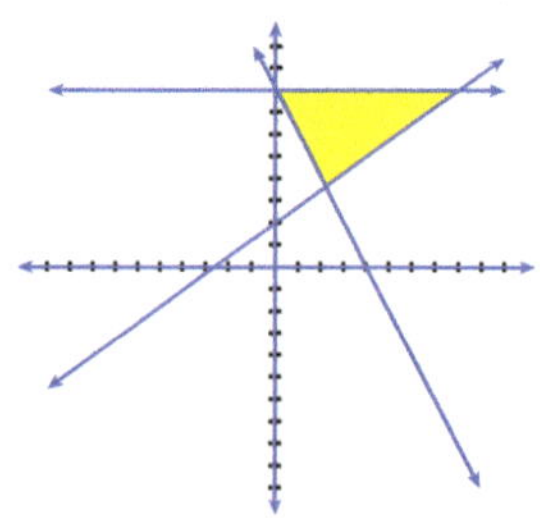

64. maximum of 9 at (3, 0)
minimum of 0 at (0, 0)

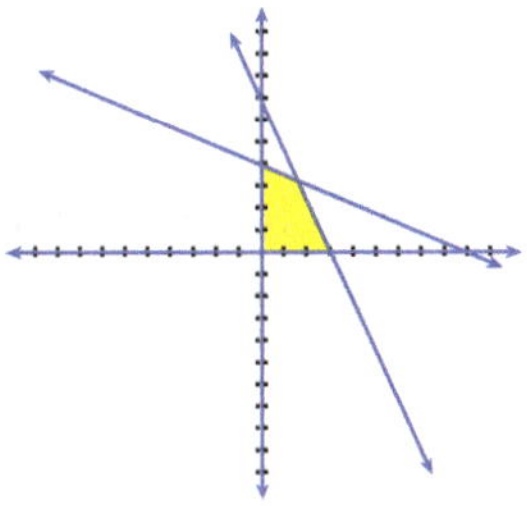

65. maximum of 15 at (3, 0)
minimum of –25 at (–5, 0)

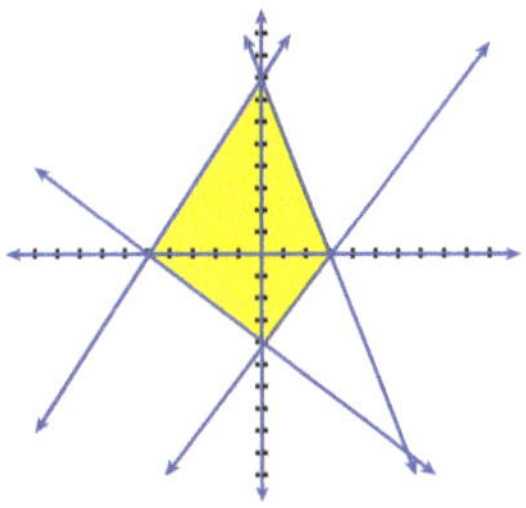

66. $325; 65 adults and 0 students

CHAPTER 2 REVIEW ANSWER KEY

1. $\begin{bmatrix} -36 & 36 \\ 20 & -36 \\ 12 & 16 \end{bmatrix}$

2. a. $\begin{bmatrix} 0 & 15 & 16 \\ 4 & 0 & 13 \end{bmatrix}$ **b.** $\begin{bmatrix} 15 & -15 & -8 \\ -2 & 30 & -19 \end{bmatrix}$

3. a. $\begin{bmatrix} 9 & 7 \\ 9 & 8 \end{bmatrix}$ **b.** $\begin{bmatrix} 0 & 6 \\ 14 & 16 \end{bmatrix}$ **c.** $\begin{bmatrix} 22 & 9 \\ 13 & 5 \end{bmatrix}$

4. $\begin{bmatrix} 12 & 2 & 0 \\ 8 & 6 & 3 \\ -3 & -9 & 1 \end{bmatrix}$

5. $x = -4, y = -1, z = -2, w = 2$

6. a. impossible
b. 3×2
c. 2×3
d. impossible
e. impossible
f. 3×2
g. 2×3
h. impossible

7. a. $\begin{bmatrix} 95 & 34 \\ 75 & 56 \end{bmatrix}$ **b.** $\begin{bmatrix} 95 & 75 \\ 34 & 56 \end{bmatrix}$

8. a. $\begin{bmatrix} 25 & 15 \\ -5 & 25 \end{bmatrix}$ **b.** $\begin{bmatrix} 20 & 10 \\ -10 & 30 \end{bmatrix}$ **c.** $\begin{bmatrix} 20 & -10 \\ 10 & 30 \end{bmatrix}$

9. a. $\begin{bmatrix} 1 & 8 & 47 \\ 0 & 1 & 15 \\ 0 & 0 & 4 \end{bmatrix}$ **b.** $\begin{bmatrix} 1 & 12 & 143 \\ 0 & 1 & 35 \\ 0 & 0 & 8 \end{bmatrix}$

10. a. $\begin{bmatrix} -24 \\ -14 \\ 5 \end{bmatrix}$
b. impossible
c. [2 3 –7]
d. $\begin{bmatrix} 2 \\ 3 \\ -7 \end{bmatrix}$

11. (–4, 2, –4)
12. N.S.
13. $(1 + 5a, 1 + a, a)$
14. (1, 3, –4)
15. (5, 0,–2)
16. $(a, 3a + 1, 5 - 4a)$

17. $A^{-1} = \begin{bmatrix} \frac{3}{5} & \frac{2}{5} \\ -\frac{1}{5} & \frac{1}{5} \end{bmatrix}$

18. $A^{-1} = \begin{bmatrix} \frac{1}{4} & \frac{1}{2} & -\frac{1}{4} \\ -\frac{1}{12} & -\frac{1}{2} & \frac{3}{4} \\ -\frac{1}{12} & \frac{1}{2} & -\frac{1}{4} \end{bmatrix}$ or $\frac{1}{12}\begin{bmatrix} 3 & 6 & -3 \\ -1 & -6 & 9 \\ -1 & 6 & -3 \end{bmatrix}$

19. $A^{-1} = \begin{bmatrix} 0 & -1 \\ \frac{1}{4} & \frac{5}{4} \end{bmatrix}$

20. not possible; A is not a square matrix
21. DNE
22. DNE

23. $A^{-1} = \begin{bmatrix} \frac{5}{18} & \frac{1}{9} & -\frac{1}{18} \\ \frac{1}{6} & -\frac{1}{3} & \frac{1}{6} \\ -\frac{2}{9} & \frac{1}{9} & \frac{4}{9} \end{bmatrix}$ or $\frac{1}{18}\begin{bmatrix} 5 & 2 & -1 \\ 3 & -6 & 3 \\ -4 & 2 & 8 \end{bmatrix}$

24. $X = \begin{bmatrix} -18 \\ 11 \\ 12 \end{bmatrix}$

25. 71 –167 77 35 –86 42 37 –112 69 –7 –14 35
26. DINNER IS READY
27. 3 of Portfolio 1, 2 of Portfolio 2 and 1 of Portfolio 3
28. 29 mg food 1, 49 mg food 2, 22 mg food 3
29. 14 oz. peanuts, 5 oz. cashews, 5 oz. almonds
30. \$18,000 from her sister, \$26,000 from her brother, and \$36,000 from her parents

CHAPTER 3 REVIEW ANSWER KEY

1. $x = \frac{\pm 5}{2}$
2. $x = -4, \frac{3}{2}$
3. $x = -2 \pm \sqrt{11}$
4. $x = -\frac{1}{3}, \frac{5}{2}$
5. $w = -1, 3$
6. $x = 3, -8$
7. $x = 11, -3$
8. $y = -6, 7$
9. **a.** (–3, –4)
   b. $x = -3$
   c. up
   d. minimum
   e. $(-3 + \sqrt{2}, 0)$ and $(-3 - \sqrt{2}, 0)$
   f.

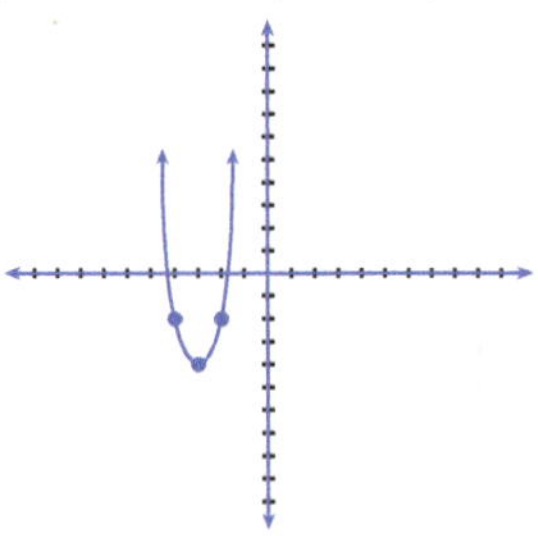

10. **a.** (5, 2)
    b. $x = 5$
    c. down
    d. maximum
    e. $(5 + \sqrt{2}, 0)$ and $(5 - \sqrt{2}, 0)$
    f.

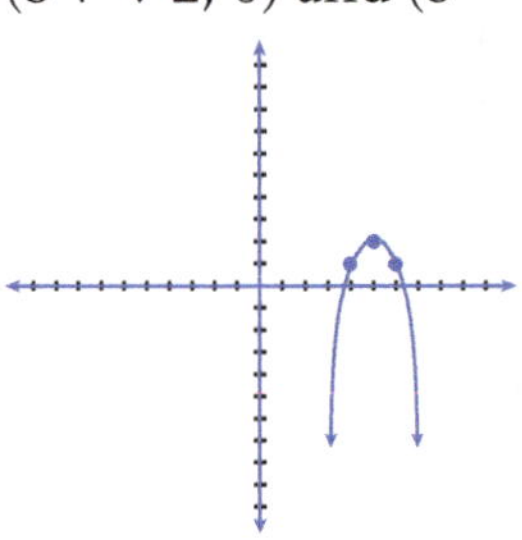

11. **a.** (2, 1)
    b. $x = 2$
    c. up
    d. minimum
    e. none
    f.

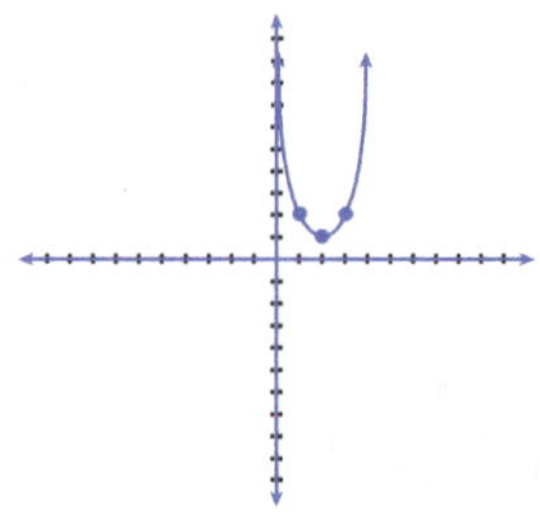

12. **a.** $\left(\frac{3}{2}, \frac{49}{8}\right)$
    b. $x = \frac{3}{2}$
    c. down
    d. maximum
    e. (–2, 0) and (5, 0)
    f.

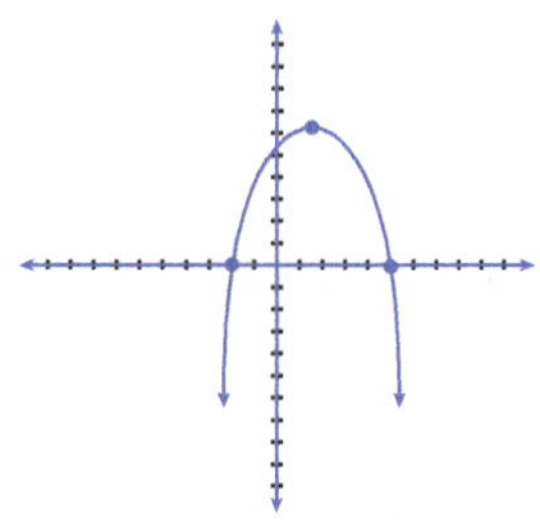

13. **a.** (–1, –1)
    b. $x = -1$
    c. down
    d. maximum
    e. none
    f.

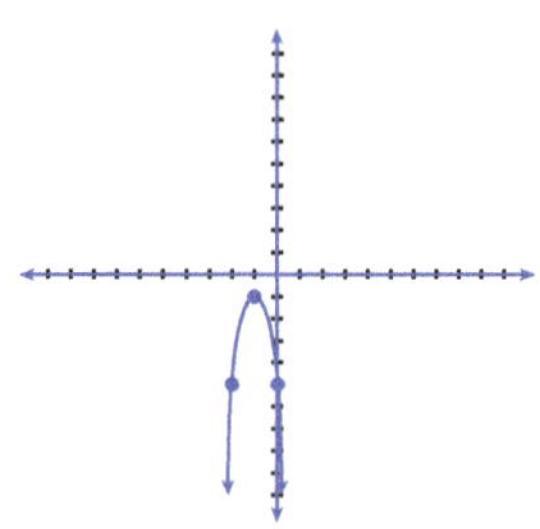

14. **a.** (3, –4)
b. $x = 3$
c. up
d. minimum
e. (1, 0) and (5, 0)
f.

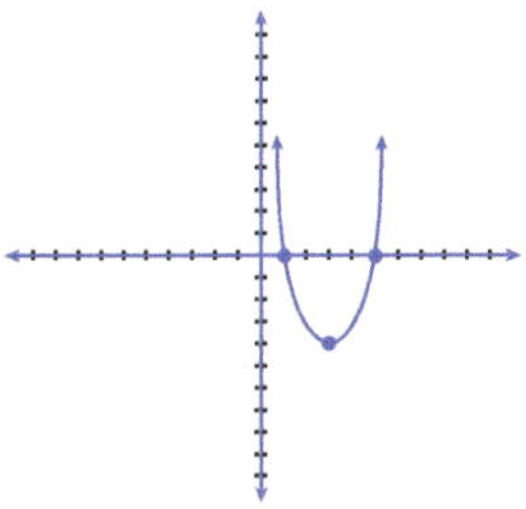

15. **a.** (0, –2)
b. $x = 0$
c. up
d. minimum
e. (3, 0) and (–3, 0)
f.

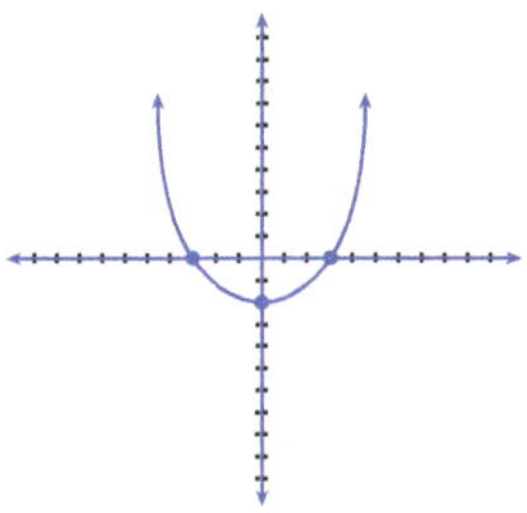

16. 92256 square feet
17. 580 boxes, $671,800
18. ~231 or 1000 units
19. 20 or 31 units
20. 14 or 28 units
21. (4, $68)
22. (7, $294)
23. (177, $39)
24. (10, $24)
25. 20; $8
26. **a.** 3
b. $x = 0, 2, -4$
c. (0, 0)
d. (0, 0), (2, 0) and (–4, 0)
e.

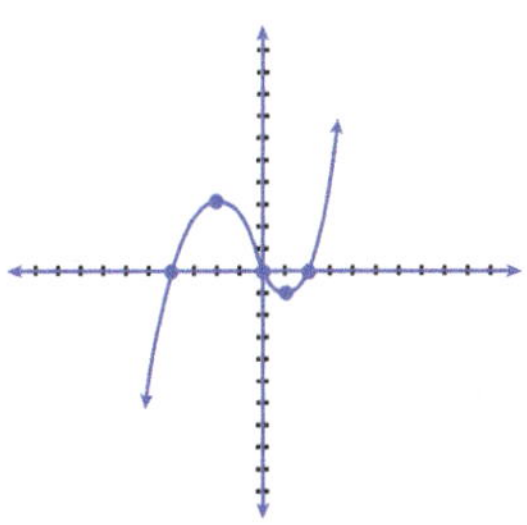

27. **a.** 4
b. $x = -1, 3$
c. $\left(0, \frac{9}{8}\right)$
d. (–1, 0) and (3, 0)
e.

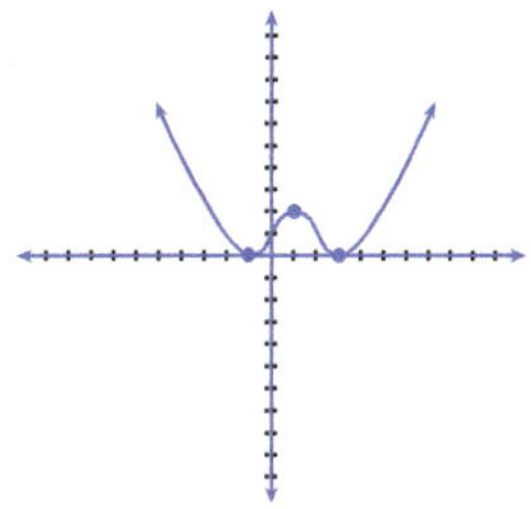

28. **a.** 3
b. $x = -4, 4$
c. (0, 4)
d. (4, 0) and (–4, 0)
e.

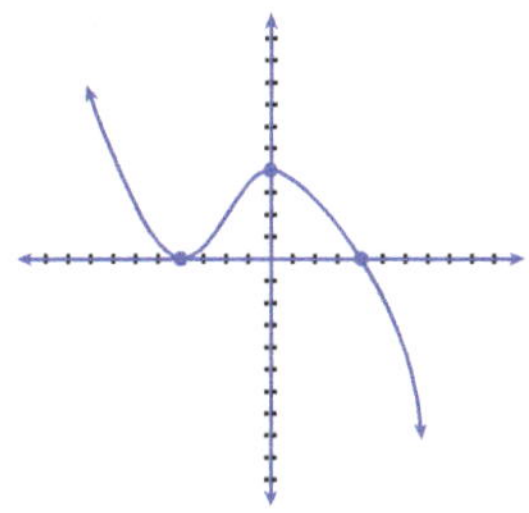

29. **a.** 4
b. $x = 2, -2, 6$
c. (0, 2)
d. (–2, 0), (2, 0) and (6, 0)
e.

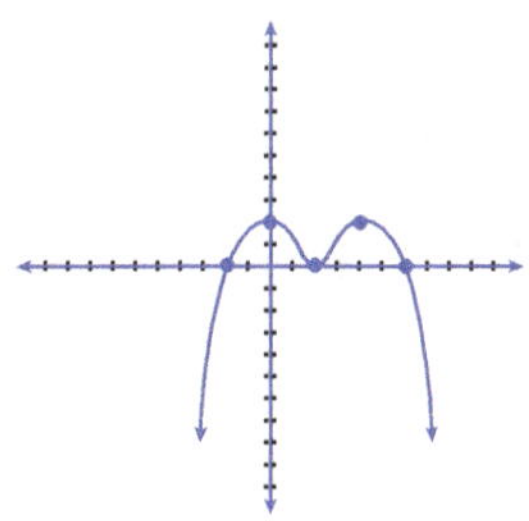

30. **a.** 5

b. $x = 0, 1, -1$

c. (0, 0)

d. (0, 0), (1, 0) and (–1, 0)

e.

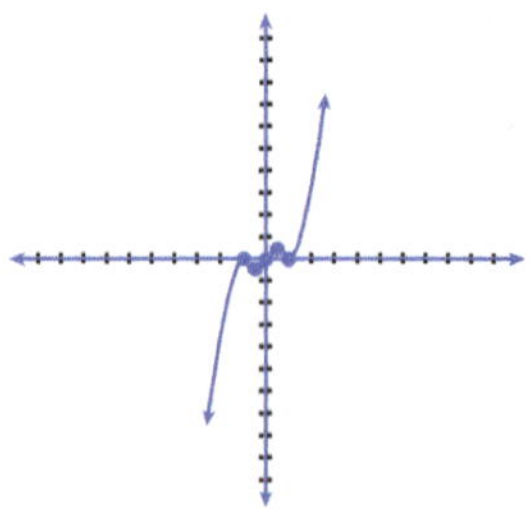

31.

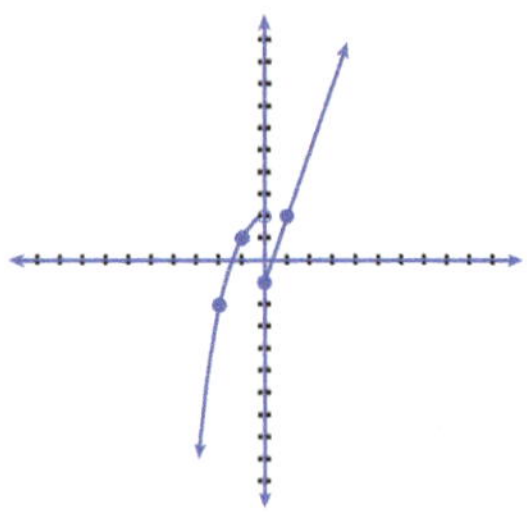

x	–2	–1	0	1	2	3
y	–2	1	–1	2	5	8

32.

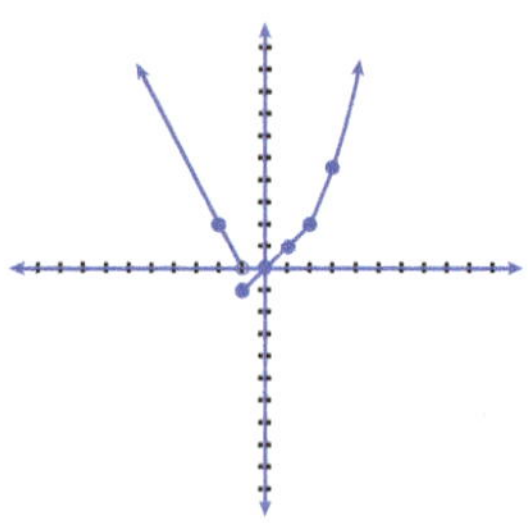

x	–2	–1	0	1	2	3
y	2	–1	0	1	2	4.5

33.

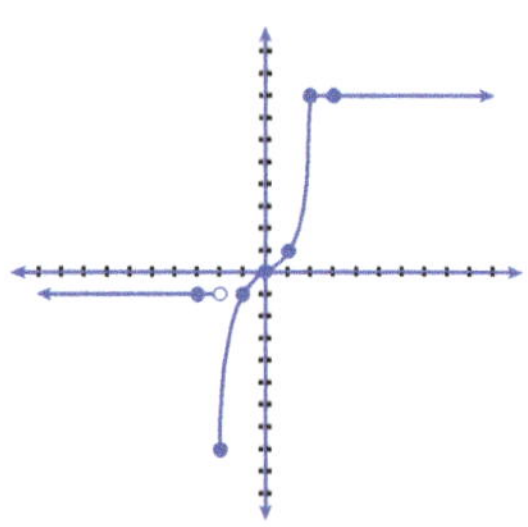

x	–3	–2	0	1	2	3
y	–1	–8	0	1	8	8

34. **a.** $x \in \mathbb{R}, x \neq 4$

b. $x = 4$

c. $y = 0$

d. (0, –0.75)

e. none

f.

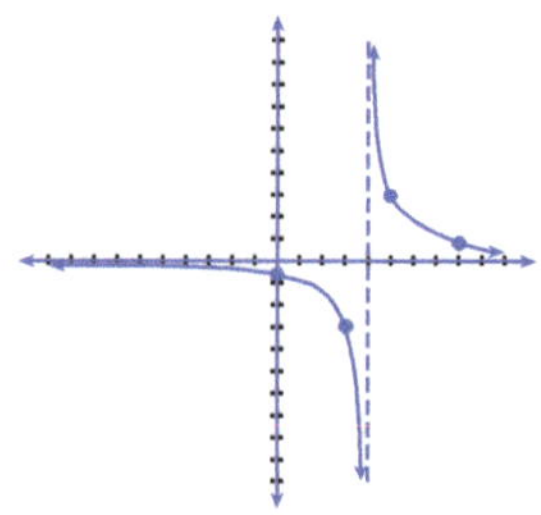

35. **a.** $x \in \mathbb{R}, x \neq 0$

b. $x = 0$

c. $y = 2$

d. none

e. (–0.5, 0)

f.

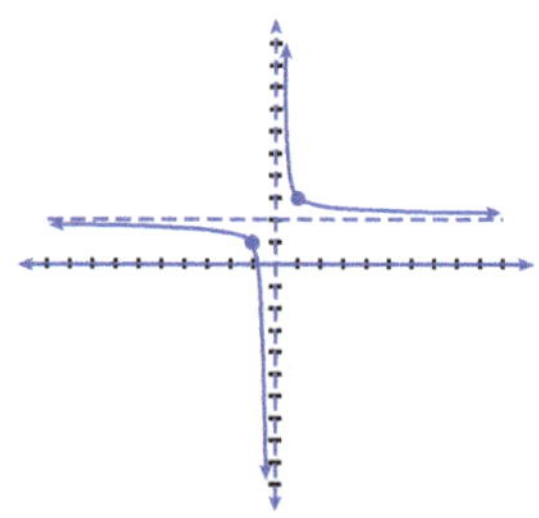

36. **a.** $x \in \mathbb{R}, x \neq 2, -1$

b. $x = -1$

c. $y = 1$

d. (0, 2)

e. (–2, 0)

f.

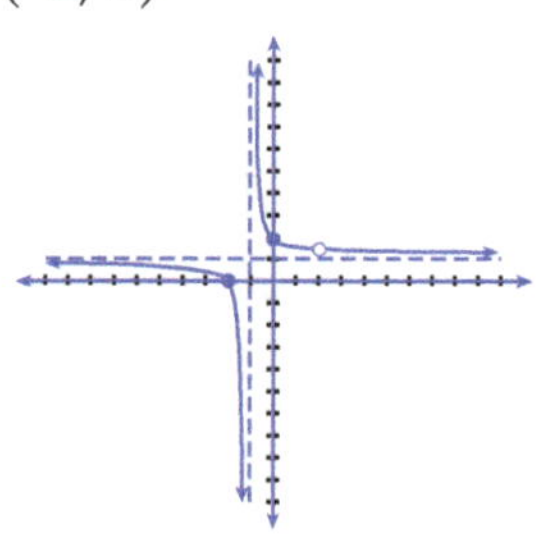

37. **a.** $x \in \mathbb{R}, x \neq \pm 2$
b. $x = 2, x = -2$
c. $y = 0$
d. (0, 0)
e. (0, 0)
f.

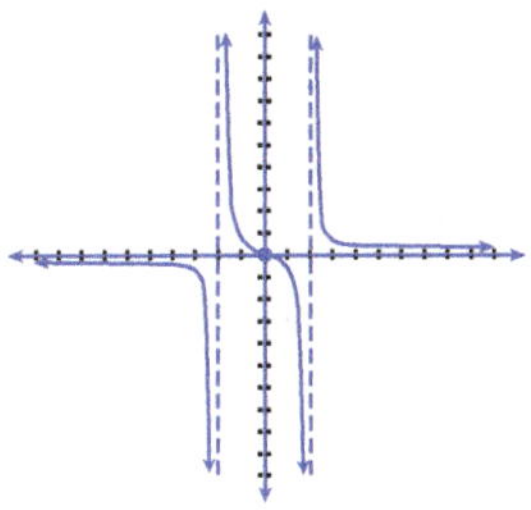

38. **a.** $x \in \mathbb{R}, x \neq \pm 3$
b. $x = 3, x = -3$
c. $y = 0$
d. $\left(0, -\frac{4}{9}\right)$
e. none
f.

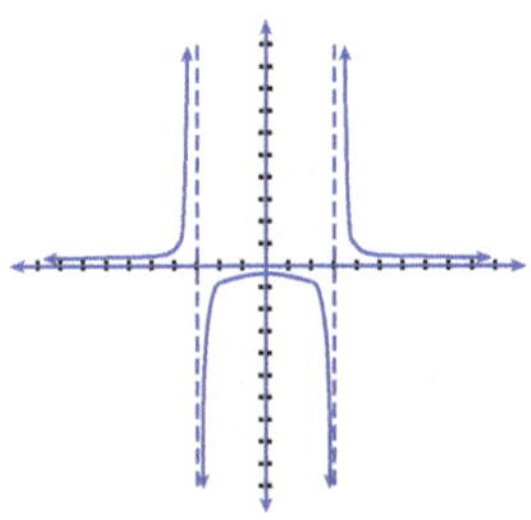

39. **a.** down 3
b. left 1
c. no
d. no
e. yes; vertical
f.

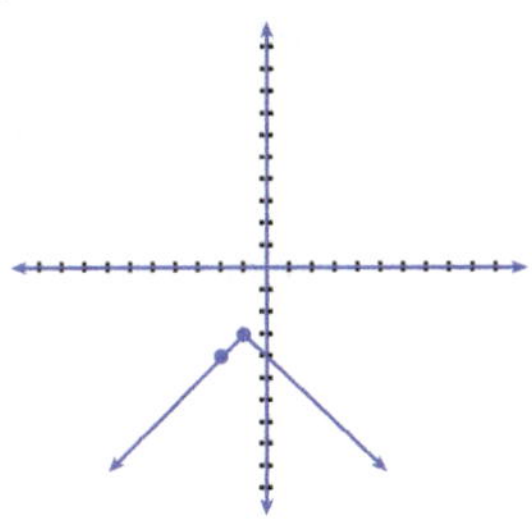

40. **a.** up 5
b. left 2
c. yes; shrunk by half
d. no
e. no
f.

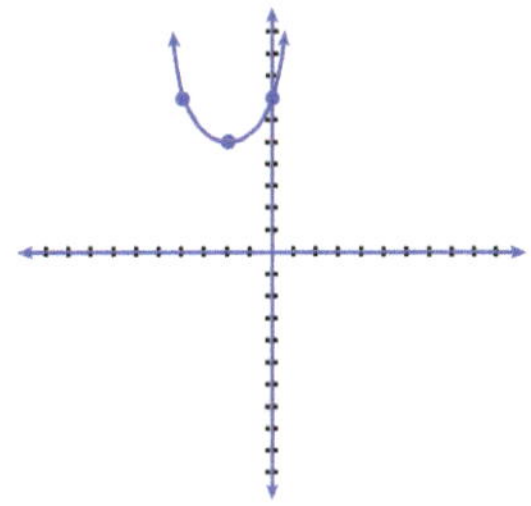

41. **a.** up 2
b. right 1
c. no
d. yes; shrunk 4
e. yes; horizontal
f.

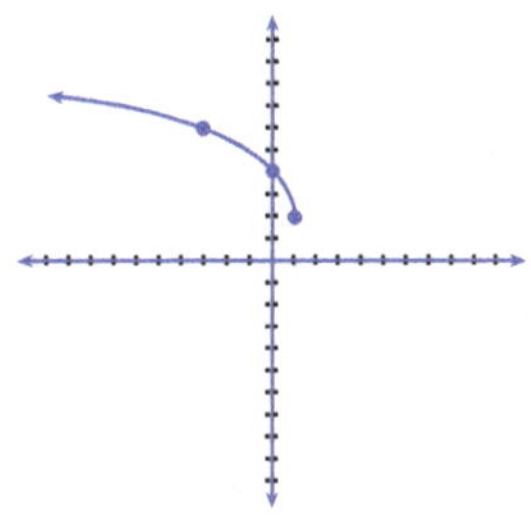

42. **a.** up 10
b. left 3
c. no
d. no
e. yes; vertical
f.

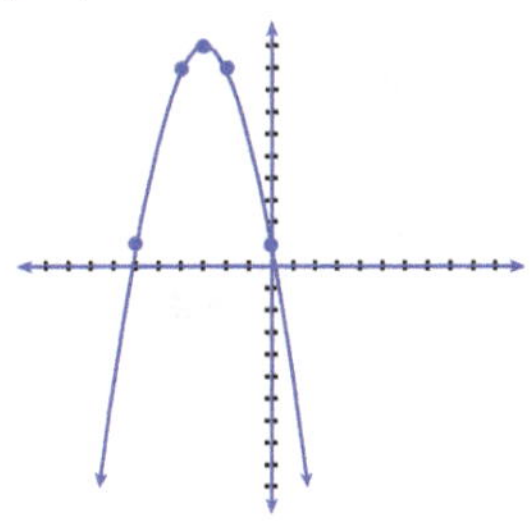

43. **a.** none
b. right 4
c. yes; stretched 2
d. no
e. no
f.

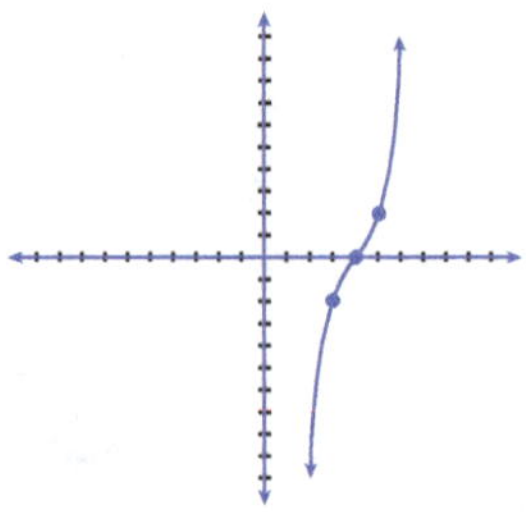

44. **a.** $\sqrt{2x+1} - x + 5$
b. $4\sqrt{3}$
c. $\sqrt{-2x+11}$
d. $5 - \sqrt{2x+1}$
e. $3 - \sqrt{5}$
f. $\dfrac{\sqrt{7}}{2}$
g. $\sqrt{2\sqrt{2x+1}+1}$

45. **a.** $x^2 + |x| + 2$
b. 0
c. $x^2 - 2|x| + 4$
d. $|x^2 + 3| - 1$ or $x^2 + 2$
e. -6
f. 6
g. $x^4 + 6x^2 + 12$

46. **a.** $x^3 + x + \dfrac{2}{x+1}$
b. 2
c. $\dfrac{2(x^2 + 2x + 5)}{(x+1)^3}$
d. $\dfrac{2}{x^3 + x + 1}$
e. $-\dfrac{28}{3}$
f. 60
g. $x^9 + 3x^7 + 3x^5 + 2x^3 + x$

47. **a.** $x^2 + \dfrac{x+5}{x-3}$
b. -3
c. $\dfrac{x^2+5}{x^2-3}$
d. $\dfrac{(x+5)^2}{(x-3)^2}$
e. 11
f. DNE
g. $\dfrac{5-3x}{x-7}$

48. **a.** $2x - 6 + \dfrac{x}{x^2 - 25}$
b. $\dfrac{1}{6}$
c. $\dfrac{2x}{x^2-25} - 6$
d. $\dfrac{2x-6}{4x^2 - 24x + 11}$
e. $\dfrac{40}{21}$
f. 0
g. $4x - 18$

49. **a.** $x^3 + \sqrt{x} - 1$
b. 0
c. $\sqrt{x^3 - 1}$
d. $x\sqrt{x} - 1$
e. $7 - \sqrt{2}$
f. $\dfrac{\sqrt{3}}{26}$
g. $\sqrt[4]{x}$

50. **a.** $x^2 - x$
b. -16
c. $-4x^2 - 12x$
d. $16x^2 - 12x$
e. 18
f. $-\dfrac{2}{3}$
g. $16x$

CHAPTER 4 REVIEW ANSWER KEY

1. $f^{-1}(x) = \sqrt[3]{\dfrac{x+4}{2}} - 3$

2. $h^{-1}(x) = \dfrac{(3x)^5 + 1}{2}$

3. $f^{-1}(w) = \dfrac{9w}{3 - 4w}$

4. $p^{-1}(y) = \dfrac{216}{(y-2)^3}$

5. $g^{-1}(x) = \sqrt[3]{\dfrac{5x-1}{x+1}}$

6. $y^{-1}(x) = \dfrac{7 + 11x^3}{2x^3 - 1}$

7. $f^{-1}(p) = -1 - \dfrac{2}{p}$

8. **a.** domain: $x \geq 1$; $f^{-1}(x) = \sqrt{\dfrac{x-3}{5}} + 1$

   b. domain: $x \geq -1$; $y^{-1}(x) = 2(x+1)^2$

   c. domain: $x \geq 0$; $h^{-1}(x) = \sqrt[4]{\dfrac{7-x}{2}}$

9.

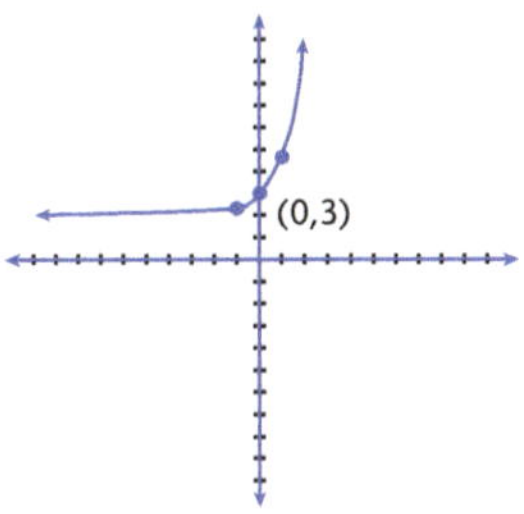

10.

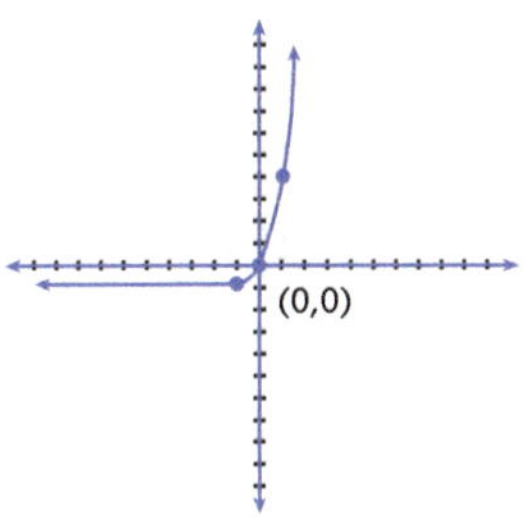

11.

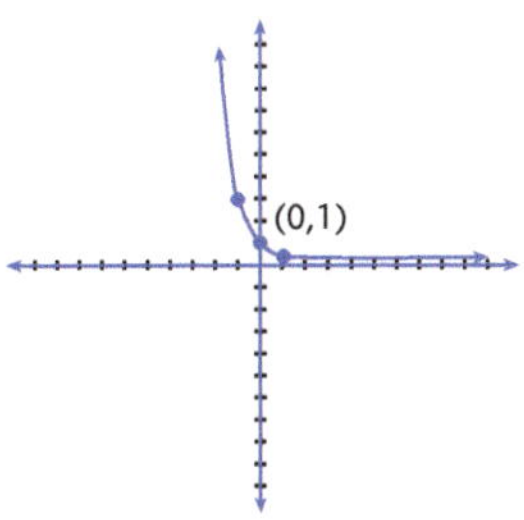

12.

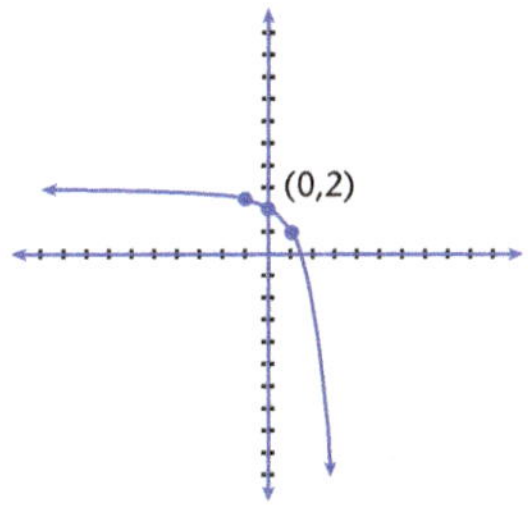

13.

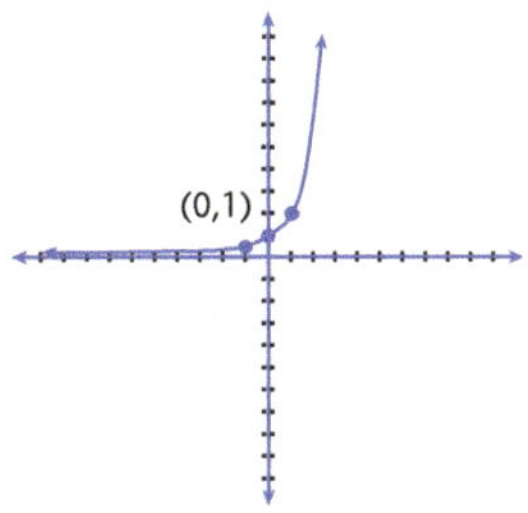

14. $3^{12x - 3x^2}$
15. 4^{2x-2}
16. π^{5x+1}
17. $(-36)^x e^x$
18. $\left(\dfrac{\pi}{e}\right)^{2x-3y}$
19. $5^2 = 25$
20. $4^{-1} = \dfrac{1}{4}$
21. $10^0 = 1$
22. $e^{-\frac{1}{2}} = \dfrac{1}{\sqrt{e}}$
23. 0
24. 4
25. –5
26. –4
27. domain: $x < 0$

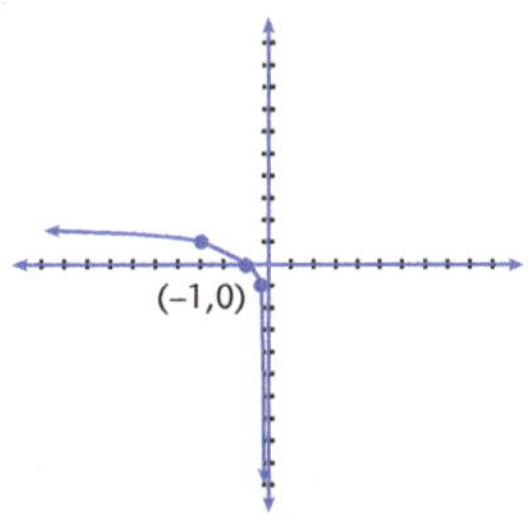

28. domain: $x > 2$

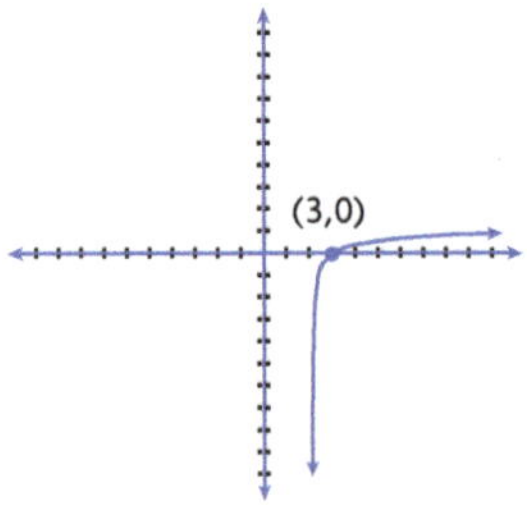

29. domain: $x > 0$

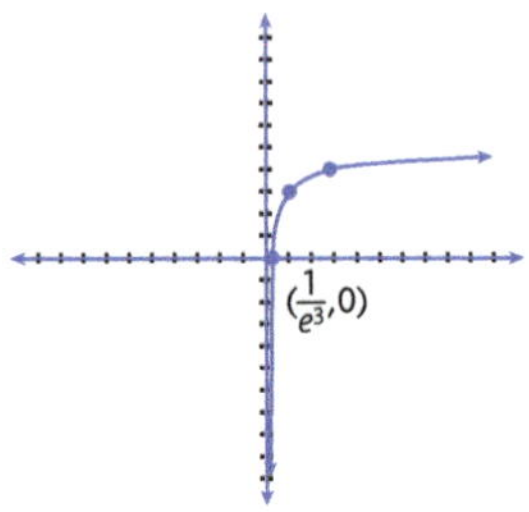

30. domain: $x > 0$

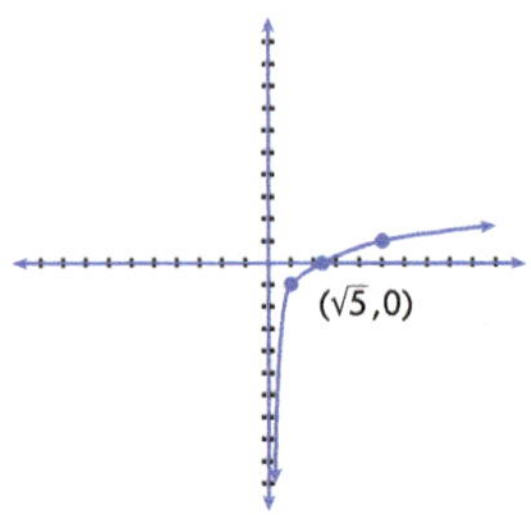

31. domain: $x > -1$

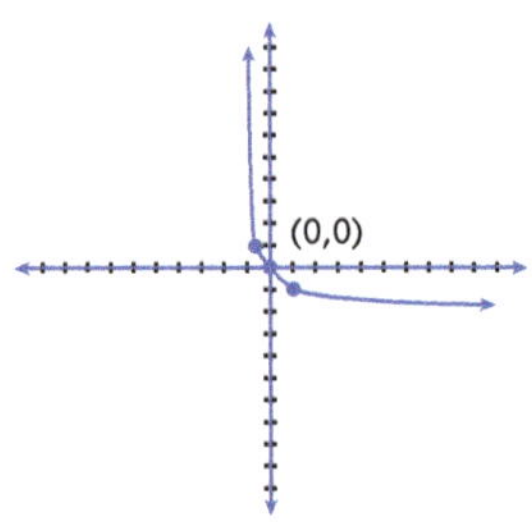

32. $4\log(x^2+5) - 12\log x$

33. $2\ln x + \frac{4}{3}\ln(x-1) - \frac{1}{3}\ln(x+9)$

34. $2\log_5(x+1) + 3\log_5(x-2) + 4\log_5(x+3) - \log_5 x$

35. $\log_2\left(\frac{9x^3}{(x+5)^4}\right)$

36. $\log_4\left(\frac{\sqrt{x-3}}{\sqrt[4]{x}}\right)$

37. $\ln\left(\frac{e^{3x}(5x-2)}{x^2}\right)$

38. $\log(20)$

39. $\log_5(2x^2(x^2-5))$

40. 7.5

41. 0

42. $\pi - 1$

43. $x = \frac{\ln 6}{5}$

44. $x = 2, -5$

45. $x = 0$

46. $w = -3$

47. $x = \ln\left(\frac{5}{2}\right)$

48. $x = 5$

49. $x = 2 \pm 2\sqrt{6}$

50. $x = 6$

51. $x = 9, -10$

52. $x = \frac{e^3 - 2}{1 - e^3}$ a -0.9476

53. $x = \frac{1}{62}$

54. $x = 0, 2$

55. **a.** 40
b. ~252
c. 400

56. **a.** 4
b. ~13 days

57. **a.** $149.62
b. ~5 units

58. 68.9%

59. ~4 years

60. ~64 years

61. **a.** $P = 14000e^{0.03t}$
b. ~14866
c. 2031

62. **a.** $6077.53
b. ~5.44 years

CHAPTER 5 REVIEW ANSWER KEY

1. **a.** $a_n = 3n - 7$; $-4, -3, 0$
b. $a_n = -4(3^{n-1})$; $-4, 12, 36$

2. **a.** $a_n = \frac{5}{2}n - 2$; $\frac{1}{2}, 3, \frac{11}{2}$

b. $a_n = \frac{1}{2}\left(\frac{5}{2}\right)^{n-1}; \frac{1}{2}, \frac{5}{4}, \frac{25}{8}$

3. a. $a_n = -2n + 27$; 25, 23, 21
   b. $a_n = 25(-2)^{n-1}$; 25, –50, 100
4. a. $a_n = 6n - 5$; 1, 7, 13
   b. $a_n = 6^{n-1}$; 1, 6, 36
5. a. 19.5
   b. 3×10^{-11}
6. a. –13.4
   b. –9,765, 625
7. a. 38
   b. $\frac{-3}{1024}$
8. 12,720
9. –17,280
10. 8,680
11. 11,184,810
12. $\frac{-4095}{512}$ or –7.998046875
13. 1559.098778
14. ~5.68 years; ~4.65 years
15. $8,588.43
16. $29,017.96
17. $57.50
18. 8.16%; 8.3%
19. $16,174.52
20. $68,654.73
21. $1,974.60
22. $2,691.65
23. $352.97
24. ~17.12 years
25. $936.68
26. $112,940.87
27. $6,013.67
28. $41,257.58
29. $2,506.91
30. $1,084.34
31. a. $429.84
    b. $25,790.14
    c. $4,790.14
32. $284,839.65
33. $222,753.81
34. $1,202.36; $151,754.41

FINAL REVIEW ANSWER KEY

1. $A^{-1} = \begin{bmatrix} 1 & \frac{2}{5} \\ 0 & \frac{1}{5} \end{bmatrix}$
2. a. $\begin{bmatrix} 2 & 3 & 6 \\ 3 & 4 & 1 \\ 6 & 1 & 6 \end{bmatrix}$
   b. $\begin{bmatrix} 29 & 25 \\ 10 & 12 \end{bmatrix}$
3. (–52, –28, –6)
4. a. 295
   b. 14,650
5. $6,069.44
6. $245,049.99
7. a. $f(x)$ domain: $x \le 1$; $g(x)$ domain: $x \in \mathbb{R}$
   b. $g(f(x)) = 2 - x$; domain $x \in \mathbb{R}$
   c. $\left(\frac{f}{g}\right)(x) = \frac{\sqrt{1-x}}{x^2+1}$
8. N.S.
9. $y = \frac{2}{3}x + \frac{5}{3}$
10. $y > \frac{-2}{3}x$
11. 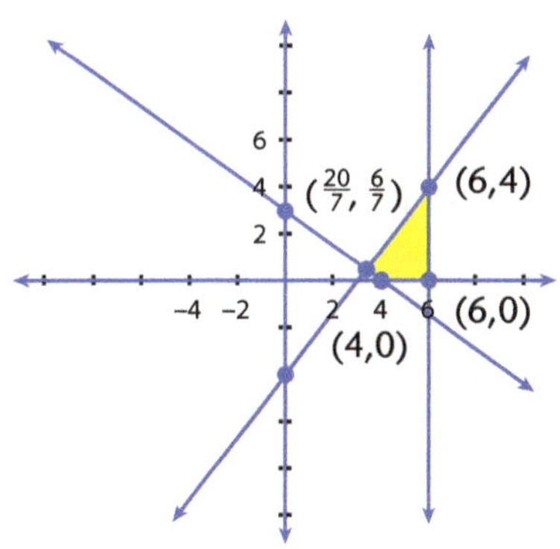

12. Max $P = 20$ at (10,0)
13. a. $P(x) = -x^2 + 19x - 84$
    b. (7, 0) and (12, 0)
14. width = 25 ft, length = 25 ft
15. a. (3, 0) and (–1, 0)
    b. (1, 4)

c.

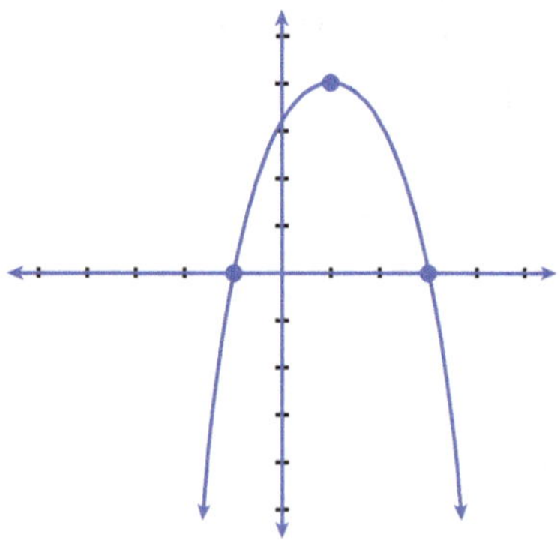

16. a. 3200 people

b. $t = \frac{\ln 2}{0.025} \approx 27.7$years

c. 3281 people

17. $2^x = 32$; $x = 5$

18. $x = 4, 2.5$

19. 330 units

20. about 720 feet

21. a. $x \in \mathbb{R}$, $x \neq -5$, $-\frac{5}{3}$

b. $x \in \mathbb{R}$

c. 0

d. $\frac{1}{25}$

e. $\frac{1}{3}$

22. y-intercept: (0, 1), x-intercepts: (1, 0) and (3, 0)

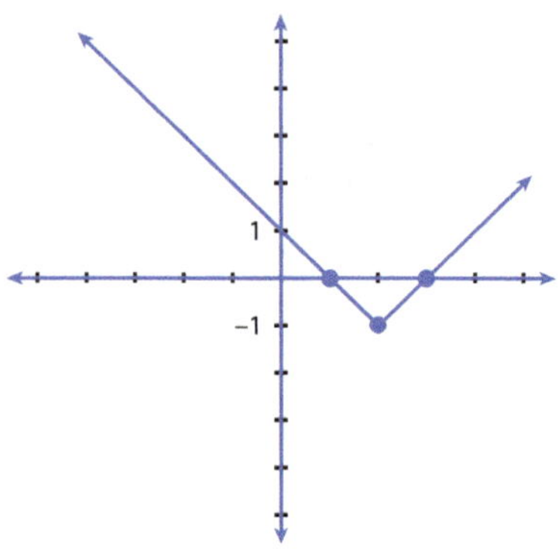

23. $p = -\frac{1}{20}q + 200$

24. $x = 5.5$

25. a. $x = -3$

b. $x = \ln 3$

c. $x = \frac{100}{3}$

26. $12,126.00

27. $701.90

28. a. $\begin{bmatrix} 3 & 0 & -4 \\ 3 & 2 & -1 \end{bmatrix}$

b. not possible; number of columns of B doesn't equal number of rows of A

c. $\begin{bmatrix} -1 & -1 & 1 \\ 2 & 2 & -1 \\ -1 & -2 & 1 \end{bmatrix}$

29. solution exists; (1, 2, 3)

30. max of 16 at (4, 0)

31. a. $x > 6$ or $x < -\frac{22}{3}$

b. $x = \frac{1}{5}$ or -11

32. 500 bicycles

33. $x = 1, 3$

34. a. $x \in \mathbb{R}$, $x \neq -1,1$

b. $4x^2 - 4x$

c. $\frac{1}{2}(x - 1)$

d. $x^2 - 2x$

35. x-intercepts: (3, 0) and (–1, 0);
y-intercept: (0, –6); vertex: (1, –8)

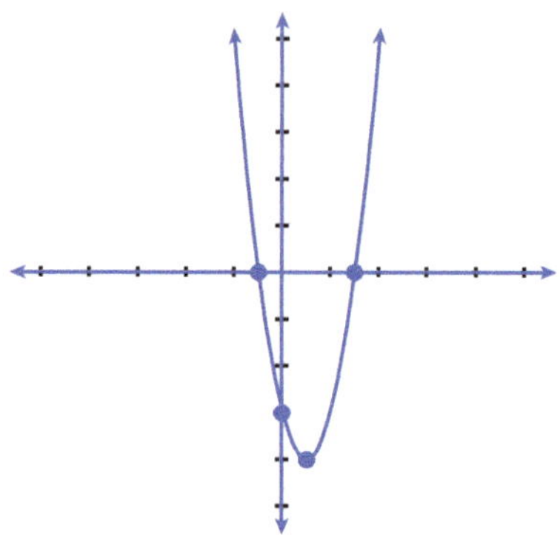

36. 18 lbs of 7% nitrogen fertilizer and 9 lbs of 13% nitrogen fertilizer

37. a. $x = 5$

b. $x = -8$

c. $x = 10$

38. about 6.8 years

39. $1,169.12

40. $89,917.83

41. a. $\begin{bmatrix} 9 & 3 & 13 & 5 \\ 6 & -4 & 6 & 6 \\ 3 & 8 & 8 & -2 \end{bmatrix}$

b. not possible; number of columns of B is not equal to number of rows of A

c. $\begin{bmatrix} -4 & 9 & -8 & -2 \\ -1 & -3 & 7 & 0 \end{bmatrix}$

42. **a.** $\begin{bmatrix} 2 & -5 \\ -1 & 3 \end{bmatrix}\begin{bmatrix} x \\ y \end{bmatrix} = \begin{bmatrix} 2 \\ 1 \end{bmatrix}$

b. $\begin{bmatrix} 3 & 5 \\ 1 & 2 \end{bmatrix}$

c. (11, 4)

43. max at (0, 12) with $P = 24$

44. N.S.

45. $y = -\frac{1}{2}x + \frac{5}{2}$

46. **a.** –16

b. $-2x^2 + 19x - 41$

47. $f^{-1}(x) = \sqrt[3]{(x-1)^5 + 5}$

48. **a.** $x \geq 0$

b. x-intercepts: (–10, 0) and (30, 0); y-intercept: (0, 150)

c. \$200

49. $q = 20$, $p = \$10$

50. 2

51. **a.** $x \approx 4.4794$

b. $x = 50$

52. in about 35 months

53. \$19,825.88

54. \$294.07

55. **a.** $\begin{bmatrix} 1 & -7 & 15 \\ 7 & -2 & 17 \\ 12 & 1 & 7 \end{bmatrix}$

b. not possible; number of columns of A not equal to number of rows of B

c. $\begin{bmatrix} 6 & 8 & 20 \\ 5 & 4 & 9 \end{bmatrix}$

56. (2, –1, 3)

57. **a.** x = number of tables, y = number of chairs;

$x \geq 0$, $y \geq 0$

$2x + 6y \leq 1800$

$x + y \leq 400$

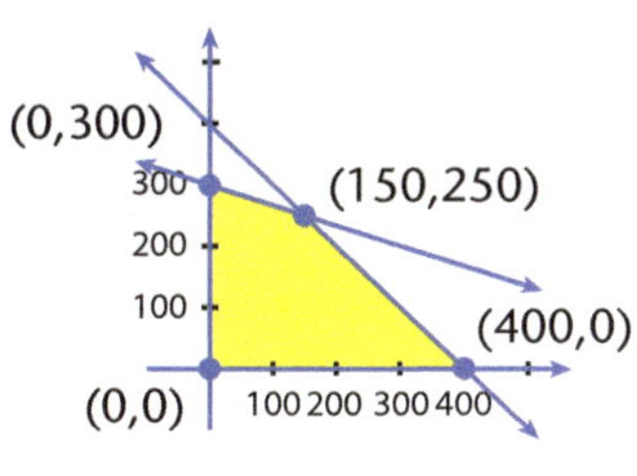

b. 400 tables and 0 chairs

58. **a.** $y = -2x - 3$

b. $y = -\frac{1}{2}x + 4$

59. **a.** $x = \frac{-5 \pm \sqrt{13}}{6}$

b. $x = 4, 9$

60. $q = 11$, $p = \$19$

61. **a.** $0.05x + 0.1y = 200$ and $x + y = 2500$

b. $\begin{bmatrix} 0.05 & 0.1 \\ 1 & 1 \end{bmatrix}\begin{bmatrix} x \\ y \end{bmatrix} = \begin{bmatrix} 200 \\ 2500 \end{bmatrix}$

c. $x = \$1000$, $y = \$1500$

62. **a.** x = # dark chocolate truffles,

y = # light chocolate truffles

$x \geq 0$, $y \geq 0$

$2x + 4y \leq 1600$

$10x + 5y \leq 5000$

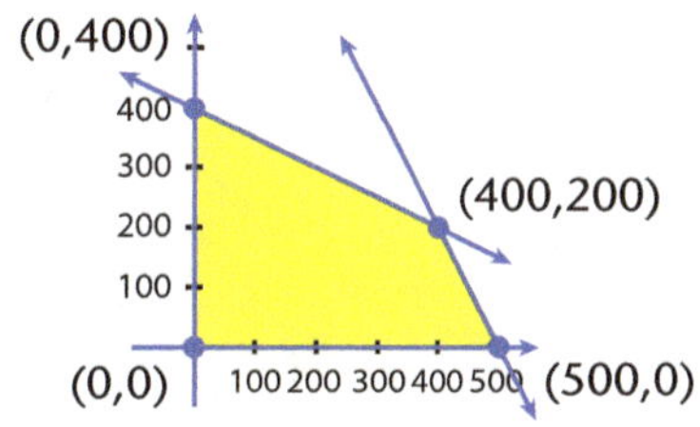

b. 400 dark chocolate truffles and 200 light chocolate truffles for \$560 revenue

63. **a.** $x = 6$

b. $x = 0, -1$

64. **a.** HA: $y = 0$

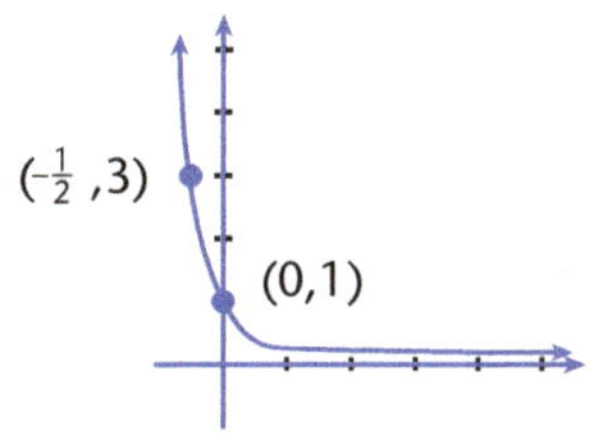

b. VA: $x = -1$

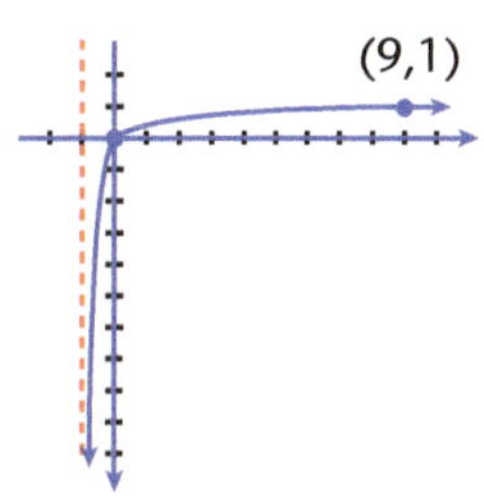

65. \$1,053.35; \$50,560.96

66. **a.** \$54,736.73

b. \$58,576.09

67. **a.** \$2 and \$3; 2000 units and 1000 units

b. max profit is \$250 for 1500 units at a price of \$2.50 per unit

68. **a.** not possible; number of columns of A is not equal to number of rows of B

b. $\begin{bmatrix} \frac{1}{4} & 0 \\ -\frac{3}{8} & -\frac{1}{2} \end{bmatrix}$

c. $\begin{bmatrix} 8 & 4 & 12 \\ -6 & -13 & -7 \end{bmatrix}$

d. $\begin{bmatrix} -1 & 1 & 10 \\ 0 & 9 & -7 \end{bmatrix}$

69. **a.** $x = 5$

b. $x = 2$

70. **a.** $814.42

b. $293,192.88

c. $143,192.88

71. **a.** $827.79

b. after 35 quarters (or 8.75 years)

Midterm Answers

MIDTERM 1

1. $x = 23$

2. $x \geq \frac{8}{5}$ or 1.6

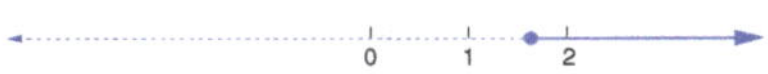

3. a.

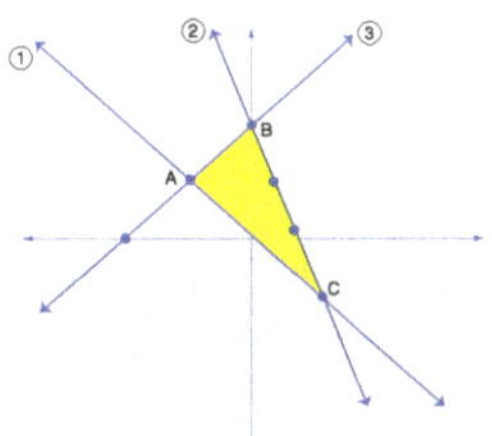

List of Vertices of the shaded region:
(0, 6) (–3, 3) (3, –3)

b. Point where max occurs:
(3, –3)
Max value of f: 51

4. $y = 4x - 19$

5. Point of intersection:
(–3, 1,4)

6. a. $R(x) = 75x$

b. $C(30) = \$2820$
Description: the cost to this company of selling 30 hours of tutoring

c. ~46.3

d. $40/hr

7. a. $\begin{bmatrix} 1 & 2 & 0 \\ 6 & -3 & 9 \end{bmatrix}$

b. 0

c. $\begin{bmatrix} -5 & 6 \\ 8 & 1 \\ 20 & -9 \end{bmatrix}$

d. We can't do this because they're not the same size

8. $(q, p) = (1250, \$13)$

9. a. Relation.
Explanation: It doesn't pass the vertical line test.

b. i $x > -2, \quad x \in \mathbb{R}$

ii $\dfrac{-1}{\sqrt{7}}$

iii $\dfrac{4 - (y + w)}{\sqrt{y + w + z}}$

c. Let x = total monthly sales

Option 1:
$1200 + 0.06(x - 1500)$

Option 2: $2500 + 0.035x$

We want to know when option 1 is better than option 2

Inequality:
$1200 + 0.06(x - 1500) \geq 2500 + 0.035x$

MIDTERM 2

1. a. $A = \begin{bmatrix} 1 & -1 & 1 \\ 1 & -1 & 2 \\ -1 & 0 & 3 \end{bmatrix}$

$X = \begin{bmatrix} x \\ y \\ z \end{bmatrix}$

$B = \begin{bmatrix} 2 \\ 1 \\ -11 \end{bmatrix}$

b. $\left[\begin{array}{ccc|c} 1 & -1 & 1 & 2 \\ 1 & -1 & 2 & 1 \\ -1 & 0 & 3 & -11 \end{array}\right]$

c. (8, 5, –1)

2. a. $= \frac{1}{24}\begin{bmatrix} 8 & 0 \\ 4 & 3 \end{bmatrix}$ OR

$\begin{bmatrix} \frac{1}{3} & 0 \\ \frac{1}{6} & \frac{1}{8} \end{bmatrix}$

b. Not possible, because D^T is 3 × 2 and C is 3 × 2; we need # columns of D = # rows of C.

c. Not possible, because they are not the same size.

d. $\begin{bmatrix} 10 & 9 \\ -14 & 29 \end{bmatrix}$

e. $\begin{bmatrix} 24 & 49 \\ 110 & 144 \end{bmatrix}$

3. a. (1) $x + y + z = 40000$

(2) $0.05x + 0.04y + 0.075z = 0.0485(40000)$

(3) $y = x + z$

b. $x = \$14{,}400$ $y = \$20{,}000$ $z = \$5{,}600$

4. a $R(x) = x(15 - 0.25x) = -0.25x^2 + 15x$

b $P(x) = -0.25x^2 + 14.5x - 80$

c 52 or 6

d 29

e \$10.25

f \$130.25

5. $A^{-1} = \begin{bmatrix} 0 & 1 & 0 \\ -1 & 1 & 0 \\ 1 & -1 & 0 \end{bmatrix}$

6. a $(3, -4)$

b $x = 3$

c up

d $\left(0, \frac{1}{2}\right)$

e $(3 + 2\sqrt{2}, 0)$ and $(3 - 2\sqrt{2}, 0)$

$\simeq (5.8, 0)$ and $(0.2, 0)$

f

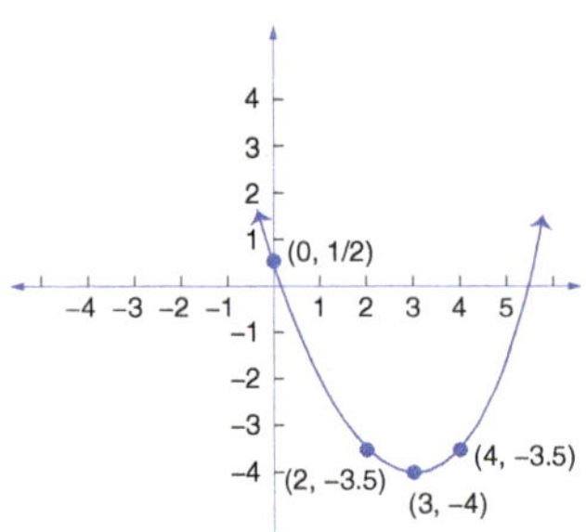

7. $(47 - 23z, 9 - 3z, z)$

MIDTERM 3

1. a. $f^{-1}(x) = \dfrac{-4x - 1}{x - 3}$

b.

	Domain	Range
$f(x)$	$(-\infty, -4) \cup (-4, \infty)$	$(-\infty, 3) \cup (3, \infty)$
$f^{-1}(x)$	$(-\infty, 3) \cup (3, \infty)$	$(-\infty, -4) \cup (-4, \infty)$

2. a. $\frac{1}{3}\log_6(x + 4) - 3\log_6 x$

b. $\log\left[\dfrac{(x+8)(x-1)^3}{x^2}\right]$

c. $5^3 = w$

d. $\log_4 L = K$

e. False

f. $3^{m+p} = 3^m \cdot 3^p$

3. $f(x) = 2^{-x}$ or $\left(\frac{1}{2}\right)^x$

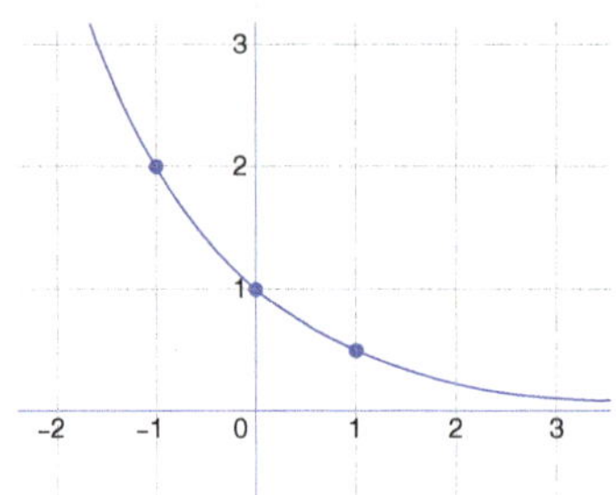

4. a. $x = \dfrac{-3}{7}$

b. $x = 4 - \log_3 6$ or

$4 - \dfrac{\ln 6}{\ln 3} = x$

5. a. (1) $d = 4$

(2) $a_n = 4n - 3$

(3) $S_{50} = 4950$

b. (1) $d = 4$

(2) $g_n = 3(4^{n-1})$ or $\frac{3}{4}(4^n)$

(3) $S_7 = 16383$

6. a. -118

b. $\left(\sqrt{x - 1}\right)^3 + 5$

c. $x \neq -5, x \in \mathbb{R}$

d. $\frac{4}{9}$

e. 21

f. $\sqrt{\frac{1}{3}}$ or $\frac{1}{\sqrt{3}}$

7. a.

x	y	
−2	1	3^0
−1	3	3^1
−3	1/3	3^{-1}

$y = 0$

Shifted left by 2

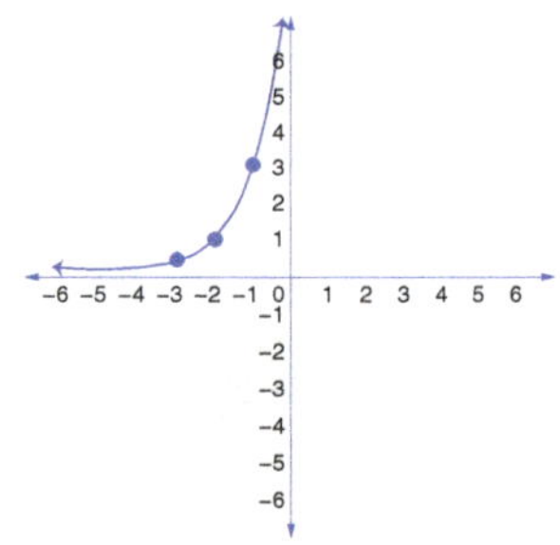

b.

(x, y) on parent graph of $y = \log_2 x$	Horizontal reflection	Shift up
(1, 0)	(−1, 0)	(−1, 3)
(2, 1)	(−2, 1)	(−2, 4)
(0.5, −1)	$\left(-\frac{1}{2}, -1\right)$	$\left(\frac{1}{2}, 2\right)$

$x = 0$

$x < 0$

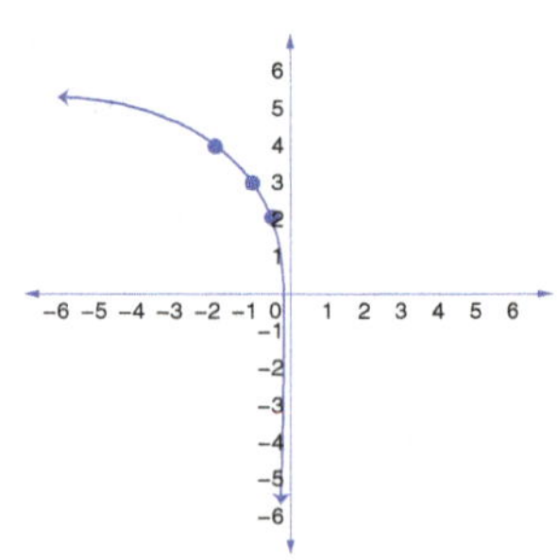

Final Exam Answers

FINAL EXAM 1

1. $N = 9$

2. $y = 2x + 1$

3. a.

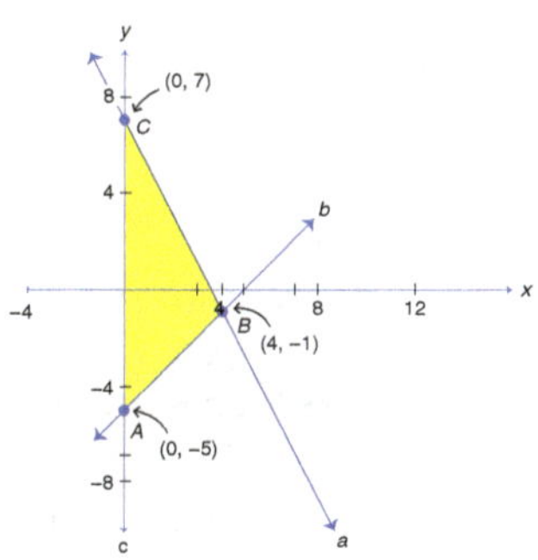

   b. Maximum: 15
   Located at: (0, −5)

   Minimum: −21
   Located at: (0, 7)

4. $(x, y, z) = (1, -5, 4)$

5. a. $\begin{bmatrix} 4 & 7 & -5 \\ 9 & -2 & 6 \end{bmatrix}$

   b. $\begin{bmatrix} -4 & -4 \\ -5 & -9 \end{bmatrix}$

   c. $-\frac{1}{8}\begin{bmatrix} 6 & -10 \\ -5 & 7 \end{bmatrix}$

   d. $\begin{bmatrix} 7 & 4 \\ 5 & 4 \end{bmatrix}$

6. a. Less
   b. $p = \$4$ $\quad q = 12$ items

7. a. $C(x) = 100 + (x + 20)\,x$
   b. Quadratic
   c. 30 services
   d. $P(x) = -2x^2 + 40x - 100$
   e. More valuable to maximize profit

8. a. $(x \in \mathbb{R} \text{ s.t. } x \geqslant 4)$
   Alternatively $x \in [4, \infty)$
   b. $\sqrt{(x^3 + 1) - 4} = \sqrt{x^3 - 3}$
   c. −4
   d. 2
   e. $\dfrac{(5 - 2x)}{3}$

9. a

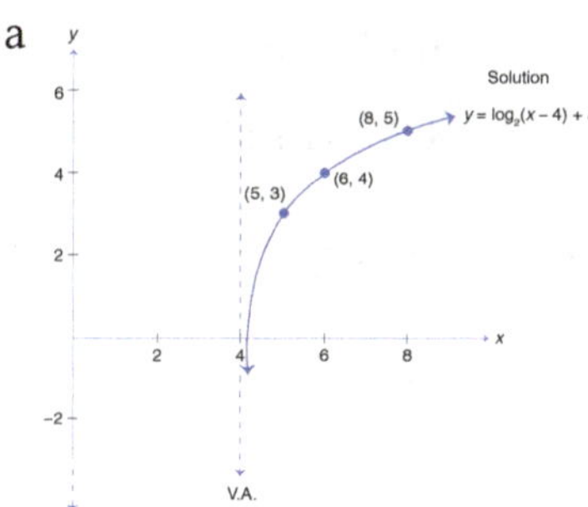

   b.

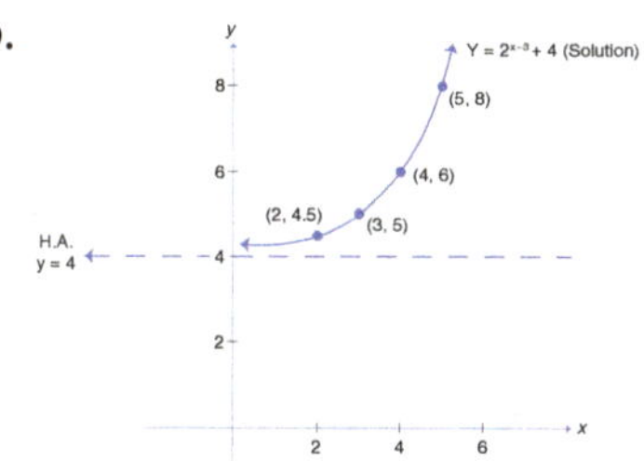

   c.

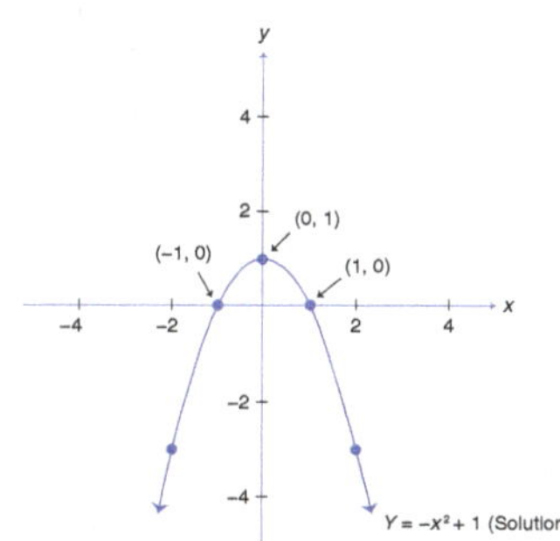

   Vertex: (0, 1)
   x-intercepts: (1, 0) & (−1, 0)

10. a. 2022
    b. $x = \dfrac{\ln(7) + 5}{2}$
    c. $x = 2$

11. 5.81 years

12. \$110.67

13. \$2353.84

14. \$17429.82

FINAL EXAM 2

1. a.

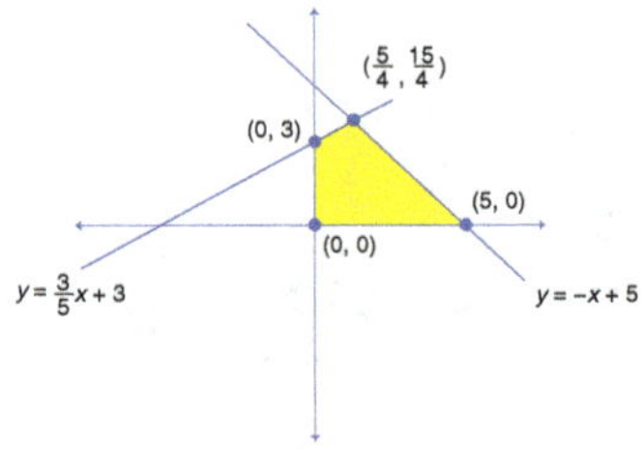

b. Max P-value: 8
Min P-value: 0

2. a. Equilibrium quantity: 50 tables
Equilibrium price: \$100

b. 200

c. 75

3. a. $\begin{pmatrix} -1 & 1 & 0 \\ 2 & 1 & -1 \\ -1 & -1 & 1 \end{pmatrix}$

b. $x = 1 \quad y = -1 \quad z = 1$

4. a. $\begin{pmatrix} -9 & 7 & -1 \\ 0 & 1 & 8 \end{pmatrix}$

b. $\begin{pmatrix} -2 & 37 & -2 \\ 0 & -11 & -3 \end{pmatrix}$

c. False
True

5. a. Axis of symmetry: $x = 1$
Vertex: (1, 4)

b. x-intercept(s): (3, 0) (−1, 0)
y-intercept: (0, 3)

c. Concave down

d.

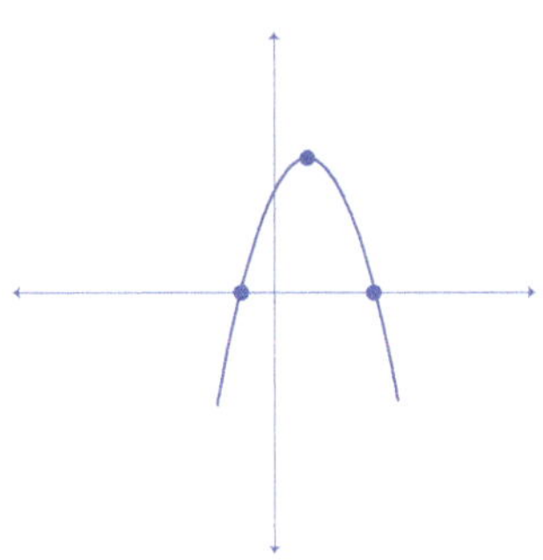

6. a. $R(x) = 100x$

b. $C(x) = (x + 20)x + 1500$

c. $P(x) = -x^2 + 80x - 1500$

d. 40

e. 100

7. a. NS

b. $\frac{1}{3}(\ln(5/2) - 1)$

8. a. 11.64 years

b. 11.55 years

9.

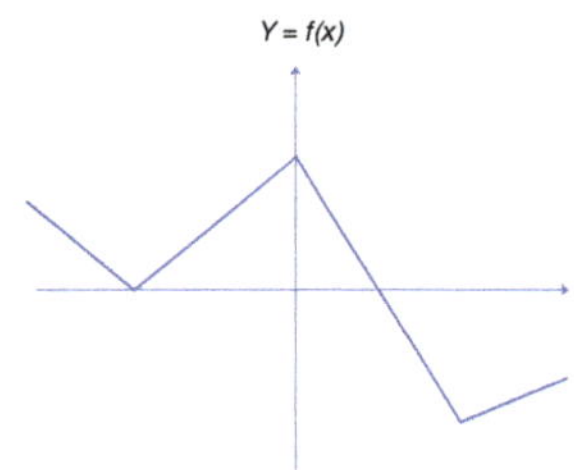

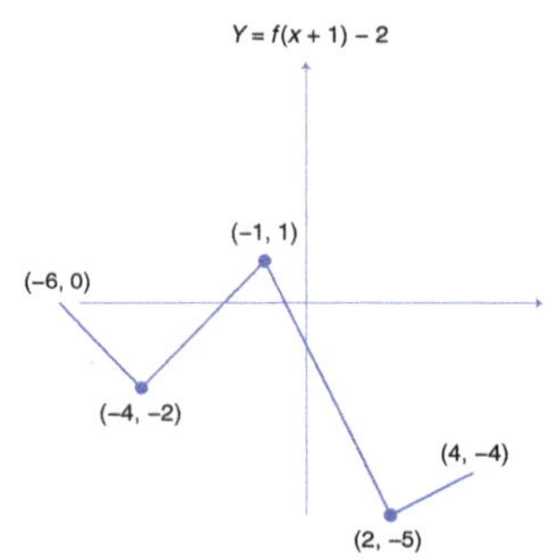

10. a. \$6658.13

b. \$213514.59

11. \$1317.25

12. a. Domain of f(x):
$x \neq 0$

Domain of $g(x)$:
$x \in \mathbb{R}$

b. $\frac{1}{x^2 + 2x - 3} + 5$

c. 1

d. $\frac{1}{x - 5}$

FINAL EXAM 3

1. a

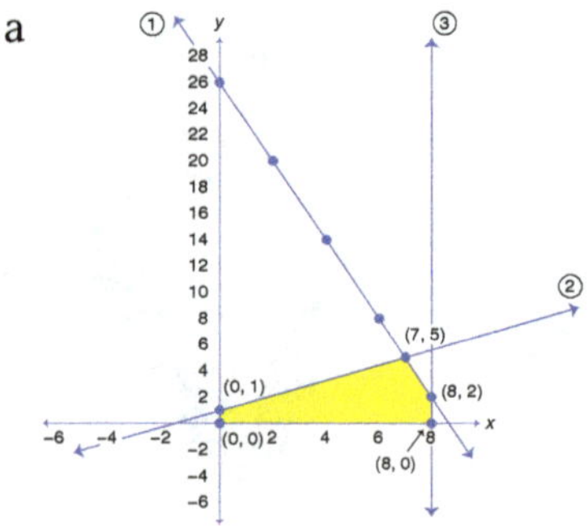

b. Maximum: 24
Located at: (8, 0)

Minimum: −2
Located at: (0, 1)

2. a. 17.5

b. 175

c. Consumers want to buy more than producers are willing to produce. Or producers are not willing to produce the quantity consumers are willing to buy.

3. Equation 1: $x + y + z = 100{,}000$

Equation 2: $0.05x + 0.09y + 0.12z = 0.08(100{,}000)$

OR $1.05x + 1.09y + 1.12z = 1.08\,(100{,}000)$

Equation 3: $10{,}000 + z = y$

4. (2, −1, 1)

5. $A^{-1} = \begin{bmatrix} 1 & -1 & 0 & 0 \\ 0 & 1 & 0 & 0 \\ 0 & 0 & 1 & 0 \\ 0 & -1 & 0 & 1 \end{bmatrix}$

6. a. $\begin{bmatrix} 2 & 4 & 6 \\ 10 & 5 & 9 \end{bmatrix}$

b. $\begin{bmatrix} 1 & 1 & 3 \\ 2 & 4 & 1 \end{bmatrix}$

c. DNE
Number of rows B is not equal to Number of columns A

d. $\begin{bmatrix} 2 & 5 \\ 3 & 5 \\ 6 & 3 \end{bmatrix}$

7. a. $\dfrac{2}{\sqrt{2x+1}}$

b. $4\left(\dfrac{2}{x}\right)^3 + 2$

c. $\left[-\dfrac{1}{2}, \infty\right)$

d. 3

e. $\dfrac{3}{2}$

f. $\sqrt[3]{\dfrac{x-2}{4}}$

8. a. Asymptote: $y = 1$

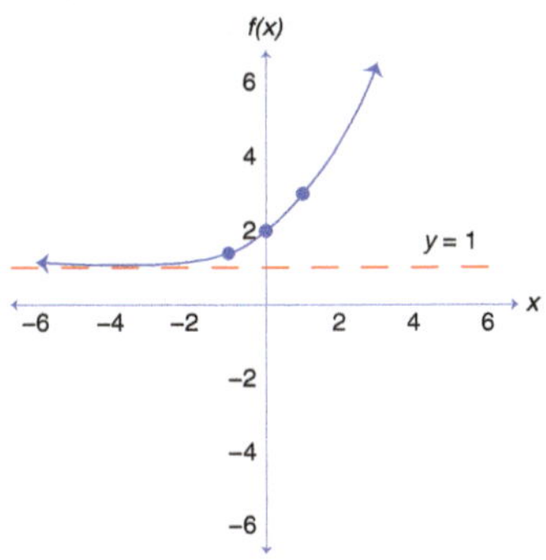

b. Asymptote: $x = 0$

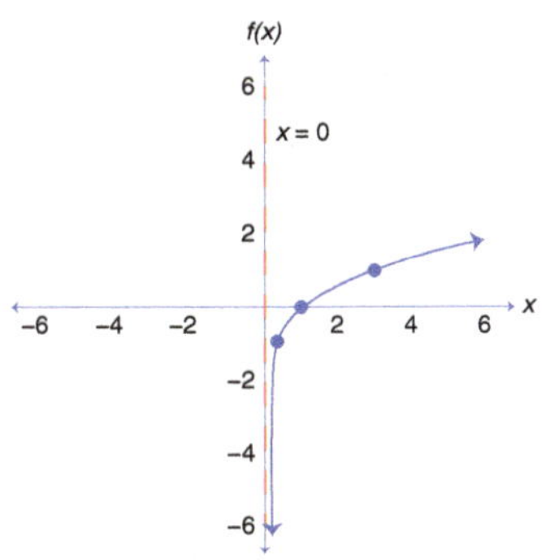

c. Vertex: (1, –2)
Axis of symmetry: $x = 1$
Concave up or down? Up
y-intercept: (0, –1)

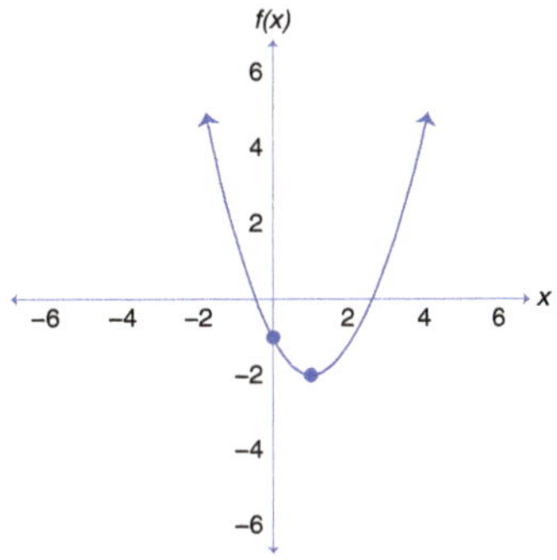

9. a. $P(x) = -30x^2 + 3300x - 85750$

b. $129,250

c. Set Revenue = Cost and solve for x. Then plug in x to solve for y.

10. a. T

b. T

c. F

d. F

e F

11. a $x = 3$

b $x = \dfrac{\log 6}{2 \log 5}$

12. a $S = R\left(\dfrac{(1+r_c)^N - 1}{r_c}\right)$

b $287,293.82

c $R = S\left(\dfrac{r_c}{(1+r_c)^N - 1}\right)$

d $575.90

e You

f Your Colleague

13. a $R = S\left(\dfrac{r_c}{1-(1+r_c)^{-N}}\right)$

b $2103.73

c $346,051.67

d $378,671.40

e $32619.73

14. a

(1) $S = 1000(1 + 0.046(5)) = \$1230$

(2) $S = 1000e^{0.041(5)} = \$1227.53$

(3) $S = 1000\left(1 + \dfrac{0.042}{12}\right)^{17(5)} = \$1233.23$

b 3

CPSIA information can be obtained
at www.ICGtesting.com
Printed in the USA
LVHW021143040822
725061LV00004B/4